SYSTEM MODELING AND RESPONSE

Theoretical and Experimental Approaches

SYSTEM MODELING AND RESPONSE

Theoretical and Experimental Approaches

Ernest O. Doebelin

Department of Mechanical Engineering The Ohio State University

JOHN WILEY & SONS

New York   Brisbane
Chichester   Toronto

WILE 1980

Library of Congress Cataloging in Publication Data:

Doebelin, Ernest O.
    System modeling and response.

    Includes bibliographies and index.
    1. Systems engineering--Mathematical models.
2. Systems engineering--Data processing. I. Title.
TA168.D593   620.7′2   79-27609
ISBN 0-471-03211-5

Printed in the United States of America

10 9 8 7 6 5 4 3 2 1

PREFACE

When introductory-level system dynamics textbooks appeared in the late 1960s and early 1970s, courses in this discipline were relatively rare in engineering curricula. Since then, more schools have introduced this material, either within individual departments or as "core" courses presented to all engineering students regardless of department. It is not too surprising that this area has not yet achieved *universal* acceptance as basic required material in the several engineering curricula where it is appropriate, because the introduction of new approaches into today's tightly packed curricula usually requires displacement, curtailment, or rearrangement of some well-entrenched "traditional" topics. Such changes in emphasis rightfully call for careful justification, which is not always quick in coming. I believe, however, that the trend in acceptance is still an upward one and that, ultimately, wide acceptance will be achieved.

The focus of this system dynamics book is at the "next higher" level after an introductory course. Whether this would be a junior, senior, or graduate course obviously depends on the particular curriculum in which it is employed. At my own school, the text is used in two graduate level courses since our undergraduate curriculum requires separate courses (with lab) in system dynamics, measurement systems, and automatic control, giving a comprehensive background sufficient for the Bachelor's degree. Beginning graduate students from schools where undergraduate system dynamics is not emphasized are able to pursue the first graduate course successfully by some intensive self-study of the basic material at the beginning of the term.

Although much of system dynamics grew out of the methods and needs of control system design, I believe these techniques are now of much wider interest and utility and thus should become an integral part of the academic preparation of almost all engineering students at both undergraduate and graduate level. However, the excessive abstraction, specialized jargon and overly mathematical approach of some treatments in this field effectively prevent the much-needed dissemination of these concepts to a wider audience. I hope this book presents this material in such a way that educators and students at present "outside" the system dynamics field will more clearly see the usefulness of these methods in a wide variety of engineering problem areas. System dynamics will then be thought of less as a specialist area and more as a basic tool for all engineers. When many graduate courses and programs today emphasize the fine details of physical phenomena, it is even more important that students not forget that engineers ultimately deal with *entire* machines and processes and need ways of understanding behavior at a system level.

While a survey of the Contents and a quick reading of Chapter 1 will give a good picture of the philosophy and content of the text, a brief discussion here is probably in order. This book is organized into three major parts: system response evaluation, generalized modeling methods, and specific applications of modeling techniques. These parts would not usually be covered in their entirety and/or sequentially in a given course and the instructor can put together a selection of topics (from any or all parts) that best meets the needs and utilizes the previous preparation of the class. Since I have written the text for use with two of our graduate courses, there is sufficient breadth of coverage to allow considerable flexibility of topic choice in designing a particular single course.

The first part, system response evaluation, includes a fairly conventional treatment of transform methods for linear system response studies followed by a heavy emphasis on frequency spectrum methods. Many recent textbooks have emphasized time-domain methods at the expense of the more classical frequency approach. I believe examination of industrial practice shows a continued heavy use of frequency domain methods in practical problems, and students consequently should be well prepared to use them. Brief sections are also included on distributed-parameter transform methods, difference equations and Z-transforms, and state-variable system description. For nonlinear systems the emphasis is on a digital simulation approach to system analysis and design. While some useful analytical methods for nonlinear systems certainly exist, the increasing availability each year of powerful, general, and economical digital simulation methods makes one wonder whether extensive student exposure to the necessarily specialized nonlinear analytical methods is of great practical utility.

Part II, generalized modeling methods, deals with methods of obtaining system descriptions, whereas Part I examined techniques for calculating response for a system whose description was given. The word "generalized" restricts the coverage to methods (both theoretical and experimental) that are applicable to broad classes of systems, not just specific physical types. Heavy concentration in this part of the book is placed on experimental test techniques, mostly for linear systems but with some attention to nonlinear ones. Of the many "system identification" methods discussed in the literature, those that are relatively well established and proven by practical use are emphasized. Theoretical questions of general interest for all kinds of models include subsystem coupling methods and problems of discretization. Coverage here is somewhat brief but sufficient to establish both basic comprehension and also some useful working tools.

The final part, specific applications of modeling techniques, emphasizes theoretical methods, both lumped and distributed, applied to specific classes of physical devices. The overall goal here is not only to develop competence in the specific problem areas but to make clear the unusually wide applicability of system analysis methods in engineering. A particular student may perhaps never need to use the specific human transfer functions developed in Chapter 13; however, the knowledge that this kind of thing can be, and has been, done is invaluable in developing the ability to tackle new problems with some self-confidence. Part III concentrates on theoretical models, but it also shows experimental test results whenever possible, thus developing appreciation for the types of simplifying assumptions that actually work. Also, comparisons between lumped and distributed models for the same physical system are made in several areas, along with comparisons between complex and simplified versions of a given model.

While I feel that an undergraduate system dynamics course definitely needs a laboratory, the need for one in a graduate course is not so clear. At present, one of our two graduate courses has a lab and while it adds some practical flavor and motivation, it certainly is not a necessity. Certain kinds of computer capability are a definite asset in this type of course, though not absolutely required. I much prefer canned programs since they allow concentration of student time on the physical and engineering aspects instead of on programming. A good selection would include a comprehensive digital simulation language (similar to CSMP), a simple-to-use general-purpose engineering language with good complex number and matrix facilities (such as SPEAKEASY), and a general-purpose statistical package (such as SAS).

Because this book presents well-established methods and not "forefront research," my contributions have been in the selection, organization, explanation, and evaluation of these techniques, rather than in their original conception. I am thus indebted to those authors from whose works I gleaned the background necessary to develop this book and have carefully referenced their original publications in every case. The support of department Chairman James E. A. John in providing manuscript production facilities is gratefully acknowledged as is the expert typing of Barbara G. Steinbrook.

Ernest O. Doebelin
Columbus, Ohio

CONTENTS

SYSTEM MODELING AND RESPONSE

Theoretical and Experimental Approaches

PART ONE

System Response
Evaluation

CHAPTER 1

The Philosophy and Methodology of the System Approach

1.1 DEALING WITH COMPLEXITY

Even though this book is aimed at intermediate-to-advanced levels of system study, and some sort of introductory level of understanding (perhaps as provided by an elementary systems text such as the author's earlier work[1]) is presumed, we will devote some initial effort to explaining the general scope, purpose, and approach of system methods in engineering. This brief introduction will serve both as a review of concepts encountered earlier and also an extension of these viewpoints in preparation for the more advanced treatment to follow in later chapters.

In clarifying the meaning of this section's title, "Dealing with Complexity," we compare and contrast the emphasis of "traditional" engineering courses such as theory of elasticity, fluid mechanics, and heat transfer with that of the more recently introduced systems studies. Our use of the word "complexity" refers specifically to the clear trend toward the design and application of machines and processes ("systems") that are made up of an ever increasing number of components, interconnected in more and more complex fashion. Of course, anyone familiar with modern physics and chemistry will realize that understanding the behavior of an apparently simple *single component* or physical effect, if rigorously pursued, *also* encounters boundless complexity. However, the complexity addressed by system methods lies not in a more detailed physical description of components but in the interconnection of a multitude of components, the behavior of each of which is modeled in a relatively simple fashion.

A few examples illustrating this distinction in the meaning of complexity may be useful at this point. In an advanced course in convection heat transfer, one might well pursue a detailed analytical study of fluid mechanics as applied to boundary-layer phenomena in an attempt to understand the physical mechanisms involved and to predict, from basic geometry and fluid properties, numerical values of heat transfer coefficients. When simplifying assumptions are

---

[1]Doebelin, E. O., *System Dynamics*, C. E. Merrill, Columbus, Ohio, 1972.

kept to a minimum in an attempt to get accurate predictions, such studies become quite complex, even for simple physical configurations. Although such calculation methods may be intricate, the end result is often a single number, the effective heat transfer coefficient in watts/(m²-C°). When analyzing (as we will in later chapters) a thermal *system* involving one or more heat exchangers, the heat transfer coefficient will be only *one* of many parameters affecting the overall system behavior. A system analyst/designer would thus of necessity consider the heat transfer phenomena from an overall operational viewpoint and presume the *existence* of a numerical value for the heat transfer coefficient without pursuing in detail the methods necessary for its calculation.

We hasten to observe that such a modus operandi is fraught with the danger that the system analyst does not maintain an adequate *physical* understanding of the basic phenomena and thus *over*-simplifies reality. This concern is particularly applicable to those trained mainly in mathematics, to whom the intricacies of system analysis are appealing, but whose background may be inadequate to allow valid judgements on physical questions. Having raised this specter, we however need to immediately reassert that both to restrain the mathematical complexity at the system level, and also in recognition of the breadth/depth limitations of human understanding, simplification at the phenomenon and component level is a *necessity* in system analysis, and one must be willing to risk the possibility of some inaccuracy in details in order to gain the benefits of overall system insight. Fortunately, if system components are actually to be *built* rather than just discussed mathematically, the system designer will surely make use of *experimental testing* to validate the assumed forms of phenomenological relations and assumed numerical values for coefficients, insofar as is practicable. In fact, for some complex and poorly understood phenomena, experimental testing of scale models or full scale prototypes may need to *precede* analytical study so as to provide information needed to formulate basic relationships. It should thus be clear that to be most effective, system engineers need to continually strive to maintain a proper *balance* of skills in the areas of physical details, system analysis tools, and experimental methods.

A final example, from the field of vehicle dynamics (an area pursued in more depth in a later chapter) may be useful in consolidating our thinking about how system methods deal with complexity. Figure 1.1 displays a schematic representation of an automobile suspension in a form useful for study of the lateral/directional or "handling" qualities of the vehicle. One purpose of the shock absorber or mechanical damper B is to control the oscillations of the body roll angle ϕ. In a Newton's Law torque equation about the roll axis, a conventional linear dynamic model represents the damping torque as $-B_t\dot{\phi}$, where B_t is the damping coefficient for both dampers and has dimensions N-m/(rad/sec). Even though different types and sizes of vehicles could employ widely varying damper designs with various geometrical connections to the frame, the *overall*

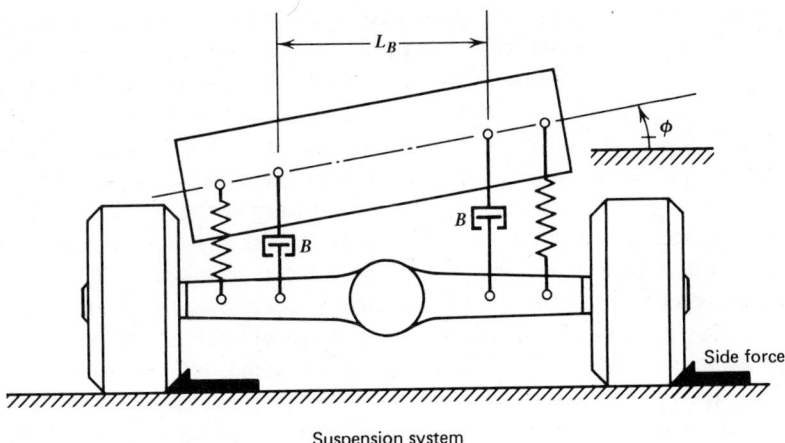

Suspension system

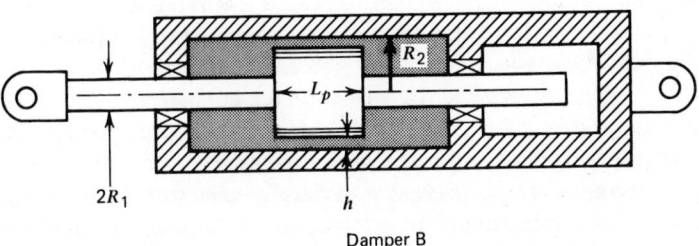

Damper B

Figure 1.1   Damper element in vehicle suspension system.

effect can always be described compactly by $-B_t\dot{\phi}$ so long as the numerical value of B_t is suitably chosen. While a system analysis does not require such detail, formulas[2] for estimating B_t for a specific construction, as in Fig. 1.1, may well be available:

$$B_t = \frac{6\pi\mu L_p L_B^2}{h^3}\left[\left(R_2 - \frac{h}{2}\right)^2 - R_1^2\right]\left[\frac{(R_2^2 - R_1^2)}{R_2 - h/2} - h\right] \qquad (1.1)$$

Even though the derivation of Eq. (1.1) *already* involves many approximations and simplifying assumptions, we note that B_t depends upon five geometrical dimensions and the fluid viscosity μ in a rather complicated way. If such details were carried through for all system elements, we would soon be overwhelmed by an incomprehensible morass of parameters. Thus a system designer/analyst needs to know of the existence of formulas (such as Eq. (1.1)) for estimating

[2]Doebelin, E. O., *System Dynamics*, C. E. Merrill, Columbus, Ohio, 1972, p. 59.

numerical values of parameters, and must understand clearly how to properly model the overall physical effect of a component in the system, but leaves the development of *detailed* component information to other members of the design team.

In Fig. 1.1 the sidewise force ("cornering force") of the road on the pneumatic tire is actually the major force involved in the control and stability of the vehicle. Even though researchers in the tire and automotive industries have expended, and continue to expend, great effort on studies of tire behavior, and even though these efforts have greatly expanded our knowledge, it is still impossible to predict from dimensions and material properties the "force coefficients" of a proposed new tire design. System analysis of vehicle dynamics thus depends on experimentally defined and measured values of tire properties obtained from laboratory tests of existing tires. Willingness to combine such experimental data with an otherwise "theoretical" treatment of a modeling study is a necessary attribute of effective system analysis.

Lest the reader be carried away by the author's enthusiasm for system methods and begins to believe that these techniques provide ready solutions for all of society's technical, economic, and social problems, we append the following disclaimer. When properly applied to well-defined engineering problems, system analysis certainly can claim many notable and well-documented triumphs. However, attempts by both engineers and nonengineers to extend these methods outside the technical realm have not generally met with conspicuous success.[3] Quoting from the reference, "In its journey from conventional engineering tasks and texts, from the realm of the military and outer space to the inner city and its cares and concerns, systems analysis has lost the methodological rigor that governed its application in accustomed fields.... Could it be that engineers are misconstruing their social mandate? With commencement exercises exhorting greater social commitment, and professors urging a deeper dedication to social values, engineers are being pushed into new roles. They are caught between the urge to do their own thing well, which in the extreme form leads to trained incapacity, and abandoning their specialized capabilities in favor of a generalized doing of the public good. A moral commitment is certainly desirable; a social conscience on the part of engineers would be commendable. But this is not to say that they should be impelled to attack social problems badly, but, instead, that they should do engineering better.... There are many useful functions that engineers can perform in a technological society such as ours. A clearer conception of "public technology" could, in fact, formalize their involvement so as to protect the public from the abuses being perpetrated in the name of systems science. Probably the greatest service the

[3]Hoos, Ida R., "Can Systems Analysis Solve Social Problems?," *Datamation*, June 1974, p. 82.

engineering profession could deliver, consistent with the Hypocratic Oath being proposed for it, would be one of disenchantment—to let society know the true limits of technology."

1.2   THE INPUT/SYSTEM/OUTPUT CONCEPT

Because of its preoccupation with the study of *interconnected* components or subsystems, it is not surprising that system analysis generally adopts a description of engineering problems that emphasizes "signals" and "systems." *Systems* are the pieces of actual hardware from which we construct the machine or process under study, while *signals* are the physical variables that "flow" in the interconnections between the systems. (Note that our use of the word "system" may seem inconsistent; we use it both for the smaller "building block" and also for the larger assemblage. While the words "component" or "subsystem" could be, and are, used to avoid this problem, many writers will use the term "system" indiscriminately in this way, relying on the context to make clear the meaning. Once we recognize this usage, it should create no problem of understanding.)

When we visualize a system as an assemblage of interconnected components, it should be clear that each component will have one or more signals, which we call *inputs*, flowing into it from other components, and one or more signals called *outputs* flowing from it to other components. Figure 1.2a[4] shows a schematic diagram for a type of steam-generation scheme as might be used in proposed nuclear power plants of the breeder reactor type, while 1.2b gives details showing the various components as they are interconnected in this system. The referenced report concentrates on an analysis of the steam generator subsystem, as shown in Fig. 1.2a, and develops a set of equations (mathematical model) that allows prediction of selected output quantities when system parameters and inputs are given. In this study, system inputs included feedwater flow rate and temperature, sodium flow rate and temperature, and intermediate spray flow rate and temperature (a total of six inputs), while the five outputs (responses) of interest were main steam flow rate and temperature, boiler outlet flow and temperature and sodium outlet temperature. System parameters (the physical characteristics that define the system) included such things as geometrical data (number, lengths, diameters, etc. of boiler tubes), material properties such as specific heats and densities, and quantities such as heat transfer coefficients, which depend both on material properties and operating conditions. Forty-four equations and some 90 parameters were necessary to model the system of Fig. 1.2c.

---

[4]Iyer, J. S., "Analog Simulation of Sodium-Heated Steam Generator," BAW-1280-42, Babcock and Wilcox Research Center, Alliance, Ohio, 1967.

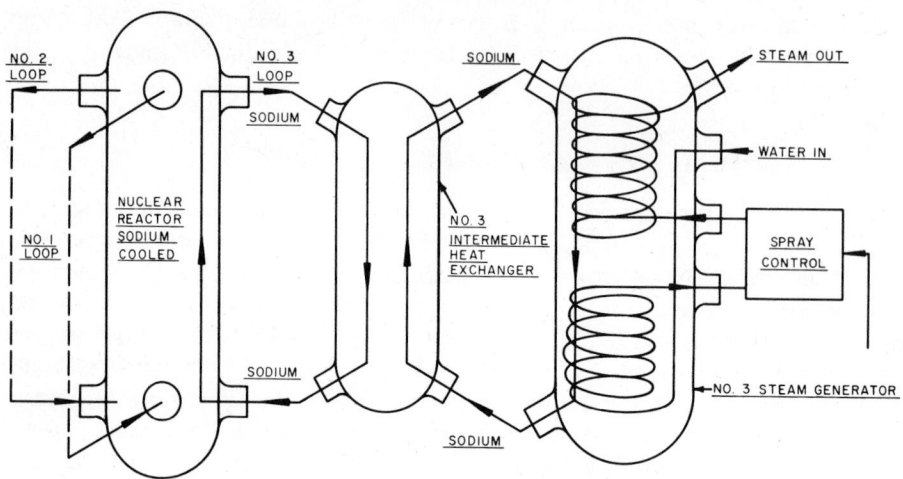

a.   A FULL SIZE LIQUID-METAL COOLED NUCLEAR POWER PLANT

(1000 MWe PLANT)

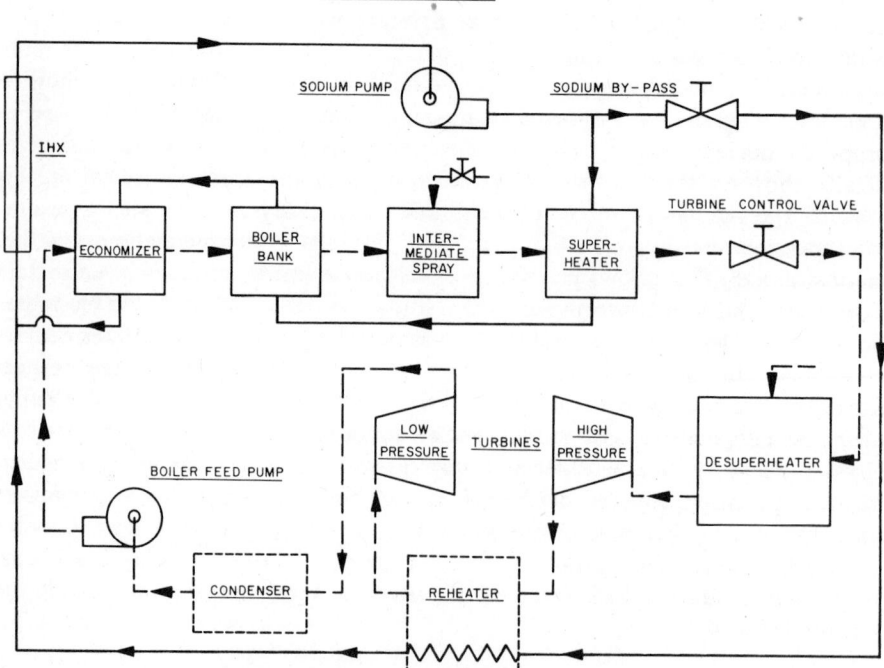

b.   DETAIL OF STEAM GENERATOR LOOP

Figure 1.2   Input/output model of nuclear power plant.

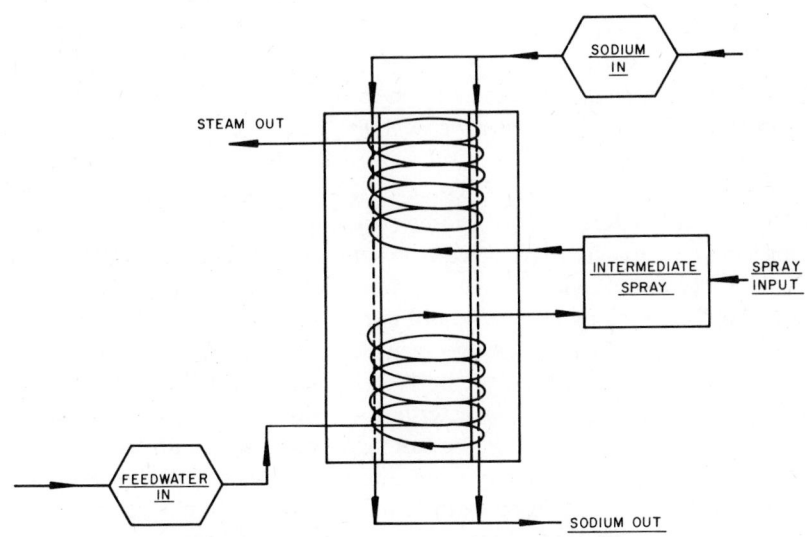

c. SCHEMATIC OF STEAM GENERATOR

Figure 1.2   *continued*

1.3   CLASSIFICATION OF TYPES OF INPUTS

In the previous section, we considered inputs as signals flowing into a system (often as *outputs* from some other system); here we take a more general view and define an input as any causative agent that can excite a system into response. Since this first chapter is intended to develop useful overviews of our subject, it is appropriate to attempt a classification of inputs into types. These types hopefully will encompass all possibilities that might arise in any specific practical application, thus enabling us to develop a coherent organization of this aspect of system analysis.

Our classification scheme will be an extension of that used in the author's earlier work.[5] The first level of classification separates inputs into three groups:

1   External driving.
2   Initial energy storage.
3   Parametric excitation.

A simple example, the electric circuit of Fig. 1.3, serves to introduce these concepts. The fixed capacitor and inductor and the variable resistor are so-called passive elements and incapable of continuously supplying energy. *External driving* consists of applying to the system an external energy input, such as

---

[5]Doebelin, *op. cit.*, p. 7.

the ideal voltage source $e_i(t)$. This will cause the system to respond with an output current i_o. If no external energy source is applied ($e_i(t)$ is replaced by a conducting wire), we can still cause the system to respond with a current i_o if the circuit is provided with some *initial stored energy*. In this electrical application (and also more generally in systems of all types), stored energy may appear in "potential" and/or "kinetic" form. If the capacitor is initially charged ("potential" energy) by some external power source (which is then *removed*), it is clear that the discharging capacitor will release its energy and cause the circuit to respond with a current. Similarly, if a current had been established in the inductor ("kinetic" energy) by application of $e_i(t)$, but then $e_i(t)$ is "shorted out" (preventing any further energy supply by e_i to the circuit), the circuit still exhibits a dynamic response in that the current i_o and the voltages across R, L, and C vary with time. Finally, if e_i is steady but we cause R to change with time, as by moving the sliding contact on R, we can again cause dynamic changes in i_o and various circuit voltages. This type of input has been called *parametric excitation*. Here the energy is still provided by the steady voltage source (which we might now wish to consider part of the system); however, the dynamic variation in energy supply is caused by externally controlled changes in an "internal" parameter of the system.

   Another useful classification scheme deals with the nature of the *time variation* of external driving and/or parametric excitation inputs (initial energy storages are by definition constant numbers, not functions of time). This mode of classification has significance both with respect to the actual operating conditions of real systems and also the methods required in the solution of the equations that mathematically model the system. One useful (though not exhaustive) scheme of this type categorizes inputs as follows:

1   Deterministic
   (a) Periodic
   (b) Transient
   (c) Special Cases (modulated and demodulated signals, etc.)
2   Random ("Stochastic")
   (a) Stationary
   (b) Nonstationary,

When we refer to types of inputs, it might be more correct to call them input *models* since, just as we cannot describe any real-world *system* precisely but must rather analyze a simplified system *model*, so we also cannot describe real inputs precisely and are forced to deal with approximations.

   A *deterministic* input model is one whose time history is prescribed by a mathematical formula, graphical curve, or table of data and that can be replicated "at will." For example, an important input to a rapid transit rail vehicle, one with significant influence on the output described as "rider comfort," is the track "roughness" as given by its vertical excursions from level. For

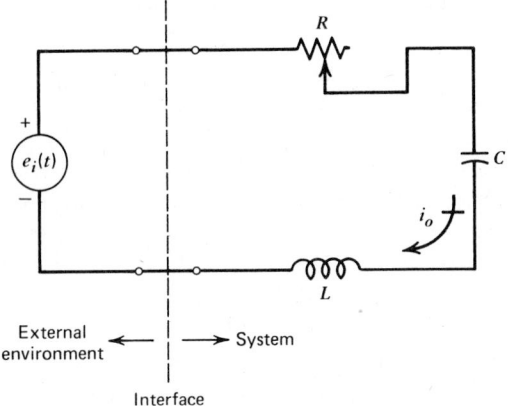

Figure 1.3   Electric Circuit with three input types.

a given stretch of track, such a track profile changes little from day to day and one could run a train over it repeatedly and be reasonably assured that the vertical input motion imparted to the vehicle was being replicated. The analogous situation for air transportation, however, gives rise to *random inputs*. Consider an aircraft flying a straight and level path through atmospheric turbulence and being subjected to "vertical gust velocity" input forces that cause vehicle vertical motions and passenger discomfort. If the aircraft is properly instrumented we can obtain a time history of the vertical air forces as we traverse a certain path; however, if we turn the vehicle around and retrace our previous path perfectly, our instruments will indicate that the air forces do *not* repeat the earlier time history, and, in fact, that time history will *never* again occur since we are dealing with a truly random phenomenon.

Elaborating now on the class of deterministic input models, the subclass of *periodic* inputs is one of great practical importance because it can be associated with the operation of a vast array of machines that run at nominally constant speed. Consider the constant-speed operation of any reciprocating and/or rotating engine, pump, turbine, compressor, and so forth. When in steady-state operation, every force, stress, velocity, temperature, pressure, flow rate, and so forth is going through some repetitive, periodic cycle. However, since all constant-speed machines must always be started up and occasionally shut down, and since such un-steady-state operations may represent critical conditions, engineers are also vitally concerned with these *transient* inputs. Transient operation is also involved in changing from one steady speed to another, as in braking or accelerating an automobile, and in emergency or failure situations such as the popping of a safety valve or blowout of an automobile tire. Also, if an occurrence is periodic but the period between events is long compared to the duration of the event, a transient input model may be appropriate. Thus, in a punch press, the effects of the previous punching impulse may have died out long before the next one occurs.

Our earlier listing of random inputs subdivided these into stationary and unstationary subtypes; we now wish to distinguish between these. In the aircraft turbulence example, while a specific time history will never be repeated, the *statistical* characteristics of the input, such as amplitude distribution, mean-squared value, and frequency content, may be reproducible from one "sample" to the next. For example, if we simultaneously fly ten airplanes, spaced at a considerable horizontal distance, through the turbulence, each time history will be different, but there can be agreement, on the *statistical* data, among the ten records. In fact, we might pool the data from all records to get more statistically reliable information to characterize the random input; this is called *ensemble averaging*. Another procedure would be to use only a single airplane but gather a ten times longer data record by flying it over a ten times longer path. If we subdivide this long record into ten subrecords of equal length, we can compute statistical data from each of these subrecords. If these statistical data are consistent among the ten records, the random input is said to be *stationary*, that is, the statistical characteristics do not vary with time. If the input is stationary we can treat the ten subrecords as one long record and pool the data for better statistical reliability (*time averaging*). If the statistical data vary significantly from subrecord to subrecord, the random input is *unstationary*. Unstationary random inputs present mathematical difficulties; however, many practical problems are dealt with satisfactorily by breaking the overall time record into judiciously selected subsections and assuming stationary behavior within each section.

In concluding this section, we remind the reader again that the input categories discussed herein refer to *model* types the analyst *chooses* to represent some physical phenomenon. All real-world inputs are nonstationary random (the most difficult type) since no real occurrence is *perfectly* predictable or reproducible. As in all other practical engineering work, the analyst must model the real-world situation in the *simplest* manner that will produce results of accuracy sufficient for the needs of the study.

1.4   CLASSIFICATION OF TYPES OF SYSTEM MODELS

Mathematical modeling of physical systems is, by definition, the description of system behavior by means of suitably chosen mathematical relations or equations. Since the mathematical description of real-world situations must *always* be, to some extent, imperfect, there is never *a* model for a given system but rather a spectrum of models. When an analysis or design is in its early stages, the engineer will often choose from this spectrum a relatively simple model so as to gain some understanding of the major determining factors without an excessive expenditure of analytical effort. The relative inaccuracy of such simple models is accepted willingly as a necessary price to be paid in order to gain the benefits of a quick overview of essential features. As basic understanding is developed from the simpler models, so also are their limitations revealed, and it becomes appropriate to add complicating features to the model to improve accuracy. This planned, gradual increase in model complexity has been found to be a logical, systematic method of dealing with difficult problems.

For those who have not earlier encountered any organized treatment of the classification of model types, we present the following brief introduction.[6] Since the difference in model types lies in the nature of the equations employed, a discussion of model types becomes essentially a discussion of equation types. To make such a discussion of practical utility, we take an engineer's (rather than a mathematician's) point of view and restrict the scope and detail so as to include only what we consider to be the major classes of equations regularly used in practical applications in system dynamics. While various types of equations have found utility in applications, our emphasis here will be on ordinary and partial differential equations.

Since physical objects extend over three-dimensional space, and since our interest generally includes dynamic response (making time an independent variable), the unknowns (outputs) in our system studies will depend on four independent variables. Thus, for example, the motion of a vibrating structure depends on *where* we look (x, y, z space coordinates) and *when* we look (t, the time variable). The "most accurate" macroscopic viewpoint employed in the modeling of problems in various fields of engineering science, such as fluid flow, heat transfer, and vibrations, considers matter as a structureless continuum; thus the unknowns vary "smoothly" over space. The physical laws that interrelate the variables generally involve rates of change with respect to space and/or time, thus we are naturally led to differential equations and these will be partial differential equations since the unknowns depend on as many as four independent variables (x, y, z, t).

The accuracy and also the degree of difficulty involved in solving the above-mentioned partial differential equations depends critically on the simplifying assumptions applied by the analyst in setting up the mathematical model of the real-world situation. As one would expect, models that more closely duplicate the actual physical behavior will involve fewer assumptions, be more complex, and thus require in their solution the application of more sophisticated and lengthy mathematical techniques, if indeed they are solvable at all. One useful method of displaying the spectrum of model types produced by various kinds of assumptions uses the two broad classifications: Nature of the Medium, and Time Variation of System Parameters, each of which is further

---

[6]Doebelin, E. O., *System Dynamics*, C. E. Merrill, Columbus, Ohio, 1972, p. 10.

TABLE 1.1 Classification of System Model Type

| MODEL TYPE NUMBER | NATURE OF THE MEDIUM, AS MODELED | | | | | | LINEARITY | | TIME VARIATION OF SYSTEM PARAMETERS | | |
| | CONTINUITY | | DIRECTIONALITY | | UNIFORMITY | | | | | | |
	CONTINUOUS	DISCRETE	ANISOTROPIC	ISOTROPIC	INHOMOGENEOUS	HOMOGENEOUS	NONLINEAR	LINEAR	RANDOM	DETERMINISTIC VARIABLE	DETERMINISTIC CONSTANT
1	X		X		X		X		X		
2	X		X		X		X			X	
3	X		X		X		X				X
4	X		X		X			X	X		
5	X		X		X			X		X	
6	X		X		X			X			X
etc.	X										
20	X			X		X	X			X	
21	X			X		X	X				X
22	X			X		X		X	X		
23	X			X		X		X		X	
24	X			X		X		X			X
25		X					X		X		
26		X					X			X	
27		X					X				X
28		X						X	X		
29		X						X		X	
30		X						X			X

Partial differential equations (rows 1–24)

Ordinary differential equations (rows 25–30)

Most realistic, hardest to solve →

Least realistic, easiest to solve →

14

subdivided, as shown in Table 1.1. The first 24 entries in this table ("Field Problems") correspond to the partial differential equation models described above, so let us now discuss this section of the table.

It may be helpful to have in mind a concrete physical example as we discuss the general significance of Table 1.1. Figure 1.4 shows a solid object of arbitrary shape, made of a material exhibiting both inertia and elasticity, and thus capable of vibration. If specified external forces (driving inputs) and/or initial deformations (energy storage) are specified, knowledge of the vibratory motions would require solution for the displacements u, v, w as functions of location x, y, z and time t. The application of appropriate physical laws and certain types of assumptions leads to a simultaneous set of partial differential equations,[7] one of which is

$$\left[\frac{E(1-\nu)}{(1+\nu)(1-2\nu)} \right] \left[\frac{\partial^2 u}{\partial x^2} + \frac{\partial^2 u}{\partial y^2} + \frac{\partial^2 u}{\partial z^2} \right] = \rho \frac{\partial^2 u}{\partial t^2} \tag{1.2}$$

This equation was obtained by using assumptions that place this system model in category 24 of the table. In the table, *directionality* refers to whether the material behavior, such as the relation between stress and strain, is different when we consider different directions in the body. For example, a composite, fiber material would exhibit rather different properties *along* the fiber direction as compared with *across* the fibers, whereas many steels would behave nearly identically in all directions. Directional materials are called *anisotropic* while nondirectional are *isotropic*.

Another descriptor of material behavior involves *uniformity* from point to point. That is, for example, is the mass density ρ the same throughout the body? Uniform media are called *homogeneous*; nonuniform *inhomogeneous*. A material or medium can be homogeneous with respect to one aspect of behavior (say, density) and inhomogeneous for another (say, modulus of elasticity, E). Finally, under Nature of the Medium, the behavior is either *linear* or *nonlinear*. This refers to the nature of the mathematical relation between variables, such as stress/strain relations. In arriving at Eq. (1.2) the *simplest* assumptions (isotropic, homogeneous, linear) were made.

The other major classification, Time Variation of System Parameters, deals with the possibility that a model might assume a known variation of certain parameters with time, as opposed to (or in addition to) the variation with direction (anisotropy) or location (inhomogeneity). For example, if we knew that the temperature of the vibrating body of Fig. 1.4 varied in a known way with time, this might be incorporated into the model as time-varying values of density ρ, modulus E, and Poisson's ratio ν. In the real world, all system parameters

---

[7]Timoshenko, S., *Theory of Elasticity*, McGraw-Hill, 1951, N.Y. p. 453.

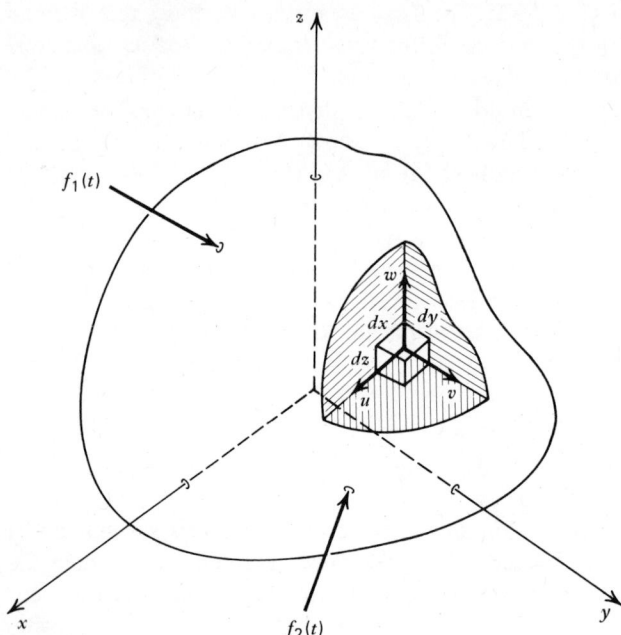

Figure 1.4 Vibrating body.

vary randomly with time due to the influence of environmental fluctuations in temperature, humidity, air pressure, and so forth; however, these random variations are often small compared with predictable (deterministic) changes or constant average values, so these simpler models are widely used. Equation (1.2) again assumes the simplest situation: that of constant values of parameters E, ν, and ρ.

While the partial differential equation ("Field Theory") models (entries 1–24) of Table 1.1 can be quite accurate, they have been found analytically solvable for only limited cases, mainly in category 24, and there only for simple geometries, inputs, and boundary conditions. To solve many practical problems it is necessary to give up the continuous medium approach and deal instead with a *discretized, lumped* or *network* model. We later discuss at some length the various schemes for implementing this concept (physical lumping, finite difference methods, finite elements); for now we draw on an assumed familiarity with *physical* lumping since it is commonly employed in many undergraduate engineering topics. Here, instead of applying the physical laws to an infinitesimal element $dx\,dy\,dz$ that shrinks to a geometrical point under the calculus limiting processes and allows smooth spacewise variation of the unknowns, we *begin* by representing the medium as a connection of discrete, finite size spring

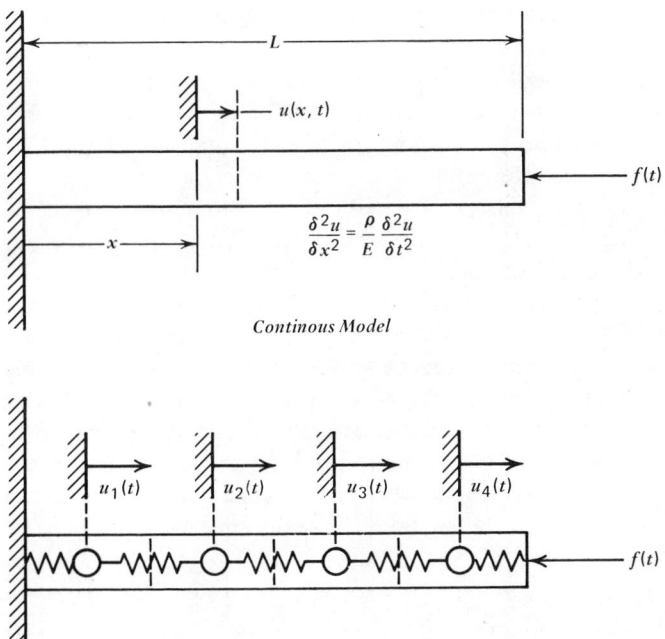

$$\frac{\delta^2 u}{\delta x^2} = \frac{\rho}{E}\frac{\delta^2 u}{\delta t^2}$$

Continous Model

Four-Lump Discrete Model

Figure 1.5   One-dimensional continuous and discrete vibration models.

and mass elements, as in the one-dimensional example of Fig. 1.5. Rather than a single partial differential equation that comprehends the spacewise variations by inclusion of the term $\partial^2 u(x,t)/\partial x^2$, we obtain a set of simultaneous *ordinary* differential equations, the spacewise variation being accommodated in discrete approximation by a finite number of unknown displacements $u_1(t), u_2(t)$, and so on.

In Table 1.1, anisotropy and/or inhomogeneity are now dealt with in simpler (through approximate) fashion by merely adjusting the *numerical values* of the individual lumped elements to agree with the *local* values of the space-wise-varying parameters. That is, if density ρ actually varies with x as $\rho = \rho_0(1 + 2x/L)$, in the discrete model, for a lump at $x = L/2$, we would just assume a *constant* density of $2\rho_0$. As in the continuous model, we may choose either linear or nonlinear relations between system variables, and random, deterministically variable, or constant system parameters. Thus all the general *features* of the real-world situation can be modeled but they are treated in a more approximate fashion. Of course the benefit obtained for this concession to inaccuracy is the generally greater *simplicity* of ordinary differential equations as compared with partial.

As a final, summary comment on Table 1.1, let us state that the theoretical foundations of most of what we call "system analysis" now lie (and probably always will) in the linear, ordinary, constant-coefficient equations of type 30, plus the corresponding partial differential equations of type 24. These (particularly the ordinary) are the only classes of equations whose complexity can be expanded to deal with large systems and for which we still are able to analytically predict behavior in a systematic, routine fashion. Of course in getting *specific* solutions to *specific* problems (as contrasted with development of a unified underlying theory), one may expect continued progress in numerical, computerized solutions. These methods can be applied to all the classes of Table 1.1; however, in essence they reduce all problems to discrete-type models. Large, fast computers and increasingly sophisticated discretization schemes can produce extremely accurate results; however, the capability for such analyses does not automatically mean that they *should* be used. Practical judgement is still necessary in deciding how accurate a result is really needed and can be economically justified.

1.5   COUPLING OF SUBSYSTEMS

In concluding this introductory chapter it is appropriate to comment briefly on the important topic of subsystem coupling. There are several reasons why the system approach emphasizes "modular" or "building block" methods in dealing with complex processes, the most obvious being that the human mind can comprehend only a limited degree of complexity at one time. Thus we must break large problems into pieces of manageable size. Even computers suffer from such limitations; therefore, efficient computer utilization often also requires a modular scheme. Aside from these problems of "comprehension," modularity also offers advantages of *efficiency*. That is, if we model components in such a way that they are easily interfaced, at both input and output ports, with other compatible devices, then such modeling effort need be expended *only once*; there is no need to repeat this work each time this component appears in a different context. Finally, the assembly of real systems often involves the bringing together of components manufactured at geographically dispersed sites by several independent firms or divisions of a larger company. If the final coupling is to involve a minimum of unanticipated problems, it is most helpful to be able to subject each component, before it leaves its site of manufacture, to experimental tests carefully designed to evaluate its compatibility with the mating subsystems. The design, implementation, and interpretation of such tests rests heavily on subsystem coupling concepts.

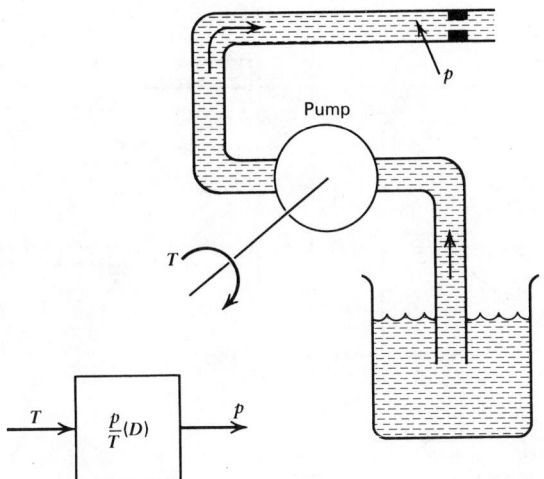

Figure 1.6   Transfer function model of hydraulic pump.

Many of the practical applications of system analysis have in the past relied on (and will most probably in the future continue to rely on) the *transfer function*[8-10] as the subsystem coupling concept. While, in keeping with the intermediate-to-advanced level of this text, a knowledge of transfer function methods by the reader is presumed, Chapter 2 will include a review of this material as an aid to those not already prepared in this area. For the purposes of this section, we will consider, in a qualitative way only, the specific examples shown in Figs. 1.6 and 1.7 as a means of introducing basic viewpoints developed in detail in Chapter 7. Figure 1.6 shows a hydraulic pump connected to a specific load and driven by a given input torque $T(t)$. By making suitable assumptions one can derive a differential equation relating input torque T to output pressure p, and thereby define a transfer function $(p/T)(D)$ and a block diagram as shown. Because of the assumptions made and the modeling technique used, this description of the pump, while useful for the specific application shown, is in reality *incomplete* and does not contain sufficient information to allow modeling of the pump's behavior when connected in ways *other* than those shown in Fig. 1.6.

To overcome the limitations of the transfer function approach, engineers have developed various modeling methods such as *impedance and mobility*,

[8]Doebelin, E. O., *Dynamic Analysis and Feedback Control*, McGraw-Hill, N.Y., 1962.
[9]Doebelin, E. O., *Systems Dynamics*, C. E. Merrill, Columbus, Ohio, 1972.
[10]Doebelin, E. O., *Measurement Systems* (Rev. Ed.), McGraw-Hill, N.Y., 1976.

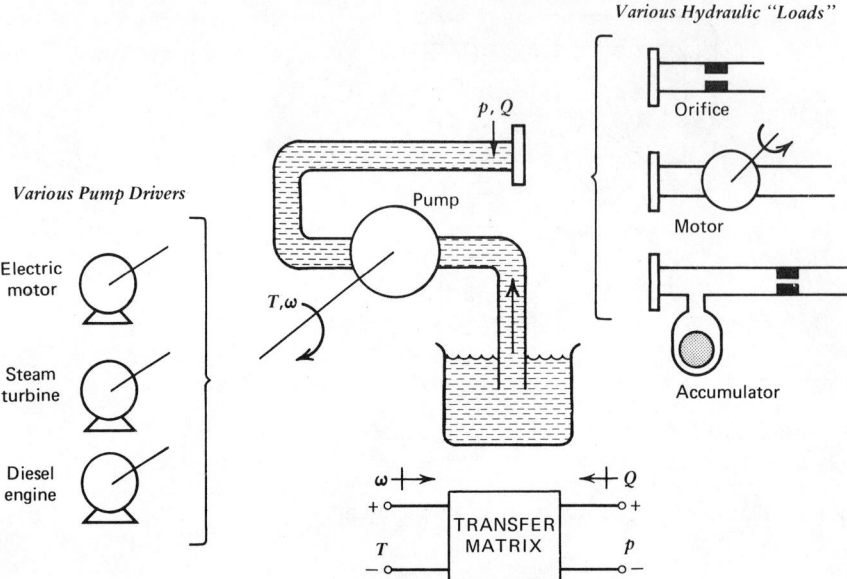

Figure 1.7 Transfer matrix model of hydraulic pump.

transfer matrix, and *bond graphs*. These methods, while differing in detail, basically recognize that the interaction between two subsystems at their joining interface cannot be rigorously described with only a *single* coupling variable, such as torque T or pressure p in Fig. 1.6. Rather, to deal with the *power transfer* that describes the energy flow between subsystems, it is necessary to consider *two* variables at each coupling port. These variables are selected so that their product is the instantaneous power being transmitted through the port. We shall show later that a modeling technique based on such concepts, while requiring greater effort to develop, allows one to interface subsystems in a much more general way than does the transfer function method. Thus in Fig. 1.7, once we have derived, by theory and/or experiment, a transfer matrix model for the pump, we can combine this with similar type models for many different kinds of "upstream" and "downstream" devices to accurately predict overall system performance.

BIBLIOGRAPHY

1 Beachley, N. H. and H. L. Harrison, *Introduction to Dynamic Systems Analysis*, Harper and Row, N.Y., 1978.

2 Berlinski, D., *On Systems Analysis*: *An Essay Concerning the Limitations of Some Mathematical Methods in the Social, Political, and Biological Sciences*, The MIT Press, Cambridge, Mass., 1976.

3   Cannon, R. H., *Dynamics of Physical Systems*, McGraw-Hill, N.Y., 1967.

4   Close, C. M. and D. K. Frederick, *Modeling and Analysis of Dynamic Systems*, Houghton Mifflin, Boston, 1978.

5   Cochin, I., *Analysis and Design of Dynamic Systems*, Harper and Row, N.Y., 1977.

6   Doebelin, E. O., *System Dynamics*, C. E. Merrill, Columbus, Ohio, 1972.

7   Haberman, C. M., *Engineering Systems Analysis*, C. E. Merrill, Columbus, Ohio, 1965.

8   Hoos, I. R., *Systems Analysis in Public Policy*, Univ. of Cal. Press, Berkeley, 1973.

9   Hoos, I. R., "Systems Analysis as Technology Transfer," *J. Dyn. Sys., Meas. and Control*, March 1974.

10   Karnopp, D. C. and R. C. Rosenberg, *System Dynamics*, John Wiley and Sons, N.Y., 1975.

11   Lilienfeld, R., *The Rise of Systems Theory*, Wiley-Interscience, N.Y., 1978.

12   Martens, H. R. and D. R. Allen, *Introduction to Systems Theory*, C. E. Merrill, Columbus, Ohio, 1969.

13   Ogata, K., *System Dynamics*, Prentice-Hall, Englewood Cliffs, N.J., 1978.

14   Paynter, H. M., *Analysis and Design of Engineering Systems*, MIT Press, Cambridge, Mass., 1961.

15   Oldenburger, R., *Mathematical Engineering Analysis*, Macmillan, N.Y., 1950.

16   Reswick, J. B. and C. K. Taft, *Introduction to Dynamic Systems*, Prentice-Hall, Englewood Cliffs, N.J., 1967.

17   Shearer, J. L., Murphy, A. T. and H. H. Richardson, *Intro. to System Dynamics*, Addison-Wesley, Reading, Mass., 1967.

18   Takahashi, Y., Rabins, M. J. and D. M. Anslander, *Control and Dynamic Systems*, Addison-Wesley, Reading, Mass., 1970, Part 2.

19   *Journal of Dynamic Systems, Measurement and Control*, Am. Soc. of Mech. Eng., N.Y.

CHAPTER 2

Transform Solution of Differential Equations

2.1 CLASSIFICATION OF PROBLEM TYPES

When we choose to represent engineering problem situations in terms of the input/system/output concept, the existence of certain classes of problem types becomes apparent. These classes are defined by enumerating all the combinations of the three entities (input, system, output) in which two are assumed "given" while the third is "to be found." The most common problem type is the so-called *analysis* problem, in which the input and system are given and we are required to solve for the system response (output). A large percentage of the problems assigned to students in undergraduate and graduate engineering curricula are of this type. At first glance this academic emphasis on analysis may seem unjustified since it is generally conceded that the major role of engineering in society revolves around the *design* of new or improved machines, processes, and services. Design or *synthesis* problems may be described in our present context as those in which certain inputs (power sources, materials, etc.) are assumed available, certain outputs (forces, motions, flow rates, etc.) must be produced to insure the desired functioning of the process, and the engineer's job is to assemble a system that will allow the required outputs to be produced from the available inputs.

For analysis problems in which both the input and system are *mathematically* described, there can be only one correct solution. If, on the other hand, only *physical* descriptions are given, then, as we saw in Chapter 1, many mathematical models, each with a different "answer," are possible. Real-world *design* problems always have many different acceptable solutions, although formal "optimum design" procedures can sometimes define isolated "best" designs by invoking mathematical performance criteria. The vast majority of practical designs, however, do not employ formal optimization methods but rather rely on an iterative process of "creative" design, design analysis, and performance evaluation. At the creative design stage a number of competitive alternative design concepts are formulated, based mainly on the designer's past experience, and familiarity with available hardware, rather than on abstract mathematical concepts. Evaluation of these competitive designs, leading ultimately to a decision to develop the "best" concept, involves mainly a succession of *analysis* problems. This explains the heavy emphasis in academic programs on analysis; much of the *design* process necessarily consists of analysis, and that part ("creative design") which does not depends so heavily on experience and intimate knowledge of local company conditions that it is difficult to present realistically outside of the actual industrial environment.

While the analysis and design types of problems already discussed are clearly of major importance, additional useful types may be defined. One such is the *system identification* or *experimental modeling* type. Here the system already exists as a physical entity, however we do not have a valid mathematical model for it. By applying suitable inputs, the system is dynamically "exercised," producing output responses. We carefully measure both input and output and from these data are able to discover the system description. Such experimental modeling methods are a necessity for those systems too complex to model accurately using theory only, and will be treated in considerable detail in Chapter 6.

Our final problem categories are called *indirect measurement* and *measurement correction* and both involve situations where system and output are known and we wish to calculate input from these given data. As an example of indirect measurement consider a case where we desire to measure a certain input quantity of a system but find that suitable sensors do not exist. If we have a valid system model that analytically relates the system variables, it may be possible to make measurements only on those quantities for which sensors are available and then combine these measured data with our system model to *calculate* those quantities that could not be directly measured.

In the case of measurement correction, we refer to a situation where a sensor for the desired measurement is available but the sensor's response is somewhat imperfect dynamically. In this case our "system" is the sensor itself and its output is of course the "indicated" value of the measured quantity which is our system input. Even if the sensor is imperfect, as long as we *know* its model we can combine the (imperfect) indicated data with our system description to analytically reconstruct the "true" value of the input quantity.

While all the above classes of system problems may be *formulated* irrespective of the type of system model selected, "guaranteed" analytical solutions by established, routine methods are available for all these classes only when the equations are ordinary linear differential equations with constant coefficients. The remainder of this chapter is devoted to presenting these solution methods in an easily usable form.

2.2 LINEARIZATION

Since the real world is always nonlinear, at least to some degree, while the most useful mathematical models are strictly linear, the concept of *linearization* is needed to help understand why linear methods have had such widespread success. While nonlinearity takes on an endless variety of forms, many relations among physical variables are in the nature of smooth curves. In considering possible methods for linearly approximating such a relation, several possibilities suggest themselves. If it is felt that accuracy "at the beginning and end points" is important, a line might be chosen as in Fig. 2.1*a*. A more accurate representation "on the average" might be obtained by use of a line as in Fig. 2.1*b*, either choosing the line "by eye" or using a formal procedure such as a least-squares fit. If small variations ("perturbations") about a chosen operating point are of interest, the local tangent to the curve gives a linear approximation that becomes "perfect" as the perturbations shrink to zero. Such an approximation can be applied at several different points over an operating range (Fig. 2.1*c*) to explore the effect on system behavior.

This last ("small-signal") type of linearization is widely used and intuitively seems to explain the success of linear methods in practice; in many real systems the slope of the nonlinear curve does not change radically and/or the perturbations of the variables from their operating points may not be excessive. A mathematical formulation of this concept is readily available from the Taylor series expansion of a function $y = f(x)$ about a chosen operating point x_0:

$$y \approx \underbrace{f(x_0) + \frac{df}{dx}\bigg|_{x=x_0}(x-x_0)}_{\text{linear approximation}} + \frac{d^2f}{dx^2}\bigg|_{x=x_0}\frac{(x-x_0)^2}{2!} + \cdots \qquad (2.1)$$

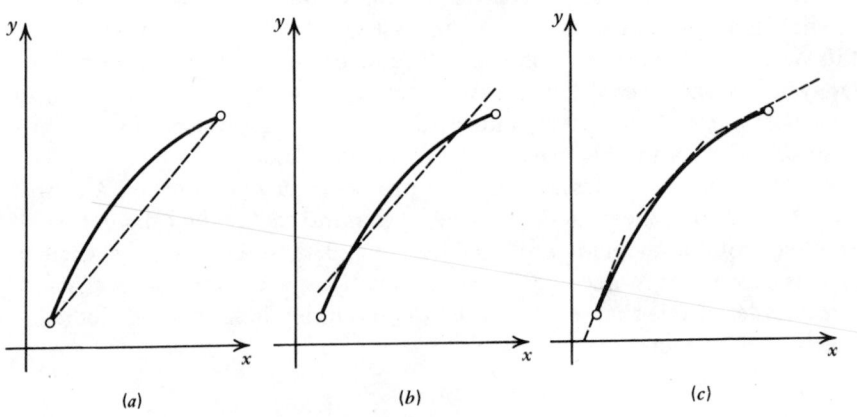

(a) (b) (c)

Figure 2.1 Three linearization schemes.

620.73 D67s
C-1

To obtain a linear approximation, we retain only the first two terms of this series, which are easily seen to give the equation of the tangent line at x_0. This linearization scheme is easily extended to nonlinear functions $y = f(x_1, x_2, x_3, \ldots)$ of several variables as follows:

$$y \approx f(x_{10}, x_{20}, x_{30}, \ldots) + \frac{\partial f}{\partial x_1}(x_1 - x_{10})$$

$$+ \frac{\partial f}{\partial x_2}(x_2 - x_{20}) + \frac{\partial f}{\partial x_3}(x_3 - x_{30}) + \cdots \qquad (2.2)$$

All the partial derivatives are *evaluated* at the operating point $x_{10}, x_{20}, x_{30}, \ldots$ so they are just numbers and y becomes a linear function of $x_1, x_2, x_3, \ldots$ in Eq. (2.2). For the case of two independent variables x_1 and x_2, a geometrical interpretation of the approximation as the tangent plane to the surface $y = f(x_1, x_2)$ is possible, as in Fig. 2.2. When there are three or more independent variables, visual geometrical interpretations are no longer possible; however, mathematicians refer to the relation among the variables as a *hyperplane*, thus retaining at least the terminology.

We wish to emphasize that *every* engineering analysis that utilizes linear mathematical models is fundamentally based on linearization concepts such as those above, *whether or not this fact is explicitly stated as part of the analysis.*

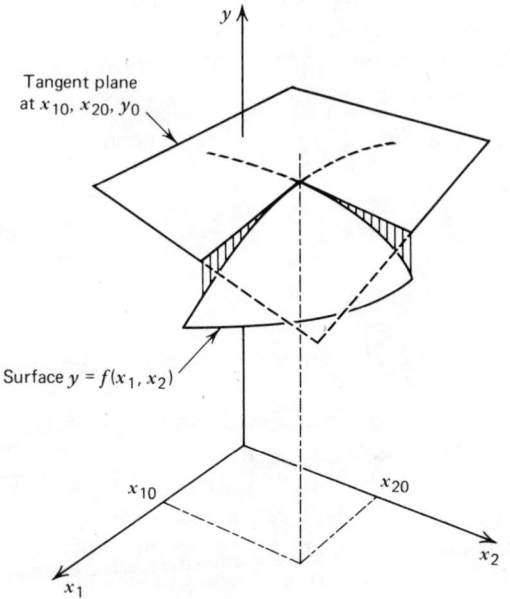

Figure 2.2 Geometrical interpretation of multivariable linearization.

Since authors often do not feel the need to continually remind readers of this situation, we feel the need to point it out forcefully here. Keep in mind, then, that results based on linear models are always subject to the limitations inherent in the small-signal concept of linearization discussed above.

2.3 REVIEW OF CLASSICAL METHOD OF LINEAR DIFFERENTIAL EQUATION SOLUTION

While this chapter emphasizes transform methods, a brief review of the classical methods of solution for linear ordinary differential equations with constant coefficients is in order. The general form of such an equation is

$$a_n \frac{d^n x}{dt^n} + a_{n-1} \frac{d^{n-1} x}{dt^{n-1}} + \cdots + a_2 \frac{d^2 x}{dt^2} + a_1 \frac{dx}{dt} + a_0 x = f(t) \tag{2.3}$$

where x is the unknown (dependent variable), t is the independent variable, the a's are constants and $f(t)$ is the "forcing function" that drives the system. In the classical method of solution a three-step procedure is followed:

1 Find the complementary solution x_c. Methods that "always work" are available for this step.

2 Find the particular solution x_p. This step may or may not be possible, depending on the nature of $f(t)$.

3 Add x_p to x_c to get the complete solution x. This complete solution will contain n arbitrary constants of integration that appeared in x_c. Numerical values may be found for these constants if n initial conditions are known. At this point an explicit equation for x in terms of t will be obtained and one can calculate x for any given t.

To find the complementary solution, Eq. (2.3) is written in operator form with the right-hand side set to zero:

$$(a_n D^n + a_{n-1} D^{n-1} + \cdots + a_2 D^2 + a_1 D + a_0) x = 0 \tag{2.4}$$

where $D \triangleq d/dt$. The system *characteristic equation* is

$$a_n D^n + a_{n-1} D^{n-1} + \cdots + a_2 D^2 + a_1 D + a_0 = 0 \tag{2.5}$$

and we now treat this as an algebraic equation in D and solve for the n roots $s_1, s_2, \ldots, s_n$. It has been shown in algebra that if $n > 4$, it is *impossible* to find these roots if the coefficients (a's) are known only as letters rather than numbers. This is unfortunate, since the a's are the physical parameters (spring constants, electrical resistances, etc.) of our system and we would much prefer to carry out our solution in *general* (letter) form so that we can examine the solution to determine the effects on system behavior of changing the parameters. Even for $n=4$, where algebra provides formulas to get the four roots in letter form, these formulas are so complex as to obscure the trends we wish to observe. For $n=3$ the root formulas become more manageable and for $n=2$ the familiar quadratic formula provides easily usable results. Thus, for $n>4$ we *must* know the a's as numbers and for $n=3$ or 4 we may *prefer* to work with numerical values. If the coefficients are known as numbers, the roots may be found by use of various numerical approximation methods or algorithms. Most computer libraries provide "canned" routines to find these roots even for large values of n; the user need only input the values of the a's. We should finally point out that this step of finding the n roots is *unavoidable* in any analytical solution method, including the transform method which we will shortly introduce.

Assuming the roots $s_1, s_2, \ldots, s_n$ have been found, one writes down the complementary solution x_c *immediately* by following certain rules that are proven in texts on differential equations but which we here simply accept as given. The roots are either real or complex; repeated or unrepeated, no other possibilities exist. For any *real unrepeated* root, such as, say, -3.78, the solution corresponding to this root is $x_c = C_1 e^{-3.78t}$, where C_1 is a constant of integration whose numerical value cannot be found until the third step of the three-step process given above, and e is the base of natural logarithms. Should we have *multiple* (repeated) *roots* (a relatively rare occurrence in practical problems) the solution takes the form

$$x_c = C_0 e^{st} + C_1 t e^{st} + C_2 t^2 e^{st} + \cdots + C_m t^m e^{st} \qquad (2.6)$$

where the root s is repeated m times.

If the a's in Eq. (2.5) are real numbers, as they usually will be since they represent real physical parameters such as spring constants and masses, then it can be shown that if any *complex roots* appear they will appear in *pairs* $a \pm ib$. The solution for such a pair will be $x_c = C e^{at} \sin(bt + \phi)$, where C and ϕ are the two arbitrary constants needed for a pair of roots. Should a pair of complex roots be repeated (again a rare occurrence in practice), the solution takes the form

$$x_c = C_0 e^{at} \sin(bt + \phi_0) + C_1 t e^{at} \sin(bt + \phi_1) + \cdots \qquad (2.7)$$

similar to the result for repeated real roots. We have now stated all the rules needed to write down x_c once the roots have been found. For example, if a computer root-finding procedure has just presented us with the following list of roots for a seventh-order equation:

$$0, -1.75, +8.62, -3.2 \pm i10.7, -8.5 \pm i4.3 \qquad (2.8)$$

we immediately write out the solution as

$$x_c = C_1 + C_2 e^{-1.75t} + C_3 e^{8.62t} + C_4 e^{-3.2t} \sin(10.7t + \phi_1)$$
$$+ C_5 e^{-8.5t} \sin(4.3t + \phi_2) \tag{2.9}$$

While the complementary solution can in principal *always* be found, the *particular solution* x_p depends on $f(t)$ and a mathematician can always concoct a sufficiently "pathological" $f(t)$ to thwart any proposed "general" method. Thus books on differential equations show *several* methods for finding x_p but do not guarantee that *any* of them will always work. Fortunately a simple method, the *method of undetermined coefficients*, works for many of the $f(t)$'s of practical interest. To check whether it *will* work, differentiate $f(t)$ repeatedly. If the higher derivatives can be seen to *go to zero* ($f(t) = 3 + 5t + 8t^2$, for example) or to *repeat themselves* ($f(t) = 6\cos 10t - 3\sin 8t$, for example), then the method will work; otherwise it will not and some other method must be tried. For those $f(t)$'s for which the method does work, the particular solution x_p is given by a sum of terms made up of each *different* type of function found in $f(t)$ and all its derivatives, each term multiplied by an undetermined constant. For the two examples given earlier we would have:

$$x_p = A + Bt + Ct^2 \tag{2.10}$$

and

$$x_p = A \cos 10t + B \sin 10t + C \cos 8t + D \sin 8t \tag{2.11}$$

respectively. These particular solutions would be substituted into the original differential equation, making it an identity. By gathering terms of like form on each side of this identity and requiring that their coefficients be identical, one can obtain a set of simultaneous algebraic equations that can be solved for numerical values of the coefficients A, B, and so on.

We have now shown how x_c, which has n unknown constants in it, and x_p, which is completely known, may be found. Now we add x_c and x_p to get the complete solution, which will contain the n unknown constants from x_c. To find these and thereby get a specific solution, we must find and impose n *initial conditions* on x and its derivatives. We should note that in the classical method the word "initial" refers to an instant of time just *after* the input is applied to the system. This requirement may lead to the need for significant additional effort in obtaining the final solution, as compared with the transform method that uses conditions just *before* the input is applied. Assuming we can obtain numerical values for the needed initial conditions, we insert these into the solution x (and/or its derivatives) and obtain n algebraic equations in the constants of integration, for which we then solve.

Only the simplest physical devices will be described by a *single* differential equation; most practical systems are modeled by a *set of simultaneous equations*. The classical solution method deals with such a set of n equations in n unknowns by using the operator form of the equations to reduce the set down to a single equation in a single unknown, which is then solved by the methods we have just explained. Once the differential equations have been rewritten in operator form, they may be treated as if they were algebraic equations, and any method learned in algebra for reducing n equations in n unknowns to one equation in one unknown may be employed. For only a few equations, substitution and elimination may be quickest; for large numbers of equations the use of determinants is usually best. An example illustrates the procedure:

$$2\frac{dx_1}{dt} + 3x_1 - x_2 = 2t \tag{2.12}$$

$$x_1 - \frac{dx_2}{dt} - x_2 = 0 \tag{2.13}$$

$$(2D+3)x_1 + (-1)x_2 = 2t \tag{2.14}$$

$$(1)x_1 + (-D-1)x_2 = 0 \tag{2.15}$$

$$x_1 = \frac{\begin{vmatrix} 2t & -1 \\ 0 & -D-1 \end{vmatrix}}{\begin{vmatrix} 2D+3 & -1 \\ 1 & -D-1 \end{vmatrix}} = \frac{-2-2t}{-2D^2-5D-2} \tag{2.16}$$

$$(2D^2+5D+2)x_1 = 2+2t \tag{2.17}$$

$$x_2 = \frac{\begin{vmatrix} 2D+3 & 2t \\ 1 & 0 \end{vmatrix}}{\begin{vmatrix} 2D+3 & -1 \\ 1 & -D-1 \end{vmatrix}} = \frac{-2t}{-2D^2-5D-2} \tag{2.18}$$

$$(2D^2+5D+2)x_2 = 2t \tag{2.19}$$

Equations (2.17) and (2.19) are of course the same form as Eq. (2.3) and are solved by the methods shown earlier. In actually carrying out these solutions, some difficulty may be encountered in applying the initial conditions to find the constants of integration. This is because the original two simultaneous equations would usually have known initial conditions for x_1 and x_2, whereas Equations (2.17) and (2.19) require initial values of x_1 and dx_1/dt and x_2 and dx_2/dt, respectively. The needed initial values of the derivatives are not obvious and must actually be *derived* from Equations (2.12) and (2.13). In the transform method for simultaneous equations, this additional complication will not arise.

2.4 LAPLACE TRANSFORM DEFINITIONS AND THEOREMS

Our presentation of Laplace transform methods will emphasize practical utility rather than mathematical rigor. Our initial interest is in presenting an alternative to the "classical" methods of solution for linear ordinary differential equations with constant coefficients. While Laplace transform methods generally will only solve equations that *are* solvable by other means, the transform method often has conspicuous advantages, particularly for the more difficult types of problems. We will also find that, in addition to its use in solving differential equations, Laplace transform techniques lead to many other useful system analysis capabilities that are difficult or impossible to duplicate without transform methods.

We will continue to use t (time) as the independent variable in our differential equations; clearly both classical and transform methods are unaffected by a change to a space (say x) independent variable if the physical problem requires it. Given a function $f(t)$, the definition of its Laplace transform $F(s)$ is:

$$\mathcal{L}[f(t)] \triangleq F(s) \triangleq \int_0^\infty f(t)e^{-st}\,dt \qquad (2.20)$$

where

$$s \triangleq \text{a complex variable} \triangleq \sigma + i\omega \qquad (2.21)$$

and e is the base of natural logarithms. The dimensions of σ, ω, and thus s are $1/\text{time}$; $F(s)$ has dimensions that are those of $f(t)$ multiplied by time (thinking of the integral of Eq. (2.20) as the "area" under the $f(t)e^{-st}$ versus t curve (e^{-st} is dimensionless)). Since in Eq. (2.20) we integrate with respect to t and then substitute numerical values ($\infty, 0$) for t, t disappears and the evaluated integral becomes a function *only* of s, $F(s)$. Conventionally, lower case letters are used for functions of t and the corresponding upper case for their Laplace transforms.

To illustrate the procedure, suppose we have a fluid pressure $p(t)$ that varies with time according to $p(t) = 6t$ pascal, t in seconds. We compute its Laplace transform $P(s)$ as follows:

$$P(s) = \int_0^\infty 6te^{-st}\,dt = \left[6\frac{e^{-st}}{s^2}(-st-1)\right]_0^\infty$$

$$= -6\left[\frac{te^{-st}}{s} + \frac{e^{-st}}{s^2}\right]_0^\infty \qquad (2.22)$$

$$P(s) = -6\left[(0+0) - \left(0 + \frac{1}{s^2}\right)\right] = \frac{6}{s^2} \text{ pascal-sec} \qquad (2.23)$$

$f(t)$	$F(s)$
at	a/s^2
a	a/s
e^{-at}	$1/(s+a)$
$(\sin \omega t)/\omega$	$1/(s^2+\omega^2)$

Figure 2.3   Some simple transform pairs.

Note that we use "ordinary" integral tables to evaluate the integrals and treat s as an "ordinary" numerical parameter, since it does *not* depend on t. It should be clear that:

1   The more general function at would have the transform a/s^2.
2   We should *tabulate* $f(t)$'s and their corresponding $F(s)$'s since there is no point in *repeating* the work of Equations (2.22) and (2.23) each time we encounter a certain function.

Figure 2.3 is a short table of this type while Appendix A gives a more comprehensive listing. In going from Eq. (2.22) to (2.23) one should, to be mathematically correct, place some *restrictions*[1] on the allowed numerical values of s. For example, the term e^{-st}/s^2 does *not* go to zero for $t=\infty$ unless $\sigma>0$, as can be seen from the following, more careful, analysis:

$$\frac{e^{-st}}{s^2} = \frac{e^{-(\sigma+i\omega)t}}{(\sigma+i\omega)^2} = \frac{e^{-\sigma t}e^{-i\omega t}}{(\sigma+i\omega)^2} = \frac{e^{-\sigma t}(\cos\omega t - i\sin\omega t)}{(\sigma+i\omega)^2} \qquad (2.24)$$

If $s=-2$, for example, e^{-st}/s^2 goes to infinity, not zero, when $t\to\infty$. Fortunately these restrictions are of no consequence in the vast majority of practical applications and we will not consider them further.

The transform definition of Eq. (2.20) allows one to compute the $F(s)$ corresponding to a given $f(t)$ for any $f(t)$ for which the integral can be evaluated (*not all* t-functions are Laplace transformable), and is called the *direct transform*. When $F(s)$ is given and we wish to compute $f(t)$, the *inverse transform* definition is available:

$$f(t) \triangleq \frac{1}{2\pi i}\int_{c-i\infty}^{c+i\infty} F(s)e^{st}\,ds \qquad (2.25)$$

The operation to be performed in Eq. (2.25) is called *contour integration* and will be meaningful only to those readers with some background in complex variable theory. Fortunately we will need to use Eq. (2.25) at only one point later in this text, elsewhere we will find the $f(t)$ corresponding to a given $F(s)$ by the simple expedient of "looking it up" in a table, such as Fig. 2.3. This is valid because of

[1]Gardner, M. F. and Barnes, J. L., *Transients in Linear Systems*, John Wiley and Sons, N. Y., 1954, p. 104.

a *uniqueness theorem*,[2] which guarantees that for each $f(t)$ there is one and only one $F(s)$.

Since in definition (2.20) the integral does not extend into the range of negative t, we would get the *same* $F(s)$ for time functions that were identical for positive t but were different for negative t. To remove this ambiguity, it is conventional to *define* our $f(t)$'s to be identically *zero* for negative t, thus ensuring that a given $F(s)$ corresponds to only one $f(t)$, even for negative time. Since the instant of physical time which we call $t=0$ is our choice, requiring $f(t)$'s to be zero for negative t is no real restriction.

In applying Laplace transform to the solution of differential equations, we will need to transform "both sides" of the equation. For example, if we have the equation

$$7\frac{d^2x}{dt^2} + 8\frac{dx}{dt} + 3x = 6e^{-5t} \tag{2.26}$$

we will generate another true equation by applying the same mathematical operation (the Laplace transform) to both sides:

$$\mathcal{L}\left[7\frac{d^2x}{dt^2} + 8\frac{dx}{dt} + 3x\right] = \mathcal{L}\left[6e^{-5t}\right] \tag{2.27}$$

To carry out the operations indicated on the left-hand side, we will need certain theorems which will now be derived.

LINEARITY THEOREM

1 $\mathcal{L}[af(t)] = aF(s)$ a is any constant (2.28)

2 $\mathcal{L}[f_1(t) + f_2(t)] = F_1(s) + F_2(s)$ (2.29)

The first part of this theorem allows us to easily deal with the numerical coefficients that always appear in our equations, while the second shows that the transform of the entire left-hand side of the equation is the same as the sum of the transforms of the individual terms. Both parts follow easily from the basic definition:

$$\mathcal{L}\left[af(t)\right] = \int_0^\infty af(t)e^{-st}\,dt = a\int_0^\infty f(t)e^{-st}\,dt = aF(s) \tag{2.30}$$

$$\mathcal{L}\left[f_1(t) + f_2(t)\right] = \int_0^\infty \left[f_1(t) + f_2(t)\right]e^{-st}\,dt$$

$$= \int_0^\infty f_1(t)e^{-st}\,dt + \int_0^\infty f_2(t)e^{-st}\,dt = F_1(s) + F_2(s) \tag{2.31}$$

---

[2]Gardner and Barnes, *op. cit.*, p. 123.

REAL DIFFERENTIATION THEOREM

$$\mathcal{L}\left[\frac{df(t)}{dt}\right] = sF(s) - f(0) \tag{2.32}$$

This theorem shows that when we transform a term of a differential equation that involves a derivative of the unknown, the numerical value $f(0)$ of the unknown at time zero (an initial condition) is automatically introduced into the transformed equation. We will see shortly that *all* the necessary initial conditions will appear *automatically* when we use the transform method, a definite advantage over the classical solution method. Another advantage resides in the interpretation of the term "initial" in the two methods. As shown in Fig. 2.4, the classical method *requires* that "initial" conditions be considered as the state of the system just an infinitesimal interval ϵ *after* an input, such as the step shown, is applied. In some systems the conditions (often zero) before the input and those after are *not* the same, and the conditions after may not be easy to find. In the transform method the meaning of "initial" conditions refers to the system's state at a time corresponding to the *lower limit* of the integral that defines the Laplace transform. Various authors choose to locate this limit at slightly different positions on the t axis; Fig. 2.4 shows our choice. The lower limit of the integral is *chosen* an interval ϵ_1 to the right of $t=0$ and ϵ_1 is allowed to shrink as close to zero as we please, but never reaches zero. The time at which the system input is applied is *chosen* at ϵ_2, where ϵ_2 is always greater than ϵ_1, and ϵ_2 also approaches as close to zero as we please. Thus we can have our inputs

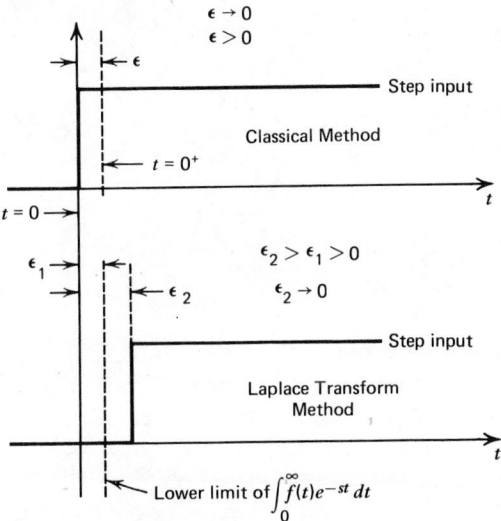

Figure 2.4 Definition of initial conditions for classical and transform methods.

"happen" as close to zero as we please but are always assured that any discontinuities (such as the step change shown) will always be *included* within the range of the Laplace transform integral. This allows the transform method to use the simpler (often zero) conditions *before* the input is applied and still properly account for any sudden changes in system state possibly caused by discontinuous inputs.

The theorem (2.32) may be proven using integration by parts and the basic definition of the Laplace transform:

$$\int_0^\infty f(t)e^{-st}\,dt = F(s), \qquad \int u\,dv = uv - \int v\,du \tag{2.33}$$

let

$$u = f(t), \qquad dv = e^{-st}\,dt$$

$$\int_0^\infty f(t)e^{-st}\,dt = -\frac{1}{s}f(t)e^{-st}\Big|_0^\infty + \frac{1}{s}\int_0^\infty \left[\frac{df(t)}{dt}\right]e^{-st}\,dt \tag{2.34}$$

$$F(s) = \frac{f(0)}{s} + \frac{1}{s}\mathcal{L}\left[\frac{df(t)}{dt}\right] \tag{2.35}$$

The theorem is easily extended to higher-order derivatives, giving:

$$\mathcal{L}\left[f^{(n)}(t)\right] = s^n F(s) - \sum_{k=1}^n f^{(k-1)}(0)s^{(n-k)} \tag{2.36}$$

where

$$f^{(k)}(t) \triangleq \frac{d^k f(t)}{dt^k} \quad \text{and} \quad f^{(0)}(t) \triangleq f(t) \tag{2.37}$$

For example:

$$\mathcal{L}\left[\frac{d^3 f}{dt^3}\right] = s^3 F(s) - f(0)s^2 - \frac{df}{dt}(0)s - \frac{d^2 f}{dt^2}(0) \tag{2.38}$$

Note that here three "initial" conditions appear, all to be interpreted as in Fig. 2.4. From here on, the notation $f(0)$, and so on for initial conditions will always be interpreted in this way; that is, $f(0) = f(\epsilon_1)$, unless explicitly stated otherwise.

REAL INTEGRATION THEOREM

Since our "differential" equations derived for physical systems may contain also terms involving integrals of the unknowns, we require a theorem to transform such terms. Again using the basic definition and integration by parts:

$$\int_0^\infty f(t)e^{-st}\,dt = F(s), \qquad \int u\,dv = uv - \int v\,du \qquad (2.39)$$

but now let

$$u = e^{-st}, \qquad dv = f(t)\,dt \qquad (2.40)$$

$$\int_0^\infty f(t)e^{-st}\,dt = e^{-st}\int f(t)\,dt\Big|_0^\infty + s\int_0^\infty \left[\int f(t)\,dt\right]e^{-st}\,dt \qquad (2.41)$$

$$F(s) = -f^{(-1)}(0) + s\mathcal{L}\left[\int f(t)\,dt\right] \qquad (2.42)$$

giving finally

$$\mathcal{L}\left[\int f(t)\,dt\right] = \frac{F(s)}{s} + \frac{f^{(-1)}(0)}{s} \qquad (2.43)$$

where $f^{(-1)}(0)$ is the initial value of $\int f(t)\,dt$. That is, if $f(t)$ were the *velocity* in a mechanical motion problem, $f^{(-1)}(0)$ would be the numerical value of the *displacement* just before the system input was applied. We may again generalize this theorem to apply to higher-order integrals:

$$\mathcal{L}\left[f^{(-n)}(t)\right] = \frac{F(s)}{s^n} + \sum_{k=1}^n \frac{f^{(-k)}(0)}{s^{n-k+1}} \qquad (2.44)$$

where

$$f^{(-n)}(t) \triangleq \int \cdots \int f(t)(dt)^n \quad \text{and} \quad f^{(-0)}(t) \triangleq f(t)$$

For example:

$$\mathcal{L}\left[\int\left[\int f(t)\,dt\right]\right]dt = \frac{F(s)}{s^2} + \frac{f^{(-1)}(0)}{s^2} + \frac{f^{(-2)}(0)}{s} \qquad (2.45)$$

2.5   SOLUTION OF INTEGRO-DIFFERENTIAL EQUATIONS: INVERSE TRANSFORMATION

While several additional useful theorems exist and will be introduced later, we now have sufficient background to begin the treatment of the transform solution method for ordinary linear integro-differential equations with constant coefficients. Since a *brief* review of some elementary system dynamics concepts may be appropriate here, we will use examples from this area as a vehicle for presenting the transform methods.

For a system with input q_i and output q_o, the differential equation of the general case takes the form:

$$a_n \frac{d^n q_o}{dt^n} + a_{n-1} \frac{d^{n-1} q_o}{dt^{n-1}} + \cdots + a_1 \frac{dq_o}{dt} + a_0 q_0 = b_m \frac{d^m q_i}{dt^m} + \cdots + b_1 \frac{dq_i}{dt} + b_0 q_i$$

$$(2.46)$$

The simplest special case, the so-called *zero-order system*, has the equation

$$a_0 q_o = b_0 q_i \tag{2.47}$$

whose *standard form* is

$$q_o = K q_i \tag{2.48}$$

where $K \triangleq b_0/a_0$ is called the system *steady-state gain* or *static sensitivity*. Such a system clearly has instantaneous dynamic response and q_o will have exactly the same shape as q_i for its time history, irrespective of the waveform of $q_i(t)$. While such instantaneous behavior appears to be an unrealistic model of *any* real system, it is often useful (and used) for modeling those components (if any) of a given system whose behavior is *much* faster than that of some other components. A good example is the electronic amplifiers often found in electro-mechanical motion control systems. While *not* instantaneous, these amplifiers are so much faster than the mechanical moving parts, that good prediction of overall system behavior is achieved using zero-order models for the electronics. The very same amplifier, when used in an *all* electronic system (*no* mechanical moving parts) might require a more sophisticated model, since now *it* might be the slowest element.

If in Eq. (2.46) we include the terms

$$a_1 \frac{dq_o}{dt} + a_0 q_o = b_0 q_i \tag{2.49}$$

we have defined the *first-order system* whose standard form is

$$\tau \frac{dq_o}{dt} + q_o = K q_i \tag{2.50}$$

where K again is the steady-state gain and τ is the system *time constant*. Figure 2.5 shows a selection of physical examples of first-order systems; in each case

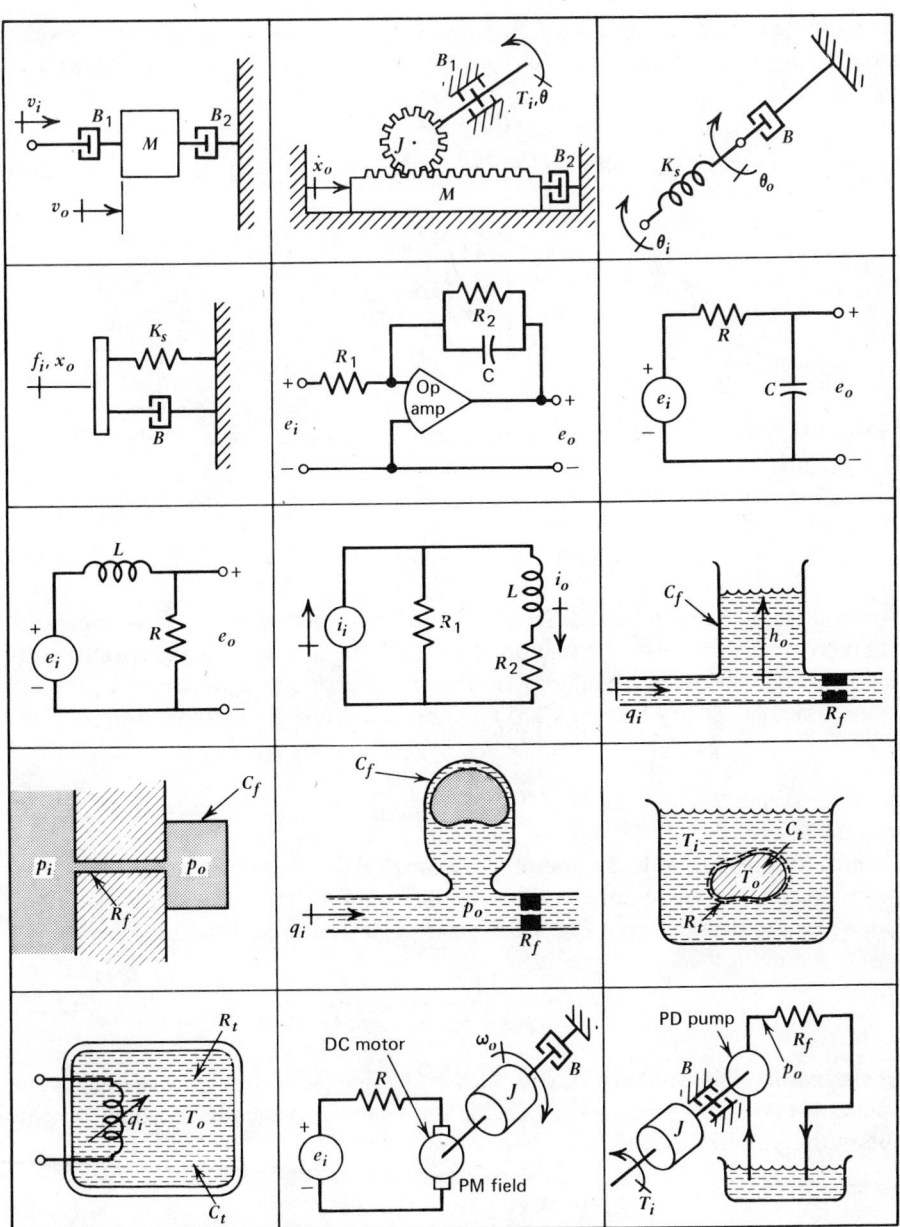

Figure 2.5 A collection of physical first-order systems.

the variable with subscript i must be considered the input while that with o is the output. Using the geared system of Fig. 2.5 to illustrate the procedure, we write

$$\Sigma T = J\alpha, \qquad \Sigma F = MA$$

$$T_i - B_1\dot{\theta} - RF_g = J\ddot{\theta}, \qquad F_g - B_2\dot{X}_o = M\ddot{X}_o \qquad (2.51)$$

$$T_i - \frac{B_1\dot{X}_o}{R} - R\left[M\ddot{X}_o + B_2\dot{X}_o\right] = J\frac{\ddot{X}_o}{R} \qquad (2.52)$$

$$\left[\frac{J}{R^2} + M\right]\dot{v}_o + \left[\frac{B_1}{R^2} + B_2\right]v_o = \left(\frac{1}{R}\right)T_i \qquad (2.53)$$

$$\tau\dot{v}_o + v_o = KT_i \qquad (2.54)$$

In this example all friction is assumed viscous; the bodies have no elasticity and the gears no backlash.

Returning to the general case of Eq. (2.50) and Laplace transforming:

$$\tau\left[sQ_o(s) - q_o(0)\right] + Q_o(s) = KQ_i(s) \qquad (2.55)$$

We note that the ordinary differential equation transforms into an *algebraic* equation; a simpler class of equation. (We shall see later that Laplace transforming a *partial* differential equation results in an *ordinary* differential equation; again a simpler class.) Solving (2.55) algebraically for the unknown $Q_o(s)$ gives

$$Q_o(s) = \frac{KQ_i(s)}{\tau s + 1} + \frac{\tau q_o(0)}{\tau s + 1} \qquad (2.56)$$

To proceed further with the solution we must now be specific about the input function $q_i(t)$ and the initial condition $q_o(0)$. Suppose $q_i(t) = A$, a step function of size A, and $q_o(0) = B$. From Fig. 2.3, $Q_i(s) = A/s$ and thus

$$Q_o(s) = \frac{KA/\tau}{s(s + 1/\tau)} + \frac{B}{s + 1/\tau} \qquad (2.57)$$

Using the linearity theorem we are allowed to inverse transform the right-hand side as two separate parts, each of which appears in Appendix A; thus we can obtain the solution directly as

$$q_o(t) = KA(1 - e^{-t/\tau}) + Be^{-t/\tau} \qquad (2.58)$$

which graphs as Fig. 2.6. Note that in *any* first-order system, K determines how *much* steady-state output is obtained per unit input while τ determines how *fast* the steady state is achieved.

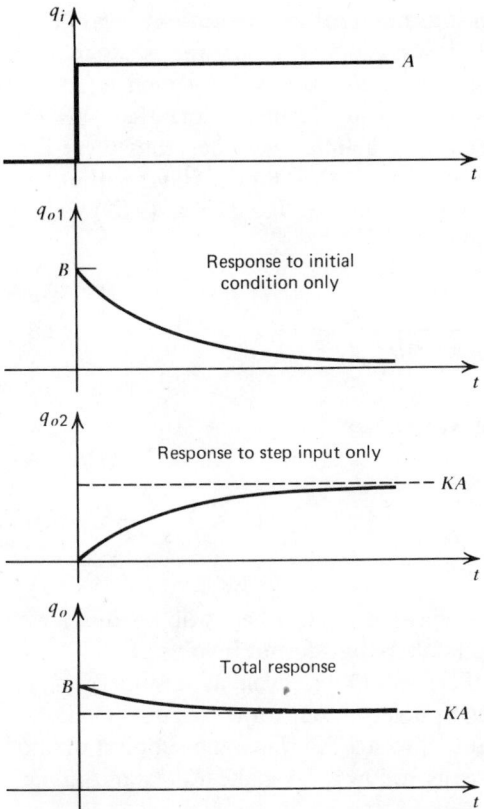

Figure 2.6   First-order system response.

In more complex problems we will need to use the *partial-fraction expansion* scheme to manipulate our s-functions into the most convenient form before inverse transforming with the tables. For the large class of practical problems modeled by equations of the form (2.46) and where the input $q_i(t)$ is limited to products and/or sums of derivatives of impulses, impulses, steps, positive integer powers of t, exponential functions and sinusoidal functions, the $Q_o(s)$ function will turn out to be a *rational algebraic function* of form

$$F(s) = \frac{N(s)}{D(s)} = \frac{B_p s^p + B_{p-1} s^{p-1} + \cdots + B_1 s + B_0}{s^n + A_{n-1} s^{n-1} + \cdots + A_1 s + A_0} \tag{2.59}$$

Usually, $n > p$ (proper fraction); if not (improper fraction), divide $D(s)$ into $N(s)$ to get

$$F(s) = \cdots + L_2 s^2 + L_1 s + L_0 + \frac{N_1(s)}{D(s)} \tag{2.60}$$

The "L terms" will inverse transform into impulses and impulse derivatives (to be discussed later) while $N_1(s)/D(s)$ is now a proper fraction and is expanded in partial fractions as shown below. The *system characteristic equation* is $D(s)=0$ and we must find its n roots before proceeding further. Note that this root finding step was also necessary in the classical solution method, thus there is no need to here repeat our earlier discussion of this topic. Having the roots $s_1, s_2, \ldots, s_n$ (real, complex, repeated, unrepeated) in hand, Eq. (2.59) can be written as

$$F(s) = \frac{N(s) \text{ or } N_1(s)}{(s-s_1)(s-s_2)\cdots(s-s_n)} \tag{2.61}$$

which, from algebra, we know can be written as

$$\frac{N(s) \text{ or } N_1(s)}{D(s)} = \frac{K_1}{s-s_1} + \frac{K_2}{s-s_2} + \cdots + \frac{K_k}{s-s_k} + \cdots + \frac{K_n}{s-s_n} \tag{2.62}$$

where we now assume there are no repeated roots (they will be dealt with shortly). The K's are real or complex numbers but do not involve s.

If we could find the K's, then $F(s)$ would be given as a sum of *simple* s-functions which we would then easily inverse transform to get $f(t)$. Let us consider first the case of an individual real root s_k. Perhaps the simplest method of finding a numerical value for K_k is as follows. To make K_k "stand alone," multiply (2.62) by $(s-s_k)$:

$$\frac{[N(s) \text{ or } N_1(s)](s-s_k)}{(s-s_1)\cdots(s-s_k)\cdots(s-s_n)} = \frac{K_1(s-s_k)}{(s-s_1)} + \cdots + K_k + \cdots + \frac{K_n(s-s_k)}{(s-s_n)} \tag{2.63}$$

Since this equation is true for any value of s, choose $s = s_k$ to "wipe out" all right-side terms except K_k, giving:

$$K_k = \frac{N(s) \text{ or } N_1(s)}{(s-s_1)\cdots()\cdots(s-s_n)}\bigg|_{s=s_k} \tag{2.64}$$

This rapid procedure will get us numerical values for each of the unrepeated real roots. As an example consider Eq. (2.57) rewritten as

$$Q_o(s) = \frac{Bs + KA/\tau}{s(s+1/\tau)} = \frac{K_1}{s} + \frac{K_2}{s+1/\tau} \tag{2.65}$$

$$K_1 = \frac{Bs + KA/\tau}{s+1/\tau}\bigg|_{s=0} = KA \tag{2.66}$$

$$K_2 = \frac{Bs + KA/\tau}{s}\bigg|_{s=-1/\tau} = B - KA \tag{2.67}$$

$$Q_o(s) = \frac{KA}{s} + \frac{B-KA}{s+1/\tau}, \qquad q_o(t) = KA + (B-KA)e^{-t/\tau} \tag{2.68}$$

To motivate our consideration of *complex* roots, let us consider now the *second-order system* model obtained by taking

$$a_2 \frac{d^2 q_o}{dt^2} + a_1 \frac{dq_o}{dt} + a_0 q_o = b_0 q_i \tag{2.69}$$

as a special case of (2.46). Here the standard form is

$$\frac{1}{\omega_n^2} \frac{d^2 q_o}{dt^2} + \frac{2\zeta}{\omega_n} \frac{dq_o}{dt} + q_o = Kq_i \tag{2.70}$$

where

$$\omega_n \triangleq \text{undamped natural frequency, rad/time} \tag{2.71}$$

$$\zeta \triangleq \text{damping ratio, fraction of critical damping, dimensionless} \tag{2.72}$$

$$K \triangleq \text{steady-state gain, output/input} \tag{2.73}$$

Figure 2.7 shows a selection of physical systems that may be modeled as second order. Taking the electrical system with current source i_i as an example, we may write

$$i_i = i_C + i_L, \qquad e_c = e_{LR}, \qquad e_o = i_L R \tag{2.74}$$

$$i_i = CDe_{LR} + i_L = CD(LD+R)i_L + i_L = (LCD^2 + RCD + 1)i_L \tag{2.75}$$

$$LC \frac{d^2 e_o}{dt^2} + RC \frac{de_o}{dt} + e_o = Ri_i \tag{2.76}$$

$$\frac{1}{\omega_n^2} \frac{d^2 e_o}{dt^2} + \frac{2\zeta}{\omega_n} \frac{de_o}{dt} + e_o = Ki_i \tag{2.77}$$

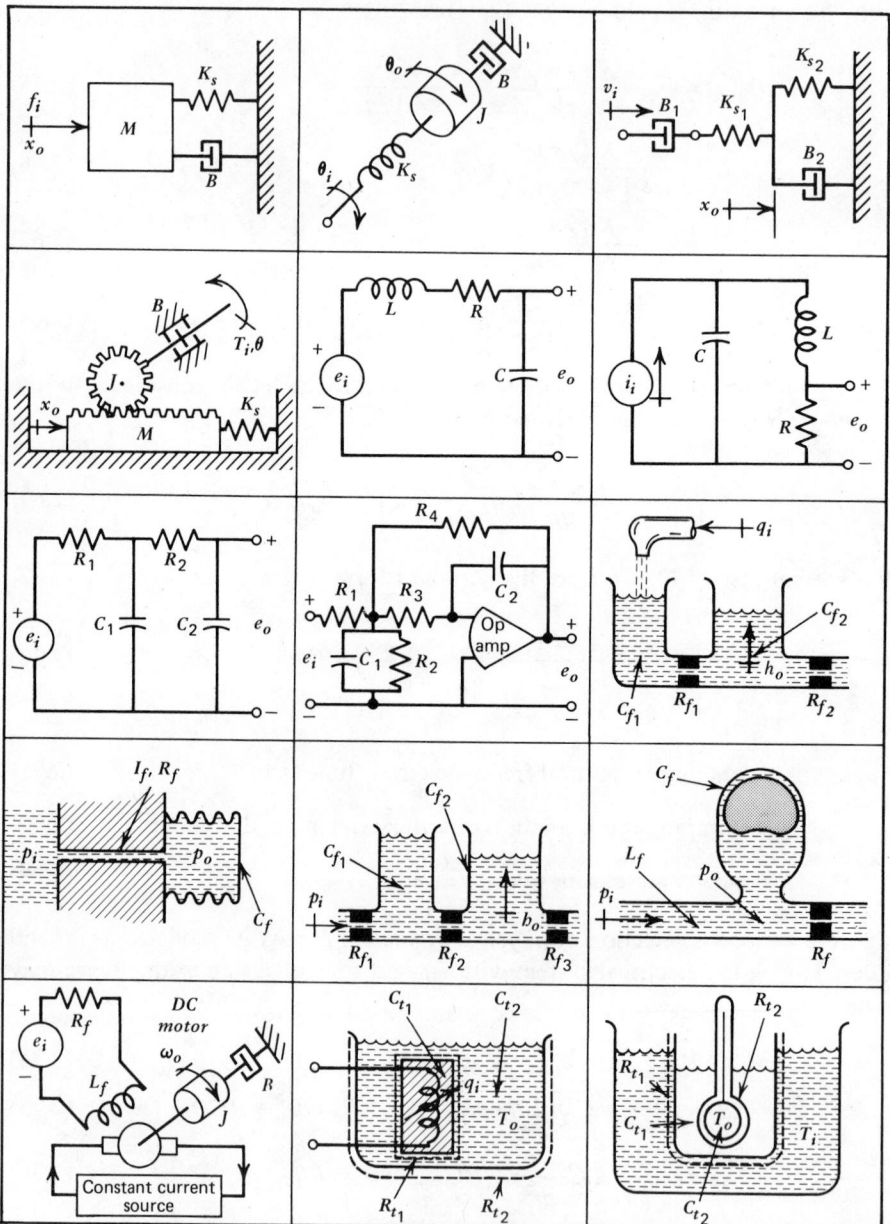

Figure 2.7 A collection of physical second-order systems.

Returning now to the general second-order system of Eq. (2.70) and transforming, we get

$$\frac{1}{\omega_n^2}\left(s^2 Q_o - s q_o(0) - \frac{dq_o}{dt}(0)\right) + \frac{2\zeta}{\omega_n}\left[sQ_o - q_o(0)\right] + Q_o = KQ_i \qquad (2.78)$$

where we now (and from here on) no longer write $Q_o(s)$ but only Q_o, relying on the use of the capital letter to remind us that we are in the s domain:

$$Q_o = \frac{KQ_i}{\dfrac{s^2}{\omega_n^2} + \dfrac{2\zeta s}{\omega_n} + 1} + \frac{\left(\dfrac{s}{\omega_n^2} + \dfrac{2\zeta}{\omega_n}\right)q_o(0) + \dfrac{1}{\omega_n^2}\dfrac{dq_o}{dt}(0)}{\dfrac{s^2}{\omega_n^2} + \dfrac{2\zeta s}{\omega_n} + 1} \qquad (2.79)$$

Let us take q_i to be a step input of size A, $q_o(0) = B$, as in the first-order system and now let $(dq_o/dt)(0) = C$. While the two separate terms of (2.79) can be inverse transformed (as they stand) using Appendix A, let us rewrite them in a form useful for explaining the partial-fraction expansion method:

$$Q_o = \frac{Bs^2 + (2\zeta\omega_n B + C)s + KA\omega_n^2}{s(s^2 + 2\zeta\omega_n s + \omega_n^2)} \qquad (2.80)$$

$$Q_o = \frac{K_1}{s} + \frac{K_2}{s + \zeta\omega_n - i\omega_n\sqrt{1-\zeta^2}} + \frac{K_3}{s + \zeta\omega_n + i\omega_n\sqrt{1-\zeta^2}} \qquad (2.81)$$

We are here taking $\zeta < 1.0$ (underdamped system) since $\zeta > 1.0$ gives *real* roots, a case we have already treated. The constant K_1 is readily found as before:

$$K_1 = \frac{Bs^2 + (2\zeta\omega_n B + C)s + KA\omega_n^2}{s^2 + 2\zeta\omega_n s + \omega_n^2}\bigg|_{s=0} = KA \qquad (2.82)$$

Proceeding in identical fashion for K_2:

$$K_2 = \frac{Bs^2 + (2\zeta\omega_n B + C)s + KA\omega_n^2}{s(s + \zeta\omega_n + i\omega_n\sqrt{1-\zeta^2})}\bigg|_{s = -\zeta\omega_n + i\omega_n\sqrt{1-\zeta^2}} \qquad (2.83)$$

Although we could go on in letter form (see Appendix A for a general result), let us, as in a practical problem, enter numerical values at this point.

Assume $B = C = 0$, $\zeta = 0.5$, and $\omega_n = 1.0$, giving:

$$K_2 = \frac{KA}{(-0.5 + i0.866)(i1.732)} = \frac{KA\,\underline{/0°}}{1.73\,\underline{/-150°}} = (KA/1.73)\underline{/150°} \qquad (2.84)$$

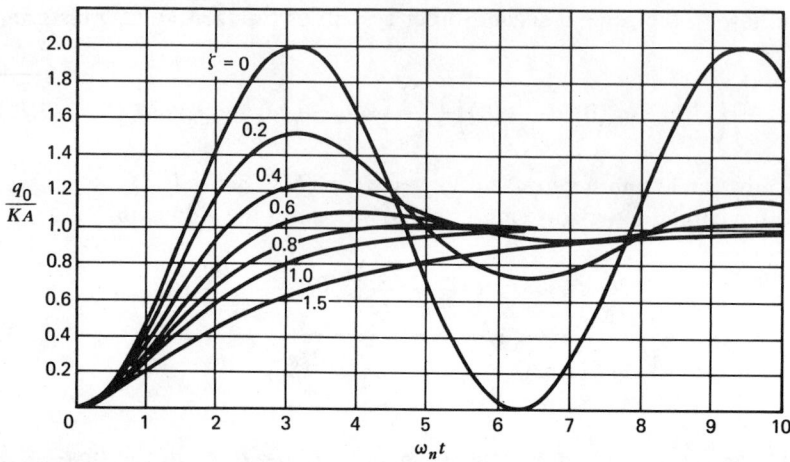

Figure 2.8 Nondimensional step response of second-order systems.

Note that K_2 is (and would *in general* be) a complex number. If we now went on to find K_3, we would see (again a *general* result) that it is the complex conjugate (same magnitude, negative angle) of K_2, that is, $K_3 = (KA/1.73)/{-150°}$. Actually there is never any need to find K_3; it can be shown[3] that once we have found K_2 we can immediately write down the inverse transform for *both* complex factors in Eq. (2.81) as:

$$q_{oc}(t) = 2|K_2|e^{-\zeta\omega_n t}\sin\left(\omega_n\sqrt{1-\zeta^2}\ t + \phi\right) \tag{2.85}$$

where $\phi \triangleq /K_2 + 90°$. Thus the complete inverse transform of (2.80) would be

$$q_o(t) = KA + 1.15 KAe^{-0.5t}\sin(0.866t + 240°) \tag{2.86}$$

Figure 2.8 shows the step response of general second-order systems. Note that, as usual, the steady-state gain K has no effect on dynamic behavior but merely determines *how much* steady-state output is achieved for each unit of input. Undamped natural frequency ω_n is a direct and proportional indicator of speed of response; holding K and ζ fixed and, say, doubling ω_n gives a speed-up of precisely two-to-one. Damping ratio ζ clearly determines the degree of oscillation; the dividing line between oscillatory and nonoscillatory systems being the critically damped case, $\zeta = 1.0$. Also, recall that the transient oscillating frequency is $\omega_n\sqrt{1-\zeta^2}$, not ω_n.

[3]D'Azzo, J. J. and C. H. Houpis, *Feedback Control System Analysis and Synthesis*, 2nd ed., McGraw=Hill, N. Y., 1966, p. 99.

The procedure just shown can be used to inverse transform *any* pair of unrepeated complex roots which might arise. If there are *several* pairs of such roots we simply do one pair at a time. To be able to use (2.85), be sure to remember that K_2 is the constant associated with the term that has the *negative* imaginary part (see (2.81)). We have now shown how to deal with the inverse transformation of any number of unrepeated real roots or complex root pairs, the most common cases in practical problems. Before completing our discussion with a treatment of repeated roots (less common, but requiring a more tedious method), let us do a more comprehensive example:

$$Q_o = \frac{1}{s(s+1)(s+2)(s+1+i1)(s+1-i1)(s+2+i2)(s+2-i2)} \tag{2.87}$$

$$= \frac{K_1}{s} + \frac{K_2}{s+1} + \frac{K_3}{s+2} + \frac{K_4}{s+1-i1} + \frac{K_5}{s+1i1} + \frac{K_6}{s+2-i2} + \frac{K_7}{s+2+i2} \tag{2.88}$$

$$K_1 = \frac{1}{(1)(2)(2)(8)} = 0.0313, \qquad K_2 = \frac{1}{(-1)(1)(1)(5)} = -0.2 \tag{2.89}$$

$$K_3 = \frac{1}{(-2)(-1)(2)(4)} = 0.0625$$

$$K_4 = \frac{1}{(-1+i)(i)(1+i)(i2)(1+i3)(1-i)} = 0.056\underline{/-26.5°} \tag{2.90}$$

$$K_6 = \frac{1}{(-2+i2)(-1+i2)(i2)(-1+i3)(-1+i1)(i4)} = 0.0041\underline{/-8.2°} \tag{2.91}$$

$$q_o(t) = 0.0313 - 0.2e^{-t} + 0.0625e^{-2t} + 0.112e^{-t}\sin(t+63.5°)$$
$$+ 0.00882e^{-2t}\sin(2t+81.8°) \tag{2.92}$$

Turning finally now to repeated roots, consider first the following example of repeated real roots:

$$Q_o = \frac{1}{s(s+1)(s+1)} = \frac{K_1}{s} + \frac{K_2}{(s+1)^2} + \frac{K_3}{s+1} \tag{2.93}$$

Note that the partial-fraction expansion contains two distinct terms for the repeated root and that if the factor $(s+1)$ had appeared *three* times we would have terms $K_2/(s+1)^3$, $K_3/(s+1)^2$, and $K_4/(s+1)$. Furthermore, if say, $(s+3)$ appeared twice in (2.93) we would merely add terms $K_5/(s+3)^2$ and $K_6/(s+3)$. Thus any number of repeated roots, each repeated any number of times may be dealt with in exactly this fashion. Let us find the K's, proceeding in our usual way:

$$K_1 = \frac{1}{(1)(1)} = 1, \qquad \frac{(s+1)^2}{s(s+1)(s+1)} = \frac{K_1(s+1)^2}{s} + K_2 + \frac{K_3(s+1)^2}{s+1} \quad (2.94)$$

If we now set $s = -1$ as usual, we get $K_2 = -1$ with no difficulty. However, if we proceed to K_3:

$$\frac{s+1}{s(s+1)(s+1)} = \frac{K_1(s+1)}{s} + \frac{K_2(s+1)}{(s+1)^2} + K_3 \quad (2.95)$$

we get

$$\infty = \infty + K_3$$

when we set $s = -1$, frustrating our attempt to find K_3. To resolve this difficulty, return to (2.94) and note that if we *differentiate* this equation with respect to s we can cause K_3 "to stand alone":

$$\frac{d}{ds}\left[\frac{1}{s}\right] = \frac{d}{ds}\left[\frac{K_1(s+1)^2}{s} + K_2 + K_3(s+1)\right] \quad (2.96)$$

Carrying out the differentiation and setting $s = -1$:

$$\frac{-1}{s^2} = -1 = \left.\frac{K_1[(s)2(s+1)-(s+1)^2]}{s^2}\right|_{s=-1} + 0 + K_3 = K_3 \quad (2.97)$$

Observe that the differentiation of the right-hand side of (2.96) (except for the term involving K_3) is a "waste of time" since in the *next* step (insertion of $s = -1$) all these terms will always go to zero. This differentiation scheme allows one to find the K's for any number of repeated real roots, each repeated any number of times; however, for roots appearing three or more times, *repeated* differentiation will be necessary to cause the desired K to stand alone. Finally, the inverse transforms of terms such as $K/(s+a)^n$ will give time functions such as $Kt^{n-1}e^{-at}/(n-1)!$ as we might expect from our experience with the classical solution method.

The treatment of repeated complex root pairs follows essentially the same pattern as for repeated real roots; the partial-fraction expansion contains terms involving *powers* of the factors and *differentiation* with respect to s is necessary to find the K's other than the one associated with the highest-power term:

$$Q_o = \frac{N(s)}{\left(s^2 + 2\zeta\omega_n s + \omega_n^2\right)^n D_1(s)}$$

$$= \frac{K_1}{\left(s + \zeta\omega_n - i\omega_n\sqrt{1-\zeta^2}\,\right)^n} + \frac{K_1'}{\left(s + \zeta\omega_n + i\omega\sqrt{1-\zeta^2}\,\right)^n}$$

$$+ \frac{K_2}{\left(s + \zeta\omega_n - \omega_n\sqrt{1-\zeta^2}\,\right)^{n-1}} + \frac{K_2'}{\left(s + \zeta\omega_n + i\omega_n\sqrt{1-\zeta^2}\,\right)^{n-1}}$$

$$+ \cdots$$

$$+ \frac{K_n}{\left(s + \zeta\omega_n - i\omega\sqrt{1-\zeta^2}\,\right)} + \frac{K_n'}{\left(s + \zeta\omega_n + i\omega_n\sqrt{1-\zeta^2}\,\right)} + \cdots \qquad (2.98)$$

We find the K's (they will be complex numbers) as shown earlier for repeated real roots. The time function takes the form

$$q_o(t) = 2|K_1|\frac{t^{n-1}}{(n-1)!}\,e^{-\zeta\omega_n t}\sin\left(\omega_n\sqrt{1-\zeta^2}\,t + \underline{/K_1} + 90°\right)$$

$$+ 2|K_2|\frac{t^{n-2}}{(n-2)!}\,e^{-\zeta\omega_n t}\sin\left(\omega_n\sqrt{1-\zeta^2}\,t + \underline{/K_2} + 90°\right)$$

$$+ \cdots + 2|K_n|e^{-\zeta\omega_n t}\sin\left(\omega_n\sqrt{1-\zeta^2}\,t + \underline{/K_n} + 90°\right) + \cdots \qquad (2.99)$$

To conclude our discussion of differential equation solution methods a final comment on simultaneous equations is appropriate. When dealing with a set of equations such as (2.12), (2.13) it is generally preferable to transform the *original* set of simultaneous equations rather than reducing the set to one equation in one unknown and *then* transforming since this latter procedure will introduce the initial conditions in a more complicated way.

2.6 STABILITY

While the question of dynamic instability is a topic of central importance in systems intentionally designed for the automatic control of some physical variable, it also arises in a number of engineering applications *not* directly related to control. Some examples of this latter type are: machine tool chatter associated with cutting and grinding processes, hydrostatic bearing oscillations, flutter of aircraft components, safety-valve pulsations, and vehicle handling instabilities. When linearized models of such systems are adequate, stability analysis is relatively straightforward since it depends entirely on the location in the complex plane of the roots of the system's characteristic equation. Roots in the right half-plane give solutions that all involve a multiplying factor of form e^{at} (a positive), which causes the system response to tend toward infinity for the slightest initial disturbance. Roots *exactly* on the imaginary axis give rise to oscillations of constant amplitude, neither building up nor dying down. Roots in the left half-plane contribute terms that die out.

Let us analyze the classroom demonstration apparatus of Fig. 2.9, which illustrates stability concepts. This simple model could represent several practical devices such as a hydrostatic gas bearing or an air cushion vehicle. It should be intuitively clear that if the air valve is opened to some fixed position, pressure p will build up until its force equals the steady load force f_0 plus the weight of mass M, which will then rise to an equilibrium height h_0 at which the air inflow rate from p_s to p just balances the outflow rate through the gap h_0. With the system in equilibrium at $p = p_0$ and $h = h_0$, we wish to predict the response of height perturbation h_p to small perturbations f_p in load force.

Our linearized dynamic analysis will deal strictly with small changes away from steady operating points for all the variables. We must express a conservation of mass equation for the volume V and also a Newton's law for the moving mass M. For a fixed inlet-valve opening the mass inflow rate depends on only

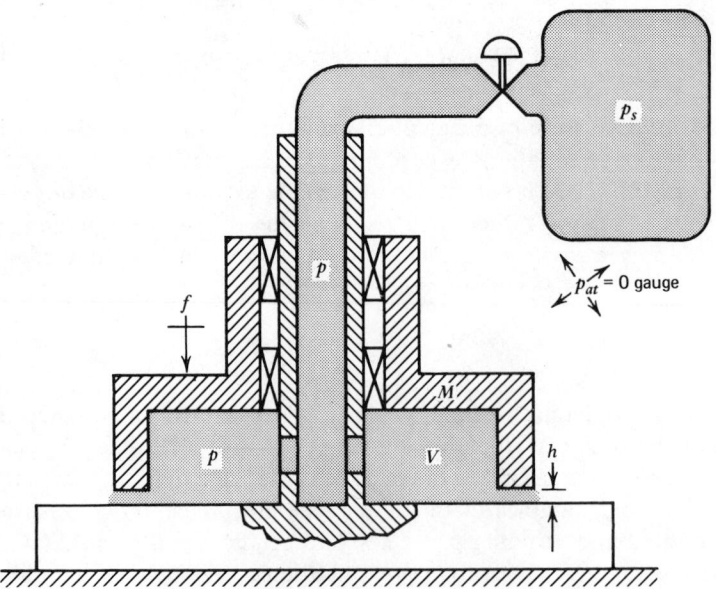

Figure 2.9 Stability demonstration model.

one variable, the pressure p. This dependency is nonlinear, but for small perturbations can be linearly approximated as

$$\dot{m}_{ip} = -K_{pi}p_p \qquad (2.100)$$

The positive coefficient K_{pi} can be estimated by linearizing a theoretical relation from steady-flow fluid mechanics or can be more accurately determined from steady-flow pressure/flow-rate experiments on the actual valve if this piece of hardware is available. One may similarly linearize the outflow relation as

$$\dot{m}_{op} = K_{po}p_p + K_{ho}h_p \qquad (2.101)$$

For the volume V the perfect gas law gives

$$\frac{dp_p}{dt} = \frac{R_o T_o}{V_o} \frac{dm_p}{dt} \qquad (2.102)$$

Since temperature changes as great as $\pm 30C°$ cause only $\pm 10\%$ changes in $T(K°)$, we have assumed T constant in 2.102. Similarly, since the volume change due to h_p is assumed small compared with the equilibrium volume, we take V as constant. For the modest pressure and temperature changes expected, R should be nearly constant. (If experiments reveal some or all of these assumptions to be of insufficient accuracy, a more complicated, though still linear, improved model would be:

$$\frac{dp_p}{dt} = \frac{R_o T_o}{V_o} \frac{dm_p}{dt} + \frac{M_o T_o}{V_o} \frac{dR_p}{dt} + \frac{M_o R_o}{V_o} \frac{dT_p}{dt} - \frac{M_o T_o R_o}{V_o^2} \frac{dV_p}{dt} \qquad (2.103)$$

Now, however, additional modeling would be needed to define dR_p/dt and dT_p/dt (dV_p/dt is easily related to h_p). This complicates the analysis considerably and a *significant* accuracy increase would have to result to justify this complication). Conservation of mass combines the above equations to produce:

$$\frac{dp_p}{dt} = \frac{R_o T_o}{V_o} \left[-K_{pi}p_p - K_{po}p_p - K_{ho}h_p \right] \qquad (2.104)$$

Turning now to Newton's law and assuming any frictional effects associated with the mass's vertical motion to be linear ("viscous"), we get

$$-f_p - B\dot{h}_p + Ap_p = M\ddot{h}_p$$

which, when combined with (2.104), leads to

$$\left\{ \frac{MV_o}{AR_oT_oK_{ho}}D^3 + \left[\frac{MR_oT_o(K_{pi}+K_{po})+BV_o}{AR_oT_oK_{ho}} \right]D^2 + \left[B\frac{(K_{pi}+K_{po})}{AK_{ho}} \right]D+1 \right\}h_p$$

$$= -\frac{K_{pi}+K_{po}}{AK_{ho}}\left[\frac{V_o}{R_oT_o(K_{pi}+K_{po})}D+1 \right]f_p \qquad (2.105)$$

While algebraic formulas exist for obtaining the roots of the cubic characteristic equation of this system in letter form, they are extremely cumbersome and rarely used for stability studies. Fortunately a method, the Routh stability criterion,[4] exists for discovering the presence of unstable roots without the need to actually get their values. We now present this criterion without proof. Assume the system characteristic equation from which any zero roots have already been removed is

$$a_ns^n + a_{n-1}s^{n-1} + \cdots + a_1s + a_0 = 0 \qquad (2.106)$$

We are to form a "triangular" array, the first two rows of which are written down, from left to right, at sight:

$$\begin{matrix} a_n & a_{n-2} & a_{n-4} & a_{n-6} \\ a_{n-1} & a_{n-3} & a_{n-5} & \cdots \end{matrix} \qquad (2.107)$$

continuing these two rows until we "run out" of coefficients. We now form a third row

$$b_1 \quad b_2 \quad b_3 \quad \cdots$$

from the first two using the following rule:

$$b_1 \triangleq \frac{a_{n-1}a_{n-2}-a_na_{n-3}}{a_{n-1}}, \qquad b_2 \triangleq \frac{a_{n-1}a_{n-4}-a_na_{n-5}}{a_{n-1}}$$

$$b_3 \triangleq \frac{a_{n-1}a_{n-6}-a_na_{n-7}}{a_{n-1}} \quad \text{etc.} \qquad (2.108)$$

and continuing in this established pattern until the b's become zero, thus completing the third row. The fourth row is formed from the second and third in exactly the same way as the third was formed from the first and second. We proceed in this fashion until all zeros are obtained, thus completing the array.

[4]Doebelin, E. O., *Dynamic Analysis and Feedback Control*, McGraw-Hill, N. Y., 1962, p. 175.

A numerical example should clarify the procedure.

$$3s^4 + 2s^3 + 5s^2 + s + 1 = 0$$

3	5	1
2	1	0
$\dfrac{10-3}{2}$	$\dfrac{2-0}{2}$	0
$\dfrac{3}{7}$	0	
1	0	
0		

Once the array is complete, we examine *only* the first column for changes of algebraic sign. *The number of roots with positive real parts is equal to the number of changes of sign*, thus the above characteristic equation represents a stable system. Two special cases should be mentioned since knowledge of them gives a quick, useful check of the validity of characteristic equations and/or saves computation time. If the characteristic equation *itself* exhibits any sign changes, there is no need to carry out the Routh test; we are guaranteed that the system is unstable. Usually, when this occurs an *error* of analysis has been made somewhere since such behavior is unusual, thus be on the lookout for such sign changes, they usually signal an earlier mistake. The second special case is "missing terms" in the characteristic equation; for example:

$$3s^4 + 2s^3 + (0)s^2 + s + 1 = 0$$

One or more missing terms signal a system with some pure imaginary roots (sustained oscillation) and/or roots with positive real parts (unstable). Again, this situation is unusual and should immediately instigate a search for modeling errors.

In addition to checking specific numerical cases, the Routh criterion can be applied in letter form to develop inequalities useful in system design. For a cubic equation $a_3 s^3 + a_2 s^2 + a_1 s + a_0 = 0$, for example, stability is assured so long as $a_0 a_3 < a_1 a_2$. Applying this to the system of Fig. 2.9, stability is assured if:

$$\frac{MV_o}{AR_o T_o K_{ho}} < \left[\frac{MR_o T_o(K_{pi} + K_{po}) + BV_o}{AR_o T_o K_{ho}} \right]\left[\frac{B(K_{pi} + K_{po})}{AK_{ho}} \right] \tag{2.109}$$

$$AK_{ho} < \left[\frac{R_o T_o}{V_o}(K_{pi} + K_{po}) + \frac{B}{M} \right](K_{pi} + K_{po})B \tag{2.110}$$

When the inequality is violated, roots in the right half-plane cause oscillations to build up. Linear theory says the amplitude of oscillation tends toward infinity, clearly impossible in a real-world system. What usually happens is that, as the amplitude builds up, one or more nonlinear effects in the real system (which were approximated or ignored in the linear model) became significant and cause the amplitude to level off. This steady oscillation is called a *limit cycle*; estimation of its amplitude requires a *nonlinear* model. Fortunately the linear stability analysis is usually sufficient for design purposes since it *does* accurately predict those combinations of parameters that will allow oscillations to *start*, a condition the designer must avoid.

Actually, inequalities such as (2.110) must not only be satisfied, they must be satisfied with some "safety margin," since left plane roots *close* to the imaginary axis, while not unstable, *do* exhibit undesirable slowly decaying oscillations. Also, roots close to instability can actually "drift over" into instability due to aging, wear, temperature change, and other "environmental" effects that cause changes in numerical values of parameters in real systems. Inequality (2.110) clearly reveals the trends of stability with each of the nine design parameters characterizing this system. Increasing A, K_{ho}, V_o, and M will decrease the stability while increasing R_o, T_o, K_{pi}, K_{po}, and B enhances stability. In the author's classroom demonstration, wiping the oil from the bearings just prior to test reduces B sufficiently to cause violent oscillations; oiling the bearings gives a stable system.

Since this is not a text on control systems, we earlier made the point that stability questions arise in noncontrol applications, such as the example just discussed. For those readers with some background in feedback control, however, it may be useful to show how noncontrol applications can be cast in feedback form, making applicable the large body of analysis and design methods developed for that area. By implementing the basic system equations in block diagram form, we can easily get the diagram of Fig. 2.10. (This figure uses *operational transfer functions*; the next section will develop the transfer function concept for those readers unfamiliar with it.) Once a system is cast in the

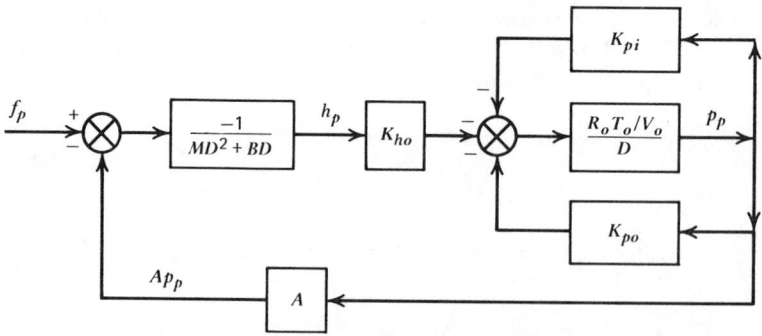

Figure 2.10   Block diagram showing feedback nature of air-cushion device.

feedback form, analytical tools such as Nyquist stability criterion[5] (an alternative to Routh criterion) and describing function[6] (an approximate method for calculating nonlinear effects, such as limit cycle amplitudes) may become applicable.

2.7 LAPLACE TRANSFER FUNCTIONS

Introductory treatments of system dynamics often use the transfer function concept with D-operator notation. If a system output q_o is related to an input q_i by an equation of form (2.46), then by definition the *operational transfer function* $(q_o/q_i)(D)$ is given by

$$\frac{q_o}{q_i}(D) \triangleq \frac{b_m D^m + b_{m-1} D^{m-1} + \cdots + b_1 D + b_0}{a_n D^n + a_{n-1} D^{n-1} + \cdots + a_1 D + a_0} \tag{2.111}$$

Thus, operationally, the output q_o is given by "multiplying" the input times the transfer function, thus allowing block diagrams like Fig. 2.10 to represent system differential equations in useful graphic form. To call the right-hand side of (2.111) a "function" of D is not quite right since D is an *operator*, not an independent variable that can take on different numerical values. A more correct terminology might call $(q_o/q_i)(D)$ a transfer *operator*; however, the transfer function usage is widespread and causes no harm, so many employ it.

When Laplace transform methods are used the transfer function concept can be defined more correctly. The *Laplace transfer function* (also called *system function*) relating a selected pair of system inputs and outputs is defined as the ratio of the output transform, divided by the input transform, when all other system inputs (including initial conditions) are set to zero. Since Laplace transfer functions are functions of s, a complex variable that *can* take on various numerical values (rather than an operator like D), the terminology transfer *function* is now correct. Let us use the system of Fig. 2.11 to develop the details:

$$f_1 - K_{s1}(x_1 - x_2) - B_1 \dot{x}_1 = M_1 \ddot{x}_1$$

$$- B_2 \dot{x}_2 + f_2 - K_{s2} x_2 - K_{s1}(x_2 - x_1) = M_2 \ddot{x}_2 \tag{2.112}$$

$$(M_1 s^2 + B_1 s + K_{s1}) X_1 + (-K_{s1}) X_2 = F_1 + M_1 \dot{x}_1(0) + (M_1 s + B_1) x_1(0)$$

$$(-K_{s1}) X_1 + (M_2 s^2 + B_2 s + (K_{s1} + K_{s2})) X_2 = F_2 + M_2 \dot{x}_2(0) + (M_2 s + B_2) x_2(0) \tag{2.113}$$

We will wish to use some elementary matrix notation, so recall that the matrix

[5]Doebelin, E. O. *Dynamic Analysis and Feedback Control*, McGraw-Hill, N. Y., 1962, p. 179.
[6]Doebelin. *Op. cit.*, p 207.

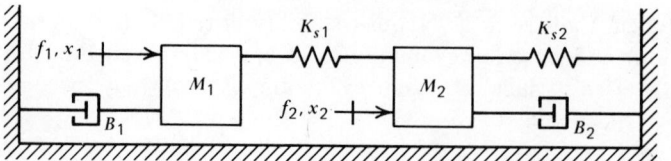

Figure 2.11 Vibrating system used for transfer function definition.

multiplication

$$\underbrace{\begin{bmatrix} a_{11} & a_{12} & a_{13} \\ a_{21} & a_{22} & a_{23} \\ a_{31} & a_{32} & a_{33} \end{bmatrix}}_{\text{square matrix}} \quad \underbrace{\begin{bmatrix} b_1 \\ b_2 \\ b_3 \end{bmatrix}}_{\text{column vectors}} = \underbrace{\begin{bmatrix} c_1 \\ c_2 \\ c_3 \end{bmatrix}}_{\text{(column matrices)}} \tag{2.114}$$

represents a set of three simultaneous equations in the unknowns b_1, b_2, and b_3:

$$\begin{aligned} a_{11}b_1 + a_{12}b_2 + a_{13}b_3 &= c_1 \\ a_{21}b_1 + a_{22}b_2 + a_{23}b_3 &= c_2 \\ a_{31}b_1 + a_{32}b_2 + a_{33}b_3 &= c_3 \end{aligned} \tag{2.115}$$

Equations (2.113) may thus be written as

$$\underbrace{\begin{bmatrix} M_1 s^2 + B_1 s + K_{s1} & -K_{s1} \\ -K_{s1} & M_2 s^2 + B_2 s + K_{s1} + K_{s2} \end{bmatrix}}_{\text{system}} \underbrace{\begin{bmatrix} X_1 \\ X_2 \end{bmatrix}}_{\text{outputs}}$$

$$= \underbrace{\begin{bmatrix} F_1 + M_1 \dot{x}_1(0) + (M_1 s + B_1)x_1(0) \\ F_2 + M_2 \dot{x}_2(0) + (M_2 s + B_2)x_2(0) \end{bmatrix}}_{\text{inputs}} \tag{2.116}$$

Since we are dealing with a set of simultaneous algebraic equations we may use determinants to solve for X_1 and X_2:

$$X_1 = \frac{\begin{vmatrix} F_1 + M_1 \dot{x}_1(0) + (M_1 s + B_1)x_1(0) & -K_{s1} \\ F_2 + M_2 \dot{x}_2(0) + (M_2 s + B_2)x_2(0) & M_2 s^2 + B_2 s + K_{s1} + K_{s2} \end{vmatrix}}{\begin{vmatrix} M_1 s^2 + B_1 s + K_{s1} & -K_{s1} \\ -K_{s1} & M_2 s^2 + B_2 s + K_{s1} + K_{s2} \end{vmatrix}} \tag{2.117}$$

$$X_2 = \frac{\begin{vmatrix} M_1 s^2 + B_1 s + K_{s1} & F_1 + M_1 \dot{x}_1(0) + (M_1 s + B_1)x_1(0) \\ -K_{s1} & F_2 + M_2 \dot{x}_2(0) + (M_2 s + B_2)x_2(0) \end{vmatrix}}{\begin{vmatrix} M_1 s^2 + B_1 s + K_{s1} & -K_{s1} \\ -K_{s1} & M_2 s^2 + B_2 s + K_{s1} + K_{s2} \end{vmatrix}} \tag{2.118}$$

The above expressions allow us to find the response of either or both x_1 and x_2 to:

a Initial conditions on x_1 (and/or x_2) only. Set F_1 and F_2 to equal to zero.
b Driving inputs on x_1 (and/or x_2) only. Set all initial conditions to zero.
c Any combination of driving inputs and initial conditions.

Our major interest presently is the transfer functions relating x_1 and x_2 to the driving inputs f_1 and f_2, thus we set all initial conditions to zero. The vast majority of transfer function applications are of this sort, thus we would usually treat the initial conditions as zero "right from the beginning" (Eq. (2.113)) to save writing and reduce confusion. Equations (2.117) and (2.118) now give

$$X_1 = \frac{(M_2 s^2 + B_2 s + K_{s1} + K_{s2})F_1 + (K_{s1})F_2}{M_1 M_2 s^4 + (M_2 B_1 + M_1 B_2)s^3 + [M_1(K_{s1}+K_{s2})+M_2 K_{s1}+B_1 B_2]s^2 + [B_1(K_{s1}+K_{s2})+B_2 K_{s1}]s + K_{s1}K_{s2}} \tag{2.119}$$

$$X_2 = \frac{(K_{s1})F_1 + (M_1 s^2 + B_1 s + K_{s1})F_2}{M_1 M_2 s^4 + (M_2 B_1 + M_1 B_2)s^3 + [M_1(K_{s1}+K_{s2})+M_2 K_{s1}+B_1 B_2]s^2 + [B_1(K_{s1}+K_{s2})+B_2 K_{s1}]s + K_{s1}K_{s2}} \tag{2.120}$$

We can now define four useful transfer functions:

$$W_{11} \triangleq \frac{X_1}{F_1}\bigg|_{F_2=0} = \frac{M_2 s^2 + B_2 s + K_{s1} + K_{s2}}{As^4 + Bs^3 + Cs^2 + Ds + E} \tag{2.121}$$

$$W_{22} \triangleq \frac{X_2}{F_2}\bigg|_{F_1=0} = \frac{M_1 s^2 + B_1 s + K_{s1}}{As^4 + Bs^3 + Cs^2 + Ds + E} \tag{2.122}$$

$$W_{12} \triangleq \frac{X_1}{F_2}\bigg|_{F_1=0} = \frac{K_{s1}}{As^4 + Bs^3 + Cs^2 + Ds + E} \tag{2.123}$$

$$W_{21} \triangleq \frac{X_2}{F_1}\bigg|_{F_2=0} = \frac{K_{s1}}{As^4 + Bs^3 + Cs^2 + Ds + E} \tag{2.124}$$

where $As^4 + Bs^3 + Cs^2 + Ds + E$ is the denominator of Equations (2.119) and (2.120). Using these transfer functions we may write:

$$X_1 = F_1 W_{11} + F_2 W_{12} \tag{2.125}$$
$$X_2 = F_1 W_{21} + F_2 W_{22} \tag{2.126}$$

which lead to the system block diagram of Fig. 2.12.

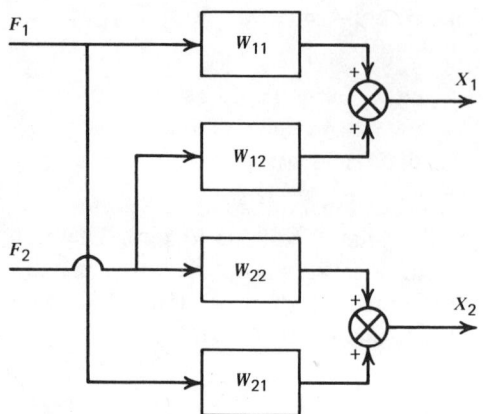

Figure 2.12 Block diagram of vibrating system.

This example illustrates several general features of system response that are worthy of note. The denominators of all the transfer functions are identical polynomials in s called the *characteristic polynomial* since, when set equal to zero, we get the system characteristic equation. Irrespective of which input/output pair is considered, we get the *same* characteristic equation; a system has only one characteristic equation. This is true for a system with any number of inputs and outputs. Also, the fact that transfer function W_{12} is identical to W_{21} is not a coincidence but a consequence of a general *reciprocity theorem*. The physical implication of $W_{12} = W_{21}$ is as follows. If we apply a force $f(t)$ of any nature whatever on M_1, it will cause a certain motion time history of x_2. If we now apply the very same force at M_2 and record the motion $x_2(t)$, it will be found to be *identical* to the $x_2(t)$ earlier produced by the same force applied at M_1. For most people, this fact is quite surprising and not physically obvious; however, it is a *general* characteristic of systems that exhibit symmetric matrices, such as our example.

Transfer functions have a number of useful features, not the least of which is the capability of utilizing block diagrams such as Fig. 2.12 to graphically visualize the interactions in a system. Since transfer functions are often combinations of simple basic elements such as first- and second-order systems, familiarity with the response of these "basic building blocks" allows one to rapidly comprehend the essential response characteristics of complex systems through the aid of the block diagram.

2.8 THE DELAY THEOREM: DEAD-TIME ELEMENTS AND DISCONTINUOUS INPUTS

Some physical processes involve a definite delay, during which time there is *no* response whatever to an input. A pressure pulse in a pneumatic transmission line travels at about 300 meters per second. If a signal is input at one end of a 600-meter tube, no response whatever will be felt at the other end until the 2-second "dead time" has gone by. In a metal rolling mill where the thickness gage is located (by necessity) 5 meters downstream from the screw-down rolls where thickness is adjusted, the thickness data reported by the gage will be delayed by 0.5 second if the sheet speed is 10 m/sec. When the front wheels of a passenger car with a 30 m wheelbase encounter a certain track irregularity at 180 km/hr train speed, the rear wheels will feel this same irregularity delayed by 0.6 sec.

   Such dead-time effects (also called transport lag) can lead, in some cases, to differential/difference equations, which present difficult mathematical problems; however, certain aspects of dead-time behavior are easily handled using Laplace transform methods. Figure 2.13 shows the relation between some arbitrary time function $f(t)$ and this same function delayed by a seconds. Using a dummy variable τ, our Laplace transform definition is

$$F(s) = \int_0^\infty f(\tau)e^{-s\tau}\,d\tau$$

We now substitute $t - a \triangleq \tau$ to get

$$F(s) = \int_a^\infty f(t-a)e^{-s(t-a)}\,dt$$

and multiplying by e^{-as} gives

$$e^{-as}F(s) = \int_a^\infty f(t-a)e^{-st}\,dt$$

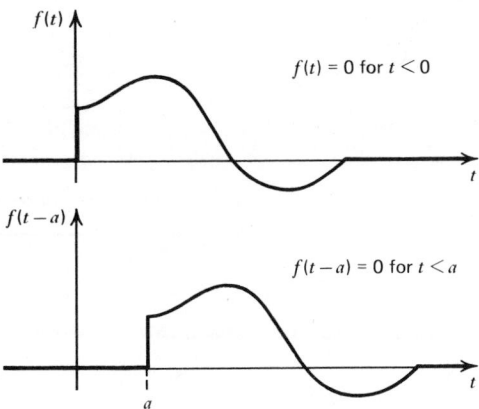

$f(t) = 0$ for $t < 0$

$f(t-a) = 0$ for $t < a$

Figure 2.13   Delayed function definition.

Since $f(t-a)$ is zero for $t<a$, we can change the lower limit to zero without changing the value of the integral, thus giving the *delay theorem* as

$$\mathcal{L}[f(t-a)u(t-a)]=e^{-as}F(s) \qquad (2.127)$$

To insure that our delayed time function is identically zero for $t<a$, we have written $f(t-a)u(t-a)$, where $u(t-a)$ is a delayed *unit step function* (see Fig. 2.14). Since the transform of the input of a dead-time element is $F(s)$ while that of the output is $e^{-as}F(s)$, its Laplace transfer function is e^{-as}.

In addition to allowing us to model the dead-time aspects of *systems*, the delay theorem is often useful for modeling *inputs* to systems, particularly those which are "discontinuous" (not representable, over their entire course, by a single smooth function). Some practical examples are shown in Fig. 2.15. The key to dealing with such inputs efficiently is to write the time functions as a sum of terms of the form $f(t-a)u(t-a)$ so that the delay theorem can be used to quickly carry out the inverse transforms. In Fig. 2.16 we would write:

$$q_i(t)=tu(t)-(t-1)u(t-1)-2u(t-2)u(t-2)$$
$$+(t-2)u(t-2)-(t-3)u(t-3) \qquad (2.128)$$

$$Q_i(s)=\frac{1}{s^2}-\frac{e^{-s}}{s^2}-\frac{2e^{-2s}}{s}+\frac{e^{-2s}}{s^2}-\frac{e^{-3s}}{s^2} \qquad (2.129)$$

$$Q_o(s)=Q_i(s)W(s)=\frac{1}{s^3}-\frac{e^{-s}}{s^3}-\frac{2e^{-2s}}{s^2}+\frac{e^{-2s}}{s^3}-\frac{e^{-3s}}{s^3} \qquad (2.130)$$

$$q_o(t)=\frac{t^2}{2}u(t)-\frac{(t-1)^2}{2}u(t-1)-2(t-2)u(t-2)$$
$$+\frac{(t-2)^2}{2}u(t-2)-\frac{(t-3)^2}{2}u(t-3) \qquad (2.131)$$

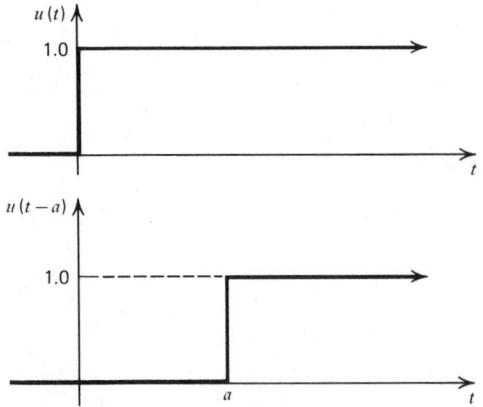

Figure 2.14 Delayed step function.

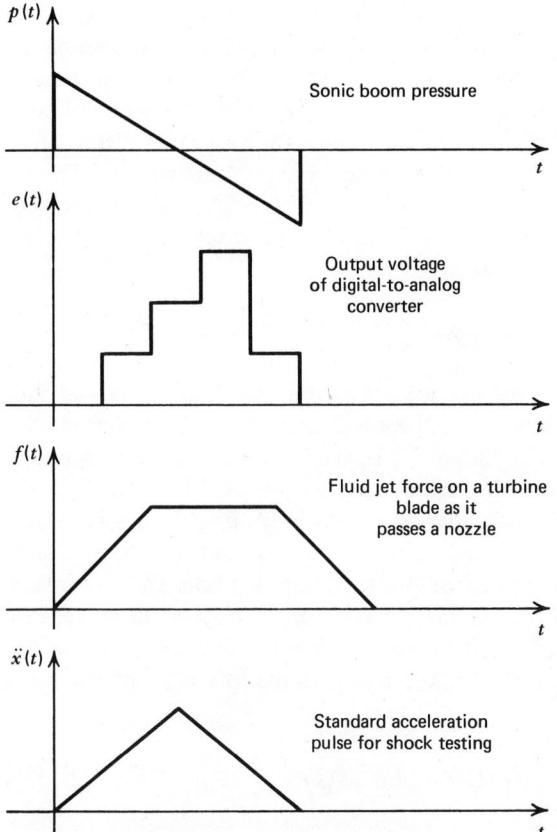

Figure 2.15 Typical "discontinuous" inputs.

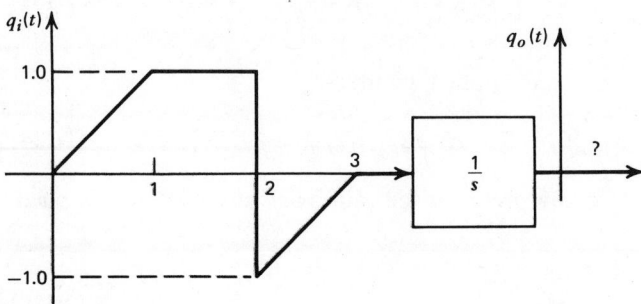

Figure 2.16 System response to discontinuous input.

In graphing expressions such as (2.131), note that a term with factor $u(t-a)$ contributes *nothing* to the total graph until $t \geqslant a$. While this particular example is also easily solved by classical methods (since the $W(s)$ is so simple), if we change the system to say, a first-order type $W = 1/(s + 1/\tau)$, the classical method becomes *very* tedious since we must sequentially solve four separate differential equations and use the final conditions of each sectional solution as the initial conditions of the next. The transform method for getting $q_o(t)$ would require very little more time than it did for $W(s) = 1/s$, in fact this would be the case for *any* $Q_o(s)$ whose terms appear in a transform table.

2.9   INITIAL-AND FINAL-VALUE THEOREMS

When we carry out the inverse transform we get a complete time history of the unknown from $t = 0$ to $t = \infty$. If we are interested *only* in the behavior at the isolated points $t = 0$ or $t = \infty$, this information is available without the expenditure of the (sometimes considerable) effort needed to perform the inverse transform. The main use of the *initial-value theorem* is in finding the system state just *after* application of the input. Recall that we earlier stated that, for certain systems and inputs, the conditions just before input application and just after are not the same. When we perform the direct transform (t to s), we interpret the "initial" conditions as those *before* the input is applied. In the initial-value theorem, however, initial means just *after* the input occurs. Stating this theorem in terms of Fig. 2.4:

$$\lim_{t \to 0} f(t) \triangleq f(\epsilon_2) = \lim_{s \to \infty} sF(s) \tag{2.132}$$

As an example, consider the system of Fig. 2.17 subjected to a unit step input voltage $e_i(t) = u(t)$. Physical analysis gives the differential equation

$$\tau_1 \frac{de_o}{dt} + e_o = K\tau_2 \frac{de_i}{dt} + Ke_i$$

$$\tau_1 \triangleq \frac{R_1 R_2}{R_1 + R_2} C, \qquad \tau_2 \triangleq R_1 C, \qquad K \triangleq \frac{R_2}{R_1 + R_2} = \frac{\tau_1}{\tau_2} \tag{2.133}$$

Suppose that e_o is zero before e_i is applied, giving:

$$\tau_1 sE_o - \tau_1 e_o(0) + E_o = K\tau_2 sE_i - K\tau_2 e_i(0) + KE_i \tag{2.134}$$

Since both $e_o(0)$ and $e_i(0)$ are interpreted as the values *before* the input is applied, they are both zero:

$$E_o = \frac{K}{s(\tau_1 s + 1)} + \frac{\tau_1}{\tau_1 s + 1} \tag{2.135}$$

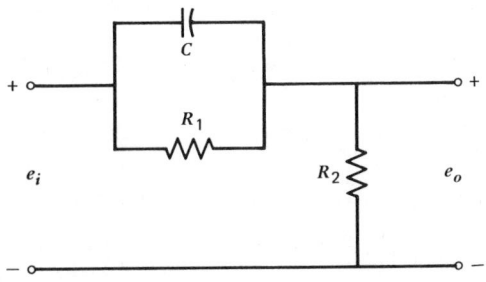

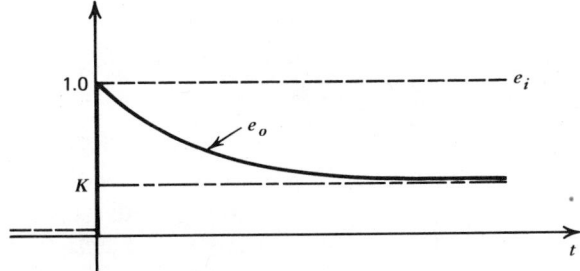

Figure 2.17   Application of initial-value theorem.

Applying the initial-value theorem:

$$\lim_{t \to 0} e_o(t) \triangleq e_o(\epsilon_2) = \lim_{s \to \infty} \left[\frac{K}{\tau_1 s + 1} + \frac{\tau_1 s}{\tau_1 s + 1} \right] = 1.0 \qquad (2.136)$$

This result can be verified by strictly physical reasoning and also by inverse transforming $E_o(s)$ and letting $t = 0$.

When the system and input are such that the output *approaches a constant value* at $t \to \infty$, then this constant value can be found easily, without the labor of the inverse transform, by use of the final-value theorem:

$$\lim_{t \to \infty} f(t) = \lim_{s \to 0} sF(s) \qquad (2.137)$$

In our present example:

$$\lim_{t \to \infty} e_o(t) = \lim_{s \to 0} \left[\frac{K}{\tau_1 s + 1} + \frac{\tau_1 s}{\tau_1 s + 1} \right] = K \qquad (2.138)$$

This theorem is not of great practical utility; its results can generally be obtained as quickly by classical methods, because the particular solution in such cases is obvious by inspection. Proofs of both these theorems are available in the literature.[7]

[7]Gardner and Barnes, *op. cit.*, p. 265.

BIBLIOGRAPHY

1 Gardner, M. F. and J. L. Barnes, *Transients in Linear Systems*, John Wiley and Sons, N. Y., 1954.

2 Kaplan, W., *Operational Methods for Linear Systems*, Addison-Wesley, Reading, Mass., 1962.

3 Lawden, D. F., *Mathematics of Engineering Systems*, John Wiley and Sons, N. Y., 1954.

4 Raven, F. H., *Mathematics of Engineering Systems*, McGraw-Hill, N. Y., 1966.

PROBLEMS

2.1 Select five first-order systems from Fig. 2.5, set up their differential equations relating input and output quantities, and define K and τ.

2.2 Select five second-order systems from Fig. 2.7, set up their differential equations relating input and output quantities, and define K, ζ, and ω_n.

2.3 In Eq. (2.100), use results from fluid mechanics to relate K_{pi} to dimensions and fluid properties, treating the inlet valve as a sharp-edge orifice of diameter d.

2.4 In Eq. (2.101), consider the flow gap h as a sharp-edge orifice and use results from fluid mechanics to relate K_{po} and K_{ho} to dimensions and fluid properties.

2.5 In Fig. P2.1, obtain a linearized expression for force F_1 in terms of displacements $X_1, X_2, X_3,$ and X_4.

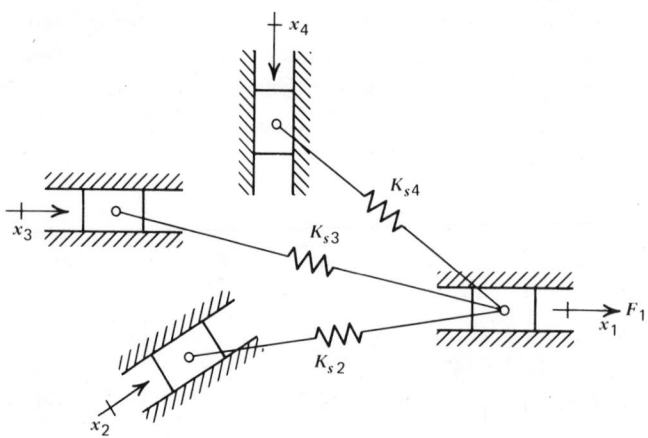

Figure P2.1

2.6 The force/velocity relation for an ideal damper (no mass, no springiness) is $(V_i/F_i)(s) = 1/B$. For the more realistic model of Fig. P2.2, find the transfer function $(V_i/F_i)(s)$.

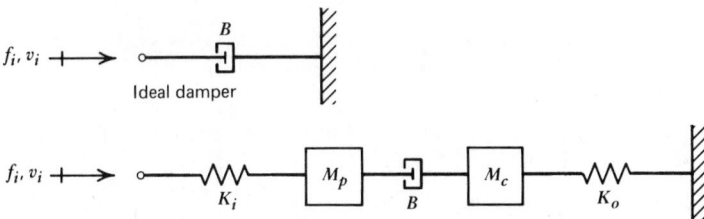

Figure P2.2

2.7 Using the delay theorem approach, write expressions for the time functions of Fig. 2.15 and then get their corresponding s functions.

2.8 Compute $q_o(t)$ in Fig. P2.3 if all initial conditions are zero.

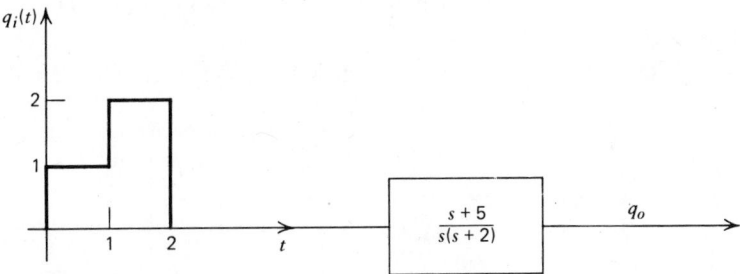

Figure P2.3

2.9 Repeat Problem 2.8 if $q_o(0) = 1$, $\dot{q}_o(0) = -1$.

2.10 In Fig. P2.4, write an expression for $f(t)$ in a form suitable for use of the delay theorem and then get $F(s)$.

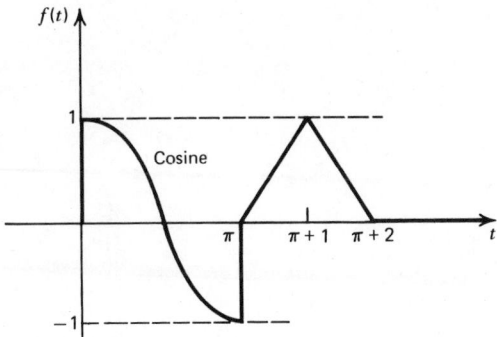

Figure P2.4

2.11 In Fig. P2.5, the electrical heater supplies $q_i(t)$ watts, the ambient air temperature is $T_a(t)$ and the R's and C's are thermal resistances and capacitances. Derive the system differential equations and obtain transfer functions $(T_2/Q_i)(s)$ and $(T_3/T_a)(s)$.

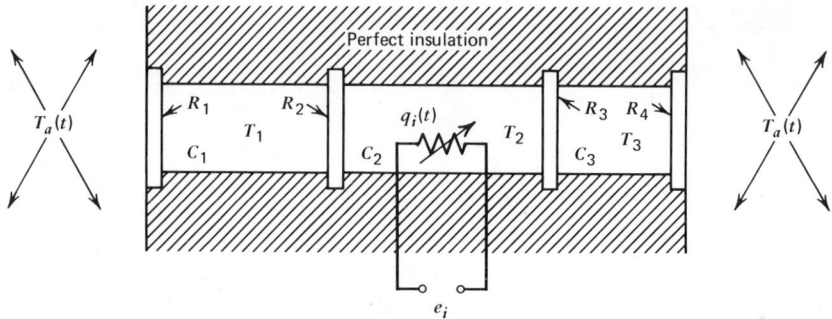

Figure P2.5

2.12 Using the partial-fraction expansion method, find $y(t)$ if $Y(s) = 5(s+3)/[s(s+1)(s+10)(s^2+100)]$.

2.13 In Fig. P2.6, let $f_o(t)$ be the force of spring K_s on mass M_2. Find $(F_o/X_i)(s)$.

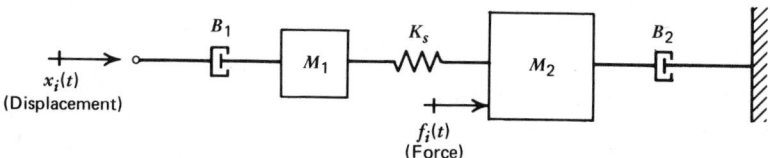

Figure P2.6

2.14 In Fig. P2.7, find $(\omega_1/T_1)(s)$, $(\omega_1/T_2)(s)$, $(\omega_2/T_1)(s)$, and $(\omega_2/T_2)(s)$. If $T_1(t) = T_2(t) = u(t)$, the unit step, use the final-value theorem to find the steady-state value of ω_1.

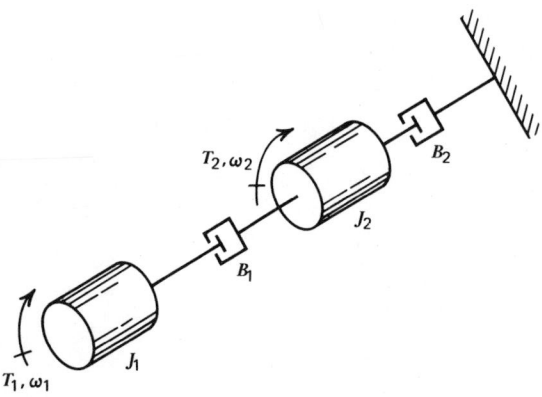

Figure P2.7

2.15 Suppose the system of Fig. 2.9 is modified as in Fig. P2.8, all other aspects remaining the same. Perform a stability analysis on this modified system.

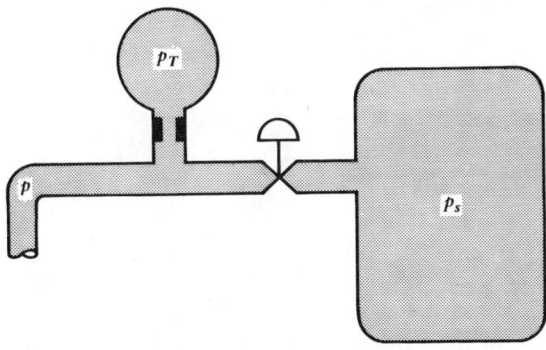

Figure P2.8

2.16 Repeat Problem 2.15, except use the modification of Fig. P2.9 instead of that in P2.8.

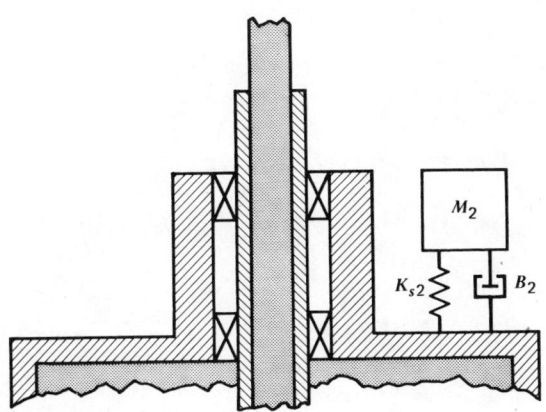

Figure P2.9

2.17 Solve by Laplace transform:

$$\frac{dx}{dt} + 3x + 2\int x \, dt = e^{-3t} - 2$$

$$x(0) = 2, \qquad x^{-1}(0) = 0$$

2.18   Use a computer root-finder plus Laplace transform to get an analytical solution for:

$$\frac{d^6q_o}{dt^6} + 5.8\frac{d^5q_o}{dt^5} + 21.36\frac{d^4q_o}{dt^4} + 44.48\frac{d^3q_o}{dt^3} + 48.92\frac{d^2q_o}{dt^2}$$

$$+ 39.0\frac{dq_o}{dt} + 18.0q_0 = 2\frac{dq_i}{dt} + q_i$$

All I.C.'s $= 0$,

$$q_i(t) = 3e^{-5t}, \quad t > 0$$
$$= 0, \quad t \leqslant 0$$

Then use CSMP (or other available digital simulation) to:
  (a) Simply graph $q_o(t)$ versus t, using the analytical solution found above.
  (b) Solve differential equation numerically.
Compare the results of a and b.

2.19   A microphone and associated amplifier are used to measure the air pressure pulse created by a sonic boom (Fig. P2.10). It is desired to investigate the dynamic accuracy of this measurement system.
  (a) For $\tau = 0.1$ sec, find $e_o(t)$ using Laplace transform to solve the differential equation.
  (b) Carefully graph $e_o(t)$, superimposed on a graph of $p_i(t)$. Find the point of maximum error by eye and express this as a percentage of the peak value.
  (c) Repeat parts a and b if an amplifier of higher input resistance is used, making $\tau = 1.0$ sec.

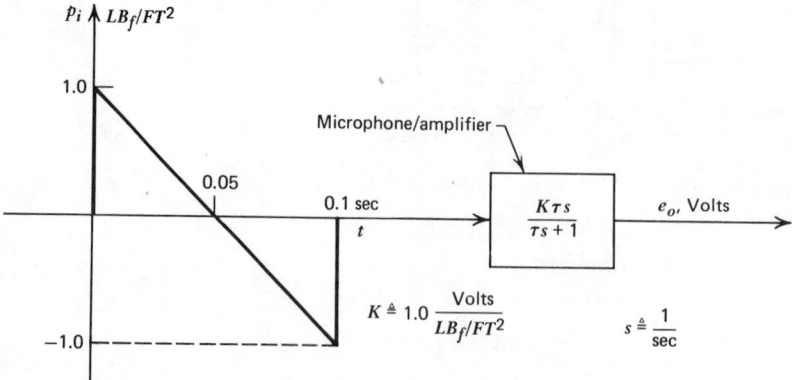

Figure P2.10

CHAPTER 3

Transform Methods for Generalized Response

We found in Chapter 2 that Laplace transform methods can considerably expedite the detailed solution for the response of a specific system to a specific input. Even more important will be the techniques developed in the present chapter since they will give us insight of a *general* nature into the system response problem. This generality refers to both the system and the input; both aspects of the situation will be treated in ways that require few restrictions. The methods are also well suited to a practical implementation that combines experimental modeling methods with powerful computer-aided analysis and design tools.

3.1 IMPULSE RESPONSE

Although the study of system response to step inputs will be familiar to most readers, impulse response methods will be less so, and because of their importance we now treat them in some detail. We must first define the *unit impulse function* $\delta(t)$. As has so often been the case in the past, impulse response concepts were originally developed by scientists and engineers in a nonrigorous fashion and then later put on a firm foundation by mathematicians. Most engineers find the nonrigorous "formal" approach quite adequate for practical work and, actually, a necessity, since the time needed to comprehend the mathematical intricacies is excessive. Even the word "function" cannot be rigorously applied to $\delta(t)$ and mathematicians have invented the concepts of "generalized functions"[1] or "distributions" to deal with these problems.

For our purposes $\delta(t)$ is defined by the statements

$$\delta(t) = 0, \qquad t \neq 0$$
$$\int_{-\varepsilon}^{+\varepsilon} \delta(t)\, dt = 1, \qquad \varepsilon > 0 \tag{3.1}$$

[1] Kaplan, W., *Operational Methods for Linear Systems*, Addison-Wesley, Reading, Mass., 1962, p. 49.

Note that this definition, unlike those for more familiar functions, does not allow one to graph $\delta(t)$ versus t. Since graphing of functions is generally an aid to understanding, it is usual to define approximating functions for $\delta(t)$ that can be graphed. Let us also at this point choose to deal with delayed functions since it will be useful to be able to make impulse functions "happen" at any desired time, not just at $t=0$. Figure 3.1 shows one of several valid ways to approximate $\delta(t-a)$:

$$\delta(t-a) = \lim_{b \to o} p(t-a) \tag{3.2}$$

It is seen that the unit impulse function has infinite height, infinitesimal width, and an area of 1. Since this behavior cannot be graphed, we "represent" impulse functions by a "spike," as in Fig. 3.2. When we multiply the unit impulse $\delta(t)$ by some number, giving, say, $3.72\delta(t)$, we create a "larger" impulse; but "larger" now means *area*, not height (as it would for ordinary functions), since the height of all impulse functions is infinite.

There is a close formal relation between the step function and the impulse function that is revealed when we approximate each of them by smooth functions, as in Fig. 3.3. As we make the approximate step function's change sharper and sharper, its slope (derivative) looks more and more like an impulse function, thus we can formally think of $\delta(t)$ as the derivative of $u(t)$, or conversely $u(t)$ as the integral of $\delta(t)$. Definition of the impulse function thus allows us to differentiate discontinuous functions such as steps, a process *not allowed* in "ordinary calculus." We can also obtain the Laplace transform of $\delta(t)$ using our earlier theorem for transforms of derivatives:

$$\mathcal{L}[\delta(t)] = \mathcal{L}\left[\frac{d}{dt}u(t)\right] = sU(s) - u(0) = s\frac{1}{s} - 0 = 1 \tag{3.3}$$

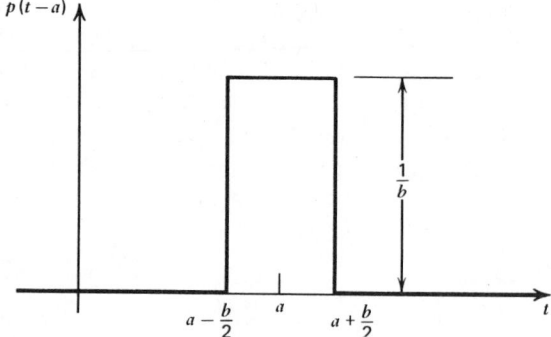

Figure 3.1 Approximating function for a unit impulse.

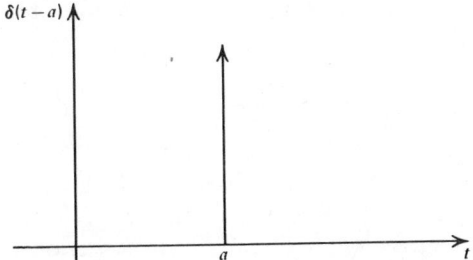

Figure 3.2   Impulse function symbol.

Impulses often appear multiplied by ordinary continuous functions within integrals:

$$\int_a^c \delta(t-b)f(t)\,dt, \qquad a < b < c \tag{3.4}$$

$$= \int_{b-\varepsilon}^{b+\varepsilon} f(t)\delta(t-b)\,dt = f(b)\int_{b-\varepsilon}^{b+\varepsilon}\delta(t-b)\,dt = f(b), \qquad \varepsilon \to 0 \tag{3.5}$$

This result, which on first sight seems as no more than a rather complicated way

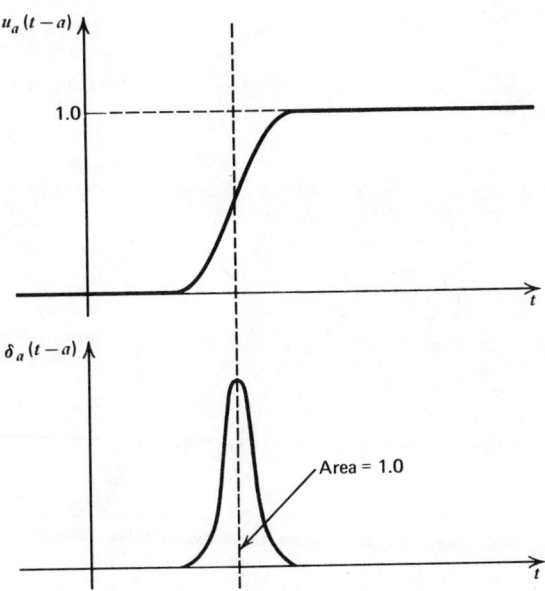

Figure 3.3   Relation between step and impulse.

of writing $f(b)$, will be useful in various developments. For example, we can confirm (3.3) by applying the basic definition:

$$\mathcal{L}[\delta(t)] = \int_0^\infty \delta(t)e^{-st}\,dt = \int_0^\varepsilon \delta(t)e^{-st}\,dt = \lim_{\varepsilon \to 0} e^{-s\varepsilon} = 1 \qquad (3.6)$$

Once we are enabled to differentiate step functions, we may be emboldened to attempt differentiation of the impulse itself. It turns out, in fact, that, using our earlier formal approach, all the derivatives of the impulse can be defined and will have Laplace transforms. For the first derivative $\delta'(t)$, let us start with a revised approximating function $\delta_a(t)$ as in Fig. 3.4, so as to easily visualize its derivative $\delta_a'(t)$. (Note that the impulse approximations in Figs. 3.1, 3.3, and 3.4 are all equally valid, even though their graphs appear different, because they all satisfy definition (3.1).) Using either the basic definition or the differentiation theorem, we can show that

$$\mathcal{L}\left[\frac{d}{dt}(\delta(t))\right] = s \qquad (3.7)$$

To visualize the next higher derivative, redraw $\delta_a'(t)$ of Fig. 3.4 as in Fig. 3.5

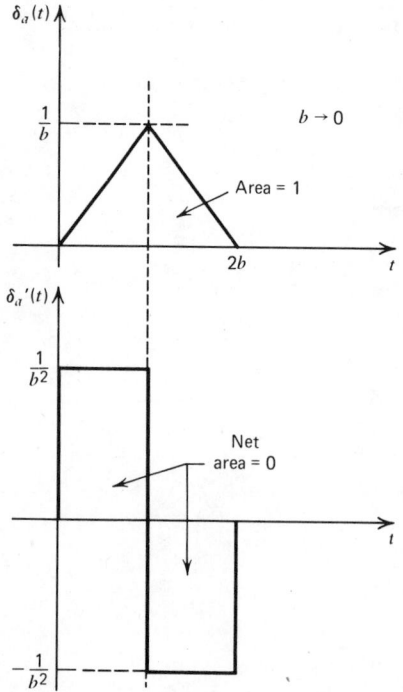

Figure 3.4 Definition of impulse first derivative.

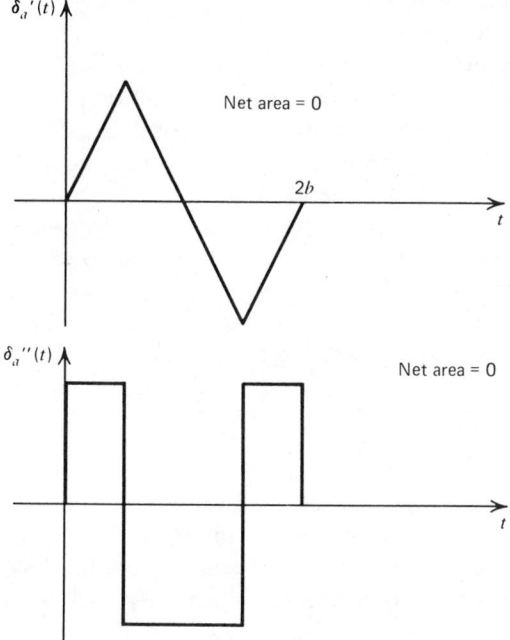

Figure 3.5 Definition of impulse second derivative.

(again valid since it fits all the definitions). This process can be continued to define derivatives of arbitrarily high order; their transforms are given by:

$$\mathcal{L}\left[\,\delta^{(n)}(t)\,\right] = s^n \tag{3.8}$$

While the derivatives of impulses have some practical applications, the impulse function itself will occupy our main interest. Suppose we apply a unit impulse as the input to a system with zero initial conditions and transfer function $W(s)$. From the definition of the transfer function:

$$Q_o(s) = Q_i(s)\,W(s) = (1)\,W(s) = W(s) \tag{3.9}$$

When the input is $\delta(t)$ the response is called the *impulse response* of the system and is denoted by $h(t)$, thus

$$h(t) = \mathcal{L}^{-1}\left[\,W(s)\,\right] \tag{3.10}$$

The function $h(t)$ is sometimes also called the *weighting function* of the system. Note that if one is trying to characterize a linear system, it can be done as well by giving $h(t)$ as by giving $W(s)$ since, given one of these two descriptions, the other can be calculated. There is often some initial objection to the concept of impulse response since it seems impractical because no *real* input voltage, force, pressure, and so on can possibly behave (go to infinity and back in "zero" time) like an impulse. People who object in this way to impulse response usually do not object to step response, yet a perfect step input is *also* not physically possible since it requires an infinite rate of change in some physical variable. Thus the "impracticality" of impulse response differs only in degree, not kind, from that of step response; they are both approximations to reality when used to model the behavior of real systems.

Let us examine some specific cases of impulse response. For an integrator, $W(s) = 1/s$ and $h(t) = u(t)$ the unit step. For a first-order system:

$$h(t) = \mathcal{L}^{-1}\left[\frac{K}{\tau s + 1}\right] = \frac{K}{\tau} e^{-t/\tau} \tag{3.11}$$

Note that this $h(t)$ is identical to the response of the system to an initial condition of size K/τ, thus one can think of an impulse input as a means of driving a system from zero initial state to some desired state in infinitesimal time. (In general, impulse *derivatives* may also be needed to achieve an arbitrary

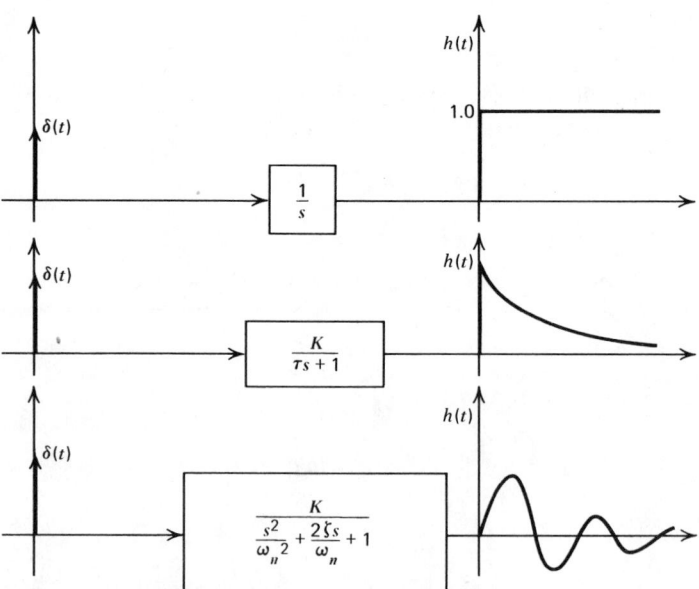

Figure 3.6 Impulse response of some basic systems.

desired initial state.) This viewpoint is analogous to the use of impulse/momentum principles in mechanics to study the behavior of impacting bodies, where "sudden" changes in velocity are caused by "impulsive" forces. For a second-order system:

$$h(t) = \mathcal{L}^{-1}\left[\frac{K}{s^2/\omega_n^2 + 2\zeta s/\omega_n + 1}\right]$$

$$= \frac{K\omega_n}{\sqrt{1-\zeta^2}}\, e^{-\zeta\omega_n t}\sin\left(\omega_n\sqrt{1-\zeta^2}\; t\right) \tag{3.12}$$

This response corresponds to that for an initial condition on $\dot{q}_o$ while $q_o(0)=0$. Figure 3.6 displays these results graphically.

3.2   CONVOLUTION INTEGRAL: RESPONSE TO ARBITRARY INPUTS WHEN IMPULSE RESPONSE IS KNOWN

Another useful Laplace transform theorem called the *convolution integral* is:

$$\mathcal{L}^{-1}[F_1(s)F_2(s)] = \int_0^t f_1(\tau)f_2(t-\tau)\,d\tau \tag{3.13}$$

where

$$F_1(s) \triangleq \mathcal{L}[f_1(t)], \qquad F_2(s) \triangleq \mathcal{L}[f_2(t)]$$

and τ is a dummy variable. Let us sketch a proof of this theorem. From Fig. 3.7, since f_1 and f_2 are zero for negative arguments:

$$\int_0^t f_1(\tau)f_2(t-\tau)\,d\tau = \int_0^\infty f_1(\tau)f_2(t-\tau)\,d\tau \tag{3.14}$$

Now,

$$\mathcal{L}\left[\int_0^\infty f_1(\tau)f_2(t-\tau)\,d\tau\right] = \int_0^\infty\left[\int_0^\infty f_1(\tau)f_2(t-\tau)\,d\tau\right]e^{-st}\,dt \tag{3.15}$$

Note that the bracketed quantity on the left-hand side of (3.15) is a function of t since we integrate with respect to dummy variable τ and then substitute numerical limits $(\infty, 0)$ for τ, thus eliminating τ and leaving t. Thus it makes sense to take the Laplace transform of the bracketed quantity. On the right-hand side of (3.15) the multiple integral is written so as to require integration first with respect to τ and then with respect to t. We wish to interchange the order of these

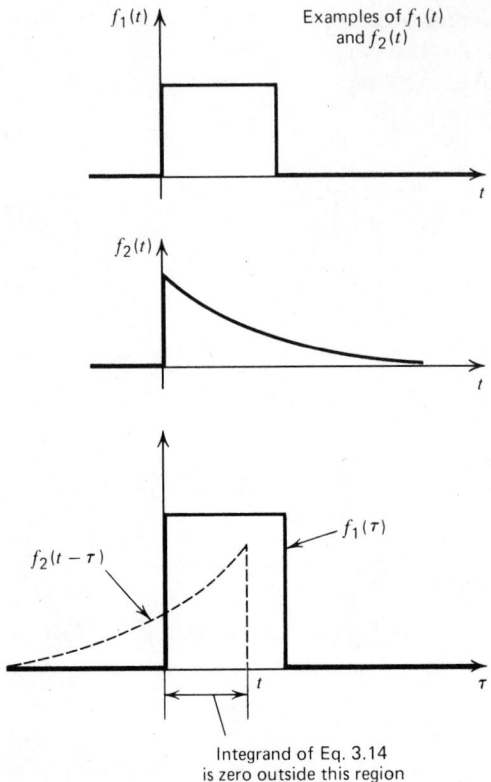

Figure 3.7 Graphical interpretation of convolution integral.

integrations, a process valid under most, but not all, conditions. For example:

$$\int_1^2 \int_0^2 xy^2 \, dy \, dx = \int_0^2 \int_1^2 xy^2 \, dx \, dy = 4 \tag{3.16}$$

Assuming the validity of the interchange:

$$\mathcal{L}\left[\int_0^\infty f_1(\tau) f_2(t-\tau) \, d\tau\right] = \int_0^\infty f_1(\tau) \int_0^\infty \left[f_2(t-\tau) e^{-st} \, dt\right] d\tau \tag{3.17}$$

From the delay theorem:

$$\int_0^\infty f_2(t-\tau) e^{-st} \, dt = e^{-s\tau} F_2(s) \tag{3.18}$$

giving

$$\mathcal{L}\left[\int_0^\infty f_1(\tau)f_2(t-\tau)\,d\tau\right] = F_2(s)\int_0^\infty f_1(\tau)e^{-\tau s}\,d\tau = F_2(s)F_1(s) \qquad (3.19)$$

which is equivalent to (3.13).

Our first application of the convolution integral will be as an alternative method of inverse transformation. If our complete s-function $F(s)$ appears as a product $F_1(s)F_2(s)$ and if $f_1(t)$ and $f_2(t)$ are known, we can use (3.13) to analytically obtain $f(t)$. Note that while $\mathcal{L}^{-1}[F_1(s) + F_2(s)] = f_1(t) + f_2(t)$, $\mathcal{L}^{-1}[F_1(s)F_2(s)] \neq f_1(t)f_2(t)$, so a method for inverse transforming *products* is of interest. Of course, if the function $F(s) = F_1(s)F_2(s)$ itself appears in a table, or is easily expanded in partial fractions, there is little incentive to use (3.13).

Our main interest in convolution integral is as the basis of a powerful method for computing the response of *any* linear system to *any* transient input when the system's impulse response is known. If an input $q_i(t)$ is applied to a system with transfer function $W(s)$:

$$Q_o(s) = Q_i(s)W(s) \qquad (3.20)$$

From (3.13) we then have

$$\mathcal{L}^{-1}[Q_i(s)W(s)] = q_o(t) = \int_0^t h(\tau)q_i(t-\tau)\,d\tau \qquad (3.21)$$

or, because of the symmetry in F_1, F_2 of (3.13), alternatively,

$$q_o(t) = \int_0^t q_i(\tau)h(t-\tau)\,d\tau \qquad (3.22)$$

where $h(t)$ is the (assumed known) impulse response of the system. The practical utility of (3.22) lies in the fact that the calculation of $q_o(t)$ can be carried out by numerical methods even if either or both $q_i(t)$ and $h(t)$ are known only as experimentally measured graphs, not as mathematical functions defined by formulas. That is, once a complex system has been experimentally impulse tested and we have an oscillograph record of $h(t)$ in hand, the system's response to any $q_i(t)$, whether a math function or another oscillograph record, can be calculated from (3.22). Using (3.22) we construct the function $q_o(t)$ "one point at a time" by choosing a specific t, say t_1, and writing

$$q_0(t_1) = \int_0^{t_1} h(\tau)q_i(t_1-\tau)\,d\tau \qquad (3.23)$$

Figure 3.8 interprets Eq. (3.23) to show how a single point $q_o(t_1)$ is calculated; by

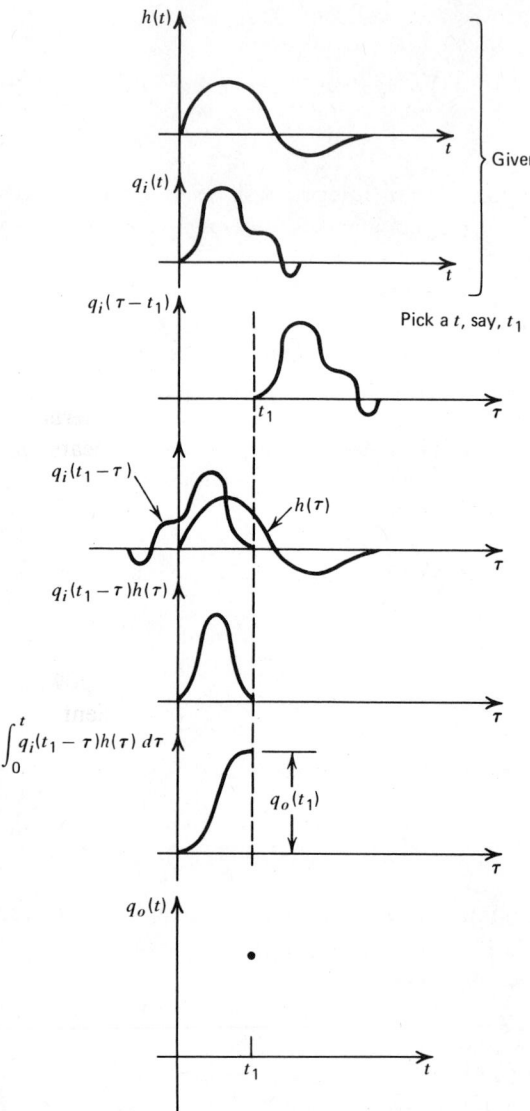

Figure 3.8 Use of convolution integral to compute system response to an arbitrary input.

repeating this process for different values of t, the entire course of $q_o(t)$ can be calculated, computer methods making the tedious operation practical.

Since we have spoken of measuring $h(t)$ experimentally and since it is clear that the real input used in such an experiment *cannot* be a perfect impulse, we must explore the conditions under which such a test is a good approximation to a true impulse test. Convolution integral will be of help in establishing a criterion for test validity; the frequency spectrum methods developed later in this chapter will give an alternative approach. Since practical test pulses of force, voltage, temperature, and so on will not usually be of unit area, let us first find the effect of impulse area on system response for a perfect impulse of area A_p. From (3.21), if $q_i(t) = A_p \delta(t)$:

$$q_o(t) = \int_0^t h(\tau) A_p \delta(t - \tau) \, d\tau = \int_{t-\varepsilon}^{t+\varepsilon} h(\tau) A_p \delta(t - \tau) \, d\tau \qquad (3.24)$$

$$q_o(t) = A_p h(t) \int_{t-\varepsilon}^{t+\varepsilon} \delta(t - \tau) \, d\tau = A_p h(t) \qquad (3.25)$$

We thus see that changing the *area* of a perfect impulse causes a proportional change in the ordinates of the system response. That is, impulses of various areas all give identical *shapes* of response, but the "sizes" are in proportion to the areas. We will find that when a *real* pulse is a good approximation to an impulse, system response magnitude will again be proportional to pulse area.

The response of any real system to perfect impulses will generally exhibit a characteristic time scale. (Note that the response of a *real* system to *perfect* impulses *exists* even though the experiment cannot be performed.) For example, an all-electronic system might operate in the region of milliseconds or microseconds while an automobile suspension vibration would be characterized by tenths of seconds or seconds. (Think of watching the response on an oscilloscope and switching the time base until the response pattern "just fills the screen" horizontally.) While it is obvious that an approximate impulse $\delta_a(t)$ becomes more and more perfect as its time duration shortens, we will now show that $\delta_a(t)$ need not be short in absolute value but only short *relative to the characteristic time of the system being tested*. Thus an all-electronic system might require nanosecond test pulses to adequately simulate impulses while pulses of several hundredths second duration would be quite adequate for vehicle suspension testing.

To put the above intuitive reasoning on a mathematical basis, consider Fig. 3.9. Here an approximate impulse $\delta_a(t)$ of area A_p is applied as input to a system whose perfect impulse response is $h(t)$. Using convolution integral:

$$q_o(t) = \int_0^t h(\tau) \delta_a(t - \tau) \, d\tau = \int_{t_1}^t h(\tau) \delta_a(t - \tau) \, d\tau \qquad (3.26)$$

We can see graphically that, since $h(\tau)$ is "slow" relative to δ_a, we can replace,

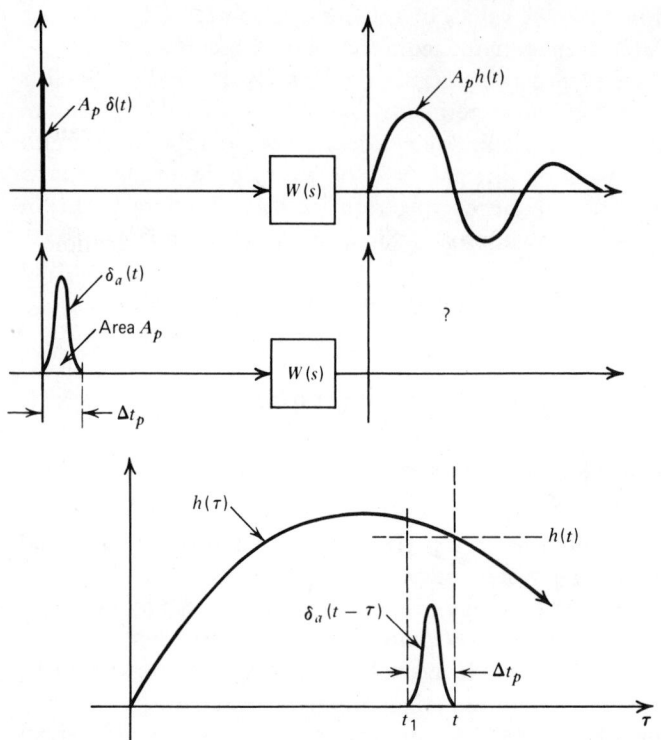

Figure 3.9   Input signal that approximates well a perfect impulse.

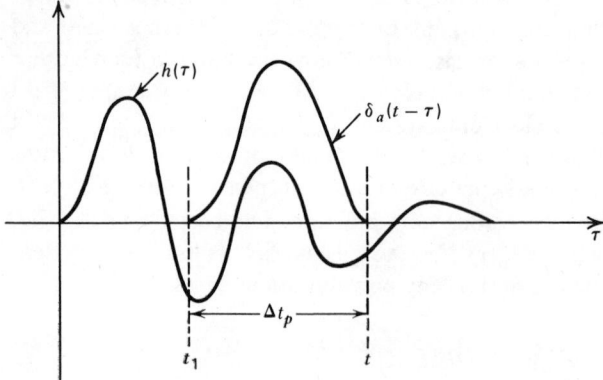

Figure 3.10   Input signal that approximates poorly a perfect impulse.

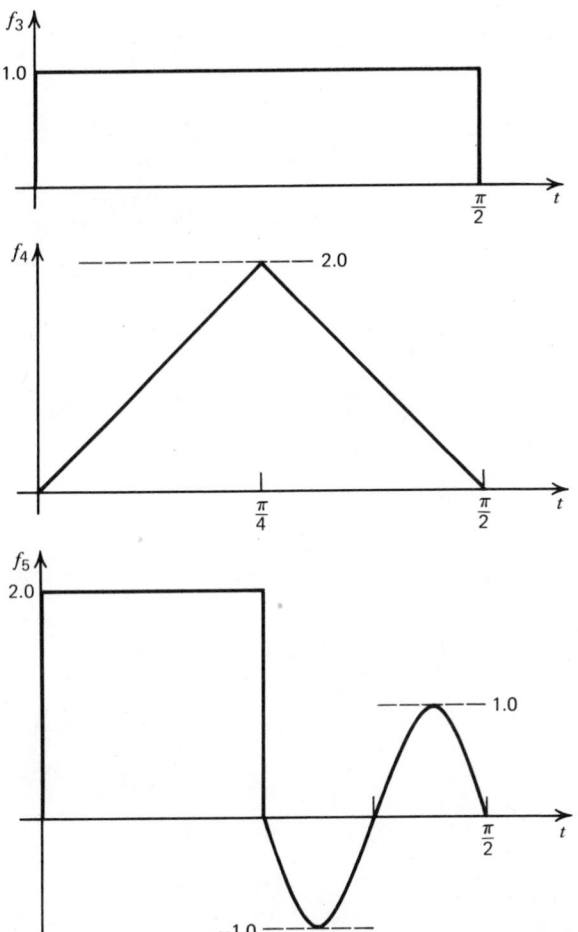

Figure 3.11   Test signals for numerical impulse response examples.

with good approximation, the function $h(\tau)$ by the number $h(t)$ over the range $t_1 \rightarrow t$, giving

$$q_0(t) \approx h(t) \int_{t_1}^{t} \delta_a(t - \tau) \, d\tau = A_p h(t) \tag{3.27}$$

We see that if δ_a's duration is "short enough," the system responds in essentially the same way it would to a perfect impulse of like area and that *the shape of δ_a makes no difference whatever*. Figure 3.10 shows clearly that the approximation of Eq. (3.27) would be very inaccurate for the "slow" δ_a shown there.

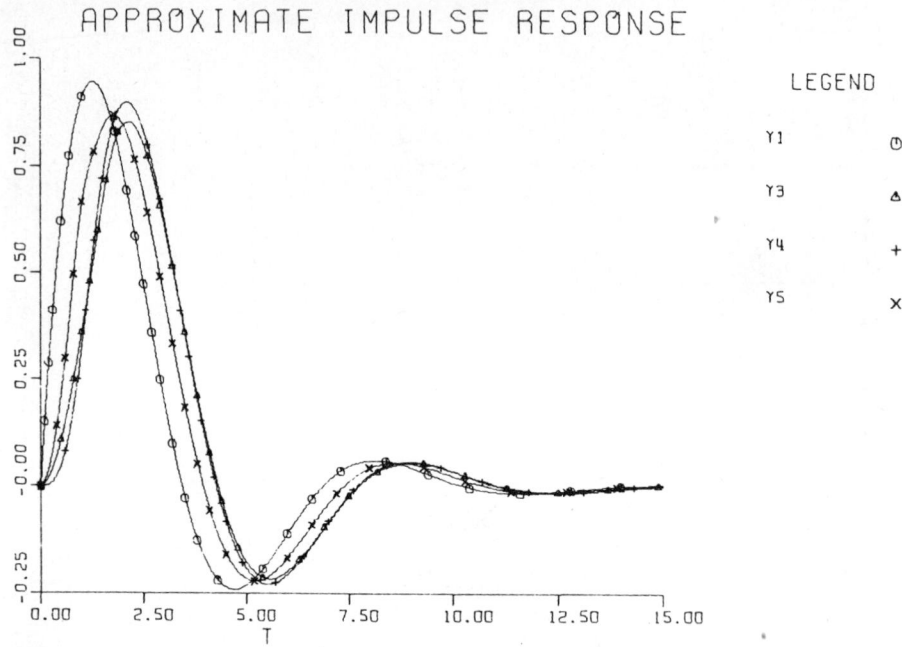

Figure 3.12 Second-order system response to test pulses: pulse duration $\frac{1}{4}$ of natural period.

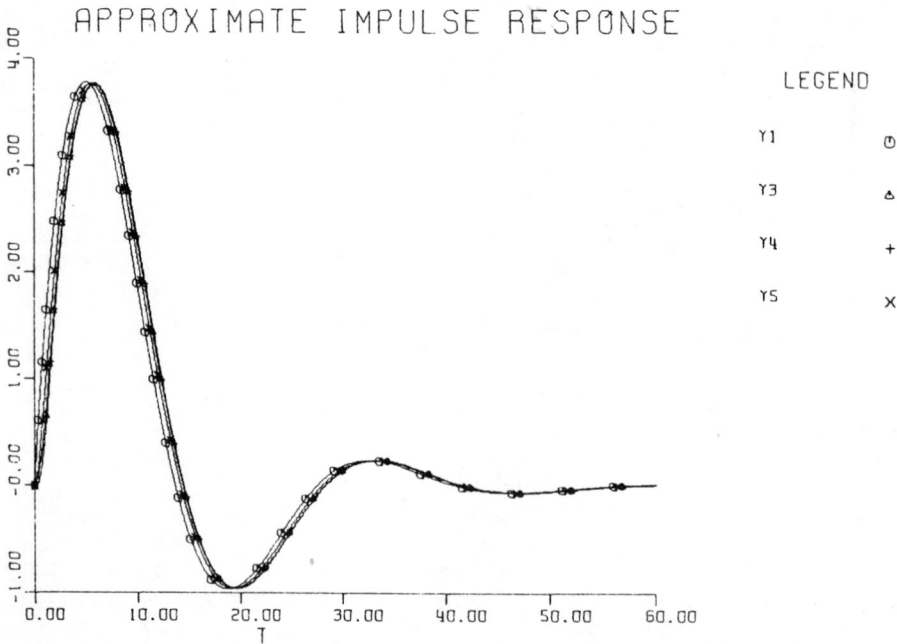

Figure 3.13 Second-order system response to test pulses: pulse duration $\frac{1}{16}$ of natural period.

80

Although Figs. 3.9 and 3.10 make qualitatively clear that system impulse response can be experimentally found using physically realizable pulse inputs, some numerical results are helpful in developing judgment as to how short Δt_p must actually be for good accuracy. Figure 3.11 shows three pulses (f_3, f_4, f_5) of different shapes but identical areas. Figure 3.12 shows the response of a second-order system with $\omega_n = 1$ rad/sec, $\zeta = 0.2$ to a perfect impulse (Y1) and to f_3, f_4, and f_5 (responses Y3, Y4, Y5). Here pulse duration is $\frac{1}{4}$ of the system's natural period 2π, the three pulses produce three noticeably different responses, and all three differ from the perfect impulse response, though not radically. In Fig. 3.13 the system's period has been changed to 8π, making Δt_p $\frac{1}{16}$ of the system's period. Now the three pulse responses are practically indistinguishable from each other and all are very close to the perfect impulse response.

3.3 FREQUENCY RESPONSE

We presume the reader is familiar with the basic concepts of the response of linear systems to sinusoidal inputs (frequency response)[2] and thus will give only a brief review treatment plus the transform aspects of the subject. Engineering interest in steady-state sinusoidal response is easily explained for mechanical engineers by the widespread use of constant-speed rotating machinery whose unavoidable unbalances produce vibration-exciting sinusoidal forces. For electrical engineers the 60-Hz power distribution system ensures a multitude of sinusoidal response problems.

By definition, frequency response refers to the sinusoidal steady-state response of a system that is observed after all transients due to both initial conditions and the "starting up" of the sinusoidal input have died out. For stable systems (roots all in left half-plane) these transients actually will eventually disappear. However, if a system model has some pure imaginary roots, their "transient" terms persist, but it is conventional in analysis to ignore these terms when speaking of the system's frequency response. Frequency response of unstable systems (right half-plane roots) is not usually of interest but can be defined and calculated if one ignores the "transient" terms. Actually, for both pure imaginary and right half-plane roots, one can find a unique set of initial conditions that will *completely suppress* all terms except the sinusoidal steady state. This scheme is, however, a mathematical rather than practical one since it requires absolute numerical perfection.

If we apply an input $q_i = A_i \sin \omega t$ to a stable system $W(s) = N(s)/D(s)$ we can calculate the frequency response from:

$$D(s)Q_o(s) + I(s) = N(s)Q_i(s) \tag{3.28}$$

[2]Doebelin, E. O., *System Dynamics*, C. E. Merrill, Columbus, Ohio, 1972.

where $I(s)$ represents terms due to initial conditions:

$$Q_o(s) = W(s)Q_i(s) - \frac{I(s)}{D(s)} = W(s)\frac{A_i\omega}{s^2+\omega^2} - \frac{I(s)}{D(s)} \tag{3.29}$$

The quantity $I(s)/D(s)$ contributes only terms that die out, so it will be ignored. Then

$$W(s)\frac{A_i\omega}{s^2+\omega^2} = \frac{K_1}{s+0-i\omega} + \frac{K_2}{s+0+i\omega} + \cdots \text{ transients} \tag{3.30}$$

These transients (due to "starting up" the sinusoidal input, rather than due to initial conditions) also disappear, so, finally, defining Q_{oss} as the sinusoidal steady state:

$$Q_{oss}(s) = W(s)\frac{A_i\omega}{s^2+\omega^2}\bigg|_{ss} = \frac{K_1}{s+0-i\omega} + \frac{K_2}{s+0+i\omega} \tag{3.31}$$

Using our earlier method for inverse transforming pairs of complex roots:

$$K_1 = W(s)\frac{A_i\omega}{s+i\omega}\bigg|_{s=i\omega} = W(i\omega)\frac{A_i\omega}{2i\omega} = \frac{W(i\omega)A_i}{2\underline{/90°}} \tag{3.32}$$

$$q_{oss}(t) = \frac{2A_i|W(i\omega)|}{2\underline{/90°}}\sin(\omega t + \underline{/W(i\omega)} + 90°) \tag{3.33}$$

$$q_{oss}(t) \triangleq A_0\sin(\omega t + \phi) = |W(i\omega)|A_i\sin(\omega t + \underline{/W(i\omega)}) \tag{3.34}$$

The output amplitude is thus $|W(i\omega)|$ times the input amplitude and the output sine wave is phase shifted by $\underline{/W(i\omega)}$ degrees. The *sinusoidal transfer function* is defined as a complex number whose magnitude is the amplitude ratio A_o/A_i and whose angle is the phase shift ϕ, thus

$$\text{Sinusoidal transfer function} \triangleq \frac{Q_o}{Q_i}(i\omega) \triangleq W(i\omega) \tag{3.35}$$

It is common to display sinusoidal transfer functions, the system frequency response, as a pair of graphs, amplitude ratio and phase shift versus frequency.

For the familiar first- and second-order systems:

$$W(i\omega) = \frac{K}{i\omega\tau + 1} = \frac{K\underline{/0°}}{\sqrt{(\omega\tau)^2 + 1}\ \underline{/\tan^{-1}\omega\tau}} = \frac{K}{\sqrt{(\omega\tau)^2 + 1}}\underline{/\tan^{-1} - \omega\tau} \quad (3.36)$$

$$W(i\omega) = \frac{K}{(i\omega)^2/\omega_n^2 + 2\zeta i\omega/\omega_n + 1} = \frac{K}{\sqrt{\left[1 - (\omega/\omega_n)^2\right]^2 + 4\zeta^2\omega^2/\omega_n^2}}\underline{/\phi} \quad (3.37)$$

$$\phi \triangleq \tan^{-1}\frac{2\zeta}{\omega/\omega_n - \omega_n/\omega} \quad (3.38)$$

Figures 3.14 and 3.15 show these curves. *Logarithmic frequency-response plotting* is also in wide use. Here the amplitude ratio is converted to decibels (db) according to

$$\text{db} \triangleq 20\log_{10}(\text{amplitude ratio}) \quad (3.39)$$

and the frequency is plotted on a logarithmic scale, giving Figs. 3.16 and 3.17 for first- and second-order systems. Advantages of logarithmic plotting include:

1 Rapid manual graphing is possible.
2 Wide ranges of amplitude ratio and frequency are conveniently displayed.

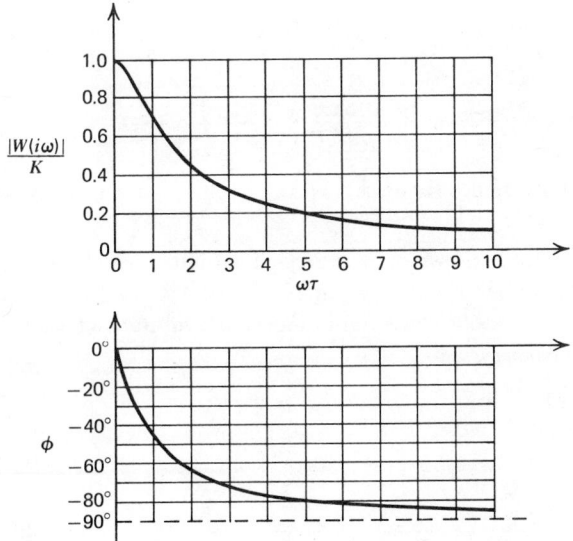

Figure 3.14   First-order system frequency response.

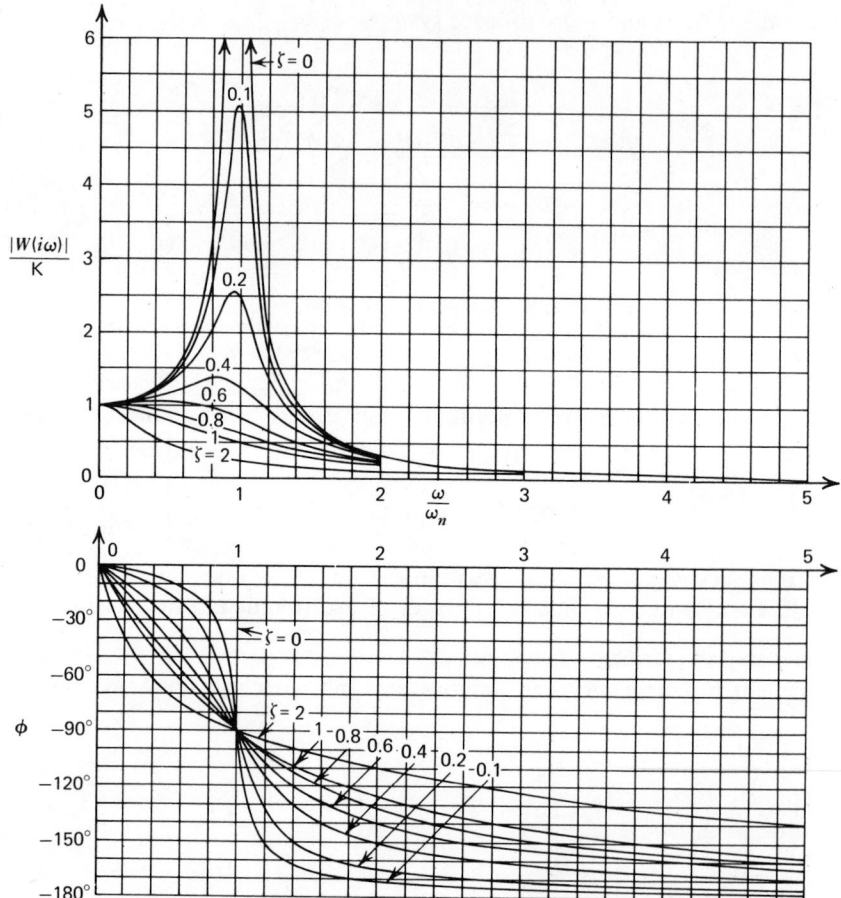

Figure 3.15   Second-order system frequency response.

3 Amplitude ratio exhibits straight-line-asymptote regions of definite slope. These are helpful in identifying model type from experimental data.

4 Complex transfer functions are easily plotted and understood as sums of simple (first-order, second-order, etc.) basic systems.

Item 4 in this list refers to the fact that the general transfer function

$$W(s) = \frac{Ks^{p}e^{-\tau s}(\tau_1 s + 1)\dots\left[\dfrac{s^2}{\omega_n^2} + \dfrac{2\zeta s}{\omega_n} + 1\right]\dots}{(\tau_a s + 1)\dots\left[\dfrac{s^2}{\omega_{na}^2} + \dfrac{2\zeta_a s}{\omega_{na}} + 1\right]\dots} \qquad (3.40)$$

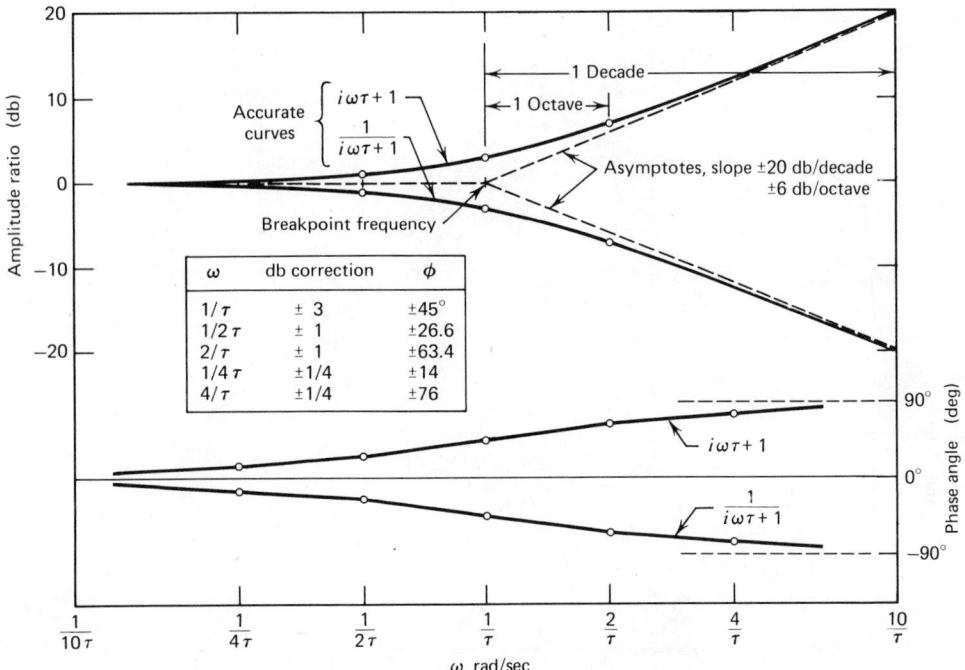

Figure 3.16 Logarithmic first-order system frequency response.

obtained by factoring $N(s)$ and $D(s)$ in the form $W(s) = N(s)/D(s)$ can be expressed in sinusoidal form as

$$W(i\omega) = W_1(i\omega) W_2(i\omega) \ldots = \left(M_1 \underline{/\phi_1} \right) \left(M_2 \underline{/\phi_2} \right) \ldots = M_1 M_2 \ldots \underline{/\phi_1 + \phi_2 + \cdots}$$

$$(3.41)$$

When we convert the amplitude ratio to decibels (logarithms) the product $M_1 M_2 \ldots$ becomes a sum of db, resulting in a graphical addition process that is rapidly carried out. The individual phase angles were already additive so no special treatment is necessary there. A numerical example may be helpful for those readers who have not previously encountered this plotting scheme:

$$W(s) = \frac{10 e^{-0.01 s} s}{(s+1)\left[s^2/10^2 + 2(0.4)s/10 + 1 \right]} \qquad (3.42)$$

A careful study of Fig. 3.18 should make clear the procedure. Note that for any dead time, $e^{-i\omega\tau} = 1.0 \underline{/-\omega\tau}$, giving a constant amplitude ratio of 1.0 and a negative phase shift proportional to frequency.

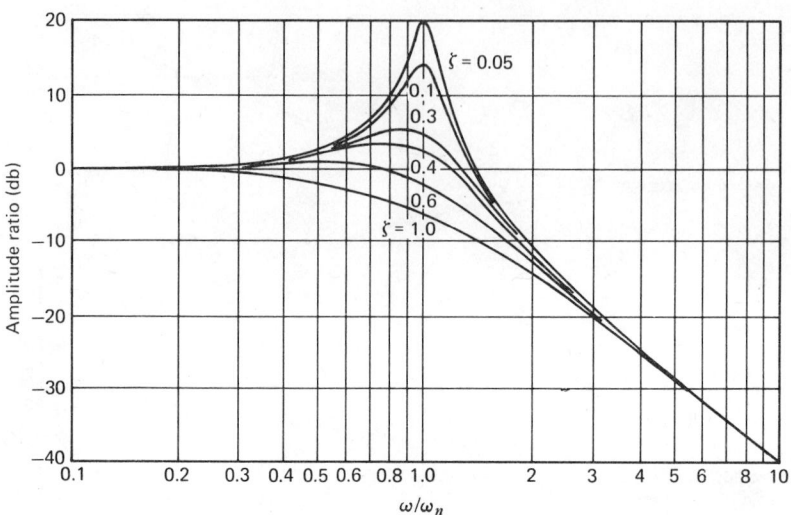

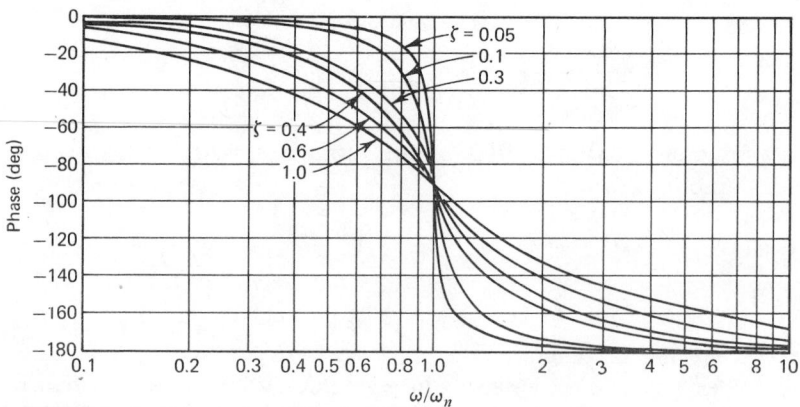

Figure 3.17   Logarithmic second-order system frequency response.

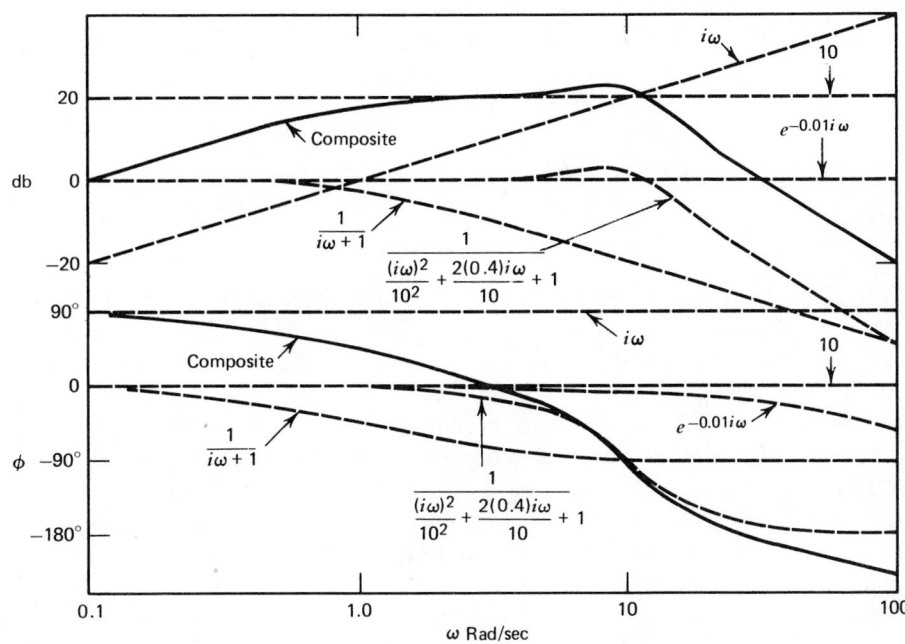

Figure 3.18 Composite system logarithmic frequency response.

3.4 RELATIONS BETWEEN IMPULSE RESPONSE AND FREQUENCY RESPONSE

In experimental modeling, both impulse tests and frequency-response tests are in practical use. The choice as to which method to employ in a specific instance depends on a number of factors that will be discussed in a later chapter devoted completely to experimental modeling. Our present interest is in showing the mathematical relations between these two alternative types of system descriptions. We will find that if, say, an impulse test has been run, there is no need to also run a frequency-response test; we can *calculate* the sinusoidal transfer function from the impulse test data. The reverse will also be true.

Assuming that the impulse response $h(t)$ of a system $W(s)$ is known, either by theoretical calculation or experimental testing, we can find the frequency-response as follows.

$$W(s) = \mathcal{L}[h(t)] = \int_0^\infty h(t)e^{-st}\,dt \qquad (3.43)$$

Since $s = \sigma + i\omega$ in general, we may take $s = i\omega$ as a special case, giving:

$$W(i\omega) = \int_0^\infty h(t)e^{-i\omega t} dt$$

$$= \int_0^\infty h(t)\cos\omega t\, dt - i\int_0^\infty h(t)\sin\omega t\, dt \qquad (3.44)$$

When $h(t)$ is given by a mathematical formula, calculation of a formula for $W(i\omega)$ requires evaluation of the two integrals in Eq. (3.44), using integral tables. If $h(t)$ is known only as an experimental graph, the following graphical/numerical procedure will produce the frequency-response graphs (but *not* any formulas) for $W(i\omega)$. The graphs of $W(i\omega)$ are produced "one point at a time" by choosing a numerical value for ω, say ω_1, making $\cos\omega_1 t$ and $\sin\omega_1 t$ definite curves, which can be plotted against t on the same graph as the given $h(t)$. One can then form, and numerically integrate, the "product curves" $h(t)\cos\omega_1 t$ and $h(t)\sin\omega_1 t$ from $t = 0$ to the point where $h(t)$ decays to zero, giving $W(i\omega_1) = a_1 - ib_1 = M_1\underline{/\phi_1}$, one point on the frequency-response curves. Figure 3.19 illustrates the procedure.

Repetition of this process for many other ω's produces the complete frequency-response curves. We of course start with $\omega = 0$; however, it is not obvious how *high* in frequency we must go, since, unlike the analytical calculus procedure possible when $h(t)$ is given by a formula, we *cannot* take ω out to infinity. This practical question is resolved by recalling our earlier statement that the impulse response of every real system occurs in some characteristic time scale. When the "time-scale" of the chosen ω is much faster than that of the system's $h(t)$, then we will find the process of Fig. 3.19 produces $|W(i\omega)| \to 0$ and we need go to no higher frequencies. Note that systems with "fast" impulse responses will have frequency response extending to "high" frequencies. Figure 3.20 should help in understanding these comments.

When $W(i\omega)$ is known and we wish to compute from it $h(t)$, then the inverse transform definition of Eq. (2.25) must be used:

$$f(t) \triangleq \frac{1}{2\pi i}\int_{c-i\infty}^{c+i\infty} F(s)e^{st} ds$$

We have avoided use of this formula up to this point; however, our desired result requires its use, so we will proceed, but in a nonrigorous fashion. For $f(t)$'s which die out to zero, the constant c may be taken as zero. We also let $s = i\omega$ to

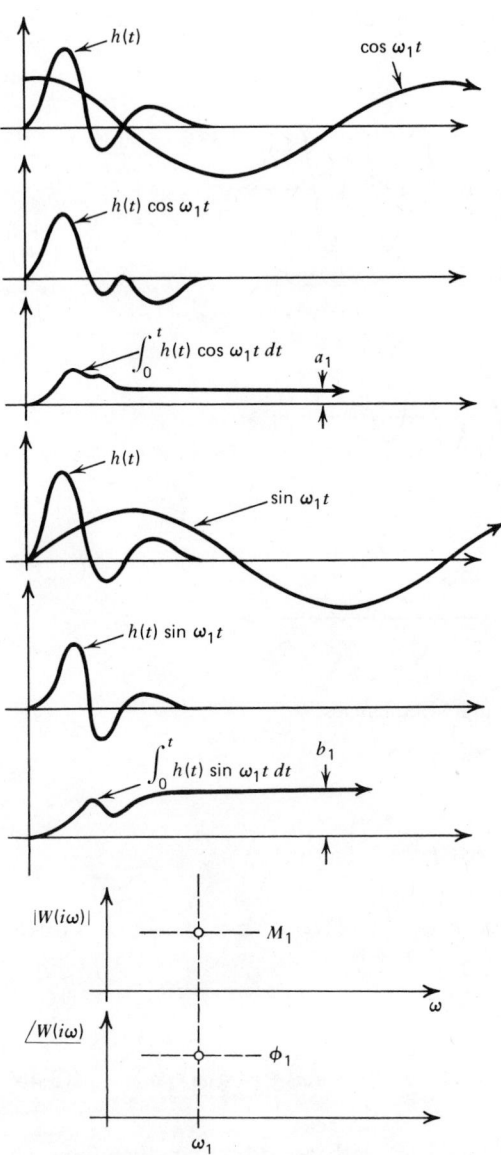

Figure 3.19 Calculation of system frequency response from known impulse response.

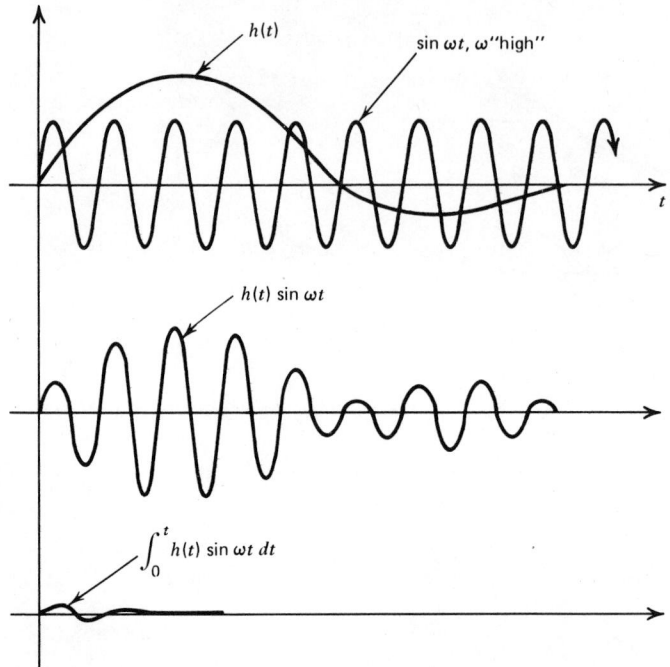

Figure 3.20 $|W(i\omega)|$ goes to zero for "high" frequencies.

get

$$f(t) = \frac{1}{2\pi i} \int_{-i\infty}^{i\infty} F(i\omega) e^{i\omega t} d(i\omega)$$

$$f(t) = \frac{i}{2\pi i} \int_{-\infty}^{\infty} F(i\omega) e^{i\omega t} d\omega = \frac{1}{2\pi} \int_{-\infty}^{\infty} F(i\omega) e^{i\omega t} d\omega \qquad (3.45)$$

$$= \frac{1}{2\pi} \int_{-\infty}^{\infty} |F(i\omega)| e^{i\phi} e^{i\omega t} d\omega$$

$$= \frac{1}{2\pi} \int_{-\infty}^{\infty} |F(i\omega)| \big[(\cos\phi + i\sin\phi)(\cos\omega t + i\sin\omega t) \big] d\omega \qquad (3.46)$$

We now examine the various terms in (3.46) with regard to whether they are even or odd functions of ω, assuming that our $F(i\omega)$'s are limited to functions whose amplitude ratio is an even function of ω and for which $\cos\phi$ is an even function of ω and $\sin\phi$ odd. (Ratios of polynomials $N(s)/D(s)$ and dead times do fit these requirements.)

The product $(\cos \omega t)(\cos \phi)$ is the product of even functions so is itself even; $(\sin \omega t)(\sin \phi)$, being the product of odd functions, is also even. The two products of odd by even functions, $(\sin \omega t)(\cos \phi)$ and $(\cos \omega t)(\sin \phi)$ are both odd. Breaking the integral of (3.46) into four separate integrals, one for each of the products just discussed:

$$\int_{-\infty}^{\infty} |F(i\omega)| \cos \omega t \cos \phi \, d\omega = \int_{-\infty}^{\infty} \text{even function} = 2 \int_{0}^{\infty} \text{even function}$$

$$\int_{-\infty}^{\infty} |F(i\omega)| \sin \omega t \sin \phi \, d\omega = \int_{-\infty}^{\infty} \text{even function} = 2 \int_{0}^{\infty} \text{even function}$$

$$i \int_{-\infty}^{\infty} |F(i\omega) \sin \omega t \cos \phi \, d\omega = \int_{-\infty}^{\infty} \text{odd function} = 0$$

$$i \int_{-\infty}^{\infty} |F(i\omega)| \cos \omega t \sin \phi \, d\omega = \int_{-\infty}^{\infty} \text{odd function} = 0$$

Equation (3.46) thus becomes

$$2(\pi)f(t) = 2 \int_{0}^{\infty} |F(i\omega)| \left[\cos \phi \cos \omega t - \sin \phi \sin \omega t \right] d\omega \qquad (3.47)$$

Finally, since $f(-t)$ must be zero (our time functions are defined to be zero for $t < 0$):

$$f(-t) = \frac{1}{\pi} \int_{0}^{\infty} |F(i\omega)| \left[\cos \phi \cos \omega t + \sin \phi \sin \omega t \right] d\omega = 0 \qquad (3.48)$$

This requires that

$$\int_{0}^{\infty} |F(i\omega)| \cos \phi \cos \omega t \, d\omega = - \int_{0}^{\infty} |F(i\omega)| \sin \phi \sin \omega t \, d\omega \qquad (3.49)$$

which gives from (3.47) our desired result as

$$f(t) = \frac{2}{\pi} \int_{0}^{\infty} |F(i\omega)| \cos \phi \cos \omega t \, d\omega \qquad (3.50)$$

This is clearly useful for transforming *any* $F(i\omega)$ into its corresponding $f(t)$. When $F(i\omega)$ happens to be a system's frequency response $W(i\omega)$, then $f(t)$ is of course its impulse response $h(t)$. When $W(i\omega)$ is given by experimental graphs, then $h(t)$ can be constructed, "one point at a time," by choosing a t, say t_1, and carrying out Eq. (3.50) graphically/numerically in a fashion analogous to that of Fig. 3.19 to obtain $f(t_1)$.

That is, $|F(i\omega)|$ is just the given amplitude ratio versus frequency curve while a curve of $\cos\phi$ versus ω can be easily plotted since ϕ versus ω is given. Choosing a numerical value of t, t_1 allows plotting of $\cos t_1\omega$ versus ω and we can now plot and then numerically integrate the "product curve" $|F(i\omega)|(\cos\phi)$ $(\cos t_1\omega)$ to get a numerical value for $f(t_1)$. (The integration need not go to $\omega = \infty$ because any real system's amplitude ratio is sure to become negligibly small beyond some *finite* ω.) Repeating this process for many t values generates the complete graph of $f(t)$.

3.5 RESPONSE TO PERIODIC INPUTS: FOURIER SERIES

As pointed out in Section 1.3, driving inputs that are periodic are of great practical interest. For linear systems, where superposition holds, response to any periodic input is easily found by expressing the input as a Fourier series (sum of sine waves), processing each sine wave through the system's frequency response, and then superimposing the individual sinusoidal responses to obtain the total response, which is also periodic. It can be shown that any periodic function $q_i(t)$, such as in Fig. 3.21, which is single-valued, finite, and has a finite number of discontinuities and maxima and minima in one cycle, may be represented by the Fourier series:

$$q_i(t) = \frac{a_0}{T} + \frac{2}{T} \sum_{n=1}^{\infty} \left(a_n \cos\left(\frac{2\pi n}{T}t\right) + b_n \sin\left(\frac{2\pi n}{T}t\right) \right) \qquad (3.51)$$

$$a_n \triangleq \int_{-T/2}^{T/2} q_i(t) \cos\left(\frac{2\pi n}{T}t\right) dt \qquad (3.52)$$

$$b_n \triangleq \int_{-T/2}^{T/2} q_i(t) \sin\left(\frac{2\pi n}{T}t\right) dt \qquad (3.53)$$

Real-world periodic forces, voltages, flow rates, and so on always meet the above-stated conditions (Dirichlet conditions). Since a perfect "curve-fit" of $q_i(t)$ requires an infinite number of terms, we can never attain this perfection; however, a suitably chosen finite number of terms will always meet practical engineering requirements. When $q_i(t)$ is given by a formula, the series is most precisely found by analytically carrying out (3.51)–(3.53). If $q_i(t)$ is given by a graph or table, then approximate numerical methods are necessary and available as "canned" programs in most computer libraries. The most efficient such programs are based on the so-called Fast-Fourier-Transform (FFT) algorithm which has revolutionized frequency spectrum analysis in recent years and which we will discuss in more detail in Chapter 6.

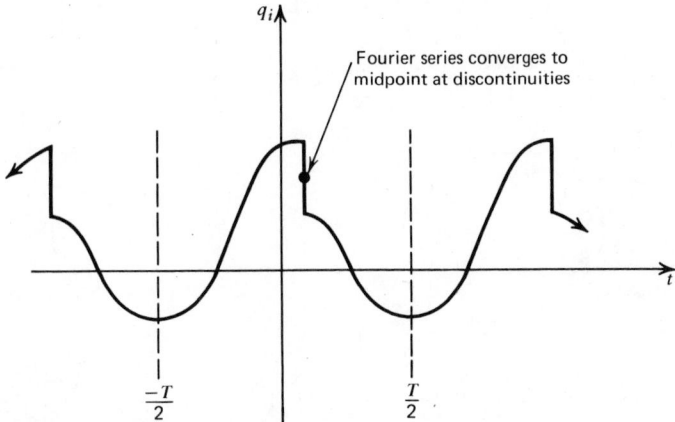

Figure 3.21 Periodic function.

The choice of location for the time origin in Fourier series calculations is ordinarily governed by mathematical convenience, since a periodic function "goes on forever," thus "absolute" time is irrelevant. In our example of Fig. 3.22 we have actually chosen an "inconvenient" origin in order to demonstrate some general features. For the constant term (average value of $q_i(t)$) in the series we compute a_0:

$$a_0 = \int_{-0.01}^{-0.0075} 0.6\,dt + \int_{-0.0075}^{0.0025} -0.4\,dt + \int_{0.0025}^{0.01} 0.6\,dt = 0.002 \qquad (3.54)$$

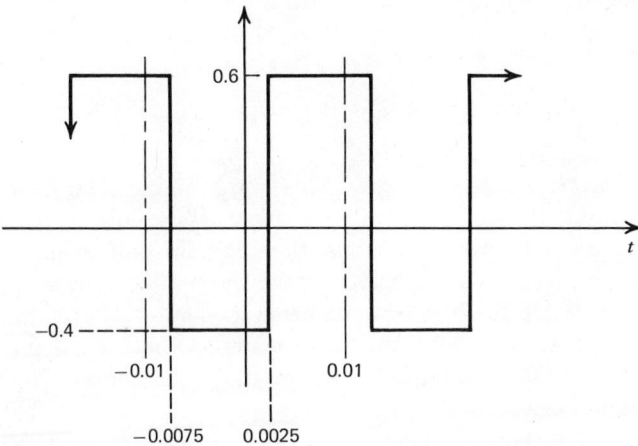

Figure 3.22 Square wave with "poor" choice of origin.

giving the average value a_0/T as 0.1, which can of course be seen by inspection.

$$a_n = \int_{-0.01}^{-0.0075} 0.6\cos\frac{2\pi nt}{0.02}\, dt + \int_{-0.0075}^{0.0025} -0.4\cos\frac{2\pi nt}{0.02}\, dt$$

$$+ \int_{0.0025}^{0.01} 0.6\cos\frac{2\pi nt}{0.02}\, dt \tag{3.55}$$

$$a_n = -\frac{1}{100\pi n}\left(\sin\frac{n\pi}{4} + \sin\frac{3n\pi}{4}\right) \tag{3.56}$$

Similarly:

$$b_n = \frac{1}{100\pi n}\left(\cos\frac{n\pi}{4} - \cos\frac{3n\pi}{4}\right) \tag{3.57}$$

The first few terms of the series are thus:

$$q_i(t) = 0.1 + \frac{\sqrt{2}}{\pi}\sin 100\pi t - \frac{\sqrt{2}}{\pi}\cos 100\pi t$$

$$-\frac{\sqrt{2}}{3\pi}\sin 300\pi t - \frac{\sqrt{2}}{3\pi}\cos 300\pi t$$

$$+\cdots \tag{3.58}$$

Since the frequency-response curves of system response are set up to deal with sine-wave (not cosine-wave) inputs, we convert (3.58) to the form of sine waves with phase angles α, using the identity $A\cos\omega t + B\sin\omega t \equiv C\sin(\omega t + \alpha)$, where $C \triangleq \sqrt{A^2 + B^2}$, $\alpha \triangleq \tan^{-1}A/B$:

$$q_i(t) = 0.1 + \frac{2}{\pi}\sin(100\pi t - 45°) + \frac{2}{3\pi}\sin(300\pi t - 135°) + \cdots \tag{3.59}$$

Figure 3.23 shows the *frequency spectrum* of $q_i(t)$. This graph makes clear at a glance which frequencies are present, and their amplitude and phase. All periodic functions display *discrete* spectra. That is, they contain only *isolated* frequencies; graphs such as Fig. 3.23 should never have the plotted points or spikes connected by a curve since there is *zero* frequency content between the spikes. The lowest frequency present is called the fundamental or first harmonic; the term corresponding to a particular value of n is called the nth harmonic. The frequency of the fundamental is always the same as the "repetition rate" of the periodic function itself. Thus we would predict from Fig. 3.22 with no calculation that the first frequency in the series would be 50 Hz or 100π rad/sec and that the possible harmonics were at 200π, 300π, and so on.

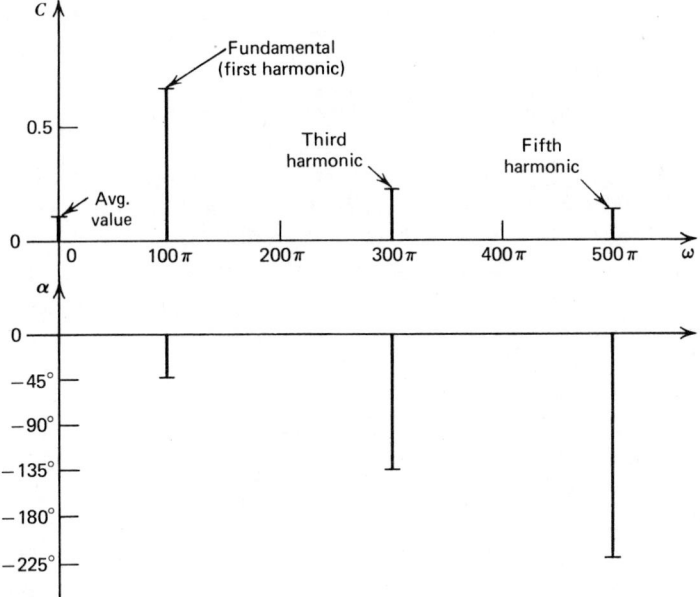

Figure 3.23   Discrete frequency spectrum of square wave.

For the waveform of Fig. 3.22 the phase angles α could all be made zero, and the effort to calculate the series greatly reduced, if we had chosen the time origin more carefully (0.0025 sec to the right). This would not, however, change the amplitude portion (C's) of the spectrum. Note that this 0.0025 sec time shift corresponds to a 45° angle shift of a 50 Hz sine wave, 135° of a 150 Hz, 225° of a 250 Hz, and so on. For waveforms of arbitrary unsymmetrical shape, such as Fig. 3.21, the phase angles α cannot be made zero for any choice of origin, thus a complete spectrum normally includes both amplitude and phase information.

An alternative expression for the Fourier series is

$$q_i(t) = \frac{1}{T} \sum_{n=-\infty}^{\infty} c_n e^{i\omega_n t} \tag{3.60}$$

$$c_n \triangleq \int_{-T/2}^{T/2} q_i(t) e^{-i\omega_n t}\, dt \tag{3.61}$$

$$\omega_n \triangleq \frac{2\pi n}{T} \tag{3.62}$$

This is equivalent to our earlier definition ((3.51)–(3.53)) as can be seen by expanding it using $e^{i\omega_n t} = \cos\omega_n t + i\sin\omega_n t$, and so on. We also find that

$$\left.\begin{array}{l} a_n = \text{real part of } c_n \\ b_n = -(\text{imaginary part of } c_n) \end{array}\right\} n = 0, 1, 2, \ldots \qquad (3.63)$$

thus relating the coefficients of the two forms. A rapid, transform-based method for computing c_n (and thus a_n, b_n) can be derived by noting that since $q_i(t)$ is periodic with period T, and $e^{-i2\pi n t/T}$ is periodic with period T/n, then

$$c_n = \int_{-T/2}^{T/2} q_i(t) e^{-i\omega_n t}\, dt = \int_0^T q_i(t) e^{-i\omega_n t}\, dt \qquad (3.64)$$

We now define

$$q_{iT}(t) \triangleq \begin{cases} q_i(t), & 0 \leqslant t \leqslant T \\ 0, & \text{elsewhere} \end{cases} \qquad (3.65)$$

giving, from (3.64),

$$c_n = \mathcal{L}\left[q_{iT}(t) \right]_{s = i\omega_n} \qquad (3.66)$$

Let us apply this new method for finding the series coefficients to the example of Fig. 3.22, however, with the time origin shifted to the more convenient position of Fig. 3.24.

$$q_{iT}(t) = 0.6u(t) - u(t - 0.01) + 0.4u(t - 0.02) \qquad (3.67)$$

$$Q_{iT}(s) = \frac{0.6 - e^{-0.01s} + 0.4e^{-0.02s}}{s} \qquad (3.68)$$

For $n = 0$, $s = 0$ and (3.68) becomes indeterminate ($0/0$); however, L'Hospital's rule gives $Q_{iT}(0) = 0.002$, giving the series average value a_0/T as 0.1. Setting $s = i100\pi n$ in (3.68) gives

$$c_n = \frac{-0.01i}{n\pi}\left[0.6 - \cos\pi n + i\sin\pi n + 0.4\cos 2\pi n + 0.4i\sin 2\pi n \right] \qquad (3.69)$$

$$c_n = \left(0 - \frac{0.02i}{\pi}\right), \left(0 - \frac{0.02i}{3\pi}\right), \left(0 - \frac{0.02i}{5\pi}\right), \ldots$$

Thus the a_n are all zero and the sine-wave amplitudes are $2/\pi$, $2/3\pi$, $2/5\pi$, and so on. Note that this corresponds to Eq. (3.59) except that now the phase angles are all zero because of our shifted origin.

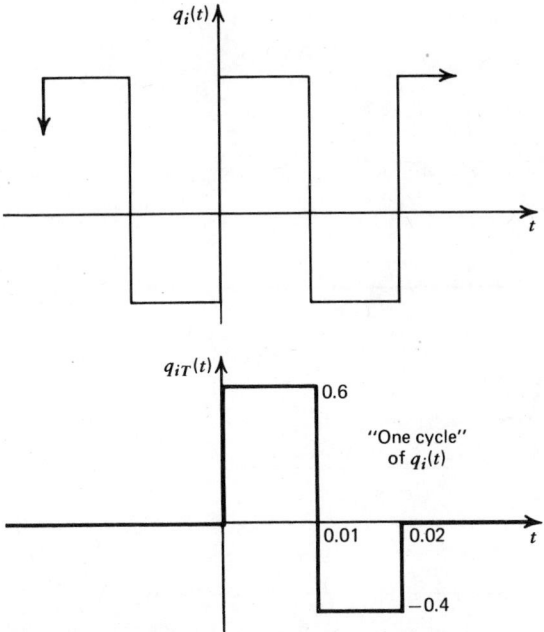

Figure 3.24 Square wave with "good" choice of origin: transform method of Fourier series calculation.

Once the Fourier series has been calculated (irrespective of the method used) the periodic steady-state response $q_{os}(t)$ of a system $W(s)$ to the periodic input $q_i(t)$ is readily obtained. If the input is given by

$$q_i(t) = A_{i0} + A_{i1}\sin(\omega_1 t + \alpha_1) + A_{i2}\sin(\omega_2 t + \alpha_2) + \cdots \qquad (3.70)$$

then

$$q_{os}(t) = A_{o0} + A_{o1}\sin(\omega_1 t + \theta_1) + A_{o2}\sin(\omega_2 t + \theta_2) + \cdots \qquad (3.71)$$

where

$$A_{ok} = A_{ik}|W(i\omega_k)|, \qquad \theta_k = \alpha_k + \underline{/W(i\omega_k)} \qquad (3.72)$$

Figure 3.25 displays this procedure and should make clear why, in general, an infinite number of series terms is unnecessary, since both $|W(i\omega)|$ for any real-world system and A_i for any real-world input tend to zero at high frequency,

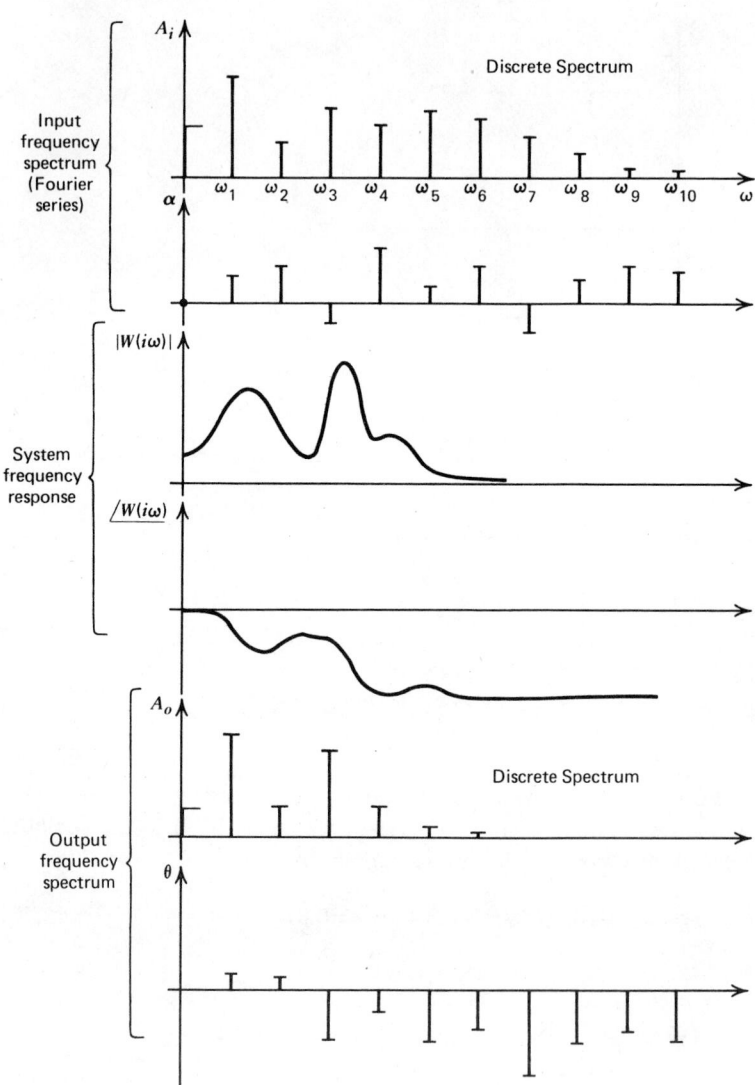

Figure 3.25 Response of arbitrary system to arbitrary periodic input.

making the response amplitude A_o go to zero "even faster." For the example of Fig. 3.25, it appears that frequencies above the fifth harmonic, even though significantly present in q_i, are of little importance since $W(i\omega)$ has become very small. Note that any coincidence between peaks of A_i and $W(i\omega)$ is a "resonance" phenomenon leading to large values of A_o at those frequencies. Also, while $q_{os}(t)$, if plotted, will be periodic with the same period as $q_i(t)$, its waveform will be in general radically different.

3.6 AMPLITUDE-MODULATED INPUTS

Amplitude-modulated and demodulated signals are intentionally employed in measurement[3] and communication systems and occur "naturally" in mechanical systems involving gearing, rolling-contact bearings, and so on where they appear as vibration and acoustic noise. While these types of signals are not usually periodic and thus do not fit the requirements for Fourier series representation, they do often have discrete frequency spectra. If we can find these spectra, then the response of linear systems to such inputs is found in exactly the same way as for periodic signals.

The general case of amplitude modulation refers to a situation where a *carrier* sine wave at frequency ω_c has an amplitude that is some function of time, $f_m(t)$:

$$q_i(t) = f_m(t)\sin(\omega_c t + \phi) \tag{3.73}$$

A simple special case that introduces the basic concepts is

$$q_i(t) = A(1 + m\cos\omega_m t)\sin(\omega_c t + \phi), \qquad \omega_c > \omega_m \tag{3.74}$$

Figure 3.26 shows a typical waveform. While this may "look to the eye" like a periodic function, it actually is not unless ω_c is an integer multiple of ω_m, which is not usually true in practical applications. Fortunately, the identity

$$\sin\alpha\cos\beta \equiv \frac{\sin(\alpha+\beta)}{2} + \frac{\sin(\alpha-\beta)}{2}$$

allows us to write:

$$q_i(t) = A\sin(\omega_c t + \phi) + \frac{mA}{2}\sin\left[(\omega_c + \omega_m)t + \phi\right] + \frac{mA}{2}\sin\left[(\omega_c - \omega_m)t + \phi\right] \tag{3.75}$$

[3]Doebelin, E. O., *Measurement Systems*, McGraw-Hill, N.Y., 1975, p. 161.

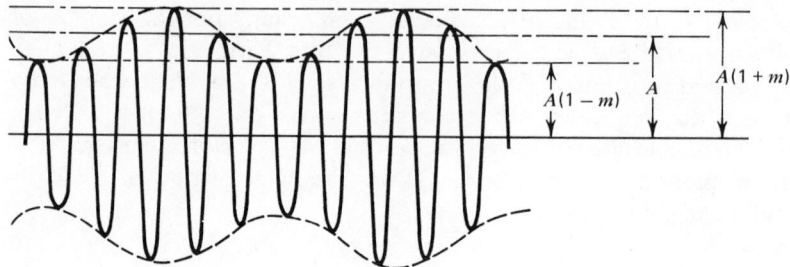

Figure 3.26 Amplitude-modulated signal.

This is clearly a sum of three distinct sine waves at frequencies $\omega_c, (\omega_c + \omega_m)$, and $(\omega_c - \omega_m)$, giving the discrete spectrum of Fig. 3.27 and allowing use of the methods of Fig. 3.25 for system response calculation.

Figure 3.28 shows some examples of mechanical systems that may exhibit amplitude-modulation effects in measured vibrations and acoustic noise. In the systems involving gears, the tooth-meshing frequency plays the role of carrier frequency while the modulation effect is caused by geometric factors or manu-facturing imperfections. When a gear wheel has spokes, its radial stiffness is not uniform around the circumference but rather varies periodically, thus effecting the meshing forces. For a four-spoke wheel with 100 teeth, rotating at 50 revolutions per second, the fundamental mesh frequency is 5000 Hz while the tooth force magnitude is modulated with a cycle of four times per revolution or

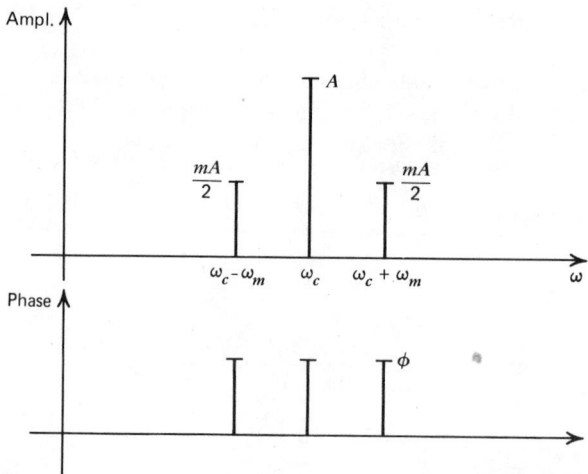

Figure 3.27 Discrete spectrum of amplitude-modulated signal.

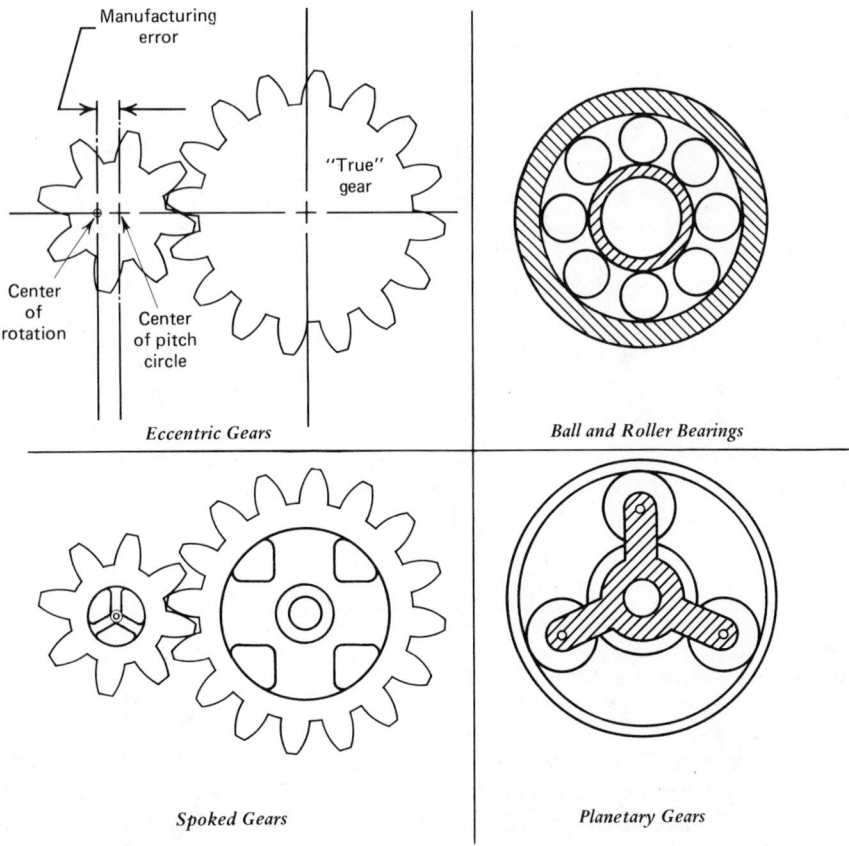

Figure 3.28   Mechanical systems that produce amplitude-modulated signals.

200 Hz. We should not be surprised to measure vibrations with components at not only the obvious shaft rotation (50 Hz) and tooth-mesh (5000 Hz) frequencies, but also at the more subtle amplitude-modulation or *sideband* frequencies of 4800 and 5200 Hz. When, as in Fig. 3.28, *both* wheels are spoked and of different diameters, even more complex modulations can occur.

Even "unspoked" gears will exhibit modulation due to unavoidable eccentricities. Figure 3.28 shows a simplified example where one gear is true and the other is eccentric. Clearly, during one rotation of the eccentric gear, the tooth force magnitudes will go through one cycle of modulation. Up to this point we have failed to mention that the gear-meshing forces for perfect gears are *not* sinusoidal but rather periodic, thus the modulation effects due to spokes, eccentricity, mounting-bolt spacing, and so on are applied, not to a *single* sine

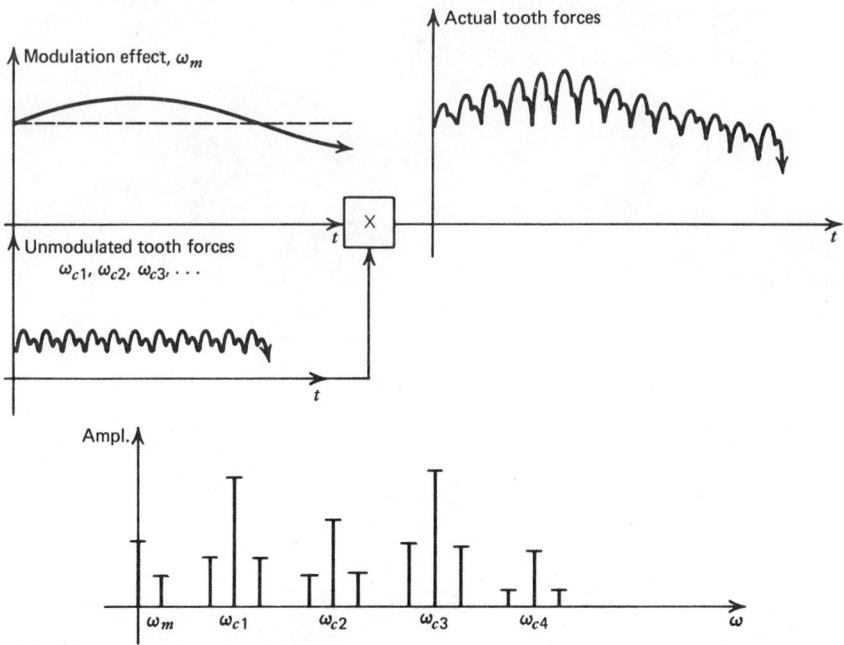

Figure 3.29 Spectrum of gear-tooth forces.

wave but to a *sum* of sine waves (Fourier series). The resulting spectrum thus contains many carriers, each with its own sidebands, as shown in Fig. 3.29. Finally, we note that the modulation effect could *also* be nonsinusoidal, requiring the multiplication of two Fourier series, and giving an even more complicated spectrum. Detailed calculations and experimental measurements for some of the situations of Fig. 3.28 may be found in various references.[4]

3.7 TRANSIENT INPUTS: FREQUENCY METHOD

In Section 3.2 we developed a method for calculating the response of any linear system to a transient input of any shape, using the assumed known impulse response of the system. We now show an alternative scheme for solving this same class of problems which utilizes the system's frequency response. Starting

[4]Remmers, E. P., "The Dynamics of Automotive Rear Axle Gear Noise," SAE Paper 710114, 1971; Mitchell, L. D. and Lynch, G. A., "Progress in Noise Analysis," *Dupont Innovation*, Vol. 1, #1, Fall 1969; Gu, A. L. and Badgley, R. H., "Prediction of Vibration Sidebands in Gear Meshes," ASME Paper 74-DET-95, 1974; Gu, A. L., Badgley, R. H., and Chiang, T., "Planet-Pass-Induced Vibration in Planetary Reduction Gears," ASME Paper 74-DET-93, 1974; Fieldhouse, K. N., "Techniques for Identifying Sources of Noise and Vibration," *Sound and Vibration*, Dec. 1970.

with the definition of transfer function (which presumes zero initial conditions) and taking as a special case $s = i\omega$:

$$Q_o(s) = Q_i(s) W(s) \tag{3.76}$$

$$Q_o(i\omega) = Q_i(i\omega) W(i\omega) \tag{3.77}$$

If $q_i(t)$ and $W(s)$ are given by formulas, and if we have no trouble finding $Q_i(s)$, then the quickest way to get $q_o(t)$ is to use (3.76) to get $Q_o(s)$ and then inverse Laplace transform. However, if $q_i(t)$ and/or $W(i\omega)$ are given only by experimental graphs or tables of data, we can still find $q_o(t)$ based on (3.77). The meaning of $W(i\omega)$ is already familiar, but that of $Q_i(i\omega)$ and $Q_o(i\omega)$ will require some discussion.

$$Q_i(s) = \int_0^\infty q_i(t) e^{-st}\, dt \tag{3.78}$$

$$Q_i(i\omega) = \int_0^\infty q_i(t) e^{-i\omega t}\, dt$$

$$= \int_0^\infty q_i(t) \cos \omega t\, dt - i \int_0^\infty q_i(t) \sin \omega t\, dt \tag{3.79}$$

The quantity $Q_i(i\omega)$ can be thought of as a Laplace transform with $s = i\omega$, but it also is the *Fourier transform* of $q_i(t)$. It extends the concept of frequency spectrum, introduced for periodic functions by the Fourier series, to transient time functions.

Even though we emphasize the utility of Eq. (3.79) for situations where $q_i(t)$ is not given by formulas, it will be instructive to first carry out some analytical cases. Considering the rectangular pulse of Fig. 3.30:

$$q_i(t) = Au(t) - Au(t - T)u(t - T) \tag{3.80}$$

$$Q_i(s) = \frac{A}{s} - \frac{Ae^{-Ts}}{s} \tag{3.81}$$

For $s = 0$, $Q_i(s)$ is the indeterminate $0/0$, but L'Hospital's rule shows the limit to be AT, the pulse area, as is verified by applying Eq. (3.79) to this example for $\omega = 0$.

$$Q_i(i\omega) = \frac{A}{i\omega}(1 - \cos \omega T + i \sin \omega T) \tag{3.82}$$

$$= \frac{\sqrt{2}\, A}{\omega} \sqrt{1 - \cos \omega T} \left/ \tan^{-1} \frac{\cos \omega T - 1}{\sin \omega T} \right. \tag{3.83}$$

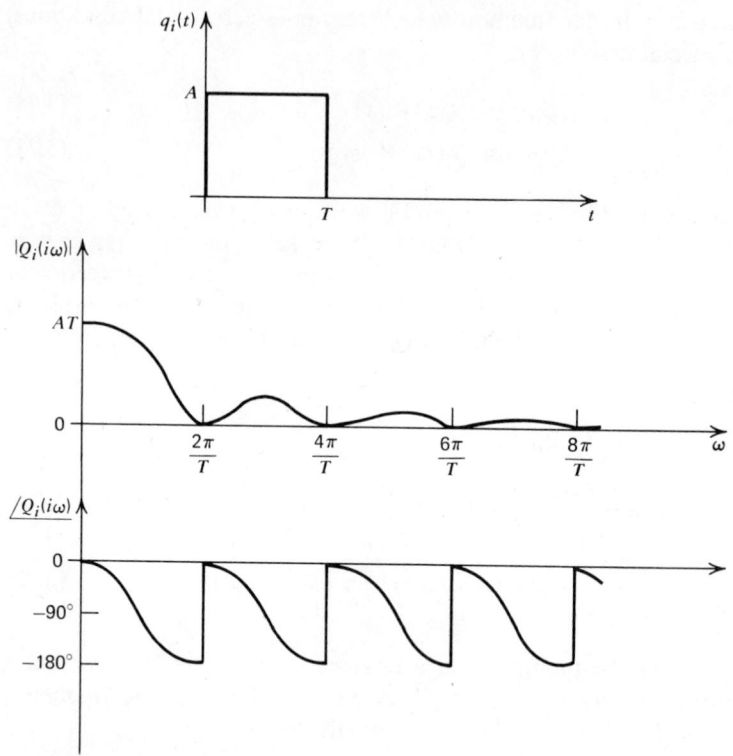

Figure 3.30   Continuous frequency spectrum of a transient.

Examination of Fig. 3.30 reveals that, for transients in general:

a The frequency spectrum is *continuous* (exists for all ω) rather than discrete as for periodic functions.
b At sufficiently high frequency the spectrum magnitude drops to negligible values, relieving us of the need to carry ω to infinity in practical calculations.
c The high-frequency "cutoff" referred to in part b is inversely related to the transient duration T; short T's lead to high frequencies.

For a perfect impulse $A_p \delta(t)$ we would have:

$$Q_i(s) = A_p, \qquad Q_i(i\omega) = A_p/\underline{0°} \tag{3.84}$$

leading to the spectrum of Fig. 3.31, which extends to $\omega = \infty$, but is, of course, the spectrum of a physically impossible transient. When $q_i(t)$ is an arbitrary transient defined by an experimental curve, since Eq. (3.79) has exactly the same

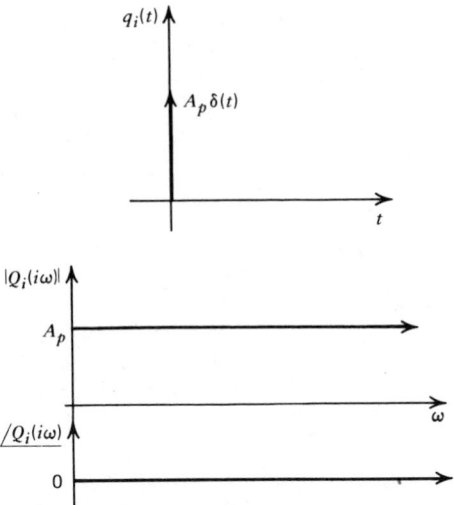

Figure 3.31 Frequency spectrum of an impulse.

form as Eq. (3.44), we can apply the graphical/numerical scheme of Fig. 3.19 to get the magnitude and phase curves for $Q_i(i\omega)$, "one frequency at a time."

Irrespective of the calculation procedure used to obtain the $Q_i(i\omega)$ curves, the meaning or interpretation of such spectra requires some discussion, since there is a temptation to carry over Fourier series concepts that, unfortunately, do not directly apply. While the $|Q_i(i\omega)|$ curves (such as Fig. 3.30) do indicate the relative importance of various frequency ranges, the ordinates are *not* the amplitudes of distinct sine waves that may be added up to "curve fit" $q_i(t)$ as was the case with the Fourier series for periodic functions. This should be clear from the fact that the *continuous* curve $|Q_i(i\omega)|$ can be "sliced" frequency-wise as fine as we please, thus a distinct, finite-size number of frequencies cannot be defined. Furthermore, the *dimensions* of $Q_i(i\omega)$ are not the same as those of $q_i(t)$. Equation (3.79) shows $Q_i(i\omega)$ to have the dimensions of $q_i(t)$ times t; if $q_i(t)$ were a pressure in pascals, $Q_i(i\omega)$ would be in pascal-seconds. This allows us to interpret $Q_i(i\omega)$ as an *amplitude density*, pascals/(rad/sec), a concept analogous to a distributed load in strength of materials. There one considers, say, a beam loaded with the gravity force of its own distributed mass. If the total beam weight is 100 newtons and its length is 4 meters, we have a distributed load of 25 N/m. At any particular beam cross section the concentrated load is *zero* newtons, but over any finite length of beam a definite load exists. Similarly for the spectrum $Q_i(i\omega)$, there is *zero* amplitude at any specific frequency even though a nonzero amplitude density exists.

Equation (3.77) tells us to find $Q_i(i\omega)$ (a process we now know how to carry out), and then to multiply this complex number, frequency by frequency, with

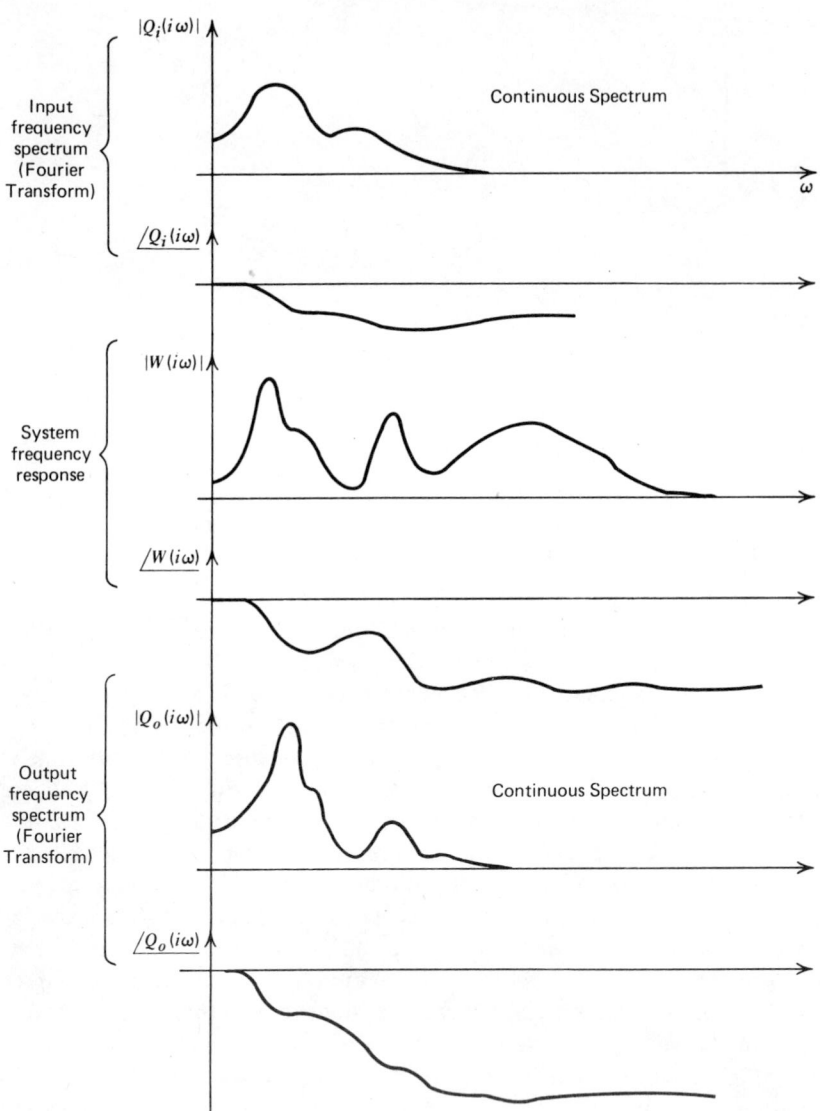

Figure 3.32 Frequency-domain response of arbitrary system to an arbitrary transient input.

the complex number $W(i\omega)$ to obtain the system response $Q_o(i\omega)$. This process is carried out to sufficiently high frequency that the product $|Q_i(i\omega)|\,|W(i\omega)|$ becomes negligibly small, as in Fig. 3.32. The Fourier transform $Q_o(i\omega)$ has the same type of interpretation as did $Q_i(i\omega)$: it is an amplitude density. Its dimensions would be those of $q_o(t)$ divided by rad/sec since $W(i\omega)$ has dimensions of q_o/q_i. While the $Q_o(i\omega)$ graph does indicate the relative importance of various frequency ranges in $q_o(t)$, we need to obtain $q_o(t)$ itself. This requires an inverse transform process for which, fortunately, we already have developed the formula. Equation (3.50) allows us to transform any given $F(i\omega)$ to its companion $f(t)$, so we apply this directly to our current problem as in Fig. 3.33. There the calculation of a single point $q_o(t_1)$ on $q_o(t)$ is explained. Repetition of this process for many other t's will generate the entire course of $q_o(t)$.

Our new response-calculation capability allows us now to usefully augment the discussion of approximate impulse testing of Section 3.2. Every approximate

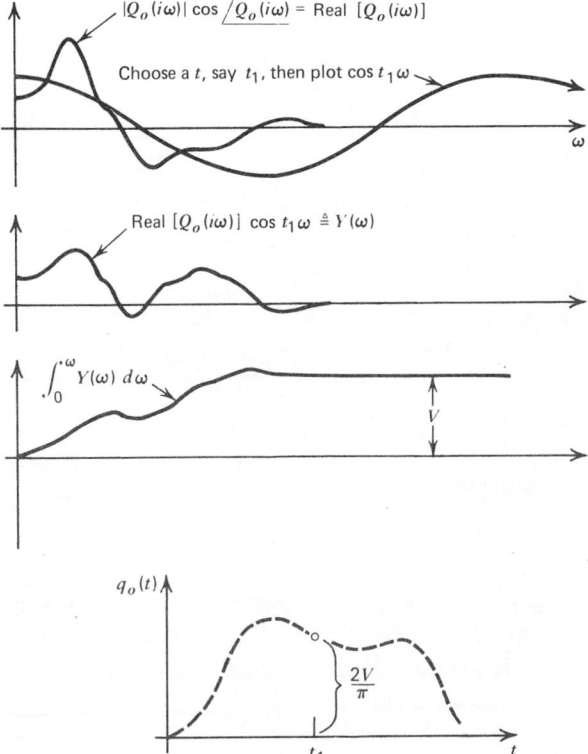

Figure 3.33 Conversion of frequency-domain response to time domain.

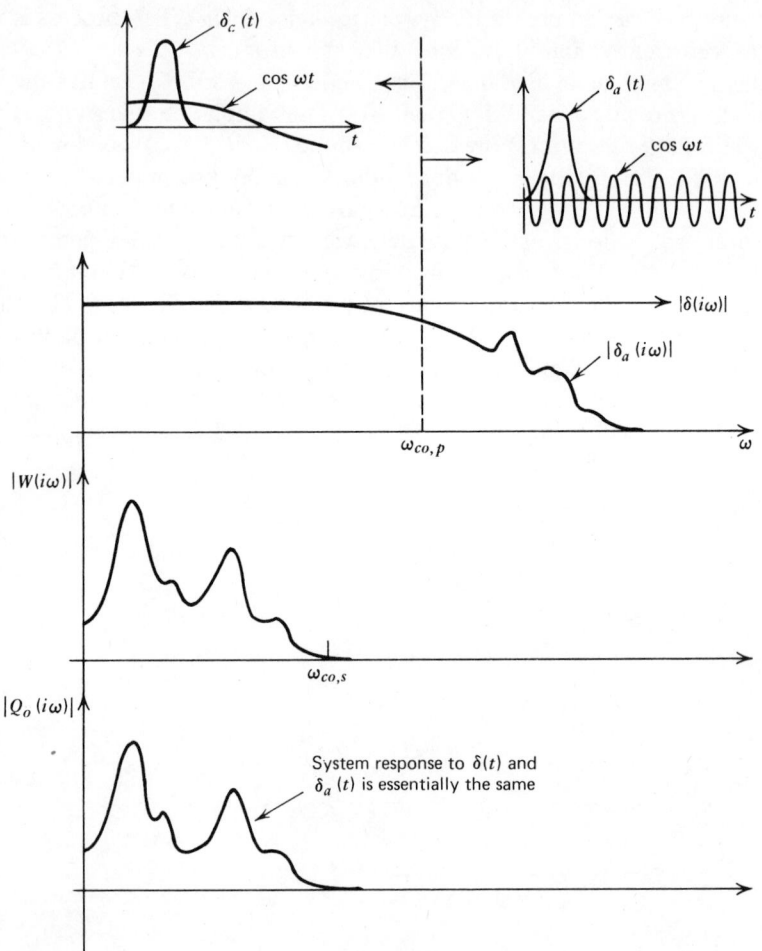

Figure 3.34 Justification of approximate-impulse testing.

impulse $\delta_a(t)$ will have a corresponding frequency spectrum $\delta_a(i\omega)$ that will behave as shown in Fig. 3.34, exhibiting a characteristic cutoff frequency $\omega_{co,p}$ beyond which its behavior gets less and less impulselike. As long as this frequency is equal to or greater than the cutoff frequency $\omega_{co,s}$ of the system being tested, the system "cannot tell" that the test pulse is not an impulse and thus responds with its "perfect" impulse response. Thus if we compute the frequency spectrum of the test pulse, we then know the limitations of our test procedure in terms of the allowable frequency response of the systems we test.

While we have concentrated on attempting to closely approximate perfect impulse testing, these requirements can actually (and sometimes must) be relaxed. Indeed, from Eq. (3.77) it would appear that $W(i\omega)$ can be computed by testing the system with *any* transient $q_i(t)$, recording $q_i(t)$ and $q_o(t)$, computing $Q_i(i\omega)$ and $Q_o(i\omega)$, and then obtaining $W(i\omega)$ from

$$W(i\omega) = \frac{Q_o(i\omega)}{Q_i(i\omega)} \qquad (3.85)$$

The difficulty in actually carrying this out in practice arises at high frequencies where $Q_i(i\omega)$ becomes small. Theoretically, the small $Q_i(i\omega)$ causes a small $Q_o(i\omega)$ and when they are divided in Eq. (3.85), the correct $W(i\omega)$ results. In practice, several factors cause the high-frequency $W(i\omega)$ to be totally unreliable. First, the measurement and recording systems for $q_i(t)$ and $q_o(t)$ are not dynamically perfect and also introduce random measurement errors. These inaccurate data must then be sampled and digitized in A/D converters whose resolution is not infinitesimal, further degrading the data. In the digital computers used to implement the transform calculations only a finite number of digits can be carried, causing additional errors. When $Q_i(i\omega)$ and $Q_o(i\omega)$ are both large, these errors are small percentages and useful results are obtained, but as the transform magnitudes get smaller and smaller the unavoidable errors make the calculation meaningless. While these measurement and numerical problems prevent us from using "just any old" $q_i(t)$, we are often able to get useful results from test pulses at frequencies significantly beyond the $\omega_{co,p}$ shown in Fig. 3.34, and we will go into this in more detail in the pulse testing section of Chapter 6.

To show some actual numerical results, $|F_3(i\omega)|$, $|F_4(i\omega)|$, and $|F_5(i\omega)|$ for the three pulses f_3, f_4, and f_5 of Fig. 3.11 were computed and plotted in Fig. 3.35. Superimposed on this same graph is the amplitude ratio of the second-order system that was pulse tested in Fig. 3.13. It should be clear now why the pulse tests of Fig. 3.13 were such good approximations to perfect impulse tests. The system's response is practically zero beyond $\omega = 1$, which is the only frequency range where the pulses' spectra differ appreciably from the perfect flatness of the impulse, thus the system "perceives" all three pulses as perfect impulses.

While this text is not one devoted to developing the details of computer programming, we definitely do wish to apprise the reader of the availability of certain computer-aided design and analysis methods that can speed the engineer's work in a cost effective manner. The graph of Fig. 3.35 was obtained in just a few minutes by use of a cathode ray tube terminal interfaced in a time sharing manner with a large central computer. Even small companies that cannot afford their own large computers may find it economic to rent terminals that can access, through telephone lines, the computing power of very large machines that provide a wide range of programs oriented to specific engineering needs.

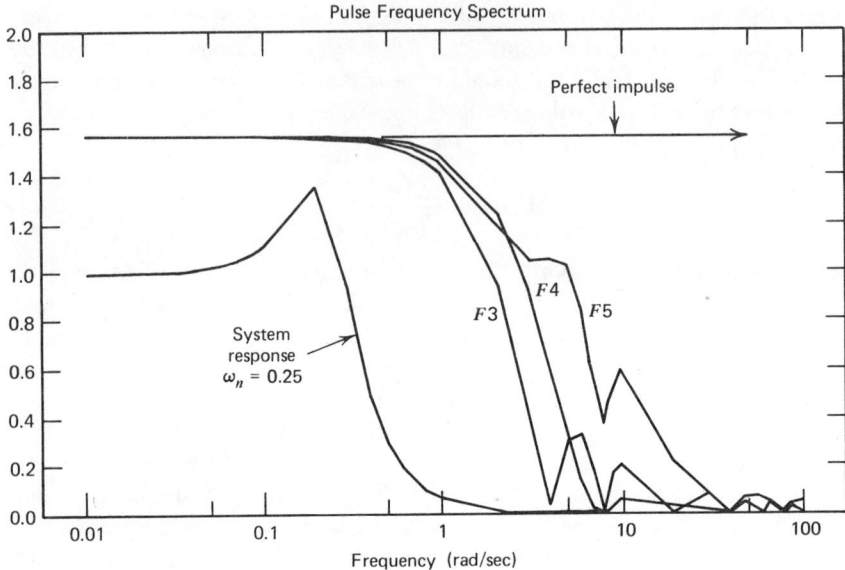

Figure 3.35   Spectra of some approximate impulses.

Figure 3.35 was produced using a general purpose scientific computing language called SPEAKEASY[5], developed at Argonne National Laboratory and presently installed at over 100 computer centers. While most engineering students begin their programming studies by learning the FORTRAN language, and most large scale, sophisticated programs used in industry also are written in FORTRAN, languages such as SPEAKEASY seem to fill a definite need for the majority of scientists and engineers who are *occasional* programmers and appreciate a language designed specifically to produce useful results quickly, with a minimum of learning time. While such languages may sometimes sacrifice *machine* efficiency, they often significantly increase *overall* efficiency (which includes the cost of the engineer's programming time) since usable results are obtained much more quickly and with fewer "false starts."

It would be inappropriate here to teach the reader SPEAKEASY; however, a brief sample program should make clear the ease of use of this type of language. We show below, with marginal comments, the complete program as it would be typed in from the terminal keyboard, for computing and plotting Fig. 3.35. In this example, SPEAKEASY is being used in a "calculator mode," where each statement is executed as it is entered into the computer, rather than waiting until the entire program has been defined.

---

[5]*The Speakeasy-3 Reference Manual*, ANL-8000 (Rev. 1), Argonne National Lab., Argonne, Ill., 1976.

DOMAIN COMPLEX   allows use of complex numbers
W1 = GRID(.01,.1,.01); W2 = GRID(.2,1,.1)   defines an array of frequency values going from .01 to .1 in .01 steps, etc.
W3 = GRID(2.1,10.1,1); W4 = GRID(20,100,10)
W = ARRAY(W1,W2,W3,W4)
S = W*1I   defines array of imaginary numbers .01i,.02i,etc.
A = 2/.785; B = 6.28/.785; C = B*B   defines some needed constants
F3 = (1 − EXP(−1.57*S))/S   calculates F3(iW) for all W values
F4 = (A*(1 − 2*EXP(−.785*S) + EXP(−1.57*S)))/S*S   same for F4(iW)
F5 = 2*(1 − EXP(−.785*S))/S + (B/(S*S + C))*(EXP(−1.57*S) − EXP(−.785*S))   same for F5(iW)
L = 1/(S*S/.0625 + 3.2*S + 1)   computes frequency response of second-order system
MS = ABS(L); M3 = ABS(F3)   computes absolute values of complex numbers (ampl. ratio of system, |F3(iω)|, etc.)
M4 = ABS(F4); M5 = ABS(F5)
USE TEKTRONIX   requests graphing routines for storage CRT terminal
NOW ISSUE THE TEKTRONIX COMMAND   computer message to user
TEKTRONIX 30   informs computer that terminal writes 30 characters/sec
SETXSCALE(4 CYCLES,FROM:.01)   requests 4-cycle log scale starting at .01
SETYSCALE(FROM:0,TO:2)   requests y scale from 0 to 2
SETTITLE('PULSE FREQUENCY SPECTRUM')   defines graph title
GRAPH(M3:W); OVERLAY; GRAPH(M4:W); OVERLAY; GRAPH(M5:W) ; OVERLAY; GRAPH(M5:W)   erases screen and plots Fig. 3.35

(In defining the array of frequency values W, the W3 statement starts with 2.1 so as to avoid a division by zero at W = 8 in F5.) The expressions for F3, F4, and F5 are obtained by using the delay theorem methods of Section 2.8. Several statements can be put on one line if we separate them by semicolons. Note that many common operations (+ , − ,*,/,etc.) and functions (EXP, etc.) use the same notation as FORTRAN. However, in SPEAKEASY, one is allowed to use complex numbers (as well as real) in almost all of these with no additional programming. Also, the evaluation of expressions for a succession of parameter values (the frequencies W in our case) is "automatic" when the parameter has been previously defined as a list or array of numbers; no DO-loops are necessary. If we had wanted a table rather than (or in addition to) a graph of our results, only one statement: TABULATE W, M3, M4, M5, MS would have been needed to produce a properly laid out table with headings.

3.8   DESCRIPTION OF RANDOM SIGNALS

Before we can deal with the response of linear systems to random inputs, we must first introduce the methods used to describe such inputs. In Section 1.3 we

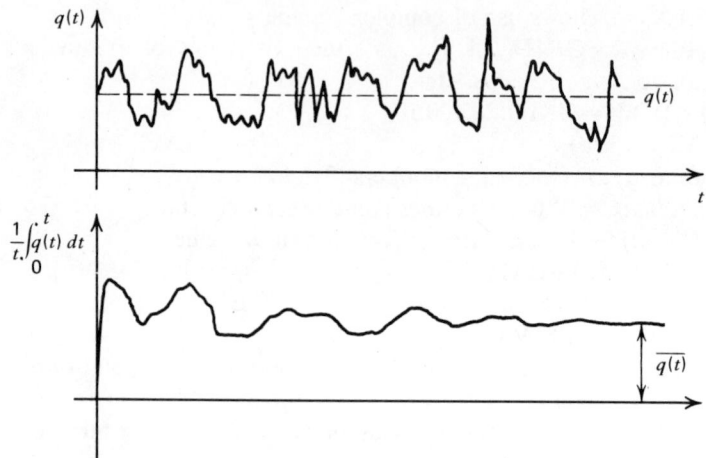

Figure 3.36   Random signal and average value.

drew the distinction between *stationary* and *nonstationary* random signals on the basis of whether the statistical properties were constant or changing with time. In this section we assume stationary behavior and proceed to define and explain the basic statistical properties needed to work with random signals in some of the more common practical applications. In the previous sections of this chapter, attention was focused on calculating the detailed time history of the response of some system to an input whose detailed time history was given. For a random input this is fundamentally impossible and we will have to content ourselves with the capability of calculating only certain statistical aspects of the system input and response.

If we consider a typical example record (Fig. 3.36) of some physical phenomenon that exhibits random behavior, the problems involved in numerically describing both its magnitude and rapidity of variation should be apparent. For a sine-wave flow velocity $V \sin \omega t$, for example, the amplitude V (m/sec) clearly tells us whether this is a large or small fluctuation while frequency ω indicates whether the fluctuation is rapid or slow. Random signals allow no such easy characterization. For a random signal $q(t)$, the average value $\overline{q(t)}$ given by

$$\overline{q(t)} \triangleq \lim_{T \to \infty} \frac{1}{T} \int_0^T q(t)\, dt \tag{3.86}$$

is a *deterministic* feature (the system response to which is easily calculated), but gives *no* indication of the size of the fluctuations, which is our main interest. Note that the calculation of the average value of a random signal follows the same procedure as for a deterministic signal except that the averaging time must be *infinite* to obtain a precise value. In practic e, thus, any numerical measurement of $\overline{q(t)}$ must be considered a statistical *estimate*, subject to uncertainty, which however can be reduced by increasing T. One of the first steps in processing random data is often to compute the mean (average) value and then subtract this from each data point to create a new variable with zero mean, *so from here on we consider zero mean variables unless otherwise stated.*

The most widely used measure of the magnitude of random fluctuations is the mean-squared value $\overline{q^2(t)}$ defined by

$$\overline{q^2(t)} \triangleq \lim_{T \to \infty} \frac{1}{T} \int_0^T q^2(t)\,dt \tag{3.87}$$

While average quantities such as $\overline{|q(t)|}$, $\overline{q^4(t)}$, and so forth could conceivably also be so used, they generally are not, since the mean-squared value (also called *variance*) leads to the simplest mathematical results that are of practical use in most of the applications. When physical data is processed (as is nowadays so common) in a digital computer, continuous time histories must be *sampled* to produce a table of discrete values (see Fig. 3.37) for entry into the computer. Definitions (3.86) and (3.87) must then be modified:

$$\overline{q} \triangleq \lim_{N \to \infty} \frac{1}{N} \sum_{k=0}^{N-1} q_s(k\,\Delta t) \tag{3.88}$$

$$\overline{q^2} \triangleq \lim_{N \to \infty} \frac{1}{N-1} \sum_{k=0}^{N-1} \left(q_s(k\,\Delta t) - \overline{q} \right)^2 \tag{3.89}$$

The denominator term $N-1$ (rather than N) in (3.89) is a statistical subtlety required to make $\overline{q^2}$ an "unbiased" estimator but has little numerical effect for the large (1000 or more) values of N used in random signal analysis. While formulas for computing the statistical variability of $\overline{q}$ and $\overline{q^2}$ are available, these uncertainties are usually negligible for the large values of N just mentioned. Sampling increment Δt is generally uniform, allowing the $k\,\Delta t$ notation of Fig. 3.37; its proper numerical value depends on the frequency content of the signal, higher frequencies requiring smaller Δt. Since $\overline{q^2}$ does not have the same physical dimensions as q, it is sometimes desirable to define and use the *root-mean-square (RMS)* value:

$$q_{\text{RMS}} \triangleq \sqrt{\overline{q^2}} \tag{3.90}$$

While the single numbers $\overline{q^2}$ or q_{RMS} are sufficient to decide whether fluctuations are large or small, or to compare the magnitudes of two random signals in a gross way, by themselves they give no capability for answering

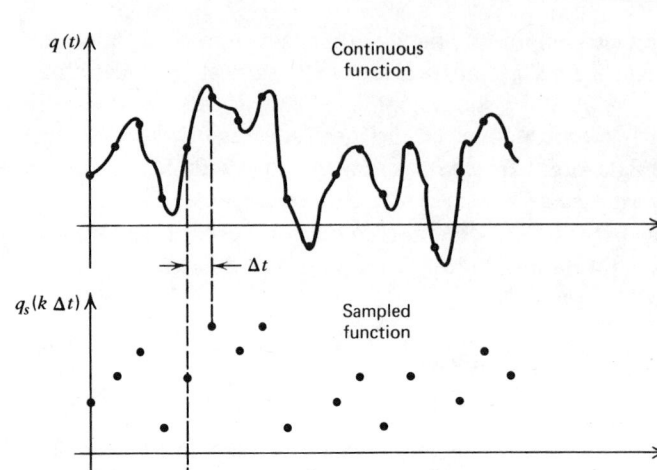

Figure 3.37   Sampling necessary for digital calculations.

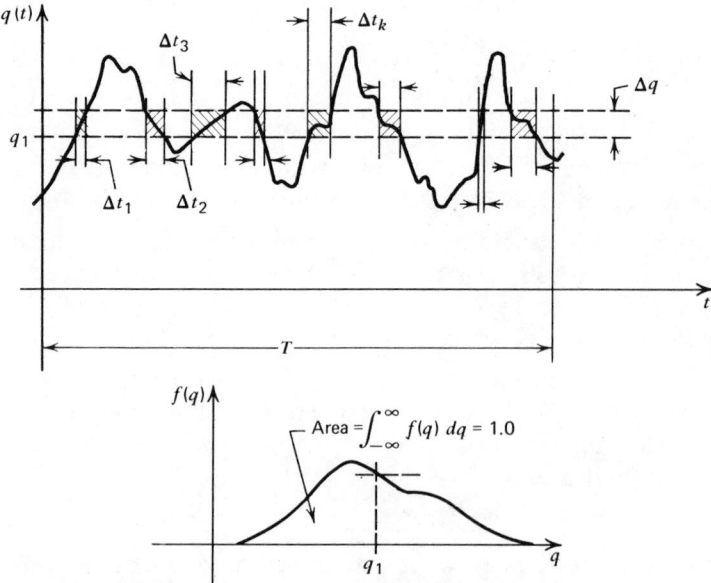

Figure 3.38   Definition of probability density function.

questions such as, "What is the probability of finding q between two selected values q_a and q_b?" To answer questions of this sort we must invest additional effort in more detailed analysis of the data, specifically in computation of the *probability density function* (also called *amplitude distribution function*). In Fig. 3.38 we define the probability P that q will be found between some specific value q_1 and $q_1 + \Delta q$ by:

$$\text{Prob. } (q_1 < q < (q_1 + \Delta q)) \triangleq P(q_1, \Delta q) \triangleq \lim_{T \to \infty} \frac{\Sigma \Delta t_k}{T} \qquad (3.91)$$

This conforms to the usual definition of probability as a decimal between 0 and 1.0 (or percentage between 0 and 100), equal to that fraction of the "total number of events" that exhibit the "desired attribute." In our case the desired attribute is that q lies between q_1 and $q_1 + \Delta q$; $\Sigma \Delta t_k / T$ is the fraction (of the total time T) that q spends in this amplitude band or "window." If q spent *all* of T in this band, P would be 1.0; if it spent *no* time, P would be zero. To define $f(q)$, the probability density function, we now let Δq squeeze down toward zero:

$$f(q) \triangleq \lim_{\Delta q \to 0} \lim_{T \to \infty} \left[\frac{\Sigma \Delta t_k / T}{\Delta q} \right] \qquad (3.92)$$

As Δq goes to zero, both the numerator and denominator of (3.92) approach zero but their ratio goes to a specific numerical value $f(q)$, which would in general be different for different values of q.

With $f(q)$ defined in this way it should be clear that

$$f(q)\, dq = \text{probability that } q \text{ lies in } dq \qquad (3.93)$$

and

$$\int_{q_a}^{q_b} f(q)\, dq = \text{probability } q \text{ lies between } a \text{ and } b \qquad (3.94)$$

If we try to implement Eq. (3.92) with actual measured data, we cannot of course let $T \to \infty$ or $\Delta q \to 0$, thus we can only estimate, rather than compute precisely, $f(q)$. When digital computer processing[6] of sampled records is used one divides the full range of q into a finite number of finite-size "bins" of size Δq and computes discrete values of $f(q)$ directly from

$$f(q_k) = \frac{1}{\Delta q} \frac{n_k}{N} \qquad (3.95)$$

[6]Otnes, R. K. and Enochson, L., *Digital Time Series Analysis*, John Wiley and Sons, N.Y. 1972, p. 371.

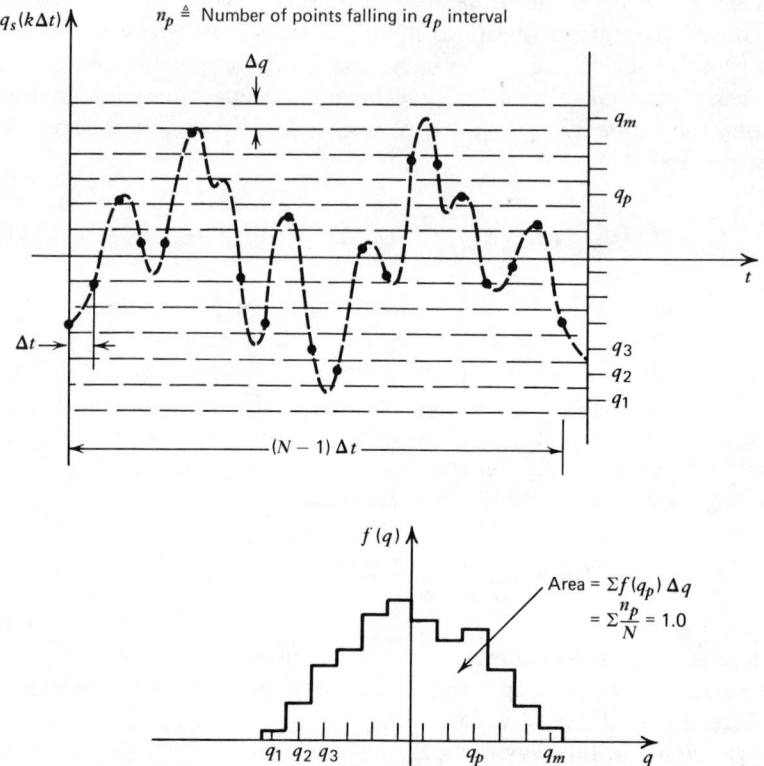

Figure 3.39 Digital calculation of probability density function.

as in Fig. 3.39. The "bar graph" display of $f(q)$ is called a *histogram*; for large numbers of points, Δq may be made quite small and the bar graph approximates a smooth curve. The *cumulative distribution function* $F(q)$, defined by

$$F(q) \triangleq \int_{-\infty}^{q} f(q)\,dq$$

gives the probability that q is less than some selected value. Figure 3.40 shows this function.

To satisfy basic statistical requirements, probability density functions must only:

1 Be nonnegative, so that increasing the range $a \rightarrow b$ in $\int_{a}^{b} f(q)\,dq$ does not *decrease* the probability,

2 Have $\int_{-\infty}^{\infty} f(q)\,dq = 1.0$ so that q is *certain* to be found somewhere between $-\infty$ and $+\infty$.

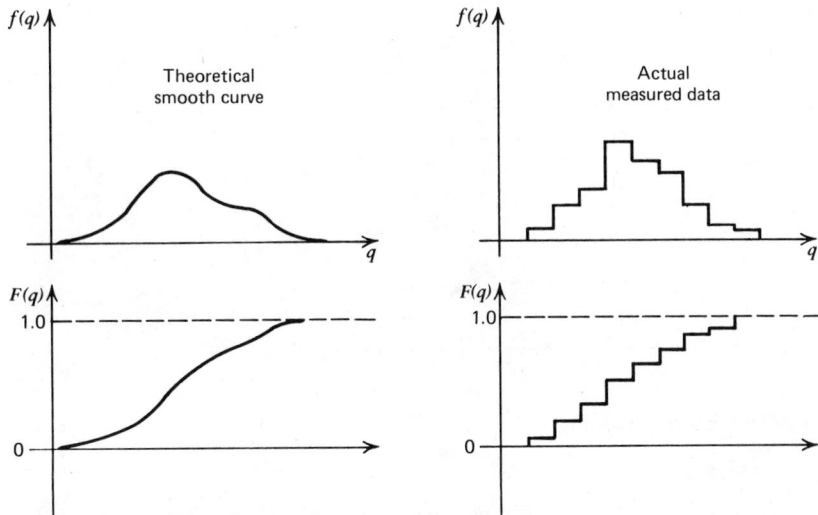

Figure 3.40   Cumulative distribution function.

Within these restrictions, *any* function could serve as a probability density function. When we process physical data as in Fig. 3.39, we of course do *not* get a formula for $f(q)$, only tabular data. To facilitate theoretical analysis and numerical computation, statistical workers have proposed various mathematical models defined by specific formulas that hopefully will be adequate "curve fits" for various types of physical random data. Some of these model density functions may be familiar to the reader: Gaussian ("Normal"), chi-square, Poisson, Binomial, *t*, *F*, and so on.

By far the most used model is the *Gaussian distribution* defined by

$$f(q) = \frac{1}{\sqrt{2\pi}\,\sigma} e^{-(q-\mu)^2/2\sigma^2}, \qquad -\infty < q < \infty \qquad (3.96)$$

where

$$\mu \triangleq \text{mean value}$$

$$\sigma \triangleq \text{standard deviation} = q_{\text{RMS}} = \sqrt{\overline{q^2}}$$

By adjusting the numerical values of σ and μ (see Fig. 3.41) one can tailor the function to get the best fit for the data; however, a perfect fit will never be possible since physical data cannot be exactly Gaussian for various reasons. One

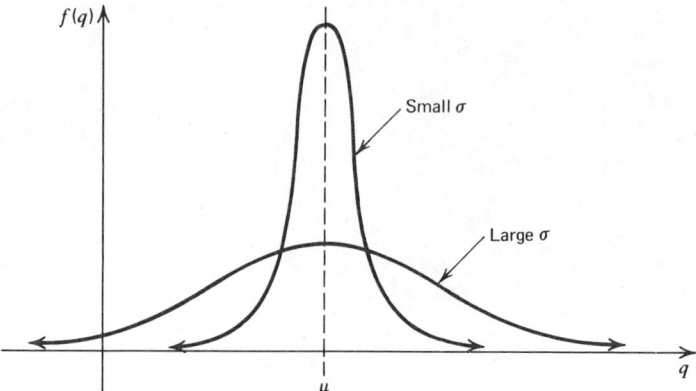

Figure 3.41 Gaussian distribution.

obvious reason is that the Gaussian distribution extends to $\pm \infty$ while a physical variable q will always have some finite limits. However, much physical data is adequately approximated as Gaussian. One can judge the adequacy visually by superimposing a graph of Eq. (3.96) (using the experimentally measured values of μ and σ) on the measured $f(q)$, or by plotting $F(q)$ on special graph paper designed to cause Gaussian data to plot as a straight line. If a numerical, rather than visual evaluation is desired, the chi-square "goodness-of-fit" test[7] is available. When the data is treated as Gaussian, many theoretical results become available. For example, 99.7% of the time, q will be within $\pm 3\sigma$ of μ, 95% of the time within $\pm 2\sigma$, 68% of the time within $\pm 1\sigma$.

Figure 3.42 shows some actual results[8] obtained in wind tunnel testing of a jet engine inlet. The variable under study is the stagnation pressure at various locations at the exit of the inlet for several operating conditions. Frequency content of the pressure transducer signals was from 0 to 200 Hz; a 20-second record length was analyzed and 24 amplitude bands ($m = 24$ in Fig. 3.39) were used to compute the histogram. Based on these measurements, it was concluded that the data could be approximated as Gaussian.

We have now developed both gross and detailed methods of describing the magnitude of a random signal but neither the mean-square value nor the probability density function give any information as to rapidity of variation. (Rapidity of variation does affect *how long it will take* to gather a sufficiently long record of $q(t)$ to compute a statistically reliable $\overline{q^2}$ or $f(q)$.) As in deterministic signals, descriptions in either the time or frequency domains are possible

[7]Otnes and Enochson, *op. cit.*, p. 378.
[8]Coltrin, R. E. and Mitchell, G. A., "Preliminary Investigation of Distortion Dynamics in a Mach 3 Mixed-Compression Inlet," NASA TM X-1706, 1968.

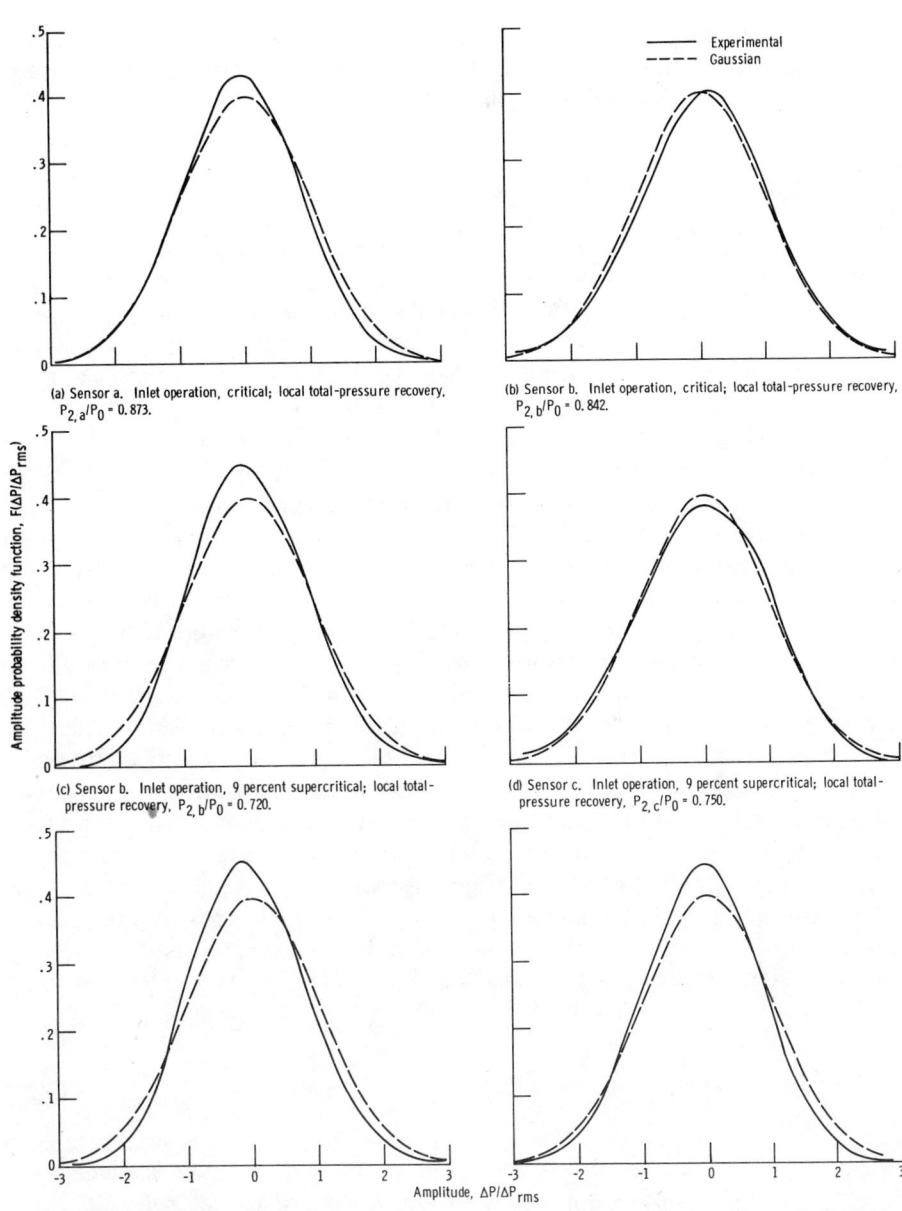

(a) Sensor a. Inlet operation, critical; local total-pressure recovery, $P_{2,a}/P_0 = 0.873$.

(b) Sensor b. Inlet operation, critical; local total-pressure recovery, $P_{2,b}/P_0 = 0.842$.

(c) Sensor b. Inlet operation, 9 percent supercritical; local total-pressure recovery, $P_{2,b}/P_0 = 0.720$.

(d) Sensor c. Inlet operation, 9 percent supercritical; local total-pressure recovery, $P_{2,c}/P_0 = 0.750$.

(e) Sensor b. Inlet operation, 13 percent supercritical; local total-pressure recovery, $P_{2,b}/P_0 = 0.654$.

(f) Sensor d. Inlet operation, 13 percent supercritical; local total-pressure recovery, $P_{2,d}/P_0 = 0.710$.

Figure 3.42 Comparison of jet-engine random pressures to Gaussian model.

and useful. In the time domain, the *autocorrelation function* is used, while the frequency domain utilizes the *power spectral density*. We define the autocorrelation function $R(\tau)$ of a random signal $q(t)$ by

$$R(\tau) \triangleq \lim_{T \to \infty} \frac{1}{T} \int_0^T q(t)q(t+\tau)\,dt \qquad (3.97)$$

To construct $R(\tau)$ "one point at a time," choose a numerical value for τ, say τ_1 seconds. If a graph of $q(t)$ is available, then $q(t+\tau_1)$ is obtained by shifting $q(t)$ to the left by τ_1 seconds. We can then multiply, integrate, and average as indicated by (3.97), though we must of course use a finite averaging time T and thus get an estimate of $R(\tau_1)$ rather than the precise value. This procedure is repeated for many values of τ, starting with $\tau = 0$, to get a graph of $R(\tau)$.

For $\tau = 0$, note that the "product curve" $q(t)q(t+\tau)$ is always positive, so that its integral will have the largest value possible for any τ, and that $R(0)$ is actually $\overline{q^2}$, the mean-squared value. When $\tau > 0$, the plus and minus portions of $q(t)$ and $q(t+\tau)$ get "misaligned," causing the product curve to have both positive and negative values, and reducing the value of $R(\tau)$. For "sufficiently large" τ values, $R(\tau)$ will approach zero since, on the average, the product curve will be as much positive as negative; a slowly varying $q(t)$ will require larger τ shifts for this to happen than will a "fast" $q(t)$. Thus fast random signals have "narrow" correlation functions while slow ones have "wide" $R(\tau)$'s. While we introduced Eq. (3.97) to deal with random signals, it can be applied to deterministic signals, such as $q(t) = A \sin(\omega t + \phi)$. For this $q(t)$ we get[9] $R(\tau) = A^2 \cos \omega \tau /2$, which we see does not go to zero as $\tau \to \infty$ but rather is periodic. One application of correlation functions makes use of this behavior to "discover" the presence of small periodic signals hidden in large random noise. For such a situation, close scrutiny of the time function itself (signal + noise) gives no hint of the presence of a periodic signal, but calculation of $R(\tau)$ makes it obvious by the periodic behavior of $R(\tau)$ for large τ, where the random signal produces zero value.

For the sampled records used in digital computer processing, the autocorrelation function is computed from[10]:

$$R(r\Delta\tau) = \frac{1}{N-r} \sum_{k=0}^{N-r-1} q_k q_{k+r} \qquad r = 0, 1, 2, \dots, m \qquad (3.98)$$

where $q_0, q_1, \dots, q_{N-1}$ are the N points of our data sample, $\Delta\tau$ is numerically equal to Δt (the sampling time interval), and r is called the *lag value*. It should be clear that here the smallest *possible* increment in which we can vary τ, and thus compute $R(\tau)$, is Δt. Note (see Fig. 3.43) that as the number of lags r is

---

[9]Otnes and Enochson, *op. cit.*, p. 25.
[10]Otnes and Enochson, *op. cit.*, p. 38.

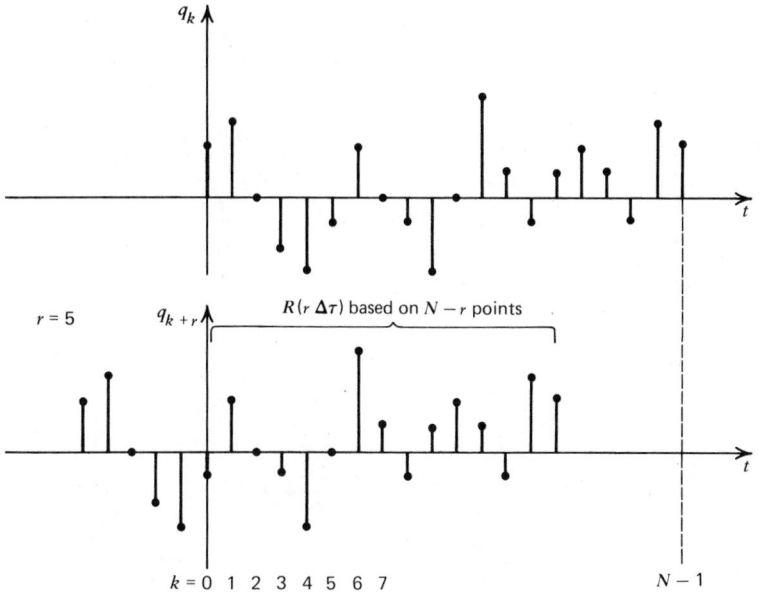

Figure 3.43   Digital computation of autocorrelation function.

increased, $R(r\Delta\tau)$ is based on fewer and fewer data points, making it statistically unreliable, thus r's maximum value m must be kept a small fraction (the order of 0.05 to 0.2) of N. A typical analysis in the field of vibration, for data with frequencies in the range 0–5000 Hz, might use $\Delta t = 0.0001$ sec, $N = 10,000$, and $m = 1000$, allowing calculation of 1001 points of $R(\tau)$ in the range $0 \leqslant \tau \leqslant 0.1$ sec.

In Fig. 3.44 we show a lightly damped resonant system with a "wide-band" (contains a wide range of frequencies) random input. Due to the filtering action of the system, the output, while still random, becomes "narrow-band" (contains only a narrow range of frequencies near ω_n). The $R(\tau)$ shapes shown are typical of wide- and narrow-band signals in general. Figure 3.45 shows $q(t)$ and $R(\tau)$ for some actual data, the electrocardiogram signal of a test pilot.[11] Here the time history exhibits some nearly periodic features, leading to the "spikes" in $R(\tau)$.

Turning our attention now to the frequency domain description of random signals (often the more useful description for many engineering applications), we begin by noting that a major use, in the past, of the autocorrelation function was as a preliminary computational step in the calculation of frequency spectra. Nowadays, frequency spectra are more often computed directly from the time data using digital Fast Fourier Transform methods, bypassing the need for correlation functions. Let us begin our discussion of frequency spectra for random signals with a definition of power spectral density (PSD) in terms of

---

[11]X-15 Data Display System, NASA CR-460, 1966.

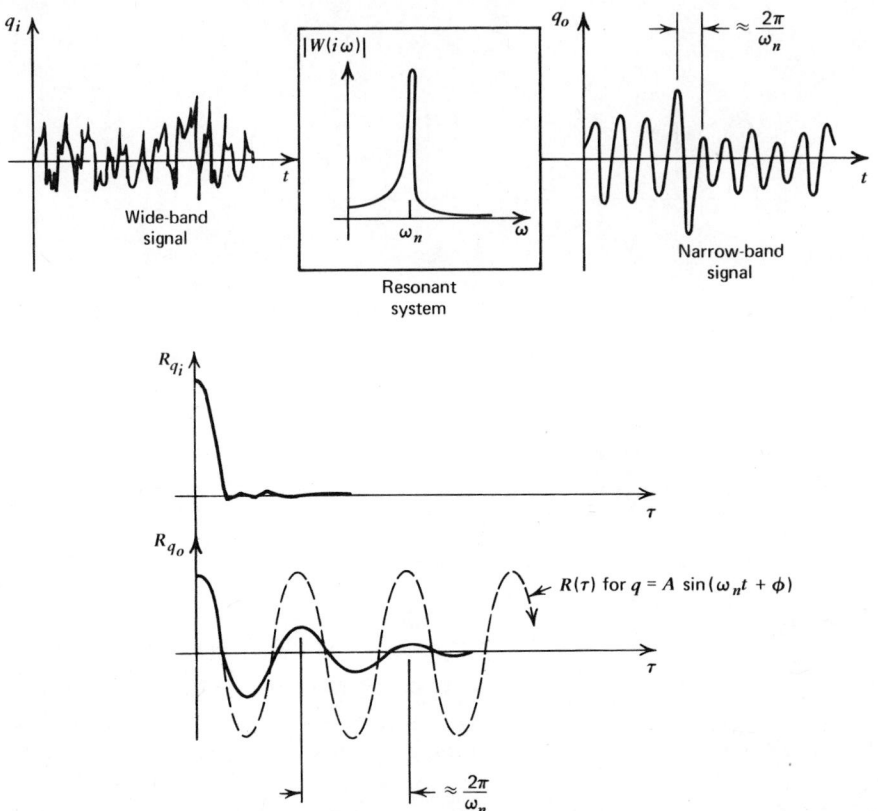

Figure 3.44 Wide-band and narrow-band random signals.

analog measurement techniques since this approach gives the clearest physical interpretation. A more descriptive name might be mean-square spectral density since our purpose is to define a per-unit-frequency quantity that, when integrated with respect to frequency between any two frequencies, will give the contribution of that frequency band to the total mean-square value.

Figure 3.46 shows the implementation of the definition:

$$\text{power spectral density} \triangleq \phi(\omega) \triangleq \lim_{\substack{T \to \infty \\ \Delta\omega \to 0}} \frac{(1/T)\int_0^T q_{\Delta\omega}^2 \, dt}{\Delta\omega} \tag{3.99}$$

The signal $q(t)$ is filtered through a narrow ($\Delta\omega$) band-pass filter with center frequency at a chosen value ω_1, thus producing the signal $q_{\Delta\omega}$ that includes only

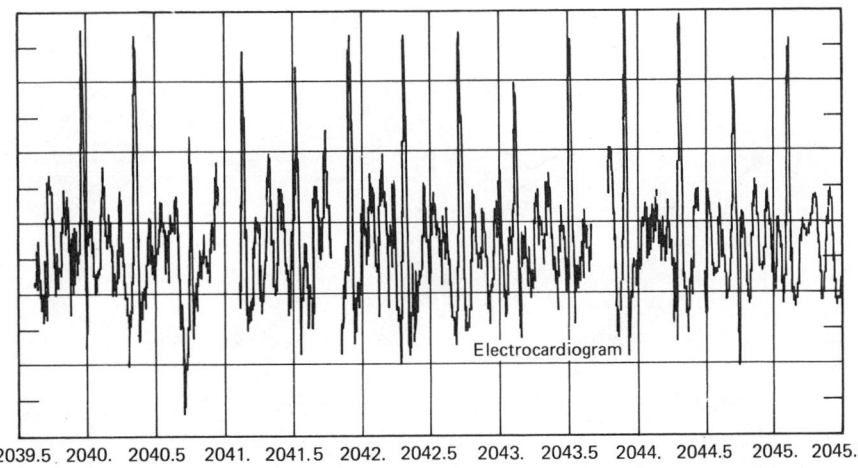

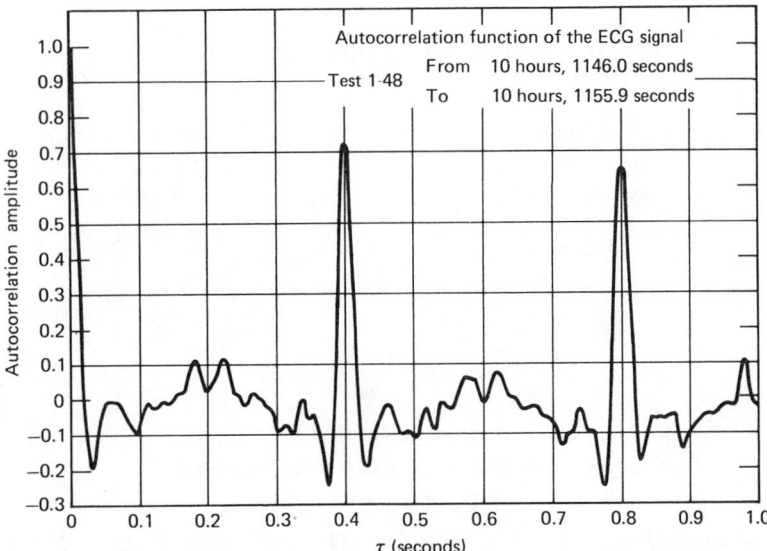

Figure 3.45 Time history and autocorrelation function for test pilot electrocardiogram.

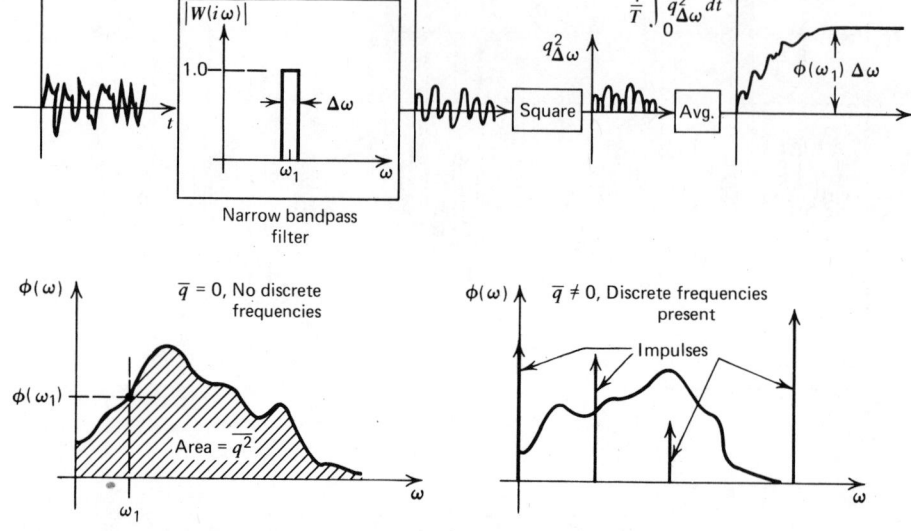

Figure 3.46 Definition of power spectral density.

the signal content in the neighborhood of ω_1. We then compute the mean-square value of $q_{\Delta\omega}$ in the usual way. If $q(t)$ were not purely random, but also contained a sinusoidal component $A\sin(\omega_1 t + \phi)$ at frequency ω_1, when we squeeze $\Delta\omega$ down toward zero, the mean-square value $\overline{q_{\Delta\omega}^2}$ would *not* go to zero, but would approach $A^2/2$.
Thus

$$\phi(\omega_1) \rightarrow \frac{A^2/2}{\Delta\omega} \rightarrow \infty \qquad (3.100)$$

showing that the power spectral density of a discrete sine wave is an impulse function (of frequency) given by $(A^2/2)\delta(\omega - \omega_1)$ since then:

$$\text{mean-square value} = \int_{\omega_1 - \epsilon}^{\omega_1 + \epsilon} \phi(\omega)\,d\omega = \int_{\omega_1 - \epsilon}^{\omega_1 + \epsilon} \frac{A^2}{2}\delta(\omega - \omega_1)\,d\omega = \frac{A^2}{2} \quad (3.101)$$

If there is *not* a discrete sine wave present at the filter center frequency ω_1, then $\phi(\omega_1)$ is some finite number and $\overline{q_{\Delta\omega}^2} \approx \phi(\omega_1)\Delta\omega$ will be proportional to $\Delta\omega$ for $\Delta\omega$'s small enough that $\phi(\omega) \approx$ constant over the range $\Delta\omega$. Thus the division of $\overline{q_{\Delta\omega}^2}$ by $\Delta\omega$ in (3.99) gives us the per-unit-frequency measure of mean-square value that we desire. Note that neither of the limiting processes of (3.99) can be achieved in practice, thus we can only estimate $\phi(\omega)$; however, in Chapter 6 we will give methods for computing the uncertainties in these estimates. If a nonzero mean value $\bar{q}$ has *not* been removed from the data previous to computing $\phi(\omega)$, then there will be an impulse function $(\bar{q})^2\delta(\omega)$ at $\omega=0$ in $\phi(\omega)$ to take care of this contribution to the total mean-square value. By using impulse functions for the mean value and any discrete frequencies present, we can write:

$$\text{total mean-square value} = \int_0^\infty \phi(\omega)\,d\omega \qquad (3.102)$$

When $\phi(\omega)$ has no impulse functions, the integral of (3.102) is just the "ordinary" area under the curve. If impulses are present we must add to this area the terms $(\bar{q})^2$, $A_1^2/2$, $A_2^2/2$, and so on to get the total mean-square value. Experimentally measured $\phi(\omega)$'s cannot exhibit impulse functions (only sharp, finite peaks) since $\Delta\omega$ cannot be zero and also because no real signal component is *perfectly* periodic.

Ideal *white noise* is defined as a random signal whose PSD is perfectly flat from $\omega=0$ to $\omega=\infty$. Since the total mean-square value of such a signal would be infinite, white noise cannot occur in nature; however, real signals can have flat PSD's over finite frequency ranges and then approach zero at high frequencies (see Fig. 3.47). Electronic apparatus used to generate random signals for laboratory test purposes often is designed to provide a flat PSD, since filtering can then be used to shape the spectrum into any curve needed to model the physical phenomenon being studied. Such electronic signal generators may utilize "naturally occurring" random physical phenomena (such as noise voltages in vacuum or gas-filled tubes or in semiconductor devices) as their principal of operation or may use digital techniques to synthesize close approximations to the flat spectrum and Gaussian probability density function usually desired. Note that frequency content and amplitude distribution are independent; a signal can be "white" and non-Gaussian or vice-versa.

Figure 3.48 shows some actual measured PSD curves; 3.48a corresponds to the electrocardiogram signal of Fig. 3.45. A vibrational acceleration signal recorded on the Apollo spacecraft[12] is shown in 3.48b, while 3.48c is for a turbulent air flow velocity signal measured with a hot-wire anemometer.[13] The prominent and regularly spaced peaks of the electrocardiogram spectrum arise from its strong near-periodic content, while the turbulent air flow's smooth spectrum testifies to pure randomness, devoid of periodicity. A mixture of periodic and random aspects appears in the Apollo structural vibration data.

[12]Watlington, R. A., Wave Analysis Applications in the Apollo Program, NASA TM X-1599, 1968.
[13]Lawrence, J. C. and Stickney, T. M., Further Measurements of Intensity, Scale, and Spectra of Turbulence in a Subsonic Jet, NASA TN 3576, 1956.

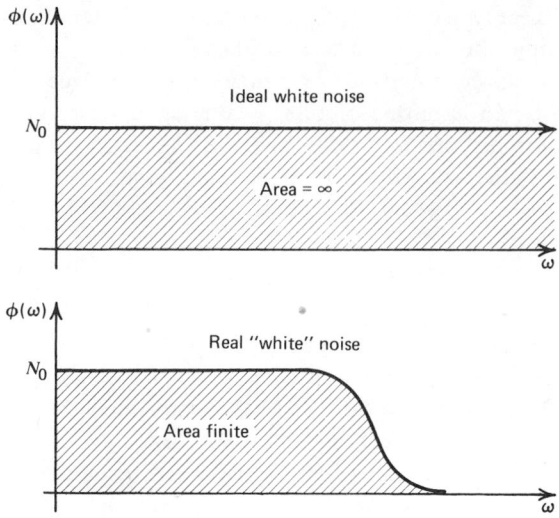

Figure 3.47   Ideal and physically realizable "white noise."

Rather than using the measurement scheme of Fig. 3.46, one can alternatively find $\phi(\omega)$ by first measuring the signal's autocorrelation function $R(\tau)$ and then using the relation[14]

$$\phi(\omega) = \frac{2}{\pi} \int_0^\infty R(\tau) \cos \omega \tau \, d\tau \tag{3.103}$$

When this is implemented digitally using the sampled data correlation function of Eq. (3.98), one gets[15] for $k = 0, 1, \ldots, m$ and $\Delta f = 1/2m\Delta t$ Hz:

$$\phi(k\Delta f) = 2\Delta t \left(R(0) + 2 \sum_{r=1}^{m-1} R(r\Delta\tau) \cos\frac{\pi k r}{m} + R(m\Delta\tau)\cos\frac{\pi k}{m} \right) \tag{3.104}$$

We are thus able to compute $m+1$ points on the PSD curve $\phi(f)$. Note we are here using frequency f in Hertz rather than ω rad/sec; one encounters formulas using either or both notations. Since the point spacing is $1/2m\Delta t$, there is no chance of resolving PSD peaks actually present in the data that are more closely spaced than this. Also, the lowest and highest frequencies computed are respectively $1/2m\Delta t$ and $1/2\Delta t$. In practice one selects the highest frequency of interest first, low-pass filters the data (before sampling) to remove

[14]Bendat, J. S., *Principles and Appl. of Random Noise Theory*, John Wiley and Sons, N.Y., 1958, p. 67.
[15]Otnes and Enochson, *op. cit.*, p. 197.

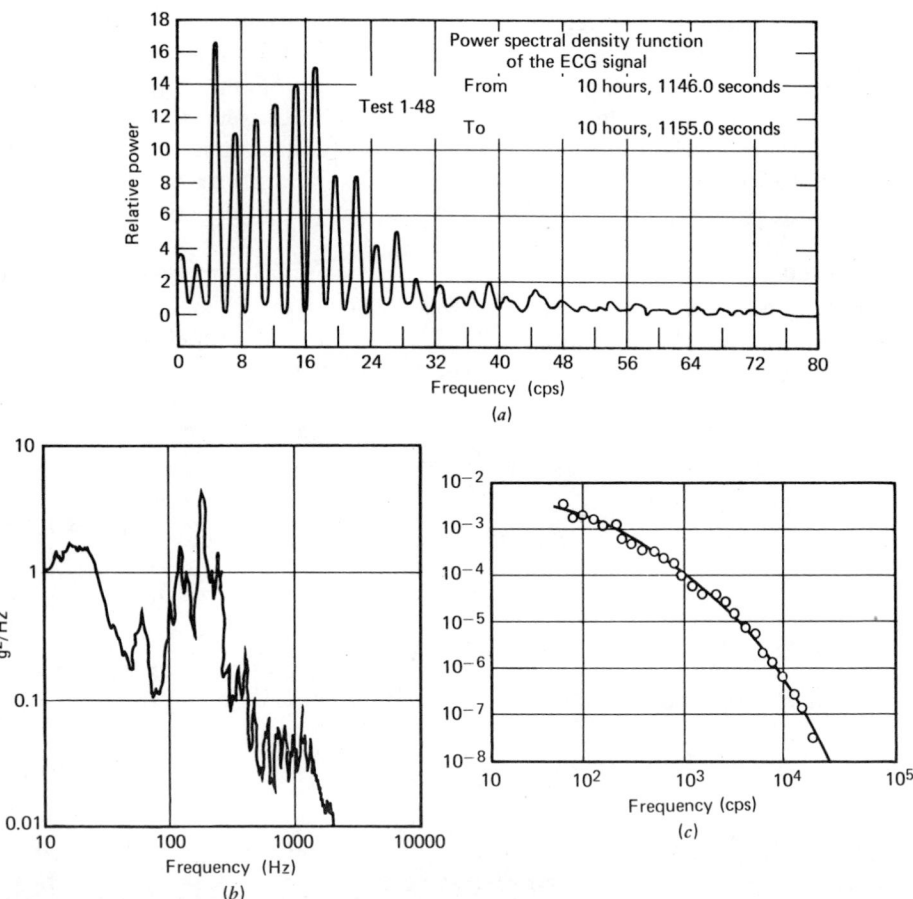

Figure 3.48 Measured power spectra for electrocardiogram, spacecraft vibration, and air-jet noise.

frequencies above this value (to prevent "aliasing"[16]), and then chooses the sampling interval Δt to meet the needs of the sampling theorem.[17] Aliasing is a phenomenon whereby a sinusoid of frequency higher than $1/2\Delta t$ Hz is sampled at a rate of $1/\Delta t$ samples per second and produces (falsely) a lower frequency component in the computed spectrum. The sampling theorem is a theoretical result which says that one must choose the sampling interval Δt small enough to provide at least 2 samples/cycle of the highest frequency to be calculated. In practice, 2.5 samples/cycle[18] has been found to give good results in many cases.

[16]Otnes and Enochson, *op. cit.*, p. 35.
[17]Otnes and Enochson, *op. cit.*, p. 28.
[18]Otnes and Enochson, *op. cit.*, p. 32.

In outlining this correlation method of computing PSD's, we have not included every detail necessary to a practical calculation; references[18, 19] should be consulted before attempting to develop a digital program. Fortunately, many computer libraries include working programs of this sort, so the practicing engineer is more often in the position of a program user than a program writer. While intelligent use of such canned programs requires some understanding of their inner workings, this level of understanding is considerably less than that needed in program development.

While the correlation method of PSD calculation is the traditional digital approach and is still in use, one is nowadays more likely to encounter programs that use Fast Fourier Transform (FFT) methods to compute the spectrum directly from the time data. Just as in the correlation method, we must first low-pass (anti-alias) filter the signal and then sample it at N points, using a Δt that gives at least 2.5 samples per cycle of the highest frequency. The Fourier transform of the sampled time data is computed from:

$$Q(k\,\Delta f) = \Delta t \sum_{p=0}^{N-1} q(p\,\Delta t)\big[\cos 2\pi k\,\Delta fp\,\Delta t - i\sin 2\pi k\,\Delta fp\,\Delta t\big]$$

$$k=0,1,\ldots,N \quad \Delta f=1/N\Delta t \quad f=0,1/N\Delta t,\ldots,1/2\Delta t \quad (3.105)$$

which is a sampled data version of Eq. (3.79). It can be shown[20] that, having the Fourier transform, one can compute the power spectral density from

$$\phi(k\,\Delta f) = \frac{2|Q(k\,\Delta f)|^2}{N\,\Delta t} \qquad (3.106)$$

If Eq. (3.106) were programmed in the most direct and naive way, a 5000-point PSD for $N = 10000$ would take several hours of computer time.[21] "Ordinary" programming cleverness can reduce this to about 12 minutes[21] while the extraordinary cleverness of an FFT algorithm cuts it to a few seconds.[21] These last speeds are 5 to 10 times faster than the traditional correlation method.[21] When we refer to the FFT algorithm we should note that while all such algorithms are based on the same principle, many detailed variations have been developed, each with its own characteristics.[22] We again refer the interested reader to the references for details and also again point out that most computer libraries have FFT programs that may be used with relative ease.

[19]Enochson, L. D. and Otnes, R. K., *Programming and Analysis for Digital Time Series Data*, Navy Publ. Office, Washington, D.C., 1968, p. 148.

[20]Bendat, *op. cit.*, p. 43.

[21]Enochson and Otnes, *op. cit.*, pp. 82, 83, 96, 168.

[22]Maynard, H. W., An Evaluation of Ten FFT Programs, R and D Tech. Rept. ECOM-5476, U.S. Army Electronics Command, Ft. Mammouth, N.J., 1973.

A final digital method used to compute PSD's is the digital filtering[23] approach. Here the analog approach of Fig. 3.46 is implemented in computer software. Actually, all the familiar analog filters (low-pass, high-pass, band-pass, band-reject) can be realized in digital form and are coming into widespread use in many areas of data processing as part of the general trend away from analog and toward digital electronics. To compute PSD's a narrow band-pass filter of desired center frequency and bandwidth is digitally synthesized, the sampled data is "passed through it" and the output squared and averaged to get one frequency point on the PSD. This is then repeated for as many different center frequencies as desired. Note that *analog* anti-aliasing filters are still required prior to the sampling operation.

In concluding this discussion of power spectral density we note that, if present trends continue, the FFT method of spectrum analysis should soon dominate the field, both in self-contained special purpose instruments and as software in general purpose computers. Portable FFT analyzers that accept analog input signals, anti-alias filter, sample and digitize the data, perform the desired analysis (autocorrelation, PSD, histogram, etc.) and present the results on a self-contained and calibrated display screen, all in less than a second, are already in use.

3.9   SYSTEM RESPONSE TO RANDOM SIGNALS

For linear, constant coefficient systems with random inputs the single most useful result[24] relating the input and output signals (see Fig. 3.49) is

$$\phi_o(\omega) = |W(i\omega)|^2 \phi_i(\omega) \tag{3.107}$$

It can also be shown[24] that if q_i is Gaussian then so also is q_o. These two results allow us to completely define the probability density function for q_o from Eq. (3.96) since we can get $\overline{q_o^2}$ from the area of the $\phi_o(\omega)$ curve. While in practice Eq. (3.107) is most often implemented graphically/numerically because ϕ_i and/or $W(i\omega)$ are experimentally measured, it can certainly be carried out analytically if formulas are available. For the familiar second-order system and with q_i a white noise with $\phi(\omega) = N_0$ we get[25]:

$$\overline{q_o^2} = \int_0^\infty \phi_o(\omega)\,d\omega = \int_0^\infty \frac{N_0 K^2}{\left[1 - (\omega/\omega_n)^2\right]^2 + 4\zeta^2\omega^2/\omega_n^2}\,d\omega = \frac{K^2\pi N_0\omega_n}{4\zeta} \tag{3.108}$$

If the second-order system is lightly damped (often the case in structural

[23]Otnes and Enochson, *op. cit.*, p. 199.

[24]Bendat, *op. cit.*, pp. 72, 124.

[25]Crandall, S., *Random Vibration*, John Wiley and Sons, N.Y., 1958, p. 83.

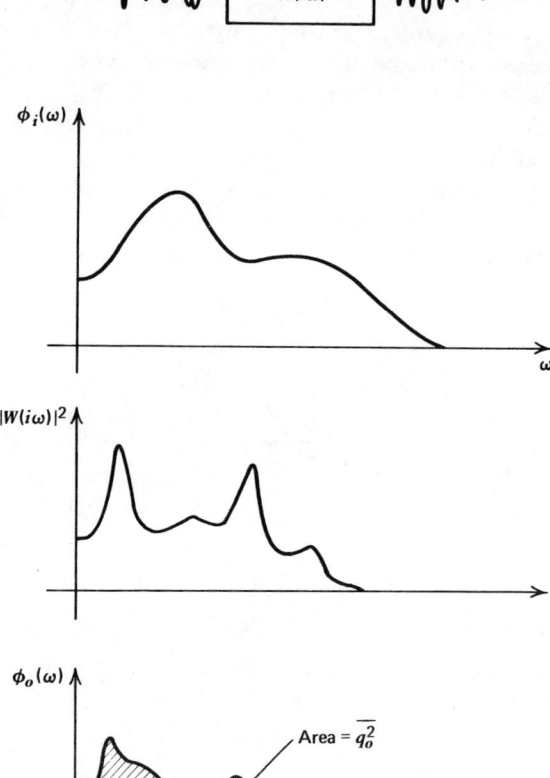

Figure 3.49 Linear-system response to random input.

vibrations), Fig. 3.50 shows that we can use (3.108) as an approximation even if $\phi_i(\omega)$ is *not* flat, so long as it does not become much larger away from ω_n than near ω_n.

Although Chapter 6 will explore in some detail the use of random signals in experimental modeling studies, we note here that measurements of ϕ_i and ϕ_o can be used together with Eq. (3.107) to find the amplitude ratio of the system under test:

$$|W(i\omega)| = \sqrt{\frac{\phi_o(\omega)}{\phi_i(\omega)}} \tag{3.109}$$

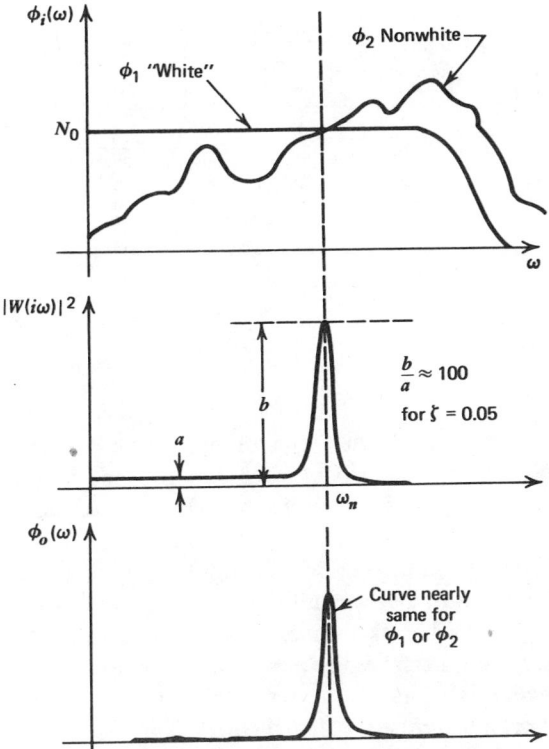

Figure 3.50 Response of lightly damped resonance to nonwhite input.

Just as in pulse testing, the spectrum of q_i must have sufficient amplitude out to the highest frequencies of interest, otherwise ϕ_o/ϕ_i will be the ratio of small, inaccurate numbers. Because PSD measurements are insensitive to phase angle, Equations (3.107) and (3.109) involve only system amplitude ratio. If phase information is necessary, cross-spectral density measurements must be made. Figure 3.51 illustrates the implementation, as an analog measurement, of the following definition of *cross-spectral density* $\phi_{1,2}(i\omega)$ for two random signals $q_1(t)$ and $q_2(t)$:

$$\phi_{1,2}(i\omega) \triangleq C(\omega) - iQ(\omega)$$

$$\text{co-spectrum} \triangleq C(\omega) \triangleq \lim_{\substack{T \to \infty \\ \Delta\omega \to 0}} \frac{(1/T)\int_0^T q_{1\Delta\omega} q_{2\Delta\omega}\, dt}{\Delta\omega} \tag{3.110}$$

$$\text{quad-spectrum} \triangleq Q(\omega) \triangleq \lim_{\substack{T \to \infty \\ \Delta\omega \to 0}} \frac{(1/T)\int_0^T q_{1\Delta\omega,90} q_{2\Delta\omega}\, dt}{\Delta\omega}$$

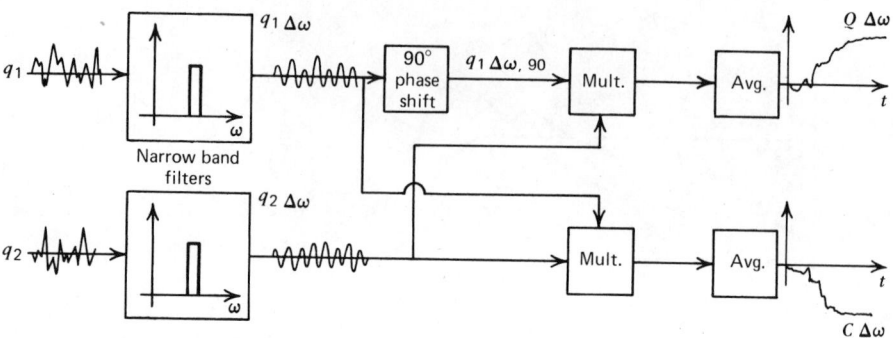

Figure 3.51   Definition of cross-spectral density.

We see that the measurement requires the same type of apparatus and operations as a PSD measurement except that two of each are needed, plus a 90° phase shifter which was not needed before. While a PSD has no phase information, $\phi_{1,2}$ is a complex number and contains both magnitude and phase information. Because of this, unintentional phase shifts in any of the apparatus components (filters, multipliers, etc.) must be carefully matched in channels 1 and 2 or else the phase angle of $\phi_{1,2}$ will be inaccurate. This applies also to the analog anti-aliasing filters necessary in otherwise digital systems. The intentional 90° shift of $q_{1\,\Delta\omega}$ is possible since this signal is "nearly a sine wave"; phase shifting an *arbitrary* signal by a certain number of degrees makes no sense.

When $\phi_{1,2}$ is computed digitally, we again have the two alternative approaches; the "traditional" method, which first computes the *cross-correlation function*[26] $R_{1,2}(\tau)$:

$$R_{1,2}(\tau) \triangleq \lim_{T \to \infty} \frac{1}{T} \int_0^T q_1(t)q_2(t+\tau)\,dt \qquad (3.111)$$

and then computes $\phi_{1,2}$ from[27]

$$C(f) = 2\int_0^\infty \left[R_{1,2}(\tau) + R_{1,2}(-\tau) \right] \cos 2\pi f\tau\,d\tau \qquad (3.112)$$

$$Q(f) = 2\int_0^\infty \left[R_{1,2}(\tau) - R_{1,2}(-\tau) \right] \sin 2\pi f\tau\,d\tau \qquad (3.113)$$

---

[26]Bendat, *op. cit.*, p. 23.
[27]Otnes and Enochson, *op. cit.*, p. 274.

and the more recent FFT method, which bypasses the cross-correlation computation. Sampled data versions of (3.111)–(3.113) and further computational details are available in the literature.[28] The direct calculation of $\phi_{1,2}$ is based on[29]

$$\phi_{1,2}(f) = \frac{2Q_1(f)^*Q_2(f)}{T} \tag{3.114}$$

where $Q_1(f)^*$ is the complex conjugate (same magnitude, negative angle) of the Fourier transform of $q_1(t)$ and $Q_2(f)$ is the Fourier transform of $q_2(t)$, both signals having a record length T. These Fourier transforms could be computed from sampled data as in Eq. (3.105), but of course an FFT algorithm gets the needed transforms much more efficiently. The reader should again consult references[30] for the details.

Our main interest in the cross-spectral density arises when q_1 is the input to a linear system $W(i\omega)$ and q_2 is the output. It can then be shown[31] that

$$W(i\omega) = \frac{\phi_{1,2}(i\omega)}{\phi_1(\omega)} \tag{3.115}$$

In terms of experimental modeling this allows us to completely determine (both amplitude ratio and phase) the frequency response of any linear constant-coefficient system by measuring one cross-spectral density and one power spectral density. We shall explore the details of this procedure in Chapter 6; however, Fig. 3.52 shows some actual results[32] for the transfer function relating the normal acceleration a_n of aircraft center of mass (as output) to the vertical gust velocity w_g of atmospheric turbulence (as input) for a large jet aircraft. A four-minute record of data was divided into two separate two-minute records for digital computation. Sampling at 0.05 sec intervals gave a total of 2400 points. Since the FFT method was not "rediscovered" until about 1965, the reference authors employed the "traditional" correlation methods, using $m = 60$ lags to compute spectral densities every $\frac{1}{6}$ Hz from $f = 0$ to 10 Hz. The solid and broken-line curves in Fig. 3.52 represent, respectively, the first and second two-minute records, plotted together to show the reproducibility of data and calculations.

---

[28]Otnes and Enochson, *op. cit.*, p. 274.
[29]Bendat, *op. cit.*, p. 64.
[30]Otnes and Enochson, *op. cit.*, p. 303.
[31]Bendat, *op. cit.*, p. 75.
[32]Coleman, T. L., Press, H., and Meadows, M. T., NASA TR R-70, 1960.

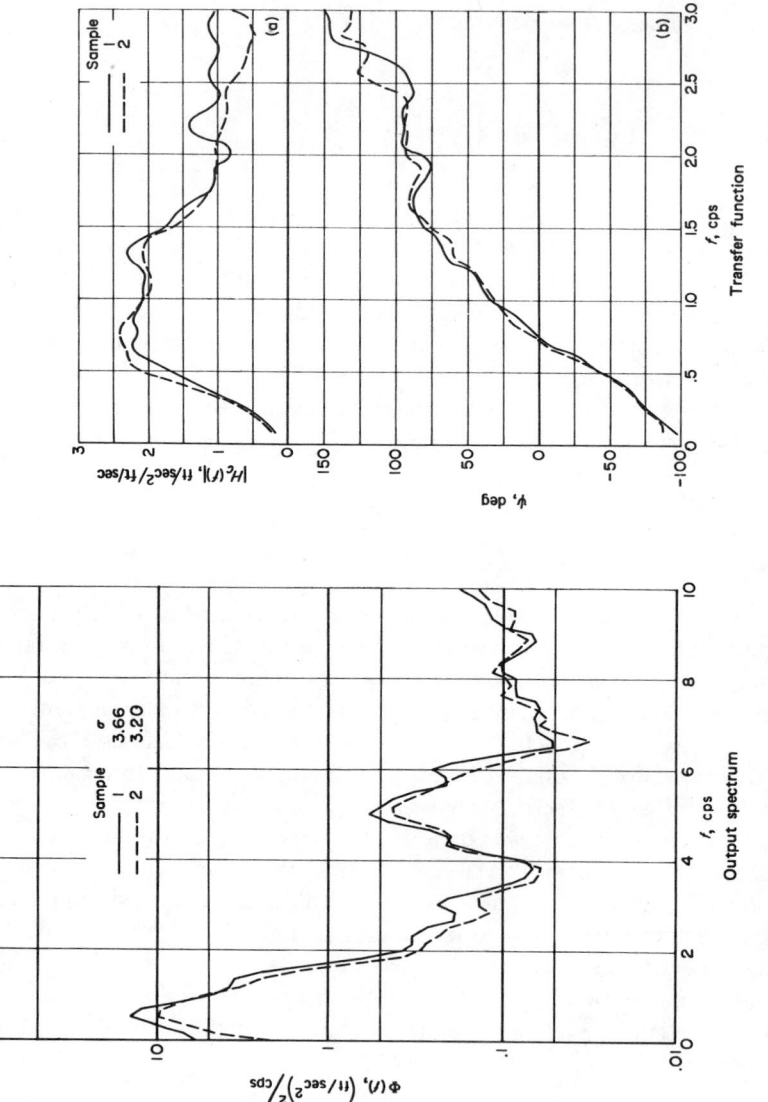

Figure 3.52 Aircraft transfer function measured using random-signal methods.

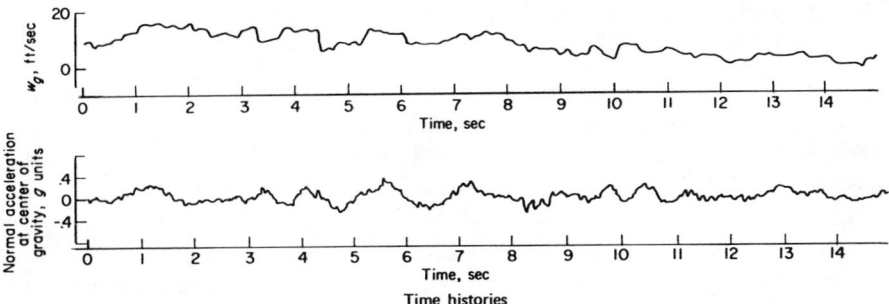

Time histories

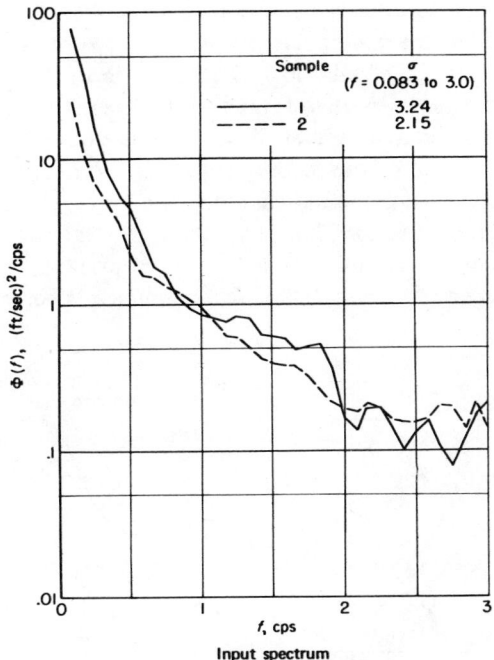

Input spectrum

Figure 3.52 *continued*

3.10   ALIASING AND LEAKAGE IN DIGITAL SPECTRUM ANALYSIS

While we have referred the reader to various references for computational details concerning certain aspects of digital spectrum analysis, a brief discussion of two of these details may be helpful at this point. In Fig. 3.53, the example time history used is a single sine wave. This simple form of data record, which makes the basic concepts most obvious, does not explore all the subtleties of the situation but will be adequate for our purposes. Note in 3.53a that when one samples the data waveform at a once-per-cycle rate and then "reconstructs" the waveform based only on the sampled points, a very misleading picture of the data is obtained. Rather than the correct sine wave we get a constant value! In 3.53b we slightly increase the sample rate to once every $\frac{3}{4}$ cycle; this is still inadequate since we get a triangular waveform of lower frequency than our original data. This creation of false low frequencies in the sampled data is called *aliasing*. In 3.53c, a twice-per-cycle sampling rate seems to give a fair approximation to the original data; however, if the sample points are not taken at the peaks (as they were in 3.53c), the reconstructed function may still be misleading.

A more sophisticated mathematical study leads to the well-known sampling theorem (mentioned in Section 3.8) that gives two samples per cycle as the theoretical lower limit for sampling rate. In actual practice, a rate of about 2.5 samples per cycle seems to be a usable lower bound. When the data record is

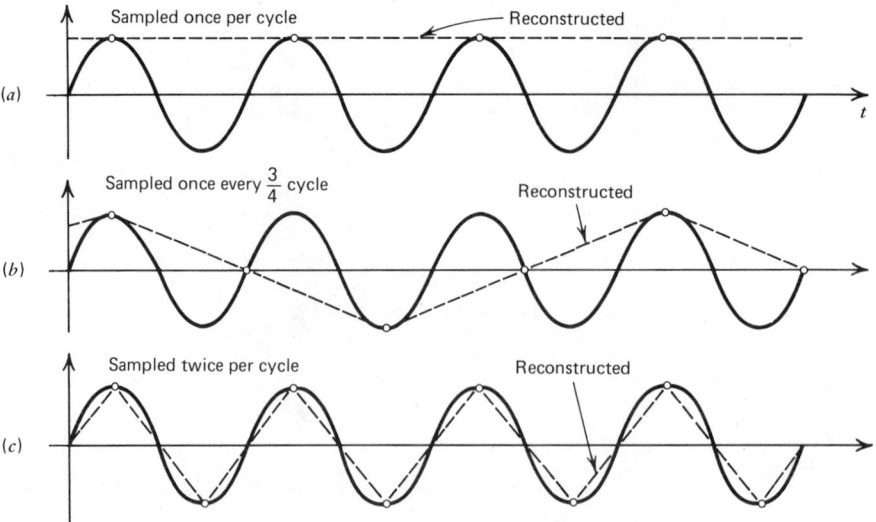

Figure 3.53   Aliasing caused by inadequate sampling rate.

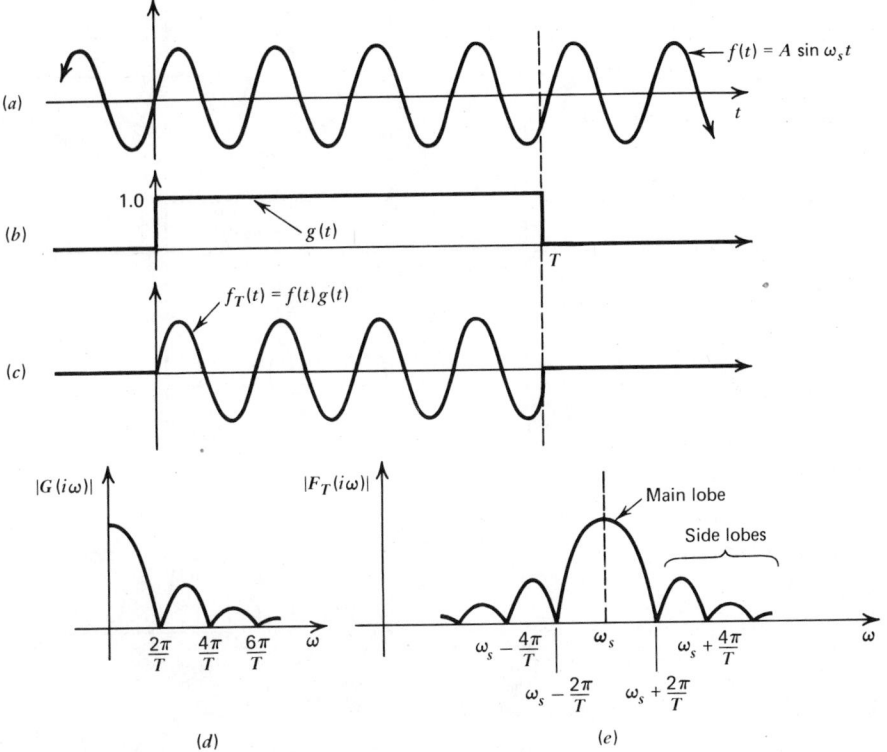

Figure 3.54 Spectrum distortion due to time-history truncation.

not a single sine wave, these values of course are applied to the *highest* frequency of interest. If one chooses to analyze a record over a certain frequency range, one must first low-pass ("anti-alias") filter the data to remove any components above the desired frequency range or else spurious low-frequency ("aliased") components will appear in the computed spectrum.

When continuous, rather than transient, signals are spectrum analyzed, a procedure called *tapering* generally is included in the Fourier transform calculations. Let us again use a simple sinusoidal data record to illustrate the need for this type of manipulation. When the ongoing sine wave of Fig. 3.54a is to be spectrum analyzed, only a finite-length portion of it, say from $t=0$ to $t=T$, can actually be dealt with; thus we spectrum analyze not the original sine wave but a truncated or "windowed" version of it, and we should not expect to get exactly the correct spectrum of the original untruncated signal. This time-windowing operation is equivalent to multiplying our original signal by a rectangular pulse function ("boxcar" function) as in Fig. 3.54b, and can be treated as an

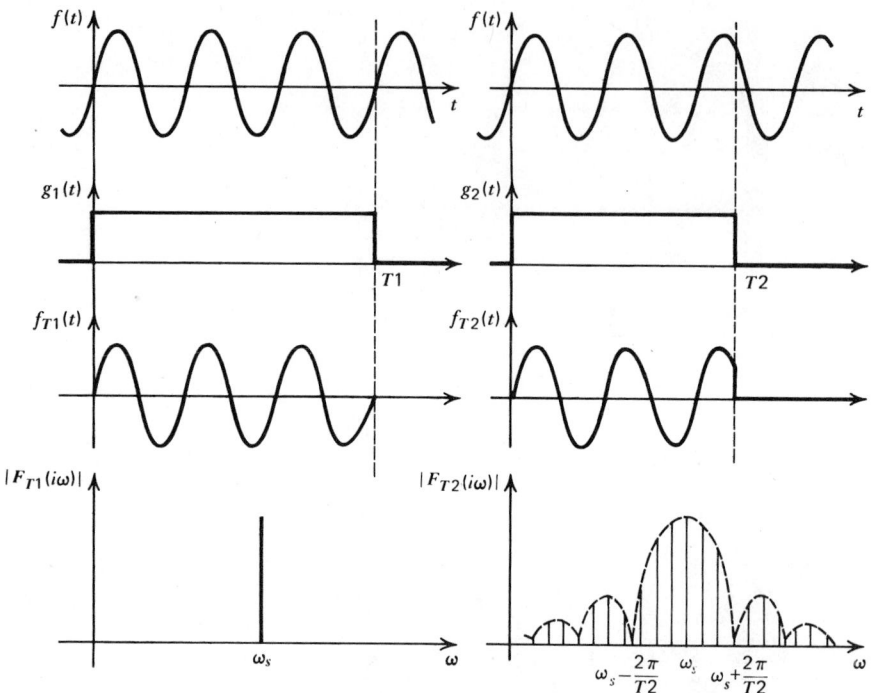

Figure 3.55 Effects of data truncation on FFT-computed spectra.

amplitude-modulation process. The spectrum of the truncated signal is obtained[33] by translating the spectrum of the rectangular pulse (see Fig. 3.30) out to the signal frequency ω_s and "folding" it back to give a symmetrical spectrum above and below ω_s (Fig. 3.54e). Clearly, this calculated spectrum is not the single "spike" at ω_s that we expect for a sine wave; however, this ideal would be approached if T became very long since the "spread" of the spectrum on either side of ω_s varies inversely with T.

When spectra are calculated numerically from sampled data by FFT methods, rather than "analytically" as in Fig. 3.54, similar defects occur, however, with the following variations. Rather than smooth, continuous curves, FFT spectra will be discrete spikes; however, the envelope of these spikes follows generally the shape shown in Fig. 3.54e. Also, if the time window defined by $g(t)$ includes *precisely* an integer number of cycles of the original sine wave, then the FFT-computed spectrum will be "perfect" in that it will show only a single spike at ω_s; however, this ideal situation is not generally realizable in practice. Figure 3.55 displays these aspects of FFT spectra.

[33]Doebelin, E. O., *Measurement Systems*, McGraw-Hill, N.Y., 1975, p. 166.

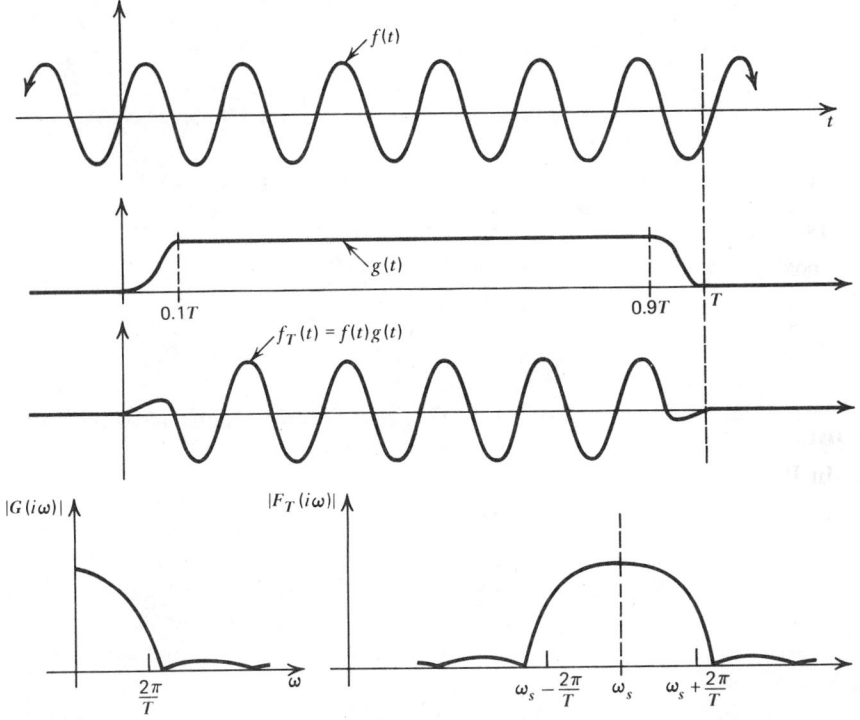

Figure 3.56 Data tapering for spectrum improvement.

To make the above-mentioned defects (called "leakage") less serious, one can "taper" the beginning and end of the time record before carrying out the FFT spectrum analysis. (Alternatively, an equivalent process can be applied in the frequency domain *after* the spectrum has been calculated from un-tapered data.) Figure 3.56 shows that the tapering process involves multiplying $f(t)$ by a $g(t)$ that rises and falls *gradually* at the beginning and end of the record. Such a $g(t)$ will have a $|G(i\omega)|$ with much less prominent side lobes, giving an improved spectrum. The tapering is generally applied to about the first and last 10% of the record and while various tapering functions can be used, a cosine squared (called "Hanning") is quite common.[34] While tapering greatly reduces the side lobes, it also slightly broadens the main lobe, reducing frequency resolution; however, this effect is not severe enough to outweigh the advantages of side-lobe suppression.

[34]Bendat, J. S. and A. G. Piersol, *Random Data*: *Analysis and Measurement Procedures*, John Wiley and Sons, N.Y., 1971, pp. 318–325.

BIBLIOGRAPHY

1 Bendat, J. S., *Principles and Applications of Random Noise Theory*, John Wiley and Sons, N.Y., 1958.

2 Bendat, J. S. and A. G. Piersol, *Random Data: Analysis and Measurement Procedures*, Wiley–Interscience, N.Y., 1971.

3 Goldman, S., *Frequency Analysis, Modulation, and Noise*, McGraw-Hill, N.Y., 1948.

4 Guillemin, E. A., *The Mathematics of Circuit Analysis*, John Wiley and Sons, N.Y., 1949.

5 Jenkins, G. M. and D. G. Watts, *Spectral Analysis*, Holden-Day, San Francisco, 1968.

6 Kaplan, W., *Operational Methods for Linear Systems*, Addison-Wesley, Reading, Mass., 1962.

7 Otnes, R. K. and L. Enochson, *Digital Time Series Analysis*, John Wiley and Sons, N.Y., 1972.

8 Papoulis, A., *The Fourier Integral and Its Applications*, McGraw-Hill, N.Y., 1962.

9 Schwartz, M., *Information Transmission, Modulation, and Noise*, McGraw-Hill, N.Y., 1959.

10 Stearns, S. D., *Digital Signal Analysis*, Hayden Book Co., Rochelle Park, N.J., 1975.

PROBLEMS

3.1 In Problem 2.6, sketch and discuss the frequency response graphs for the two damper models given there.

3.2 For the periodic function shown in Fig. P3.1, use transform methods to get expressions for a_n and b_n in the Fourier series.

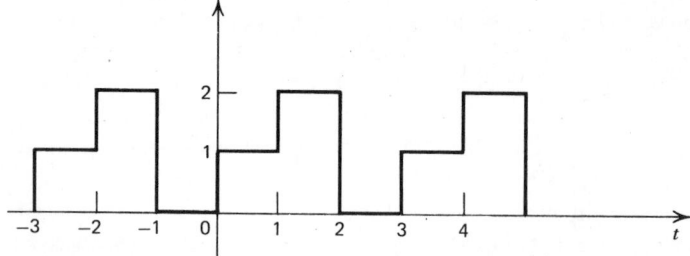

Figure P3.1

3.3 If $(q_o/q_i)(s) = 1/(s+a)$, find and sketch the impulse response $h(t)$. Then find $q_o(t)$ if $q_i(t) = ce^{-bt}$ and initial conditions are zero. If b is made much greater than "a," what value should c have to get a good approximation to perfect unit impulse response? For this value of c, check the area of $q_i(t)$ and comment on its significance.

3.4 If $(q_o/q_i)(s)=1/(s+1)(s+2)$ and $q_i(t)=1\sin t$, what must initial conditions $q_o(0)$ and $\dot{q}_o(0)$ be so that the system goes instantly into the sinusoidal steady state with no transient whatsoever? Solve this problem two ways; somewhat intuitively using frequency response methods, and rigorously, using Laplace transform methods of solving differential equations.

3.5 Use convolution integral to find $\mathcal{L}^{-1}[1/s][1/(s+a)]$ and then check the result from the transform table for $1/s(s+a)$.

3.6 For $a=1$ in Problem 3.5, use convolution integral graphically/numerically (as in Figs. 3.7 and 3.8) to compute points at $t=0$, 1, 2, and 3 for $f(t)$, where $F(s)=[1/s][1/(s+a)]$, and check the numerical results using a transform table.

3.7 A system has the transfer function

$$\frac{q_o}{q_i}(s)=\frac{0.001}{(0.1s+1)(0.02s+1)\left(\dfrac{s^2}{(100)^2}+\dfrac{(2)(0.3)s}{100}+1\right)}$$

Choose the *longest* duration triangular pulse (Fig. 3.11) you feel would be a good approximation to a perfect impulse input for this system. We wish to check this result by comparing the response to the triangular pulse with the perfect impulse response, using CSMP (or other available) digital simulation (see Chapter 5). Since we cannot digitally simulate a perfect impulse input (it has *infinite* magnitude) we must determine the initial conditions that are exactly *equivalent* to an impulse input. Use the initial-value theorem to find the numerical values of $q_o(0)$, $q_o'(0)$, $q_o''(0)$, and $q_o'''(0)$ that are needed in the simulation. Then run the simulation to compare the system responses for perfect and approximate impulse inputs. If your triangular pulse is inadequate, adjust it and rerun until a "good" response is obtained.

3.8 Sketch the approximate logarithmic frequency response curves for the systems given and then check your results using any digital computer frequency response program available to you.

(a) $\quad W(s)=\dfrac{5(0.1s+1)e^{-s}}{s(0.5s+1)(8s+1)}$

(b) $\quad W(s)=\dfrac{25(0.01s+1)(0.005s+1)}{\left(\dfrac{s^2}{(100)^2}+\dfrac{(2)(0.1)s}{100}+1\right)\left(\dfrac{s^2}{(500)^2}+\dfrac{(2)(0.2)s}{500}+1\right)}$

3.9 Using Eq. (3.44), find $W(i\omega)$ for the following $h(t)$'s, using integral tables where needed:

(a) e^{-5t} (b) $1-e^{-t}$ (c) $e^{-t}-e^{-2t}$
(d) $(\sin 5t)/5$ (e) $e^{-t}\sin t$

3.10 Using Laplace transform methods, find the Fourier series for the function of Fig. P3.2. Discuss, using sketches, the difference between the frequency spectra for $T=20\pi$ and $T=200\pi$.

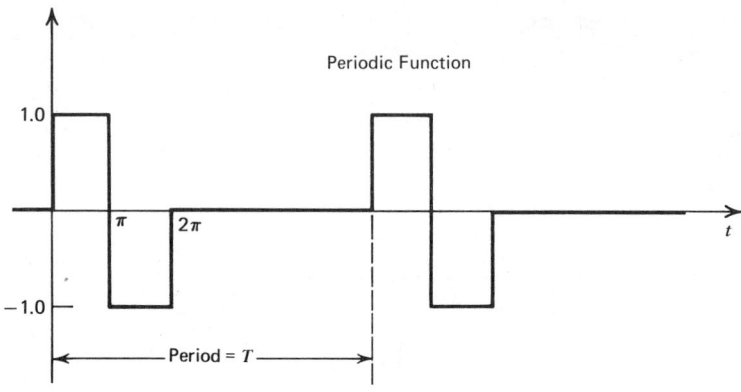

Figure P3.2

3.11 For the transient in Fig. P3.3, find $F(i\omega)$ and sketch its magnitude and phase versus frequency.

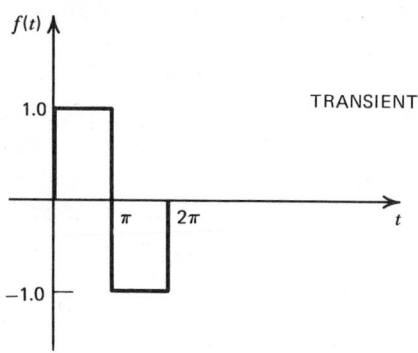

Figure P3.3

3.12 In Fig. P3.4, find the Fourier series for displacement x_i using transform methods. Calculate numerical values for the average value and first harmonic term in both x_i and x_o.

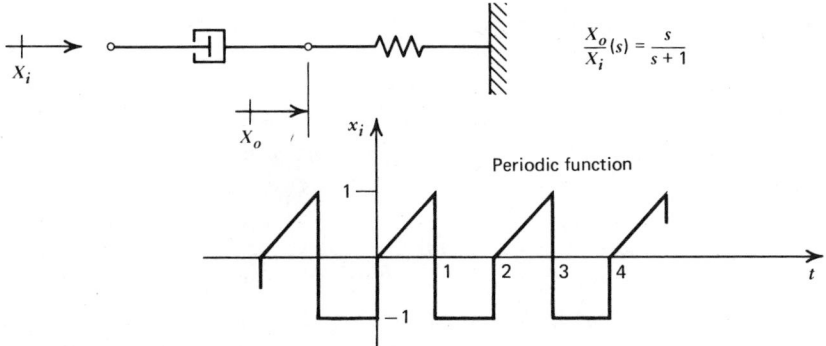

Figure P3.4

3.13 Using CSMP (or other available digital simulation) get a graph of $x_o(t)$ in Fig. P3.4, starting system from rest at $t=0$ and running for five complete cycles of x_i.

3.14 A vibrating system with transfer function

$$\frac{X}{F}(s)=\frac{2.6(0.1s+1)}{(0.004s+1)(0.04s^2+0.04s+1)(0.0025s^2+0.01s+1)}\frac{\text{cm}}{\text{N}}$$

is subjected to the periodic input force of Fig. P3.5.

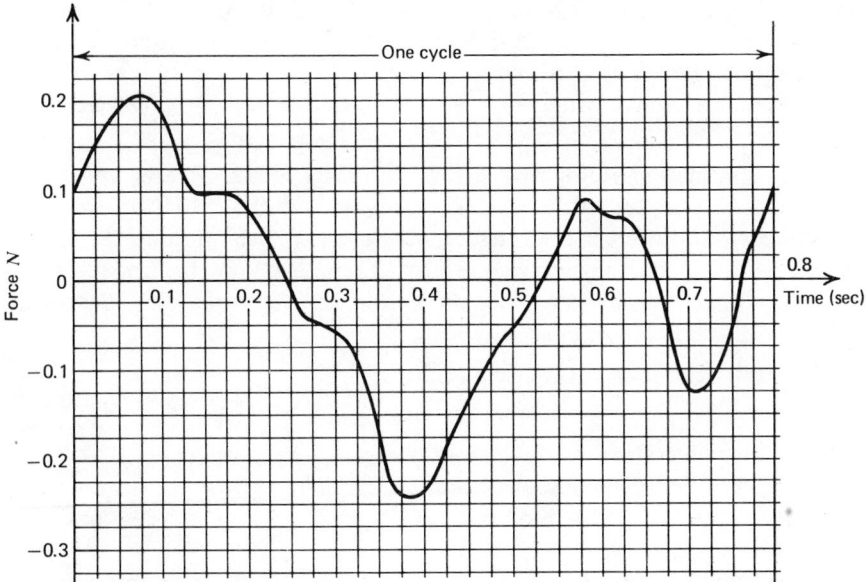

Figure P3.5

(a) Using logarithmic plotting methods, sketch the straightline asymptotes only for the amplitude ratio $(X/F)(i\omega)$.

(b) Using available frequency-response computer program, fill in the details of the amplitude ratio of part a and add the phase angle graph.

(c) Use available Fourier series computer program to get the Fourier series for the input force. Plot the frequency spectrum (magnitude and phase) of this force.

(d) Using the results of parts b and c, compute and plot the frequency spectrum of motion x and give an expression for $x(t)$.

3.15 In Problem 2.19, use SPEAKEASY (or other available computer language) to find and graph $P_i(i\omega)$. If this sonic boom strikes a window whose natural frequency is ω_n, comment on what values of ω_n might be particularly dangerous, based on your results for $P_i(i\omega)$. To check these predictions more carefully, let us use CSMP (or other available digital simulation) to find the time response of a window to $p_i(t)$. Assume that the transfer function relating window glass stress s_w to sonic boom pressure $p_i(t)$ is:

$$\frac{S_w}{P_i}(s) = \frac{6000}{s^2/\omega_n^2 + 0.1s/\omega_n + 1} \frac{\text{psi}}{\text{lb}_f/\text{ft}^2}$$

and that the failure stress is 10,000 psi. Make one CSMP run in which the stress in each of five different windows ($\omega_n = 30, 40, 50, 60, 70$ rad/sec) is computed and graphed. Use PAGE GROUP to get a composite graph showing all five stress/time histories on one sheet, for easy comparison. Comment on which window is the most critical and whether any are broken.

3.16 The DC power supply for the magnet of a large electrodynamic vibration shaker (see Chapter 10) contains harmonics of the 60-Hz power line frequency due to imperfect filtering. If the shaker armature coil is driven by a periodic, nonsinusoidal current of fundamental frequency 1000 Hz, use amplitude modulation concepts to predict the frequencies present in the magnetic force produced by the shaker.

3.17 Analytically compute $F_4(i\omega)$ for $f_4(t)$ in Fig. 3.11.

3.18 Repeat Problem 3.17 for $f_5(t)$ in Fig. 3.11.

3.19 The pneumatic system of Fig. P3.6 operates at a mean pressure of 200 psig; however, $p_i(t)$ exhibits Gaussian random fluctuations that have a constant spectral density of 0.314 psi^2/(rad/sec) from 0 to 1000 rad/sec and zero elsewhere.

(a) Set up and run a CSMP simulation (see Section 5.9) to compute the time history of the fluctuating component of p_o(PO) and also the mean-square value of p_o(POMSQ), using $\omega_{co} = 1000$. In calculating the

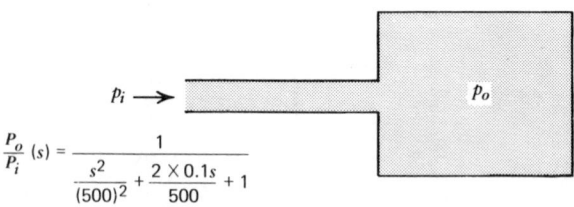

$$\frac{P_o}{P_i}(s) = \frac{1}{\dfrac{s^2}{(500)^2} + \dfrac{2 \times 0.1 s}{500} + 1}$$

Figure P3.6

mean-square value, use a "trick" to avoid dividing by zero at TIME = 0. Use the following run-control cards to define a total sample length of 0.5505 sec, divided into three ranges. The first and last ranges use plotting increments small enough to clearly show the waveforms of p_i and p_o while the middle range uses a larger increment to save paper.

TIMERьDELT = .0001101, FINTIM = .05505, OUTDEL = .0005505
OUTPUTьPI, PO, POMSQ
ENDьCONTINUE
TIMERьFINTIM = .4955, OUTDEL = .004404
ENDьCONTINUE
TIMERьFINTIM = .5505, OUTDEL = .0005505
END

When you get your computer-plotted graphs, use colored markers to distinguish the curves for PI, PO, and POMSQ clearly. Why are the PI and PO graphs misleading for the time interval .05505 to .4955?

(b)   Use Eq. (3.108) to estimate the RMS value of p_o and compare with your CSMP result. Should these results agree exactly? Discuss.

(c)   When theoretical results such as that of part b are not available (as they would not be if the system dynamics were only slightly more complicated, or experimentally defined), we can fall back on the general formula (3.107) to plot $\phi_o(\omega)$ and then integrate to get $\overline{p_o^2}$. Plot a few points of $|W(i\omega)|^2$ superimposed on Fig. 5.19 to show that $\phi_i(\omega)$ can be replaced, with good accuracy, by the constant 0.314. Then compute $\phi_i(\omega)|W(i\omega)|^2$ (for the appropriate range of frequencies) using SPEAKEASY and employ the INTEGRAL command to get the area under the curve. Compare the RMS value of p_o obtained this way with the two earlier values and discuss the reasons for any differences observed.

CHAPTER 4

Miscellaneous Transform Topics

4.1 DISTRIBUTED-PARAMETER MODELS

Although in later chapters we deal at some length with distributed-parameter models (partial differential equation models), we here wish to prepare for these specific applications by presenting some common general background. To motivate and illustrate this material, consider the system of Fig. 4.1, which might be used to study dynamic effects in a drilling machine in which the drill is stationary and the workpiece rotates, cutting forces at the drill tip producing the torque $T(L,t)$.

We wish to obtain a relation between the input torque $T(L,t)$ and the output twist angle $\theta(y,t)$ at any location y along the rod. Since the unknown θ depends both on y (where we look) and t (when we look), it is a function of two variables, and any derivatives of θ we might need to use to describe the motion will have to be partial derivatives. The differential equation relating θ to T can be obtained by applying Newton's law to a shaft slice of thickness dy. Neglecting all internal and external frictional effects, the only torques acting on this slice are those due to torsional shearing stresses on the two faces. From strength of materials we know that the elastic torque exerted by the upper rod section on the upper face of the element at any instant of time t is:

$$T(y,t) = - I_p G \frac{\partial \theta}{\partial y} \qquad (4.1)$$

where $I_p = \pi R^4/2$ is the polar area moment of inertia of the (circular) cross section and G is the material modulus of elasticity in shear. At the same time instant, since T is a function of location y, the torque of the lower rod section on the lower face of the element is:

$$T(y+dy,t) = -\left(T(y,t) + \frac{\partial}{\partial y} T(y,t)\,dy \right) = I_p G \frac{\partial \theta}{\partial y} + I_p G \frac{\partial^2 \theta}{\partial y^2}\,dy \qquad (4.2)$$

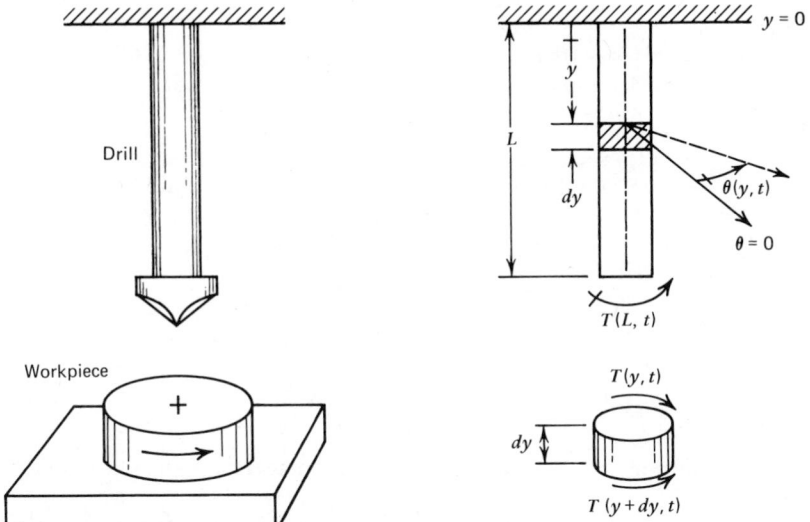

Figure 4.1   Distributed-parameter model of drilling operation.

Newton's law now gives:

$$\sum T = J\alpha$$

$$-I_p G \frac{\partial \theta}{\partial y} + I_p G \frac{\partial \theta}{\partial y} + I_p G \frac{\partial^2 \theta}{\partial y^2} = \frac{\rho \pi R^2 \, dy \, R^2}{2} \frac{\partial^2 \theta}{\partial t^2} = \rho I_p \frac{\partial^2 \theta}{\partial t^2} \qquad (4.3)$$

$$\frac{\partial^2 \theta}{\partial y^2} = \frac{\rho}{G} \frac{\partial^2 \theta}{\partial t^2} \qquad (4.4)$$

This partial differential equation is one of the classical equations of physics, the one-dimensional wave equation.

To solve Eq. (4.4) by Laplace transform methods we will need some additional theorems dealing with the transforms of functions of several variables. In transforming a function $\theta(y, t)$ we must choose whether the independent variable y or t will be transformed to s; either could be, the other being treated as a constant parameter during the transformation:

$$\mathcal{L}_y \theta(y, t) \triangleq \int_0^\infty \theta(y, t) e^{-sy} \, dy = \theta(t, s) \qquad (4.5)$$

$$\mathcal{L}_t \theta(y, t) \triangleq \int_0^\infty \theta(y, t) e^{-st} \, dt = \theta(y, s) \qquad (4.6)$$

(In system analysis, we often choose to transform t to s since then if we later set $s = i\omega$, we obtain the system frequency response, which the last chapter showed to be the key to general response studies.) An example of (4.5) and (4.6) for $\theta(y,t) = \cos(yt)$ gives:

$$\mathcal{L}_y(\cos yt) = \theta(t,s)$$

$$= \int_0^\infty \cos yt \, e^{-sy} \, dy = \frac{s}{s^2 + t^2} \tag{4.7}$$

$$\mathcal{L}_t(\cos yt) = \theta(y,s)$$

$$= \int_0^\infty \cos yt \, e^{-st} \, dt = \frac{s}{s^2 + y^2} \tag{4.8}$$

Note that the transforms can be obtained from our usual tables since the second variable is treated as a constant parameter.

When we transform a partial derivative, say $\partial\theta/\partial y$, we are again faced with choosing either

$$\mathcal{L}_y\left(\frac{\partial\theta}{\partial y}\right) = \int_0^\infty \frac{\partial\theta}{\partial y} e^{-sy} \, dy \tag{4.9}$$

or

$$\mathcal{L}_t\left(\frac{\partial\theta}{\partial y}\right) = \int_0^\infty \frac{\partial\theta}{\partial y} e^{-st} \, dt \tag{4.10}$$

since $\partial\theta/\partial y$ is a function of both y and t. Because, according to the usual definition of partial derivatives, $\partial\theta/\partial y$ considers t as a constant, the transform (4.9) follows the same rules as an ordinary derivative, giving

$$\mathcal{L}_y\left(\frac{\partial\theta}{\partial y}\right) = s\theta(t,s) - \theta(0,t) \tag{4.11}$$

where $\theta(0,t)$ is a boundary condition, $\theta(y,t)$ at $y = 0$. Higher derivatives follow the same pattern:

$$\mathcal{L}_y\left(\frac{\partial^2\theta}{\partial y^2}\right) = s^2\theta(t,s) - s\theta(0,t) - \theta'(0,t) \tag{4.12}$$

where $\theta'(0,t)$ is another boundary condition, $(\partial\theta/\partial y)(y,t)$ at $y = 0$. For transforms of form (4.10), consider the example $\theta(y,t) = y^2 e^{-t}$, where (4.10) directly applied would give:

$$\mathcal{L}_t\left(\frac{\partial\theta}{\partial y}\right) = \int_0^\infty 2ye^{-t}e^{-st}\,dt = 2y\frac{1}{s+1} \tag{4.13}$$

If the order of partial differentiation and integration in (4.10) were *interchanged* (legal under most circumstances), we could write:

$$\mathcal{L}_t\left(\frac{\partial\theta}{\partial y}\right) = \frac{\partial}{\partial y}\int_0^\infty y^2 e^{-t}e^{-st}\,dt$$

$$= \frac{\partial}{\partial y}\left[y^2\frac{1}{s+1}\right] = 2y\frac{1}{s+1} \tag{4.14}$$

This example, though it is *not* a proof, should make plausible the following theorem:

$$\mathcal{L}_t\left(\frac{\partial\theta}{\partial y}\right) = \frac{\partial}{\partial y}\theta(y,s) \tag{4.15}$$

Similarly:

$$\mathcal{L}_t\left(\frac{\partial^2\theta}{\partial y^2}\right) = \frac{\partial^2}{\partial y^2}\theta(y,s) \tag{4.16}$$

The reader is asked to accept (4.11), (4.12), (4.15), and (4.16) as theorems that may be applied to the transform solution of partial differential equations. Let us illustrate this application by returning to Eq. (4.4).

Since we are interested in obtaining the frequency response $\theta(y,i\omega)/T(L,i\omega)$, we choose to transform t to s:

$$\mathcal{L}_t\left(\frac{\partial^2\theta}{\partial y^2}\right) = \mathcal{L}_t\left(\frac{\rho}{G}\frac{\partial^2\theta}{\partial t^2}\right) \tag{4.17}$$

$$\frac{\partial^2}{\partial y^2}\theta(y,s) = \frac{\rho}{G}\left[s^2\theta(y,s) - s\theta(y,0) - \theta'(y,0)\right] \tag{4.18}$$

Note that $\theta(y,0)$ and $\theta'(y,0)$ are initial (time$=0$) conditions, and if we choose to obtain the transfer function $\theta(y,s)/T(L,s)$, we would, as usual, take all initial conditions as zero. We then get from (4.18):

$$\frac{d^2\theta}{dy^2} - \frac{\rho s^2}{G}\theta = 0 \qquad (4.19)$$

where $\partial^2\theta/\partial y^2$ is now written as an *ordinary* derivative (since the second *variable* t has been transformed to the *parameter* s) and $\theta(y,s)$ is written simply as θ. Note that the transform process has reduced a partial differential equation to an ordinary one (a simpler class), which we could now solve using a *second* Laplace transform with a new transform variable, say w. However, since (4.19) is simply solved by the classical operator method, we choose that approach since the multiple transform scheme is needlessly confusing.

Defining $D \triangleq d/dy$:

$$\left(D^2 - \frac{\rho s^2}{G}\right)\theta = 0$$

$$\theta(y,s) = C_1 e^{+s\sqrt{\rho/G}\,y} + C_2 e^{-s\sqrt{\rho/G}\,y} \qquad (4.20)$$

At the fixed end of the rod, $\theta(0,t) \equiv 0$, giving $\theta(0,s) \equiv 0$, which we can use in (4.20) to help find C_1 and C_2:

$$\theta(0,s) = 0 = C_1 + C_2, \qquad C_1 = -C_2 \qquad (4.21)$$

At $y = L$, Eq. (4.2) gives:

$$T(L,t) = I_p G \frac{\partial\theta}{\partial y}(L,t), \qquad T(L,s) = I_p G \frac{\partial\theta}{\partial y}(L,s) \qquad (4.22)$$

From (4.20) and (4.21):

$$\frac{\partial\theta}{\partial y} = s\sqrt{\rho/G}\ C_1 e^{s\sqrt{\rho/G}\,y} + s\sqrt{\rho/G}\ C_1 e^{-s\sqrt{\rho/G}\,y} \qquad (4.23)$$

which gives at $y = L$

$$\frac{\partial\theta}{\partial y}(L,s) = C_1 s\sqrt{\rho/G}\left[e^{s\sqrt{\rho/G}\,L} + e^{-s\sqrt{\rho/G}\,L}\right] \qquad (4.24)$$

and thus

$$C_1 = \frac{1}{I_p Gs\sqrt{\rho/G}}\left[\frac{1}{e^{s\sqrt{\rho/G}\,L} + e^{-s\sqrt{\rho/G}\,L}}\right]T(L,s) \qquad (4.25)$$

$$\frac{\theta(y,s)}{T(L,s)} = \frac{e^{s\sqrt{\rho/G}\,y} - e^{-s\sqrt{\rho/G}\,y}}{I_p\sqrt{\rho G}\,s\left(e^{s\sqrt{\rho/G}\,L} + e^{-s\sqrt{\rho/G}\,L}\right)} \qquad (4.26)$$

$$\frac{\theta(y,s)}{T(L,s)} = \frac{1}{I_p\sqrt{\rho G}\,s}\frac{\sinh\left(s\sqrt{\rho/G}\,y\right)}{\cosh\left(s\sqrt{\rho/G}\,L\right)} \qquad (4.27)$$

To obtain the frequency response for the angular motion at $y = L$, let $s = i\omega$:

$$\frac{\theta(L,i\omega)}{T(L,i\omega)} = \frac{\sin\left(\sqrt{\rho/G}\,L\omega\right)}{\omega I_p\sqrt{\rho G}\,\cos\left(\sqrt{\rho/G}\,L\omega\right)} \qquad (4.28)$$

As seen in Fig. 4.2, this frequency response exhibits an infinite number of natural frequencies given by

$$\omega_{nk} = \frac{k\pi}{2L}\sqrt{\frac{G}{\rho}}, \qquad k = 1,3,5,7,\ldots \qquad (4.29)$$

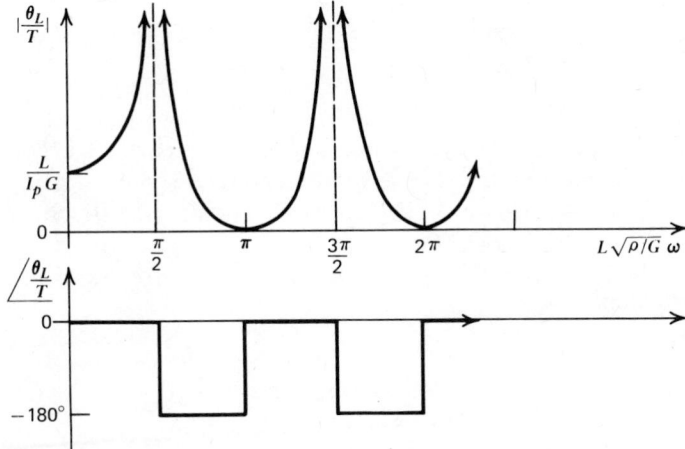

Figure 4.2 Frequency response of distributed-parameter torsional vibration model.

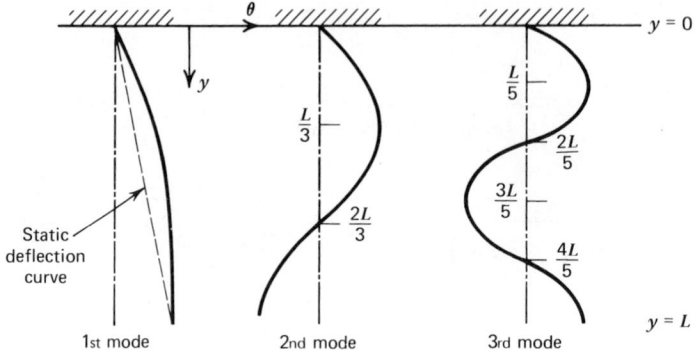

Figure 4.3   Mode shapes of torsional vibration.

When a structure is vibrating at one of its natural frequencies the dynamic deflection curve is called the *mode shape*. We can easily get the mode shapes from Eq. (4.28) by forming the ratio

$$\frac{\theta(y,i\omega_{nk})}{\theta(L,i\omega_{nk})} = \frac{\theta(y,i\omega_{nk})}{T(L,i\omega_{nk})} \div \frac{\theta(L,i\omega_{nk})}{T(L,i\omega_{nk})} = \sin\left(\frac{k\pi}{2}\frac{y}{L}\right) \qquad (4.30)$$

Figure 4.3 shows the first three mode shapes.

If the applied torque were a step input $T(L,t) = T_s u(t)$, then $T(L,s) = T_s/s$ and Eq. (4.26) gives

$$\theta(y,s) = \frac{T_s}{I_p\sqrt{\rho G}\,s^2}\frac{e^{s\sqrt{\rho/G}\,y} - e^{-s\sqrt{\rho/G}\,y}}{e^{s\sqrt{\rho/G}\,L} + e^{-s\sqrt{\rho/G}\,L}} \qquad (4.31)$$

This s function is not a ratio of polynomials in s and thus the partial-fraction expansion method is not applicable to the inverse transformation. In fact, many of the s functions obtained from partial differential equations are difficult or impossible to inverse transform in closed form. Occasionally, however, a suitable trick will yield a solution. A trick that works here is based on a transform theorem relating to periodic functions:

$$\mathscr{L}f(t) = \frac{1}{1 - e^{-Ts}}\mathscr{L}f_T(t) \qquad (4.32)$$

where $f(t)$ is a periodic function with period T and $f_T(t)$ is $f(t)$ for $0 \leqslant t \leqslant T$ but zero elsewhere, that is, $f_T(t)$ is "one cycle" of $f(t)$. Using the delay theorem we can prove (4.32) as follows:

$$f(t)=f_T(t)+f_T(t-T)u(t-T)+f_T(t-2T)\mu(t-2T)+\cdots \qquad (4.33)$$

$$F(s)=F_T(s)\left[1+e^{-Ts}+e^{-2Ts}+\cdots\right]\triangleq F_T(s)G \qquad (4.34)$$

$$Ge^{-Ts}=e^{-Ts}+e^{-2Ts}+e^{-3Ts}+\cdots \qquad (4.35)$$

Subtraction gives $G-Ge^{-Ts}=1$, $G=1/(1-e^{-Ts})$ and thus

$$F(s)=\frac{1}{1-e^{-Ts}}F_T(s) \qquad (4.36)$$

To get a term of form $1-e^{-Ts}$ into the denominator of (4.31) we multiply top and bottom by $(e^{-s\sqrt{\rho/G}\,L}-e^{-3s\sqrt{\rho/G}\,L})$ to get:

$$\theta(y,s)=\frac{T_s}{I_p\sqrt{\rho G}\;s^2}\left[e^{-[(L-y)/c]s}-e^{-[(L+y)/c]s}\right.$$

$$\left.-e^{-[(3L-y)/c]s}+e^{-[(3L+y)/c]s}\right]\frac{1}{1-e^{-(4L/c)s}} \qquad (4.37)$$

where $c\triangleq\sqrt{G/\rho}$ must (and does) have the dimensions of a velocity since the exponent of e must be dimensionless. In fact c is the *velocity of propagation* of disturbances (deflection, strain, torque, etc.) in the rod, analogous to the "speed of sound" in fluid mechanics. Equation (4.37) is clearly of form (4.36) with the period $T=4L/c$, thus if we can inverse transform the $F_T(s)$ portion for $0\leqslant t\leqslant 4L/c$, the time function will repeat cyclically after that. Let us find the angular displacement at $y=L$:

$$\theta_T(L,s)=\frac{T_s}{I_p\sqrt{\rho G}\;s^2}\left[1-2e^{-(2L/c)s}+e^{-(4L/c)s}\right] \qquad (4.38)$$

Using the delay theorem to inverse transform:

$$\frac{\theta_T(L,t)}{T_s/I_p\sqrt{\rho G}}=t-2\left(t-\frac{2L}{c}\right)u\left(t-\frac{2L}{c}\right)$$

$$+\left(t-\frac{4L}{c}\right)u\left(t-\frac{4L}{c}\right) \qquad (4.39)$$

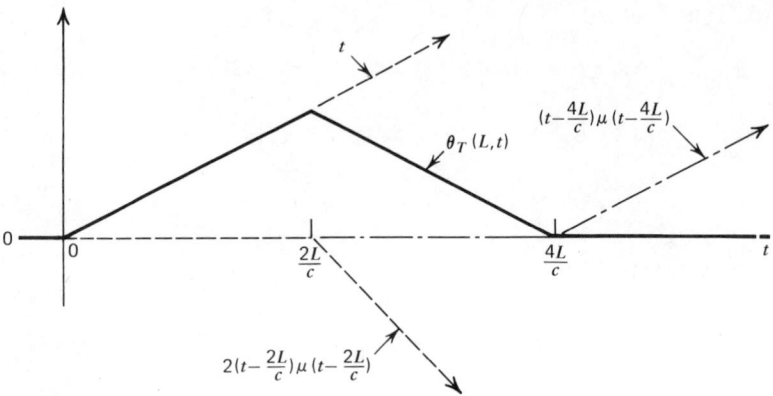

Figure 4.4 One cycle of step response at L.

The waveform shown between $t=0$ and $t=4L/c$ in Fig. 4.4 repeats periodically to give $\theta(L,t)$.

For the motion at $y = L/2$ we get:

$$\frac{\theta_T(L/2,t)}{T_s/I_p\sqrt{\rho G}} = \left(t - \frac{L}{2c}\right)u\left(t - \frac{L}{2c}\right) - \left(t - \frac{3L}{2c}\right)u\left(t - \frac{3L}{2c}\right)$$

$$-\left(t - \frac{5L}{2c}\right)u\left(t - \frac{5L}{2c}\right) + \left(t - \frac{7L}{2c}\right)u\left(t - \frac{7L}{2c}\right) \qquad (4.40)$$

giving for $\theta(L/2,t)$ the periodic waveform of Fig. 4.5. The significance of c as propagation velocity should be clear from the behavior between $t=0$ and $t=L/2c$. Even though the torque is applied at $t=0$, *nothing* happens at $L/2$ until a time interval equal to $(L/2)/c$ passes; clearly this interval is the time required for a disturbance moving with speed c to traverse a distance $L/2$. The changes that occur at $t=3L/2c$, $5L/2c$, and so on may similarly be associated with the arrival (at the station $y = L/2$) of *reflected* waves.

The main defect in the model of Eq. (4.4) is the complete lack of energy dissipation (frictional effects). In a real rod the peaks in Fig. 4.2 would be of finite height and the motions of Figs. 4.4 and 4.5 would not be periodic but would gradually damp out to *steady* deflections corresponding to the steady applied torque. These modeling errors do not completely negate the practical utility of the model, however, since:

1 In Fig. 4.2 the numerical values of the natural frequencies *are* accurately predicted for the light damping usually present.

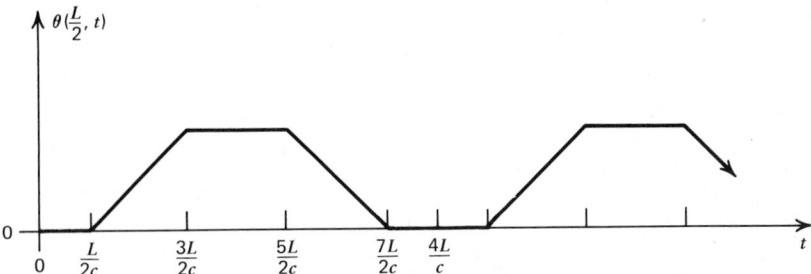

Figure 4.5 Step response at $L/2$.

2 In Fig. 4.4 the peak $(t=2L/c)$ value of θ (and associated stresses), which turn out to be *twice* those produced by static application of the torque T_s, are again accurately predicted.

3 The general nature of the response (an infinite number of natural frequencies, the wave propagation effects, etc.) is revealed.

4.2 DIFFERENCE EQUATIONS AND Z TRANSFORMS

A short section on these extensive subjects can only introduce basic concepts and make the reader aware of the existence of working tools for certain classes of problems. Our main motivation is to develop some understanding of the characteristics of sampled-data systems that employ digital computers as components in measurement and control applications. Figure 4.6 shows a typical feedback control system of this type. The hardware external to the computer and its interface devices is usually of an analog or continuous nature and, if linear, would be described by Laplace transfer functions. Since the digital computer accepts as input, and produces as output, sequences of discrete numbers, we periodically (every T seconds) *sample* the continuous signal $e(t)$ and then hold this value until the next sampling instant. During the interval T, the computer performs its calculation and then, at the sampling instant, outputs this number through the D/A converter, where it is held until the next sampling instant. Information in three forms is thus present in such a system: "smooth" signals, number sequences, and sampled-and-held signals.

Let us first consider the digital computer itself. Its operation is often characterized by a linear, time-invariant *difference equation*:

$$m(kT)=b_0e(kT)+b_1e[(k-1)T]+\cdots+b_ne[(k-n)T]$$

$$-a_1m[(k-1)T]-a_2m[(k-2)T]-\cdots-a_nm[(k-n)T] \quad (4.41)$$

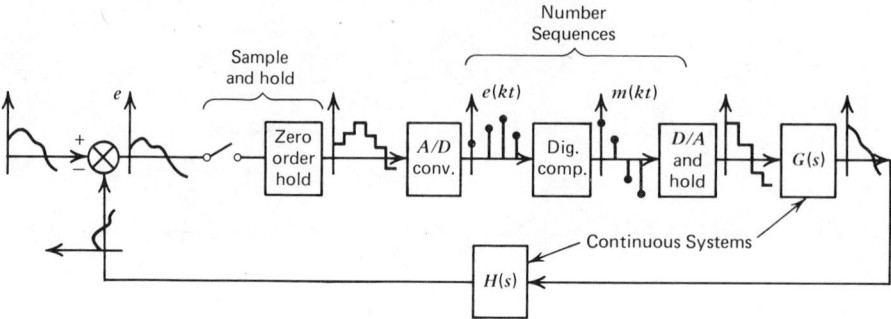

Figure 4.6 Sampled-data system.

Here m and e are, respectively, the output and input number sequences for the computer; $m(kT), e(kT)$ denoting the values at the kth sampling interval, that is, at time kT, where T is the sampling interval. The symbols $e[(k-1)T]$, $m[(k-1)T]$, and so on represent previous values of e and m that might be stored in the computer. The coefficients $a_n,\ldots,a_1$ and $b_n,\ldots,b_0$ are fixed numbers whose presence and numerical value determine the input/output relation for the computer.

As an example of (4.41), suppose we wish to use a digital computer to implement the popular proportional plus integral plus derivative (PID) mode of control.[1] For a continuous (analog) controller we would have:

$$\frac{M}{E}(s) = K_D s + K_P + \frac{K_I}{s} = \frac{1}{s}\left(K_D s^2 + K_P s + K_I\right) \tag{4.42}$$

$$\frac{dm}{dt} = K_D \frac{d^2e}{dt^2} + K_P \frac{de}{dt} + K_I e \tag{4.43}$$

To convert this to difference-equation form we need to approximate the derivatives by finite differences. Of the various schemes possible, the simplest uses

$$\frac{de}{dt}(kT) \approx \frac{e(kT) - e[(k-1)T]}{T}$$

which can be applied successively to approximate the higher derivatives:

$$\frac{d^2e}{dt^2}(kT) \approx \frac{\dfrac{de}{dt}(kT) - \dfrac{de}{dt}[(k-1)T]}{T}$$

$$\approx \frac{e(kT) - 2e[(k-1)T] + e[(k-2)T]}{T^2} \tag{4.44}$$

[1]Doebelin, E. O., *Dynamic Analysis and Feedback Control*, McGraw-Hill, N.Y., 1962, p. 262.

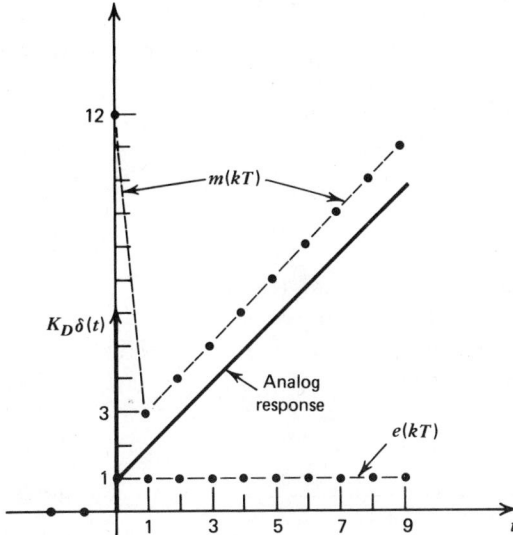

Figure 4.7 Comparison of analog and digital versions of PID controller.

Applying this to Eq. (4.43) we get:

$$m(kT) = m[(k-1)T] + \left(\frac{K_D}{T} + K_P + K_I T\right)e(kT)$$

$$- \left(\frac{2K_D}{T} + K_P\right)e[(k-1)T] + \left(\frac{K_D}{T}\right)e[(k-2)T] \qquad (4.45)$$

To see how equations of form (4.41) are used to construct the sequence $m(kT)$ when input $e(kT)$ is given, let us take in Eq. (4.45) the numerical values $K_I = K_P = T = 1$, $K_D = 10$ and let $e(kT)$ be a "unit step input." That is, let $e(kT)$ be zero for $k < 0$, and 1 for $k \geqslant 0$; also let $m(kT)$ be zero for $k < 0$ as an initial condition:

$$m(kT) = m[(k-1)T] + 12e(kT) - 21e[(k-1)T] + 10e[(k-2)T] \quad (4.46)$$

Substitution of $k = 0, 1, 2, \ldots$ into (4.46) generates the $m(kT)$ sequence shown in Fig. 4.7. Note that while the digital response does not (and cannot) duplicate the analog, the digital response does clearly exhibit the correct *characteristics* of a PID response to step input:

1   A large initial "derivative" response that dies away.
2   A sustained "proportional" response.
3   A growing "integral" response.

We next introduce the concept of the z transform as applied to number sequences. This transform operates on a number sequence $f(kT)$ and produces a function of the complex variable $z = a + ib$. The one-sided (number sequence defined for positive k only) z transform is defined[2] by:

$$\mathcal{Z}[f(kT)] \triangleq F(z) \triangleq \sum_{k=0}^{\infty} f(kT)z^{-k} \tag{4.47}$$

As defined by (4.47), $F(z)$ is a power series in z^{-1} and may or may not converge, depending on the sequence $f(kT)$ and the region of the complex plane where z falls. For most applications the convergence question is not a problem and we will pursue it no further. When $F(z)$ is convergent it can be put into a closed form

$$F(z) = \frac{b_0 z^m + b_1 z^{m-1} + \cdots + b_m}{z^n + a_1 z^{n-1} + \cdots + a_n} \tag{4.48}$$

By factoring the polynomials in z we can put (4.48) in the form

$$F(z) = b_0 \frac{(z-z_1)(z-z_2)\cdots(z-z_m)}{(z-p_1)(z-p_2)\cdots(z-p_n)} \tag{4.49}$$

where the numbers z_i are called the zeros of $F(z)$ and the p_i the poles. When z takes on the value of a zero, $F(z) = 0$ and when z takes on the value of a pole, $F(z) = \infty$. Note the similarity in form to Laplace transforms which also appear as ratios of polynomials (in s) and also have poles and zeros.

Let us apply definition (4.47) to determine the *digital transfer function* of the digital computer characterized by Eq. (4.41). In z transforming both sides of (4.41), a linearity theorem allows us to transform each term separately and also take multiplying constants (b_0, b_1, etc.) outside the transform. The terms $m(kT)$ and $b_0 e(kT)$ transform directly to give $M(z)$ and $b_0 E(z)$. To deal with the term $b_1 e[(k-1)T]$, note that if the sequence $e(kT)$ is, say, $3, 4, 2, -1, 6, \ldots$ for $k = 0, 1, 2, 3, 4, \ldots$ (and zero for $k < 0$), then $e[(k-1)T]$ is $0, 3, 4, 2, -1, 6, \ldots$. This gives:

$$\mathcal{Z}[e(kT)] = 3 + 4z^{-1} + 2z^{-2} - z^{-3} + 6z^{-4} + \cdots$$

$$\mathcal{Z}[e[(k-1)T]] = 3z^{-1} + 4z^{-2} + 2z^{-3} - z^{-4} + 6z^{-5} + \cdots$$

[2]Cadzow, J. A. and Martens, H. R., *Discrete-Time and Computer Control Systems*, Prentice-Hall, Englewood Cliffs, N.J.

and shows that $\mathscr{Z}[e[(k-1)]] = z^{-1}\mathscr{Z}[e(kT)]$. We can now write

$$M(z) = b_0 E(z) + b_1 z^{-1} E(z) + \cdots + b_n z^{-n} E(z)$$
$$- a_1 z^{-1} M(z) - a_2 z^{-2} M(z) - \cdots - a_n z^{-n} M(z) \qquad (4.50)$$

which gives

$$M(z) = \frac{b_0 + b_1 z^{-1} + \cdots + b_n z^{-n}}{1 + a_1 z^{-1} + a_2 z^{-2} + \cdots + a_n z^{-n}} E(z) \qquad (4.51)$$

This is in the form of a transfer function relationship $M(z) = D(z)E(z)$, where

$$D(z) \triangleq \frac{b_0 + b_1 z^{-1} + \cdots + b_n z^{-n}}{1 + a_1 z^{-1} + a_2 z^{-2} + \cdots + a_n z^{-n}} \qquad (4.52)$$

is called the *digital transfer function*. For a computer programmed for the PID algorithm as in Eq. (4.46) we would have

$$D(z) = \frac{12 - 21z^{-1} + 10z^{-2}}{1 - z^{-1}} \qquad (4.53)$$

We now need to show how the z transform is applied to the analysis of the remaining portions of systems such as that of Fig. 4.6. Consider the sample-and-hold portion of the system shown in Fig. 4.8. We assume the sampling

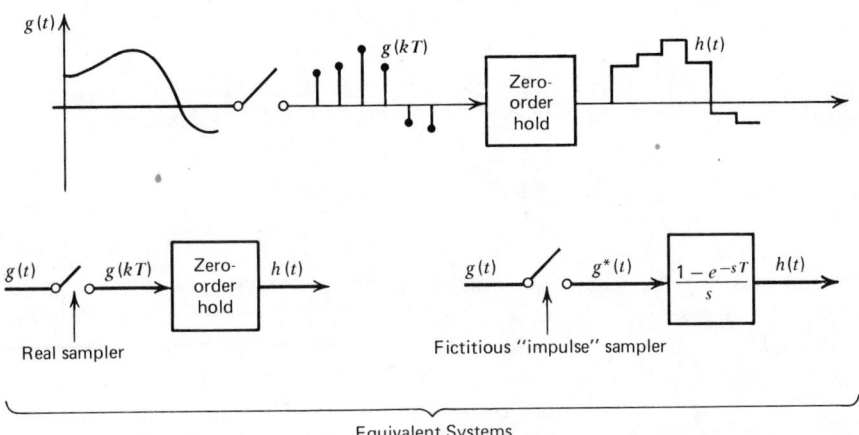

Figure 4.8 Definition of impulse sampler.

switch picks off an instantaneous value of $g(t)$ every T seconds and the hold device than holds this value constant until the next sampling instant, at which time it is updated. (Other types of hold schemes are possible but this *zero-order hold* is the simplest and most common.) This action is described mathematically (using delayed step functions) by:

$$h(t) = \sum_{k=0}^{\infty} g(kT) \left[u(t-kT) - u(t-kT-T) \right] \qquad (4.54)$$

$$\mathcal{L}[h(t)] = H(s) = \sum_{k=0}^{\infty} g(kT) \left[\frac{e^{-skT} - e^{-s(k+1)T}}{s} \right] \qquad (4.55)$$

$$H(s) = \frac{(1-e^{-sT})}{s} \sum_{k=0}^{\infty} g(kT) e^{-skT} \qquad (4.56)$$

This last result takes the form of a transfer function relation, output function = (system function) (input function):

$$H(s) = G_m(s) G^*(s)$$

if we define

$$G_m(s) \triangleq \frac{1-e^{-sT}}{s} \qquad (4.57)$$

$$G^*(s) \triangleq \sum_{k=0}^{\infty} g(kT) e^{-skT} \qquad (4.58)$$

If we have $G^*(s)$ we should be able to find its $g^*(t)$; actually,

$$g^*(t) = \sum_{k=0}^{\infty} g(t) \delta(t-kT) \qquad (4.59)$$

which can be verified by Laplace transforming to get (4.58). The function $g^*(t)$ is a train of impulses of size (area) equal to $g(kT)$, occurring at the sampling instants $t = kT$. Even though these impulses do not physically occur anywhere in the system, we may think of, and analyze, the system in this way since the overall effect is identical with that of the real system. This allows us to treat the continuous ("analog") parts of the system by studying their response to impulse trains. Let a continuous system with transfer function $W(s)$ be subjected to an impulse-train input

$$r^*(t) = \sum_{k=0}^{\infty} r(t)\delta(t - kT)$$

If $w(t)$ is the system's impulse response, superposition gives for the system output $c(t)$:

$$c(t) = \sum_{k=0}^{\infty} r(kT)w(t - kT) \tag{4.60}$$

To use z-transform methods we must deal with number sequences, not continuous functions; so think of $c(t)$ as being sampled at intervals T and let us inquire as to the value of $c(t)$ at an arbitrary instant nT:

$$c(nT) = \sum_{k=0}^{\infty} r(kT)w(nT - kT) \tag{4.61}$$

We can now take the z transform of both sides of (4.61), which can be shown[3] to give

$$C(z) = R(z)W(z) \tag{4.62}$$

where $R(z)$ is the z transform of the sequence $r(kT)$ and $W(z)$ is the z transform of $w(kT)$.

To obtain $W(z)$ when $W(s)$ is given (we do not just substitute z for s!) we evaluate $w(t)$ at the sampling instants and then use (4.47) to get $W(z)$. Let us take $W(s) = a/s(s + a)$ as an example; $w(t) = (1 - e^{-at})u(t)$. This gives $h(kT) = 1 - e^{-akT}$ for $k \geq 0$ and 0 for $k < 0$. This sequence has the z transform:

$$W(z) = \sum_{k=0}^{\infty} (1 - e^{-akT})z^{-k}$$

$$= \sum_{k=0}^{\infty} z^{-k} - \sum_{k=0}^{\infty} z^{-k}e^{-akT} \tag{4.63}$$

By long division:

$$\frac{z}{z-1} = 1 + z^{-1} + z^{-2} + z^{-3} + \cdots$$

$$\frac{z}{z - e^{-aT}} = 1 + e^{-aT}z^{-1} + e^{-2aT}z^{-2} + e^{-3aT}z^{-3} + \cdots$$

---

[3]Cadzow and Martens, *op. cit.*, p. 210.

so we get

$$W(z) = \frac{z}{z-1} - \frac{z}{z - e^{-aT}} = \frac{z^{-1}(1 - e^{-aT})}{(1 - z^{-1})(1 - e^{-aT}z^{-1})} \tag{4.64}$$

The use of z^{-1} in the right-hand side of 4.64 is preferred usage since it eases the transition to iterative equations when this is necessary. Since tables such as Table 4.1 are available, we can often go from $W(s)$ to $W(z)$ with little effort.

We are now in a position to discuss the analysis of complete systems such as that of Fig. 4.6, which we redraw as in Fig. 4.9. While the hardware of a real system includes two hold devices as shown in Fig. 4.6, these are synchronized and only *one* "delay" of T seconds is involved since the computer receives a new sampled input value at the beginning of T and outputs a new value through the

Table 4.1   Short Table of z Transforms.

$F(s)$	$\mathscr{Z}[F(s)]$
$\left(\dfrac{1 - e^{-sT}}{s}\right)F(s)$	$(1 - z^{-1})\mathscr{Z}\left(\dfrac{F(s)}{s}\right)$
$\dfrac{1}{s}$	$\dfrac{1}{1 - z^{-1}}$
$\dfrac{1}{s^2}$	$\dfrac{Tz^{-1}}{(1 - z^{-1})^2}$
$\dfrac{1}{s^3}$	$\dfrac{T^2(1 + z^{-1})z^{-1}}{2(1 - z^{-1})^3}$
$\dfrac{a}{s(s+a)}$	$\dfrac{(1 - e^{-aT})z^{-1}}{(1 - z^{-1})(1 - e^{-aT}z^{-1})}$
$\dfrac{a}{s^2(s+a)}$	$\dfrac{Tz^{-1}}{(1 - z^{-1})^2} - \dfrac{(1 - e^{-aT})z^{-1}}{a(1 - z^{-1})(1 - e^{-aT}z^{-1})}$
$\dfrac{ab}{s(s+a)(s+b)}$	$\dfrac{1}{1 - z^{-1}} + \dfrac{(be^{-bT} - ae^{-aT})z^{-1} - (b - a)}{(b - a)(1 - e^{-aT}z^{-1})(1 - e^{-bT}z^{-1})}$
$\dfrac{a^2 b^2}{s^2(s+a)(s+b)}$	$\dfrac{abTz}{(z-1)^2} - \dfrac{(a+b)z}{z-1} - \dfrac{b^2 z}{(a-b)(z - e^{-aT})} + \dfrac{a^2 z}{(a-b)(z - e^{-bT})}$
$\dfrac{a^2 + b^2}{s\left[(s+a)^2 + b^2\right]}$	$\dfrac{z}{z-1} - \dfrac{z^2 - ze^{-aT}\sec\phi\cos(bT + \phi)}{z^2 - 2ze^{-aT}\cos bT + e^{-2aT}}$
	$\phi \triangleq \tan^{-1}(-a/b)$

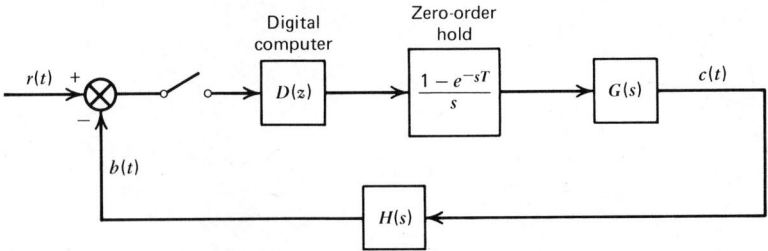

Figure 4.9 Computer control system.

D/A converter and output hold at the end of the *same* interval. Thus in Fig. 4.9 only one hold transfer function $(1-e^{-sT})/s$ is appropriate. It can be shown[4] that the closed-loop transfer function is given by

$$\frac{C(z)}{R(z)} = \frac{D(z)\mathscr{Z}\left[\dfrac{1-e^{-sT}}{s}G(s)\right]}{1+D(z)\mathscr{Z}\left[\dfrac{1-e^{-sT}}{s}G(s)H(s)\right]} \tag{4.65}$$

Note that the evaluation of the $\mathscr{Z}$ transforms of the two bracketed quantities is facilitated by the first entry of Table 4.1. A numerical example will be of help at this point. Suppose we wish to control a system $G(s)=1/s(s+1)$ using a measuring device $H(s)=1$ and a PID controller with $K_D=1.8$, $K_P=2.6$, and $K_I=2$; however, the PID controller is to be realized with a digital computer, using $T=0.1$ sec. From (4.45) and (4.52) we find

$$D(z)=\frac{20.8-38.6z^{-1}+18.0z^{-2}}{1-z^{-1}} \tag{4.66}$$

which gives, with the aid of Table 4.1:

$$\frac{C(z)}{R(z)} = \frac{\dfrac{20.8z^2-38.6z+18}{z^2-z}\left[\dfrac{0.005(z+0.9)}{(z-1)(z-0.905)}\right]}{1+\dfrac{20.8z^2-38.6z+18}{z^2-z}\left[\dfrac{0.005(z+0.9)}{(z-1)(z-0.905)}\right]} \tag{4.67}$$

$$\frac{C(z)}{R(z)} = \frac{0.104z^3-0.0995z^2-0.084z+0.081}{z^4-2.801z^3+2.71z^2-0.989z+0.081} \tag{4.68}$$

[4]Cadzow and Martens, *op. cit.*, p. 223.

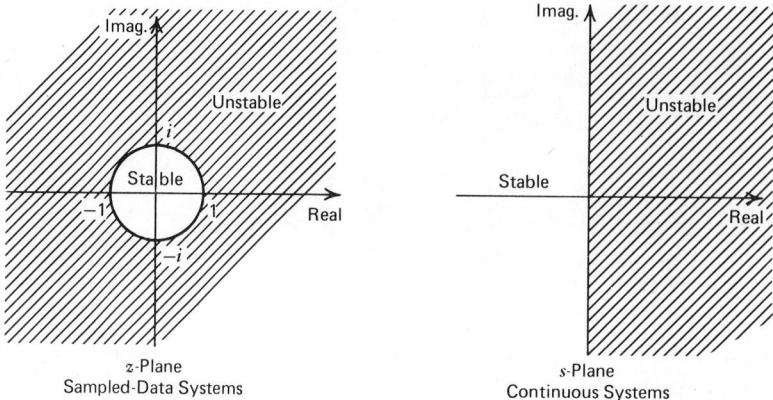

Figure 4.10 Comparison of continuous and sampled-data stability criteria.

Equation (4.68) has a number of practical uses, two of which we will explore. The first consideration in all control systems is absolute stability; until this is achieved there is no point in going further. It can be shown[5] that the characteristic equation

$$z^4 - 2.801z^3 + 2.71z^2 - 0.989z + 0.081 = 0 \tag{4.69}$$

formed from the denominator of (4.68) must have all its roots inside the unit circle $|z| = 1.0$ if the closed-loop system is to be stable. This is analogous to requiring, for continuous systems, that all the roots of the characteristic equation in s be in the left half-plane (see Fig. 4.10). While methods analogous to Routh criterion for detecting the presence of unstable roots without actually finding their numerical values are available for sampled-data systems, they are tedious and the ready availability of computer root-finding routines makes this direct approach preferable. SPEAKEASY discussed in Chapter 3 has such a root finder; only three statements are needed to find the roots of 4.69:

DOMAIN COMPLEX
COFS = .081, − .989, 2.71, − 2.801, 1
POLYROOT(COFS)

whereupon the computer returns the list of roots $0.11293, 0.82354, 0.93227 \pm i0.04243$. These are all inside the unit circle so the system is absolutely stable. Its *relative* stability (whether any large, though decaying, oscillations are present) is now of interest.

---

[5]Cadzow and Martens, *op. cit.*, p. 223.

If the roots are all close to the origin of the unit circle in Fig. 4.10, then the system response will generally not exhibit excessive oscillation; however, if some roots are close to the circle itself, the system *may or may not* be oscillatory, thus the root positions themselves are not always a reliable guide to relative stability. This is particularly true if the sampling rate is high (*T* small) since then the roots tend to be close to the unit circle while relative stability may actually be excellent. Intuitively, when *T* becomes small compared to system time constants, the sampled-data behavior should approach that of the continuous system obtained by ignoring the sample-and-hold operation and replacing the computer controller by its continuous equivalent. This does happen but the z-transform stability analysis is misleading under these circumstances since the roots will be close to the instability boundary.

To get a more correct indication of system behavior, the actual time response to a specific input *r* (usually a step input) should be calculated. Fortunately z-transform methods allow this, although the basic method will find *c(t) only* at the sampling instants, behavior *between* these instants remaining unknown. While an *inverse z-transform* method using partial-fraction expansions similar to those of Laplace transform is available, we here show only the simpler "long division" approach. For a step input $r(t) = u(t), r(kT)$ would be the sequence $1, 1, 1, \ldots$ and

$$R(z) = \sum_{k=0}^{\infty} 1 + z^{-1} + z^{-2} + z^{-3} + \cdots$$

$$= \frac{z}{z-1} \tag{4.70}$$

From (4.68):

$$C(z) = \frac{0.104z^3 - 0.0995z^2 - 0.084z + 0.081}{z^4 - 2.801z^3 + 2.71z^2 - 0.989z + 0.081}\left(\frac{z}{z-1}\right) \tag{4.71}$$

$$C(z) = \frac{0.104z^4 - 0.0995z^3 - 0.084z^2 + 0.081z}{z^5 - 3.80z^4 + 5.51z^3 - 3.70z^2 + 1.07z - 0.081} \tag{4.72}$$

An "ordinary" division of denominator into numerator produces:

$$C(z) = 0.104z^{-1} + 0.295z^{-2} + 0.463z^{-3} + \cdots \tag{4.73}$$

which gives the values of $c(kT)$ for $k = 0, 1, \ldots$ from Eq. (4.47) as

$$c(0, 0.1, 0.2, 0.3, \ldots) = 0, 0.104, 0.295, 0.463, \ldots \tag{4.74}$$

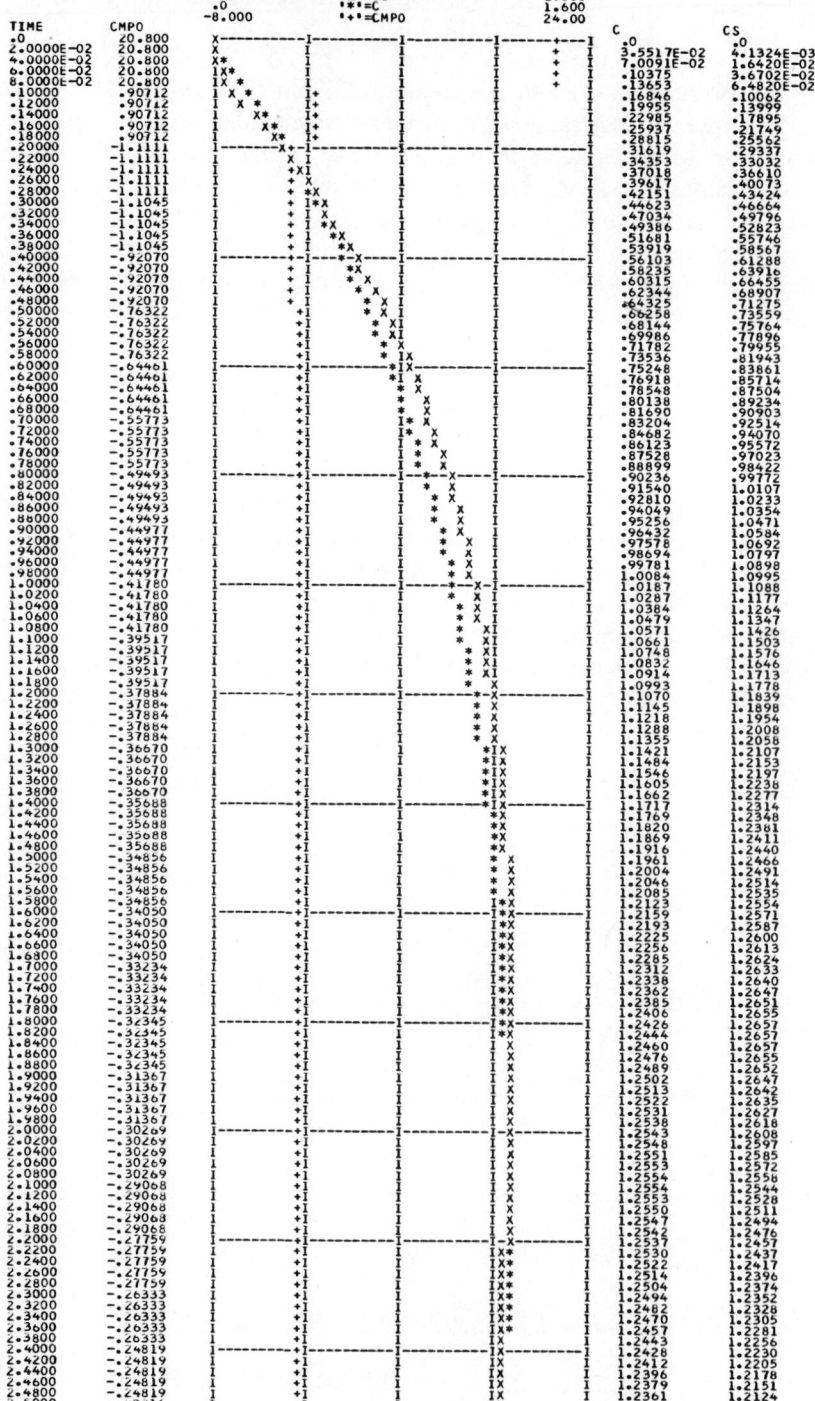

Figure 4.11 Digital simulation results for sampled-data control system.

While the division can in principle be carried on indefinitely to produce $c(kT)$ values as far into time as we wish, this becomes both tedious and inaccurate. Partial-fraction expansion methods avoid these problems by giving an expression for the *general* term $c(kT)$, and the modified z transform even will get values of c *between* sampling instants. However, these analytical methods will nowadays often be passed over in favor of a *digital simulation* of the system. We will give the details of such a simulation in Chapter 5, showing only the results here. In Fig. 4.11, *CS* is the symbol for $c(t)$ in the sampled-data system we have been using as a numerical example, *C* is the same signal in the purely continuous version of this system, and *CMPO* is the output of the PID digital algorithm of Eq. (4.66). Note that the response of the sampled-data ($T = 0.1$) system and its continuous counterpart are almost identical (because the sampling rate is fast relative to the system time constant of 1 sec) and that both are well damped with overshoots of about 25%. Values of *CS* at $t = 0, 0.1, 0.2$, and so on check those of Eq. (4.74) (the simulation values are more correct since the computer carries more digits than we did "manually" in (4.74)).

In most sampled-data feedback systems an increase in sampling interval T has a destabilizing effect. By changing T to 1 sec in our present example, the new characteristic equation (corresponding to Eq. (4.69)) now has the roots $0.17071, 1.0565, -0.43359 \pm i1.12951$ indicating absolute instability. A rerun of the simulation with $T = 1$ bears this out. While our treatment of sampled data systems has been very brief and has avoided proofs of certain results, the tools given, especially when combined with an easy-to-use simulation language, provide a basic level of analysis capability that is of practical use.

4.3   SYSTEM DESCRIPTION USING STATE VARIABLES

Starting around 1960 and continuing today, many papers and books have appeared promoting a so-called "modern control theory" based on state-variable methods. While scrutiny of actual industrial practice reveals that most practical problems continue to be adequately treated using largely classical methods, the state-variable techniques are being applied to some problems with good results, particularly in complex multiple-input/multiple-output systems such as jet engines. While originally applied mainly to feedback control system problems, state-variable techniques and nomenclature also appear with some frequency in general system dynamics studies, thus it may be appropriate to include a brief treatment in this text. Since a detailed treatment[6,7] is of necessity lengthy, our

---

[6]DeRusso, P. M., Roy, R. J., and Close, C. M., *State Variables for Engineers*, John Wiley and Sons, N.Y., 1965.

[7]Schultz, D. G. and Melsa, J. L., *State Functions and Linear Control Systems*, McGraw-Hill, N.Y., 1967.

main purpose here will be to present basic definitions and nomenclature. This is often adequate for understanding reports and papers dealing with practical applications; comprehension of the more esoteric research oriented publications requires an investment of learning time whose practical payoff may be questionable for most working engineers.

Let us use the mechanical system of Fig. 2.11 as an initial example. The application of simplifying assumptions and the appropriate physical laws are always necessary to obtain basic system equations such as 2.112, whether classical or state-variable methods are applied thereafter. (There may, of course, be more than one correct way to arrive at the identical basic equations. In Fig. 2.11 one might for example choose an energy approach through Lagrange's equations in preference to a summation-of-forces approach through Newton's law.) Having obtained Equations (2.112), in Chapter 2 we proceeded to manipulate these equations so as to produce four output/input relations, the transfer functions relating the various output motions to the input forces. While these transfer functions are unquestionably of great practical utility, a proponent of state-variable methods might argue that their introduction tends, in a sense, to *obscure* the internal details of interaction in the system. To keep these details more visible instead of *combining* the second-order equations (2.112) to obtain the fourth-order transfer functions (2.121)–(2.124), the state-variable method requires that (2.112) be rewritten as a set of *first-order* equations.

Leaving, for the moment, our example, the general form of the state-variable description of a linear system is of the form:

$$\dot{\mathbf{x}}(t) = \mathbf{A}\mathbf{x}(t) + \mathbf{B}\mathbf{u}(t)$$

$$\mathbf{y}(t) = \mathbf{C}\mathbf{x}(t) \tag{4.75}$$

These are *matrix differential equations,* a compact way of writing the set of

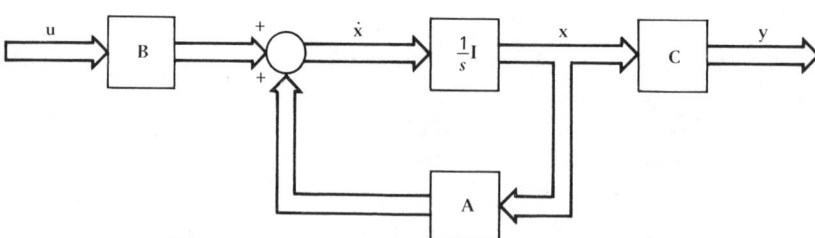

Figure 4.12   State-variable diagram for multiple-input/multiple-output system.

simultaneous equations:

$$
\begin{bmatrix} \dot{x}_1(t) \\ \dot{x}_2(t) \\ \vdots \\ \dot{x}_n(t) \end{bmatrix} = \begin{bmatrix} a_{11} & a_{12} & \cdots & a_{1n} \\ a_{21} & a_{22} & \cdots & a_{2n} \\ \vdots & & & \\ a_{n1} & a_{n2} & \cdots & a_{nn} \end{bmatrix} \begin{bmatrix} x_1(t) \\ x_2(t) \\ \vdots \\ x_n(t) \end{bmatrix}
$$

$$
+ \begin{bmatrix} b_{11} & b_{12} & \cdots & b_{1r} \\ b_{21} & b_{22} & \cdots & b_{2r} \\ \vdots & & & \\ b_{n1} & b_{n2} & \cdots & b_{nr} \end{bmatrix} \begin{bmatrix} u_1(t) \\ u_2(t) \\ \vdots \\ u_r(t) \end{bmatrix} \tag{4.76}
$$

$$
\begin{bmatrix} y_1(t) \\ y_2(t) \\ \vdots \\ y_m(t) \end{bmatrix} = \begin{bmatrix} c_{11} & c_{12} & \cdots & c_{1n} \\ c_{21} & c_{22} & \cdots & c_{2n} \\ \vdots & & & \\ c_{m1} & c_{m2} & \cdots & c_{mn} \end{bmatrix} \begin{bmatrix} x_1(t) \\ x_2(t) \\ \vdots \\ x_n(t) \end{bmatrix} \tag{4.77}
$$

The column vector **x** is an *n*-dimensional *state vector* in the *state variables* $x_1, x_2, \ldots, x_n$, **u** is an *r*-dimensional *control vector*, and **y** is an *m*-dimensional *output vector*. Matrix **A** is called the *system matrix* and is $n \times n$, **B** is the $n \times r$ *control matrix* and **C** is the $m \times n$ *output matrix*. Figure 4.12 shows the conventional block diagram used to represent such a multiple-input/multiple-output system; the broad arrows (as contrasted to the single lines used in classical block diagrams) denote the flow of *multiple* signals. The symbol **I** represents the *identity matrix* while $1/s$ is our usual Laplace operator for integration:

$$
\frac{1}{s} \begin{bmatrix} 1 & 0 & \cdots & 0 & 1 \\ 0 & 1 & \cdots & 0 & 0 \\ \cdots & \cdots & & \cdots & \cdots \\ 0 & 0 & \cdots & 1 & 0 \\ 0 & 0 & \cdots & 0 & 1 \end{bmatrix} = \begin{bmatrix} 1/s & 0 & \cdots & 0 & 1/s \\ 0 & 1/s & \cdots & 0 & 0 \\ \cdots & \cdots & & \cdots & \cdots \\ 0 & 0 & \cdots & 1/s & 0 \\ 0 & 0 & \cdots & 0 & 1/s \end{bmatrix}
$$

$$
\underbrace{\qquad\qquad}_{\mathbf{I}} \qquad \underbrace{\qquad\qquad\qquad}_{(1/s)[\mathbf{I}]}
$$

Thus $[\mathbf{x}] = (1/s)[\mathbf{I}][\dot{\mathbf{x}}]$ merely shows how the *x*'s are obtained by integrating the $\dot{x}$'s.

The "feedback system appearance" of Fig. 4.12 is due solely to the form of Equations (4.76); the actual physical system may or may not be a feedback control system. Note also that the *actual* unknowns are the x's; the y's are merely linear combinations ($y_1 = c_{11}x_1 + c_{12}x_2 + \cdots + c_{1n}x_n$, etc.) of the x's that are easily found once the x's have been solved for. In a given practical problem it may or may not be necessary to define the y's. Let us now interpret our example in state-variable terms. Since the physical equations are second-order while only first-order equations are permitted by the state-variable format, we simply define $x_1 \triangleq x_1$, $x_2 \triangleq x_2$, $x_3 \triangleq v_1 \triangleq dx_1/dt$, and $x_4 \triangleq v_2 \triangleq dx_2/dt$. It is also clear that the physical forces f play the role of the "controls" u, so we let $u \triangleq f_1$ and $u_2 \triangleq f_2$. We can now write:

$$
\begin{aligned}
\dot{x}_1 &= (0)x_1 & + (0)x_2 & & + (1)x_3 & + (0)x_4 \\
\dot{x}_2 &= (0)x_1 & + (0)x_2 & & + (0)x_3 & + (1)x_4 \\
\dot{x}_3 &= \left(\frac{-K_{s1}}{M_1}\right)x_1 + \left(\frac{K_{s1}}{M_2}\right)x_2 & & + \left(\frac{-B_1}{M_1}\right)x_3 + (0)x_4 & + \left(\frac{1}{M_1}\right)u_1 \\
\dot{x}_4 &= \left(\frac{K_{s1}}{M_2}\right)x_1 + \left(\frac{-K_{s1}-K_{s2}}{M_2}\right)x_2 + (0)x_3 & & + \left(\frac{-B_2}{M_2}\right)x_4 + \left(\frac{1}{M_2}\right)u_2
\end{aligned}
$$

$$(4.78)$$

or, in matrix form:

$$
\begin{bmatrix} \dot{x}_1 \\ \dot{x}_2 \\ \dot{x}_3 \\ \dot{x}_4 \end{bmatrix} =
\begin{bmatrix}
0 & 0 & 1 & 0 \\
0 & 0 & 0 & 1 \\
-\dfrac{K_{s1}}{M_1} & \dfrac{K_{s1}}{M_2} & \dfrac{-B_1}{M_1} & 0 \\
\dfrac{K_{s1}}{M_2} & \dfrac{-K_{s1}-K_{s2}}{M_2} & 0 & \dfrac{-B_2}{M_2}
\end{bmatrix}
\begin{bmatrix} x_1 \\ x_2 \\ x_3 \\ x_4 \end{bmatrix} +
\begin{bmatrix}
0 & 0 \\
0 & 0 \\
\dfrac{1}{M_1} & 0 \\
0 & \dfrac{1}{M_2}
\end{bmatrix}
\begin{bmatrix} u_1 \\ u_2 \end{bmatrix}
$$

$$(4.79)$$

If, now, say, our interest in this vibration comes down to a concern about the *stresses* in the springs, we might wish to define two outputs as stresses y_1 and y_2:

$$y_1 = L_1(x_1 - x_2), \qquad y_2 = L_2(x_2)$$

where L_1 and L_2 are known physical constants with dimensions stress/deflection. Equations (4.77) would then be:

$$
\begin{bmatrix} y_1 \\ y_2 \end{bmatrix} =
\begin{bmatrix} L_1 & -L_1 & 0 & 0 \\ 0 & L_2 & 0 & 0 \end{bmatrix}
\begin{bmatrix} x_1 \\ x_2 \\ x_3 \\ x_4 \end{bmatrix}
$$

$$(4.80)$$

We choose to terminate our discussion of this topic at this point with the following comments. A logical next step from a general system dynamics viewpoint would be to develop the analytical matrix methods needed to manipulate, solve and interpret equations of form (4.76). This material constitutes a classical mathematical topic not dependent on "modern control theory" or "state-variable" concepts per se and we will return to it at appropriate points later in the text. The development of the control system applications of state-variable methods (optimal control, state-variable feedback, etc.) is not felt to be appropriate to this more general text and we will not pursue it. Actually, the practical implementation of equation solution techniques by state-variable methods usually requires computer assistance. When, in the next chapter, we study digital simulation methods, we shall see that these "naturally" work with equations of form (4.76). In fact, analog computer methods (on which the digital simulations are based) also integrate the equations in the first-order format. Both analog and digital simulation methods were originally developed without consideration of state-variable methods and remain effective analysis tools that are easily comprehended and used by practicing engineers. They also inherently obtain the time response of *all* the system variables (not just the "input" and "output" of some transfer function) and thus share, with the state-variable methods, the advantage of not obscuring internal system details.

BIBLIOGRAPHY

1 Cadzow, J. A. and H. R. Martens, *Discrete-Time and Computer Control Systems*, Prentice-Hall, Englewood Cliffs, N.J., 1970.

2 Churchhill, R. V., *Operational Mathematics*, McGraw-Hill, N.Y., 1958.

3 Oldenburger, R., Theory of Distributed Systems, ASME Paper 69-FE-15, 1969.

4 Raven, F. H., *Mathematics of Engineering Systems*, McGraw-Hill, N.Y., 1966.

5 Takahashi, Y., Rabins, M. J. and D. M. Auslander, *Control and Dynamic Systems*, Addison-Wesley, Reading, Mass., 1970, Part 2.

6 Timothy, L. K. and B. E. Bona, *State Space Analysis: An Introduction*, McGraw-Hill, N.Y., 1968.

PROBLEMS

4.1 Using the methods of Section 4.1, find $U(L, i\omega)/F(L, i\omega)$ for the longitudinal rod vibration in Fig. P4.1.

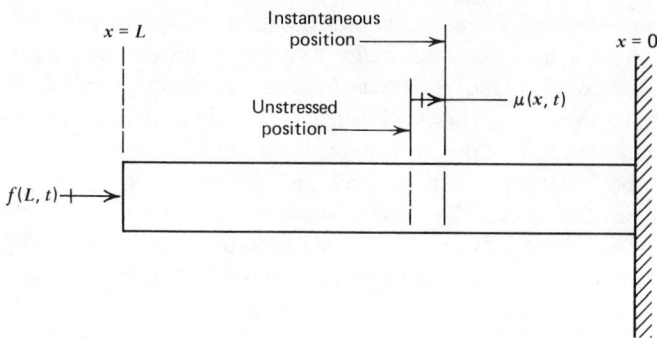

Figure P4.1

4.2   Check the correctness of Eq. (4.28) for $\omega=0$.

4.3   Use Equations (4.1) and (4.27) to find $T(y,s)/T(L,s)$.

4.4   In Fig. 4.1 let the rod be free at $y=0$ and find $\theta(y,s)/T(L,s)$. Also find the natural frequencies.

4.5   Repeat Problem 4.1 for a rod free at $x=0$.

4.6   Modify the curves of Fig. 4.7 so that they correspond to a system with no derivative control.

4.7   Compute $C(z)/R(z)$; check stability using computer root finder; and calculate (for a unit step input $r(t)$) $c(kT)$ for $k=0$, 1, and 2 for the feedback systems with digital PID controllers:

(a) $G(s)=1/(s(s+1))$, $H(s)=1$, $K_D=1.8$, $K_P=2.6$, $K_I=6$, $T=0.1$.

(b) $G(s)=1/s$, $H(s)=1$, $K_D=1.8$, $K_P=2.6$, $K_I=2$, $T=0.1$.

(c) $G(s)=1/(s(s+1))$, $H(s)=1$, $K_D=0$, $K_P=0$, $K_I=2$, $T=0.1$.

CHAPTER 5

Digital Simulation Methods

5.1 LIMITATIONS OF ANALYTICAL METHODS

The analytical methods of earlier chapters provide a sound framework for understanding the general character of the dynamic response of the many practical engineering systems that behave in roughly linear time-invariant fashion. For the simpler systems they also allow relatively quick and easy calculation of numerical results for specific applications. As system complexity increases, however, response calculations become increasingly time consuming and subject to human error. Furthermore, when nonlinear and/or time-invariant features of behavior are included in system models, effective analytical techniques are to a large extent nonexistent. In such circumstances the versatile capabilities of a comprehensive digital simulation language have become a powerful tool for the engineer.

5.2 ANALOG, DIGITAL, AND HYBRID SIMULATION

Computer simulation methods have, since their inception, found application in two broad areas: mathematical problem solving and coupled machine/computer simulators. In the first of these the computer is used in a "stand alone" fashion as a general purpose differential equation solver and, while speed of operation is desirable, there is no necessity that the solution be generated at any particular time rate. Machine/computer simulators, on the other hand, interface the computer with an operating machine or process to create an assemblage that "mimics" the behavior of some real system, the computer portion allowing versatile adjustment of overall system behavior for test or training purposes. Possibly the best known of these applications are the flight simulators used in the aerospace industry for pilot training and system development studies (see Fig. 5.1). Here the computer simulates the aircraft's flight by continuously solving its equations of motion in real time, thereby providing voltages that vary with time in the same way as do the vehicle's motion variables such as pitch angle, vertical velocity, and so on. The computer can simulate *any* aircraft (including ones not yet built) simply by changing its programming. Input commands to the simulated aircraft come from transducers attached to the

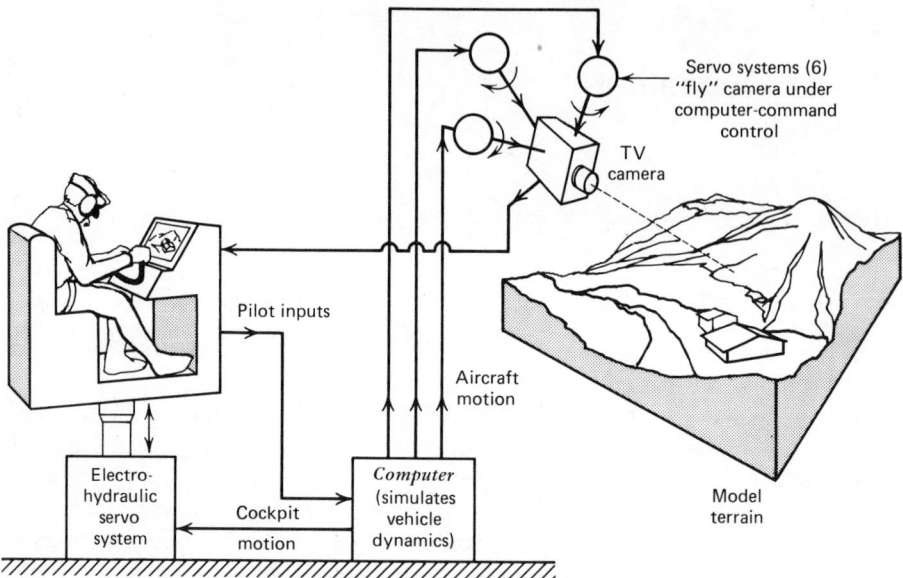

Figure 5.1   Machine/computer simulator.

pilot's control wheel, rudder pedals, and so forth. From these inputs the computer calculates the resulting aircraft motions, voltages proportional to these being sent as commands to hydraulic systems that move the cockpit with a replica of aircraft motion, and to electromechanical servos that cause a television camera to move with six degrees of freedom over a model terrain as would the real airplane. The pilots thus see a realistic moving image on their TV screen and feel realistic motions as they "fly" the simulator. It should be clear that in such applications, whatever type of computer is used, it must be fast enough to "keep up" with the real-time behavior of the simulated system.

Historically, electronic *analog computers* (developed in the 1940s) first dominated both the above areas of computer simulation and were vital tools in the analysis and design of a wide variety of engineering systems. Electronic digital computers (also introduced in the 1940s) have always had the basic capability for solving differential equations (and thus serving as simulators), but two factors delayed their widespread application in this area until more recent times. The first of these was the relatively poor speed of digital simulation (compared with analog); this ruled out digital methods for most real-time computer/machine simulator applications. Second, the programming of a digital computer (even in a user-oriented language such as FORTRAN) for simulation purposes was very time consuming and required programming and numerical analysis expertise beyond the capabilities of most engineers, while analog computer methods were much easier to learn and use.

In the 1960s both the above situations began to change in favor of digital methods. Although analog and hybrid (combined analog/digital) machines continue to be used, the small number of manufacturers of general purpose analog or hybrid equipment surviving today can be taken as an indication of the ascendency of digital methods. (This is clearly a trend in all areas of electronics, not just computers.) For real-time simulator applications this has come about because of the increased speed of digital methods; for general purpose problem solving the appearance of user oriented digital simulation languages that are *easier* to use than analog methods has made the difference. Since these simulation languages sacrifice speed for ease of use, they usually are not appropriate for any but the simplest real-time applications. Sophisticated real-time applications require careful theoretical treatment and assembler language programming[1]; however, this is a specialty field of interest to a relatively small number of engineers, so we will not pursue it here but rather concentrate on the simulation languages, a topic useful to a very large number of engineers. Also, our coverage will not include the *design* of such languages (again a narrow specialty field), but rather concentrate on the *use* of available languages.

5.3   FEATURES OF DIGITAL SIMULATION LANGUAGES

We first restrict ourselves to *continuous system simulation languages* as contrasted with *discrete system simulation languages*. The latter concentrate on the study of discrete serial processes and are of more interest to operations researchers than to engineers concerned with physical hardware design. Our interest also centers on general purpose languages rather than those designed for specific problem areas such as electrical circuit analysis. Within these constraints the first and overriding feature such languages share is their close conceptual relation to the analog computer methods that historically preceded them. Since basic analog techniques[2] are nowadays presumed an established part of undergraduate curricula, we will assume familiarity and utilize a simple example to relate digital simulation to them. Consider the equation

$$(1 - At)M\ddot{x} + B(\dot{x})^3 + F_s(x) = G(t) \tag{5.1}$$

[1]Rosko, J. S., *Digital Simulation of Physical Systems*, Addison-Wesley, Reading, Mass., 1972.
[2]Doebelin, E. O., *System Dynamics*, C. E. Merrill, Columbus, Ohio, 1972, p. 251.

which has a nonconstant coefficient (variable mass $(1 - At)M$), nonlinear damping $B(\dot{x})^3$, and a nonlinear spring force given by an experimentally measured force/deflection relation $F_s(x)$. It is unlikely that equations of this form will *ever* be analytically solvable, yet analog methods deal with them easily in principle (but may become tedious to set up and scale, and inaccurate in execution), while digital simulation, implementing the analog concept in digital form, solves the equation quickly, easily, and accurately. (We should remind the reader that *both* analog and digital simulation require numerical (not literal) values for all parameters and initial conditions; that is, we must have, say $(1 - 5.2t)8\ddot{x}$, not $(1 - At)M\ddot{x}$.)

As in analog simulation, we first arrange the equation to make the highest derivative of the unknown "stand alone":

$$\ddot{x} = \big(G(t) - B(\dot{x})^3 - F_s(x)\big)/(1 - At)M \tag{5.2}$$

and then use numerical integration (choosing a suitable algorithm from the several available) rather than analog electronic (op-amp) integration to compute $\dot{x}$ and then x. Having $\dot{x}$ and x we can produce $B(\dot{x})^3$ with multipliers and $F_s(x)$ with an arbitrary function generator. The driving force $G(t)$ is a known function of time, as is $1 - At$, so these are directly expressible in some language such as FORTRAN. The division by $(1 - At)M$ and the introduction of initial conditions at each integrator completes the simulation, as shown in Fig. 5.2. We see that the pattern is identical to that of analog computation. Note that when summing and integration are done digitally (rather than with op-amps) there of course is no need for the sign change inherent in analog methods.

Just as op-amp integrators are the heart of an analog simulation, numerical integration algorithms are the heart of digital simulation. An important aspect of numerical integration is the choice of the computing increment Δt; too coarse a step size leads to inaccuracy and/or instability, too fine increases computing time (cost) and may lead to inaccuracy from accumulation of many small round-off errors. Sophisticated variable-step-size algorithms perform built-in accuracy checks at every Δt step and will change the step size automatically during a problem solution to optimize the trade-off between computing speed and accuracy. Some simulation languages give the user a choice among several algorithms to suit the needs of particular classes of problems.

If a simulation language allows use of a general purpose language such as FORTRAN within it, many problems can be programmed using only the integrating capability plus the standard arithmetic operations $(+, -, *, /)$. (In Eq. (5.2)

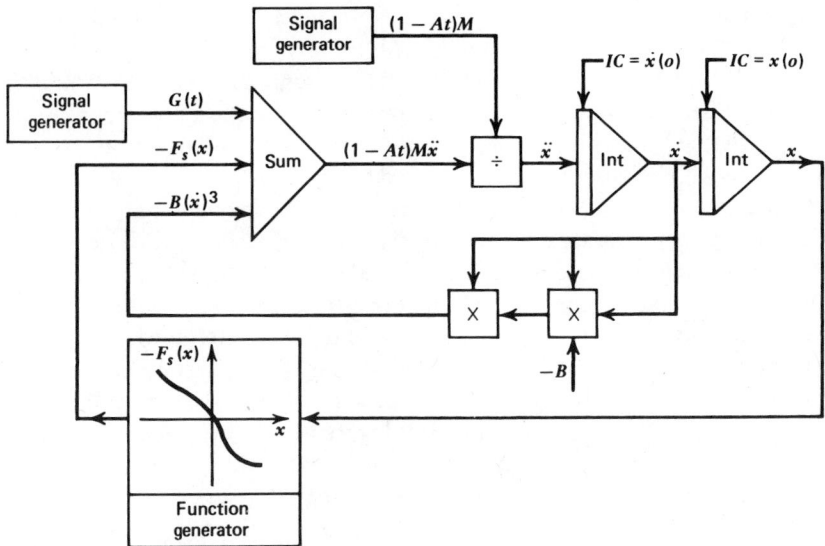

Figure 5.2   Simulation diagram.

these would handle everything except the spring force.) Most languages however are much more versatile than this, offering a large "menu" of useful operations from which the user can efficiently assemble the system model. A typical list of such modules might include first- and second-order systems, general Laplace transfer functions, arbitrary function generators (either one or two independent variables), on–off switches, dead space, dead time (fixed or variable), saturation, quantizers, zero-order hold, pulse generators, logic functions, random signal generators, and so forth. These are all easily invoked, often with only a single statement, and generally without regard for the time and magnitude scaling problems that are the most annoying features of analog simulation.

Most simulation languages allow versatile and simple "run control." A run can be terminated either at a specified problem time or when a selected variable reaches a certain magnitude. A simple listing of multiple values for a selected parameter causes a sequence of runs, one for each value of the parameter. A run can be halted at a chosen condition, parameters changed, and the run continued from there. Some languages group statements into INITIAL, DYNAMIC, and TERMINAL sections. The INITIAL section is used for computations that need be done only once, such as computing the spring constant of a cantilever beam from given dimensions and material properties. In the DYNAMIC segment the differential equations are stated and integrated, usually from time equals zero to the chosen finish condition. Once the solution is complete, the TERMINAL section can be used to summarize or manipulate the final results, possibly as part of a design optimization scheme that causes successive reruns of the problem until some performance criterion is satisfied.

Tabular and/or graphical output is usually very convenient; the formatting of tables and scaling of graphs is completely automatic, one need only state what is wanted. If the usual printer plots are too inaccurate (resolution$\approx$0.1 in.) or not of satisfactory visual quality, punched decks for driving a graphic plotter off line can be easily called for or plotters can be directly interfaced to produce high-quality on-line graphs. Interactive use with a time-sharing computer system and remote CRT graphics terminals duplicates the close man/machine synergism characteristic of analog operation, but high costs may preclude this mode of operation for many potential users, at least at present.

A final feature of many of these languages is their "nonprocedural" nature. A language such as FORTRAN or BASIC is called procedural since the statements of a program are executed in the exact order they are written and thus extreme care must be taken to get the proper sequence. With few exceptions, the order of statements in a simulation language program is immaterial since the simulation processor includes a *sorting algorithm* that sorts the statements into a proper sequence for computation, relieving the programmer of this burden. In addition to the sorting function a processor also will include translation and execution functions. In the CSMP language, which we will discuss in some detail, the simulation language statements are translated first into FORTRAN, then into assembler, and finally into machine language. Fortunately these operations are "behind the scenes" and normally of no concern to the simulation language user.

Before leaving this section we should note that while the number of *new* installations of analog computers is modest, many earlier installations continue in useful service since analog simulation may be more cost effective in certain kinds of situations, particularly if equipment costs have already been amortized. For example, the calculation of the frequency response of lightly damped nonlinear systems requires run times of many cycles (to allow transients to die out) and may be prohibitively expensive with digital methods. Also the immediate computer/engineer interaction present in even the simplest analog facility (but presently costly to achieve with digital methods) can provide savings due to more efficient utilization of engineering personnel. The difficulty faced in justifying *new* analog facilities is that modern engineering operations *require* digital computers for many nonsimulation tasks and once the digital machine is cost-justified on this basis, the simulation language software is easily added, often at a nominal lease cost of about 100 dollars per month.

5.4   A REPRESENTATIVE DIGITAL SIMULATION LANGUAGE: CSMP III

An historical account of the development of digital simulation languages together with a description of their operation is available in the literature.[3] Current developments appear in a number of journals, one[4] being devoted solely to the general area of simulation. Our purpose here is to present a fairly complete view of the capabilities and method of use of such languages by presenting details on a particular one, CSMP, the Continuous System Modeling Program[5] of IBM Corporation. This language is felt to be a good one for our purposes since it is one of the more versatile and complete, it is in widespread use, it may be "intermixed" with FORTRAN (the most common engineering language) and a recent textbook[6] devoted entirely to it is available as a reference for further details. While it would of course be desirable that the reader have CSMP (or some similar language) actually available for personal use when studying this chapter, the presentation is structured so that a good appreciation of the utility of simulation languages can be obtained without "hands-on" experience. This is accomplished by presenting the language in terms of specific engineering examples chosen to exercise and demonstrate its main features.

Let us take for our first example the "air bearing" of Fig. 2.9, whose stability was earlier analyzed using a linearized model. An important class of simulation problems involves the validation of approximate models by comparing their predictions with simulation results for a more correct model. In the present case, available stability theory (Routh criterion) allows us to come up with a useful design criterion (Eq. (2.110)) in letter form, but requires a linearized model whose accuracy is open to question. To gain confidence in the linear model we can compare its predictions with those of a more correct nonlinear one for some specific numerical values. If the two results compare well, it is likely that the linear model will be useful as a design tool since it gives us relations in letter form, whereas the nonlinear model (while more accurate) can only be "solved" by simulation for specific numerical parameters. In earlier days engineers had to rely on laboratory experimentation to validate their

[3]Chu, Y., *Digital Simulation of Continuous Systems*, McGraw-Hill, N. Y., 1969, p. 8.
[4]*Simulation*, The Society for Computer Simulation, 1010 Pearl St., La Jolla, Cal., 92037.
[5]Document SH19-7001-2, IBM Corp.
[6]Speckhart, F. H. and Green, W. L., *A Guide to Using CSMP*, Prentice-Hall, Englewood Cliffs, N. J., 1976.

approximate models. While lab testing will always be needed in practical engineering, we can now eliminate some of this with "computer testing" (simulation).

The "more correct" model for the air bearing is obtained by replacing the linearized flow equations (2.100) and (2.101) by the standard isentropic flow relations[7] for compressible flow through a flow restriction, using discharge coefficients (as for orifice flow) to correct for deviations from theoretical conditions:

$$\dot{m}_i = \left(\frac{A_i p_s C_{di}}{g\sqrt{T_s}} \right) \left(\sqrt{\frac{2\gamma g}{R(\gamma-1)}} \right) \left[\sqrt{\left(\frac{p}{p_s}\right)^{2/\gamma} - \left(\frac{p}{p_s}\right)^{(\gamma+1)/\gamma}} \right] \tag{5.3}$$

$$\dot{m}_o = \left[\frac{C_{ir} hp C_{do}}{g\sqrt{T_s(p/p_s)^{(\gamma-1)/\gamma}}} \right] \left(\sqrt{\frac{2\gamma g}{R(\gamma-1)}} \right) \left(\sqrt{\left(\frac{p_a}{p}\right)^{2/\gamma} - \left(\frac{p_a}{p}\right)^{(\gamma+1)/\gamma}} \right) \tag{5.4}$$

In the perfect gas law our nonlinear model can now let temperature and volume vary:

$$\dot{p} = \left[\frac{RT_s(p/p_s)^{(\gamma-1)/\gamma}}{V + A_p h} \right] (\dot{m}_i - \dot{m}_o) g \tag{5.5}$$

while in Newton's law we now include the gravity force (weight) since total forces (not perturbations away from equilibrium) are needed:

$$\ddot{h} = \frac{-B\dot{h} - f + A_p(p - p_a) - W}{W/g} \tag{5.6}$$

Definitions, and numerical values corresponding to the author's physical apparatus are:

A_i $\triangleq$ inlet valve area, 0.00285 in.^{2}

p_s $\triangleq$ supply pressure, 27.0 psia

C_{di} $\triangleq$ inlet value discharge coefficient, 0.7

T_s $\triangleq$ supply temperature, 530°R

---

[7] Andersen, B. W., *The Analysis and Design of Pneumatic Systems*, John Wiley and Sons, N. Y., 1967, p. 19.

γ $\triangleq$ ratio of specific heats, 1.40 for air

g $\triangleq$ acceleration of gravity, 386 in./sec^2

R $\triangleq$ gas constant, 639.6 in./°R

C_{ir} $\triangleq$ chamber outer circumference, 7.85 in.

C_{do} $\triangleq$ outflow discharge coefficient, 0.7

p_a $\triangleq$ atmospheric pressure, 14.7 psia

V $\triangleq$ chamber volume when $h=0$, 2.3 in.^{3}

A_p $\triangleq$ pressure-force area, 4.48 in.^{2}

W $\triangleq$ weight of moving part, 0.448 lb$_f$

B $\triangleq$ damping coefficient, lb$_f$/(in./sec), numerical value unknown

f $\triangleq$ external force, lb$_f$

h $\triangleq$ outflow gap height, in.

$\dot{m}_i, \dot{m}_o$ $\triangleq$ mass flow rates, lb$_f$-sec/in.

p $\triangleq$ chamber pressure, psia

Equations (5.3)–(5.6) define a nonlinear model of the air bearing that defies analytical solution but is relatively easy to solve numerically using CSMP. To compare the predictions of this model with those of the more approximate linear model, we must first establish an equilibrium operating point at which the linearized parameters are computed. Suppose we take external force f as zero; Eq. (5.6) then gives ($\ddot{h} = \dot{h} = 0$ at equilibrium) the equilibrium value of p as 14.8. Using this in (5.3) we can compute $\dot{m}_i$, and since at equilibrium $\dot{m}_i = \dot{m}_o$, (5.4) now allows us to calculate h as 0.00359 in. In the linearized stability analysis we need to compute:

$$K_{pi} \triangleq \frac{\partial \dot{m}_i}{\partial p} = 1.25 \times 10^{-7} \quad \text{(small; flow nearly choked)}$$

$$K_{po} \triangleq \frac{\partial \dot{m}_o}{\partial p} = 1.61 \times 10^{-5}$$

$$K_{ho} \triangleq \frac{\partial \dot{m}_o}{\partial h} = 8.98 \times 10^{-4}$$

These could be calculated from Equations (5.3) and (5.4); available tables[8] ease this task. Since all numerical values other than friction B are given, we can use Eq. (2.110) to compute the value of B that puts the system on the borderline of

---

[8]Andersen, *op. cit.*, p. 57.

instability by replacing the inequality sign with an equal sign. Substituting numbers:

$$(4.48)(8.98 \times 10^{-4}) = \left(\frac{(386 \times 639.6 \times 530)}{2.3} (0.0125 \times 10^{-5} + 1.61 \times 10^{-5}) \right.$$

$$\left. + \frac{B(386)}{0.448} \right)(1.62 \times 10^{-5})B$$

$$13.96B^2 + 14.85B - 4.02 = 0, \qquad B = 0.222$$

For this value of B the characteristic equation should have two pure imaginary roots; use of a digital computer root finder gives the roots -1108, $-1.27 \pm i418$. The -1.27 should really be zero but is not exact since we carried only three digits in our earlier calculations. If one "plots" $-1.27 \pm i418$ on, say $8\frac{1}{2} \times 11$ paper, it is clear that one can not tell it from $0.00 \pm i418$ "by eye," so our result is certainly usable. We thus predict an oscillation frequency of 418 rad/sec = 66.7 Hz for this marginally stable case.

We can now check the predictions of the linearized model by writing a CSMP program for the more correct nonlinear model of Equations (5.3)–(5.6). While certain basic operations would be common to all possible programs for this model, details can vary, so the program we now display should be thought of as *a* program, not *the* program.

```
TITLE NONLINEAR GAS BEARING
INIT
PARAM  PA=14.7,GAM=1.4,AP=4.48,V=2.3,PS=27.
PARAM  TS=530.,G=386.,R=639.6,CIR=7.85
PARAM  AI=.00285,HO=.00335,P0=14.8,CDI=.7,CD0=.7
PARAM  W=.448,B=1.0
       K1=2./GAM
       K2=(GAM+1.)/GAM
       K3=(GAM-1.)/GAM
       K4=SQRT(2.*GAM*G/((GAM-1.)*R))
       K5=(AI*PS*CDI/SQRT(TS)*K4
       M=W/G
DYNAM
       MIDOT=K5*SQRT((P/PS)**K1-(P/PS)**K2)
       MOD=(CIR*H*CDO*P)/SQRT(TS*(P/PS)**K3)
       MODOT=K4*MOD*SQRT((PA/P)**K1-(PA/P)**K2)
       PDOT=(R*(TS*(P/PS)**K3)/(V+(AP*H)))*(MIDOT-MODOT)
       P=INTGRL(PO,PDOT)
       FI=0.0
```

$$HDOT2 = (-B*HDOT1 - FI + (P - PA)*AP - W)/M$$
$$HDOT1 = INTGRL(0.0, HDOT2)$$
$$H = INTGRL(HO, HDOT1)$$
TIMER FINTIM = .05, DELT = .0001, OUTDEL = .001
OUTPUT H,P,MIDOT,MODOT
END
PARAM B = (.22,.25,.28)
TIMER FINTIM = .08, DELT = .0001, OUTDEL = .0002
OUTPUT H
PAGE MERGE,GROUP
OUTPUT TIME,H
PAGE MERGE,GROUP, XYPLOT, HEIGHT = 5.5, WIDTH = 6.5
LABEL AIR BEARING STABILITY
END

The CSMP language allows segmentation of a program into INITIAL, DYNAMIC, and TERMINAL portions; the present program requires only the first two, a subsequent example will illustrate the use of TERMINAL. The statements between the cards INIT and DYNAM could be in any order. Those cards labeled PARAM belong to the class called *data statements* and are used to enter numerical values of system parameters; the others are FORTRAN calculations of constants needed in the DYNAM section. Any statements between INIT and DYNAM need be (and will be) calculated only once, but the results are available to the rest of the program at any stage of the calculation. The cards between DYNAM and TIMER define the system equations and are called *structure* statements. They also can be given in any order; however, it is "natural" to first write in FORTRAN the expressions for the highest derivatives of the unknowns and then use the INTGRL (integral) statements to compute the lower derivatives and the unknowns themselves. Each INTGRL statement includes in parentheses the initial condition (symbol or number) and the name of the quantity to be integrated. Symbolic names may have one to six alphanumeric characters but must begin with an alphabetic character and cannot contain embedded blanks or "special" characters (+, −, /, =, etc.).

The TIMER card is considered a *control* statement; it normally specifies:

1 FINTIM—the length of time the independent variable TIME is allowed to run, starting from zero.
2 DELT—the time interval to be used in the numerical integration algorithm.
3 PRDEL—the time increment to be used in printing tabular output.
4 OUTDEL—the time increment to be used in producing printer plots.

Intelligent choice of FINTIM requires some knowledge of system response from previous experience or estimates from approximate models, such as the linearized air bearing model. We choose to run the problem first with damping

$B = 1.0$, a value the linearized model predicts to be thoroughly stable and should thus cause the system to settle to a steady state that we can check against our earlier hand calculations. Assuming any oscillations present will have a frequency near 66.7 Hz and that well-damped systems rarely require more than a few cycles to settle down, a FINTIM of 0.05 sec is a reasonable first try. To get the system to react dynamically we take the initial pressure as the predicted equilibrium value of 14.8, but start h off at 0.00335 rather than the calculated steady state of 0.00359. (Any other off-equilibrium combinations of HO and PO could be used to excite a system response that should then settle to equilibrium; the numbers just given have no special significance.)

A discussion of the choice of DELT leads us also to the subject of integrating algorithms. CSMP III provides a choice of numerical integration methods ranging from the simple rectangular, trapezoidal, and Simpson's rule techniques familiar to most calculus students, to sophisticated variable-step-size methods that adjust DELT each step of integration to strike an optimum balance between accuracy and computing time. Unless a problem requires "synchronization" with some strictly timed event (such as a sampling switch in a sampled-data system), the default method of integration RKS (a variable-step-size Runge–Kutta method) usually is a good first choice. This is automatically selected if no card labeled METHOD appears in the program, and produces good solutions for most problems without any further consideration by the user. When a *fixed* step-size method is desired to achieve the synchronization mentioned above, a card METHOD RKSFX inserted at the beginning of the program selects a Runge–Kutta algorithm that usually gives good results. This method should also be tried if the variable-step RKS method "shuts itself off" due to DELT having been reduced to excessively small values (FINTIM $\times 10^{-7}$) in an attempt to "keep up" with rapidly changing variables.

When the default method of integration RKS is used, if no value of DELT is stated on the TIMER card, the system will initially set DELT to $\frac{1}{16}$ of the smaller of PRDEL or OUTDEL and then continuously adjust it to meet accuracy criteria throughout the run. It is probably better for the user to specify an initial DELT. Reasonable values are FINTIM/2000 or, if oscillations of a known frequency are expected, $\frac{1}{200}$ of the oscillation period. These trial values can also be used with the fixed-step-size algorithms. If a fixed-step algorithm runs successfully, one should repeat it with larger and smaller DELT's. If the results do not change significantly, this is a good indication (though not infallible) that a proper range of values for DELT is being used. Systems in which the speed of response varies widely during the problem duration (for linear systems this happens when the characteristic equation has roots with a wide spread (say 1000 to 1) of values) are called *stiff systems* and present particularly difficult integration problems. CSMP III includes an integration scheme designed specially for such problems; we will discuss its use later.

Returning to our example, Fig. 5.3 shows the printer plot produced by the statement OUTPUT H, P, MIDOT, MODOT; we see that the system is stable and that h settles to 0.0035901 (very close to our earlier hand-calculated steady-state value) in 0.05 sec while MIDOT and MODOT finally balance out in the steady state. To check the linearized model stability predictions we now want to rerun the problem with three different values of B that span the stable/unstable boundary. This is accomplished by defining a new "case" through use of the END card, which resets the model back to time zero and allows us to modify data and control statements (but *not* structure statements) and rerun with these new conditions. (If the model *structure* must be changed (such as adding a spring), then the END card must be followed by an ENDJOB card and a *complete* set of model structure statements (*not* just the *new* statements for the spring effect).) The card PARAM B = (.22, .25, .28) calls for three "runs" of the model with all conditions as in the previous case except that B now takes on successively the values .22, .25, and .28. Up to 50 values of a parameter may be called for within the parentheses; if the increments are equally spaced one can use B = (1.0, 6*.1) to get the values 1.0, 1.1, 1.2, ..., 1.6. Only *one* multiple-value parameter can be used at a time (B = (.22, .25, .28), W = (.448, .6, 1.3) is *not* allowed); however, we can accomplish the desired result with:

```
PARAM B = (.22,.25,.28),W = .448
END
PARAM W = .6
END
PARAM W = 1.3
```

We are also allowed to change the TIMER card (it is a control statement) if we wish. In the present problem we decide a longer FINTIM and smaller OUTDEL are desirable since we want to produce a high-quality (many points) off-line graph on the drum plotter and we want to see several cycles of oscillation so as to clearly detect build-up or decay. Since stability will be apparent from the trend of *any* of the unknowns, we choose to plot only h, using the card OUTPUT H. This produces a *printer* plot of H versus TIME which allows us to see what the drum plot will look like before actually running the punched cards (produced by CSMP) through the off-line plotter. The statement PAGE MERGE causes the three graphs for B = .22, .25, and .28 to be superimposed on one sheet, whereas, GROUP causes them to all be plotted to the same scale. To actually generate the punched deck for the off-line plotter we use:

```
OUTPUT TIME,H time will be horizontal axis,h vertical
PAGE MERGE,GROUP,XYPLOT,HEIGHT = 5.5, WIDTH = 6.5
LABEL AIR BEARING STABILITY
```

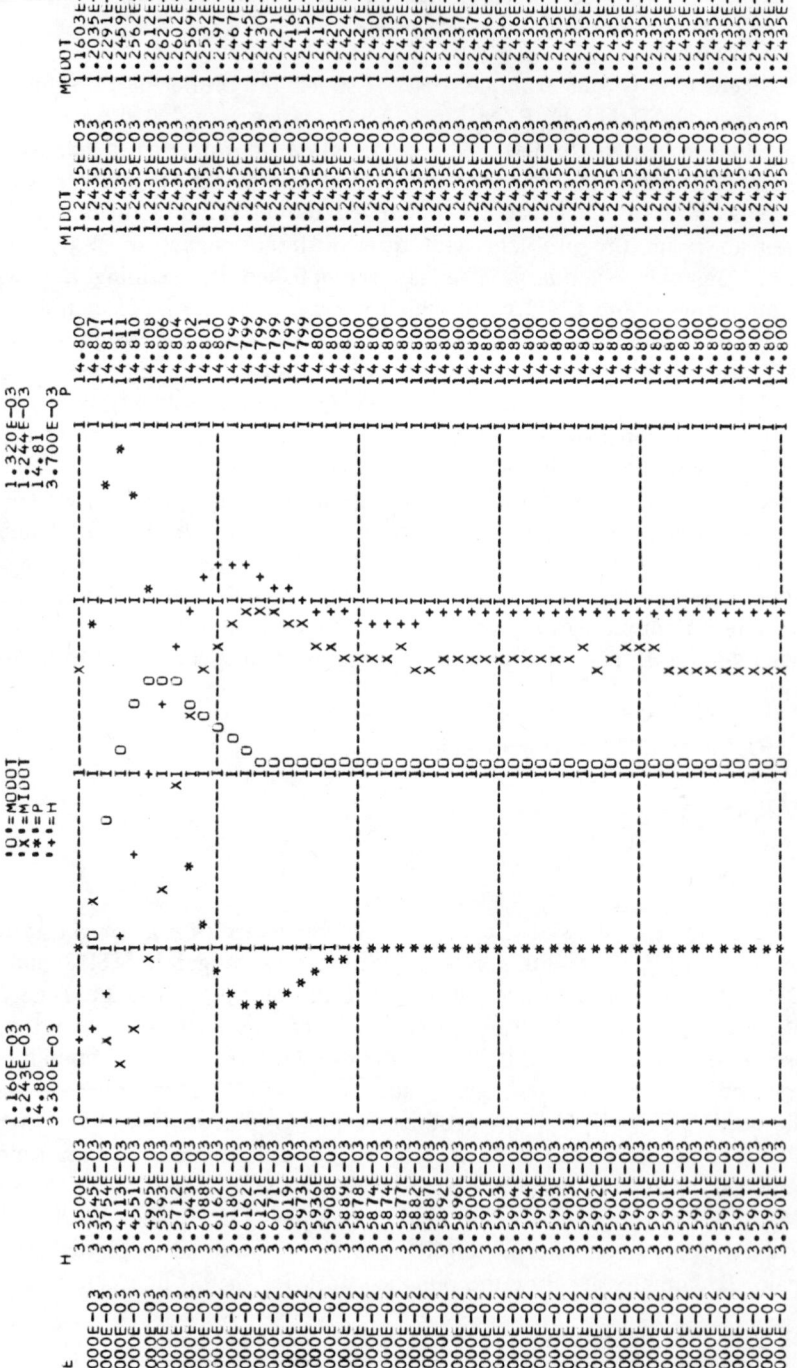

Figure 5.3 Air bearing response.

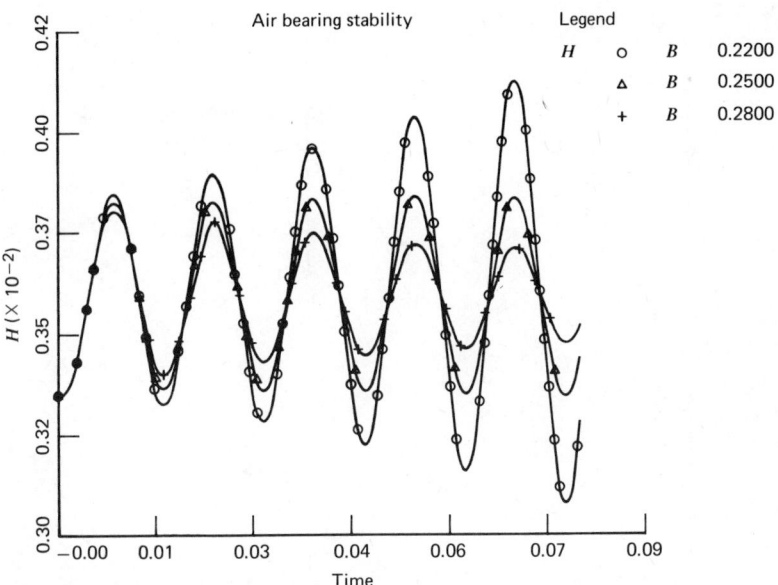

Figure 5.4 Air bearing stability check.

The word XYPLOT causes a deck for the plotter to be produced, HEIGHT and WIDTH determine its size in inches, and LABEL specifies the title to be printed on the graph. Figure 5.4 shows the resulting plot, which confirms the stability predictions of the linearized model since B = .25 seems to be the boundary between stability and instability and the oscillation frequency is about 64 Hz.

5.5 ARBITRARY FUNCTION GENERATORS; USE OF TERMINAL SEGMENT IN DESIGN OPTIMIZATION

While the availability of standard FORTRAN operations in CSMP allows direct and simple treatment of many nonlinear and variable-coefficient effects in equations (as we saw in the previous example), CSMP also includes a number of useful special purpose function generators, as shown in Fig. 5.5. These are invoked by simple single statements that define the input and output variables and assign numerical values to parameters. *Arbitrary function generators* are also available to model functional relations defined by experimental graphs or tables; our next example illustrates their use.

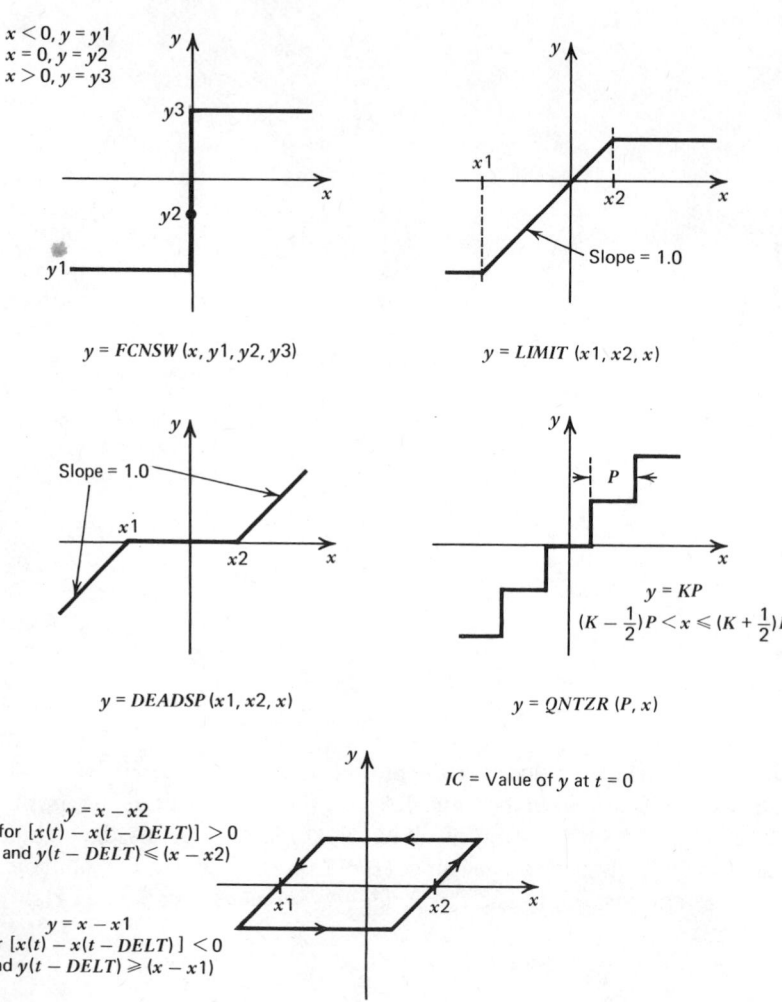

$y = FCNSW\ (x, y1, y2, y3)$

$y = LIMIT\ (x1, x2, x)$

$y = DEADSP\ (x1, x2, x)$

$y = QNTZR\ (P, x)$

$y = x - x2$
for $[x(t) - x(t - DELT)] > 0$
and $y(t - DELT) \leqslant (x - x2)$

$y = x - x1$
for $[x(t) - x(t - DELT)] < 0$
and $y(t - DELT) \geqslant (x - x1)$

otherwise, $y = y(t - DELT)$ $y = HSTRSS\ (IC, x1, x2, x)$

Figure 5.5 Special-purpose function generators.

The mass/spring/damper system of Fig. 5.6 has a nonlinear spring whose design has been finalized and whose force/deflection characteristic is known only as a table of experimental data. The damper is also nonlinear with experimentally defined force/velocity characteristics; however, its design can be "scaled up or down" to produce a range of numerical values, but always with the same *shape* for the force/velocity curve. We need to choose from this family a specific damper that will give a 10% overshoot of displacement when a 1000 Newton step force is applied to the mass. We can use CSMP's arbitrary function generator for both the spring and damper characteristics while the choice of the proper damper can be implemented using the TERMINAL sections's capabilities. For arbitrary functions of one independent variable, CSMP offers a choice of three schemes; linear interpolation (straight-line segments connecting the given data points), parabolic interpolation, and a "general" routine that allows the user to select the interpolation polynomial degree from one (straight lines) to five. If the known data points have reasonably close spacing, the simplest (linear) interpolation often gives satisfactory results, and we use it in the program below.

```
TITLE COMPUTER-AIDED DESIGN
PARAM   M=.6
PARAM   B=1.0
        FI=1000.*STEP(0.0)
        FS=AFGEN(CURVE1, XS)
AFGEN CURVE1= -5., -1100., -4., -800., -3., -540., -2., -310., -1.,
        -160.,... 0.0, 0.0, 1., 160., 2., 440., 2.5, 680., 3.0, 1000., 3.5, 1500.,
        3.7, 1800.,...5., 4000
        FD=AFGEN(CURVE2, XSDOT)
AFGEN CURVE2= -150.,-800.. -125., -700., -100., -580., -50., -320.,
        0.0, 0.0, ... 25., 200., 50., 450., 62.5, 650., 75., 820., 87.5, 1250.,
        100., 1820.
```

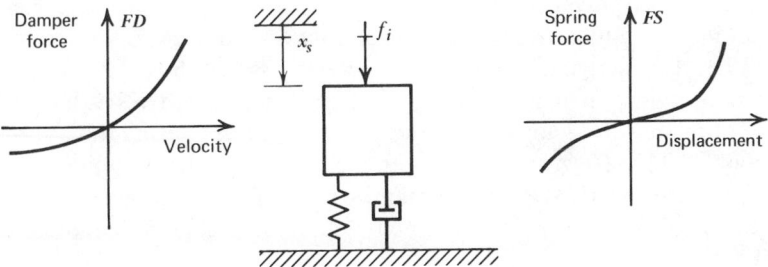

Figure 5.6 Nonlinear mechanical system.

```
        XSDOT2 = ( − FS − B*FD + FI)/M
        XSDOT = INTGRL(.01,XSDOT2)
        XS = INTGRL(0.0,XSDOT)
FINISH XSDOT = 0.0
TIMER FINTIM = 1.0,DELT = .001,OUTDEL = .002
OUTPUT XSDOT2,XSDOT,XS
RERUN B
TERMINAL
        IF(ABS(XS − 3.3) − .03)50,50,40
    40  B = B + B*(XS − 3.3)/3.3
        CALL RERUN
        GO TO 60
    50  CONTINUE
    60  CONTINUE
END
```

In the Newton's law statement XSDOT2 = ... we write the damper force as − B*FD, where FD is the basic nonlinear characteristic and B is the numerical scale factor that determines the "size" of the damper. Our problem is to find the proper value of B to meet the 10% overshoot requirement. In the TERMINAL section we provide an algorithm for adjusting B to meet this specification; however, we need a "starting value," given as 1.0 on a PARAM card. The mass M is fixed and known at 0.6 Kg. The input force FI makes use of the CSMP signal generator STEP(TS), which produces a step change of size 1.0 at time TS, in our case at $T = 0$. This, and the ramp generator RAMP(TS), which produces a ramp of slope 1.0 starting at time TS, are very useful in "manufacturing" input signals of various forms. For example, the input in Fig. 2.16 would be given by:
QI = RAMP(0.0) − RAMP(1.0) − 2.*STEP(2.0) + RAMP(2.0) − RAMP(3.0)

Each arbitrary function generator is implemented with two statements: one that names the variables and a second that tabulates the numerical "X − Y" data points which specify the particular function. The statement FS = AFGEN(CURVE1, XS) identifies spring force FS as a function of displacement XS and states that the numerical values will be found at an AFGEN statement called CURVE1. This statement, AFGEN CURVE1, simply lists the displacement, force data points in pairs, starting at the most negative FS value and working toward the positive values. Thus the first point on the spring force curve is at − 5 cm and − 1100 Newtons; since there are too many pairs of points to fit on one card, we use the CSMP continuation symbol ... to continue them on successive cards. The damper force/velocity curve is similarly specified in AFGEN CURVE2.

Since we are interested in controlling the size of the first overshoot of displacement, we can detect this occurrence and stop the calculation by using a FINISH statement on the velocity XSDOT. Clearly, XSDOT will be zero when XS first reaches a peak and "turns around," thus FINISH XSDOT = 0.0 will halt the calculation at the peak of the first overshoot. The FINTIM = 1.0 in the TIMER statement is not really necessary; just a safety precaution if FINISH should somehow fail to "shut off" the run. (Rough estimates showed the first peak to occur before TIME = 1.0.) The statement RERUN B causes the current value of B to be printed as part of each run. The TERMINAL segment of the program comes into play only at the end of each run. It checks to see whether the desired percent overshoot (10%) has been achieved and modifies B up or down as needed; CALL RERUN then reruns the problem with the new B. This cycle continues until the proper B value is found. Since a "perfect" 10% overshoot is impossible, we need to allow a small "error band" in our algorithm. The standard FORTRAN IF statement checks whether the peak value of XS falls within 3.3 ± 0.03. If it does not, a new trial value of B is computed from:

$$B = B + B*(XS - 3.3)/3.3$$

This adjustment algorithm changes B by an amount proportional to the deviation of XS from the goal of 3.3. While this approach seems reasonable, it is certainly not the only possible scheme and it is also not obvious what the "best" proportionality factor might be. If B is changed too rapidly, the process may oscillate and even diverge; if too slowly, we use excessive computer time. Since the convergence process is not amenable to mathematical analysis, we will simply have to use trial-and-error on the computer.

The program was run as shown and went through the sequence of B values 1.00, 1.17, 1.34, 1.50, 1.64, 1.76, 1.86, 1.94, 2.01, 2.06, 2.11, 2.14, 2.17 and stopped at 2.19, which value gave a peak XS of 3.3264, which meets our requirements. Figure 5.7 shows the printer plot for the last of these runs. Since convergence was rather slow (14 trials), any further programs of this kind might try a more rapid adjustment of B.

While we will not give an example of its use, we should at least briefly discuss CSMP III's arbitrary function generator for functions of two variables. Figure 5.8 shows how $Z(X, Y)$ is represented by a set of functions $Z_1 = f_{X1}(Y)$, $Z_2 = f_{X2}(Y)$, $Z_3 = f_{X3}(Y)$, and so on obtained by "slicing" through the surface $Z = Z(X, Y)$ at specified values of X. Each such function is defined by listing pairs of Y, Z points and is "labelled" with its respective X value. To calculate the Z corresponding to a given X and Y, CSMP first notes the two X values that bracket the given X, computes one Z value for each of these curves (at the given Y), and then interpolates between these two Z values to get Z. The statements

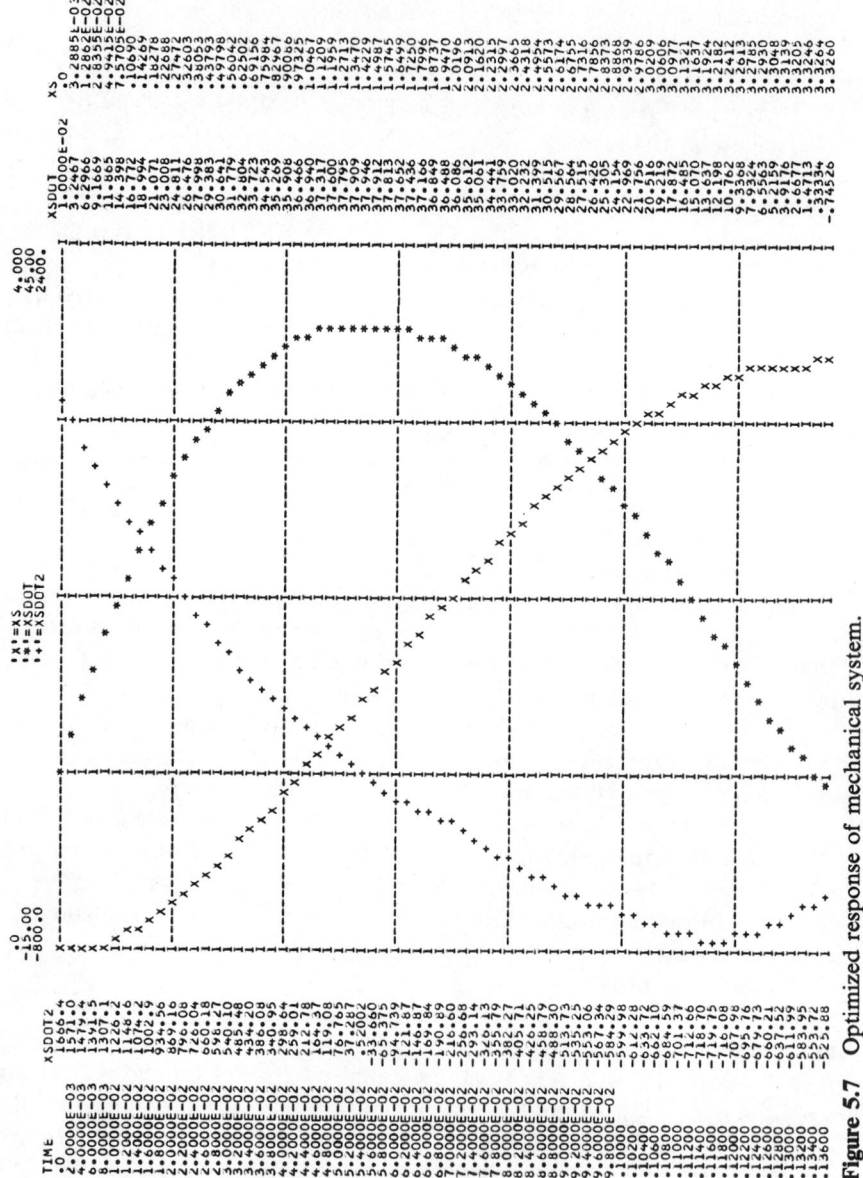

Figure 5.7 Optimized response of mechanical system.

192

needed to accomplish this would be, for example:

$$Z = \text{TWOVAR(FUNCT,Y,X)}$$
FUNCTION FUNCT,0.0 = (0.0,3.),(2.0,4.2),(4.0,5.1),...
(6.0,5.6),(8.0,5.4),(10.0,4.9),(12.0,3.4)
FUNCTION FUNCT,2.0 = (0.0,3.5),(2.0,4.0),(4.0,4.3),...
(6.0,4.4),(8.0,4.2),(10.0,3.7),(12.0,2.6)
FUNCTION FUNCT,4.0 = (0.0,3.2),(2.0,3.1),(4.0,3.0),...
(6.0,2.8),(8.0,2.5),(10.0,2.0),(12.0,1.4)

etc.

The number (0.0, 2.0, 4.0,...) after FUNCT, in each FUNCTION statement is the X value for that "slice," while the paired values (0.0, 3.) and so on after the equal sign are the respective Y, Z points which define that curve. Neither the X increments nor the Y increments need be equal, but the FUNCTION statements must be ordered with X increasing monotonically and Y must increase monotonically in each such statement.

5.6 DEAD TIMES, CONSTANT AND VARIABLE

While the dead-time effects of Section 2.8 are easily handled by analytical methods when the dead time is simply cascaded with some other transfer

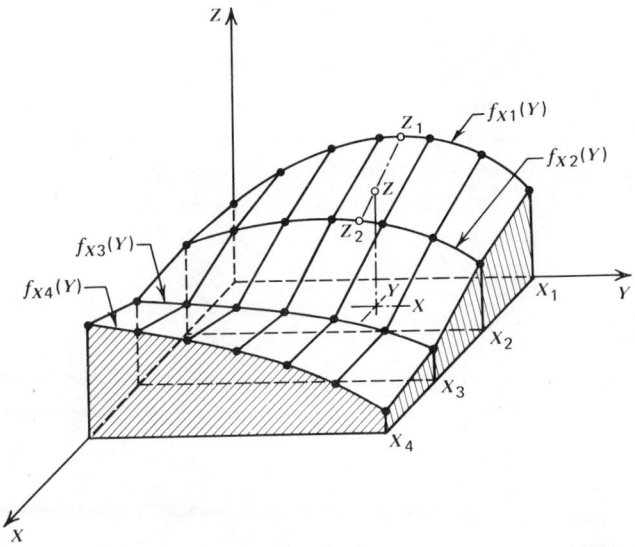

Figure 5.8 Function generator for functions of two variables.

function, for example, $(e^{-\tau_{dt}s})(K/(\tau s+1))$, there are situations, such as when the dead time is embedded in a feedback system, where analytical methods become tedious or fail entirely. Consider the system of Fig. 5.9 when R is a unit step input:

$$(R-e^{-\tau_{dt}s}C)\frac{K}{\tau s+1}=C \tag{5.7}$$

$$C=\frac{K}{\tau s+Ke^{-\tau_{dt}s}+1}R=\frac{K}{s(\tau s+1)+Kse^{-\tau_{dt}s}} \tag{5.8}$$

Since the denominator of (5.8) is not a polynomial in s, our usual partial-fraction expansion method for finding $c(t)$ fails. In this particular case, however, division of denominator into numerator produces a sequence of terms that can be inverse transformed and evaluated using the delay theorem:

$$C(s)=\frac{K}{s(\tau s+1)}-\frac{K^2e^{-\tau_{dt}s}}{s(\tau s+1)^2}+\frac{K^3e^{-2\tau_{dt}s}}{s(\tau s+1)^3}-\frac{K^4e^{-3\tau_{dt}s}}{s(\tau s+1)^4}+\text{etc.} \tag{5.9}$$

$$c(t)=K(1-e^{-t/\tau})-K^2\left[1-\left(1+\frac{(t-\tau_{dt})}{\tau}\right)e^{-(t-\tau_{dt})/\tau}\right]u(t-\tau_{dt})+\text{etc.}$$
$$\tag{5.10}$$

While this is an infinite sequence, the usual questions of series convergence do not arise since the delayed terms are each zero until their "starting times" (τ_{dt}, $2\tau_{dt}$, $3\tau_{dt}$, etc.) are reached. We can thus evaluate and plot $c(t)$ starting from $t=0$ and proceeding toward $t=\infty$ knowing that at any specific t we have the *total* contribution of the series at that point in time. Of course the behavior of $c(t)$ *beyond* the currently computed t-value is not always obvious in such a situation, thus questions such as stability may be obscured.

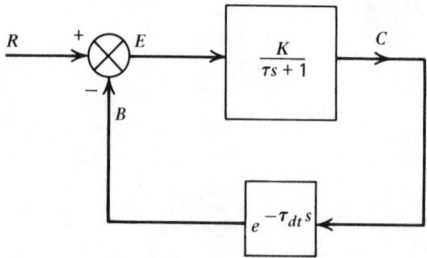

Figure 5.9 Feedback system with dead time.

While the above or other analytical methods sometimes yield usable results, it is clear that the very simple example just shown leads to quite tedious computations. To avoid these, and more importantly, to deal with those many situations where no analytical approach exists, CSMP provides the DELAY statement. In analog computers, simulation of dead times is very difficult; however, digital simulation handles it easily and accurately since we need merely store past values and then "re-create" them later. A program for exploring the effect of system gain K in the example of Fig. 5.9 might go as follows:

```
PARAM   TAU=1.0
        R=STEP(0.0)
        E=R-B
PARAM K=(.5,4*.8)
        C1=K*E
        C=REALPL(0.0,TAU,C1)
        B=DELAY(100,1.0,C)
TIMER FINTIM=10.,DELT=.01,OUTDEL=.2
OUTPUT C
PAGE MERGE,GROUP,WIDTH=100
OUTPUT C
PAGE MERGE,GROUP,WIDTH=100,NTAB=0
```

The REALPL statement models a first-order system with gain=1.0 and time constant TAU between input C1 and output C. The initial value of C is the first entry inside the parenthesis; in our case it is 0.0. If a second-order system $1/(s^2+2\zeta\omega_n s+\omega_n^2)$ with input X, output Y, $Y(0)=YD0$, and $Y(0)=Y0$ were needed, CSMP provides the statement:

$$Y=CMPXPL(Y0, YD0, ZETA, OMEGAN, X)$$

In the DELAY statement the integer 100 is the number of DELT's in the dead time, 1.0 is the dead time, C is the input, and B the output. One needs to choose a DELT such that the desired dead time can be expressed as an integer multiple. In the first PAGE statement, WIDTH=100 restricts the printed output to 100 characters, giving a convenient $8\frac{1}{2}\times 11$ inch format for reports. The second PAGE statement requests a repeat of the earlier OUTPUT C data, but NTAB=0 suppresses the usual marginal tables, giving more space for the graph. If we wanted the first n variables of a multi-curve graph to be both plotted and tabulated, with the remaining variables only plotted, NTAB=n would be used. Figure 5.10, showing the printout for this program, makes clear the importance of K; some values make the system unstable.

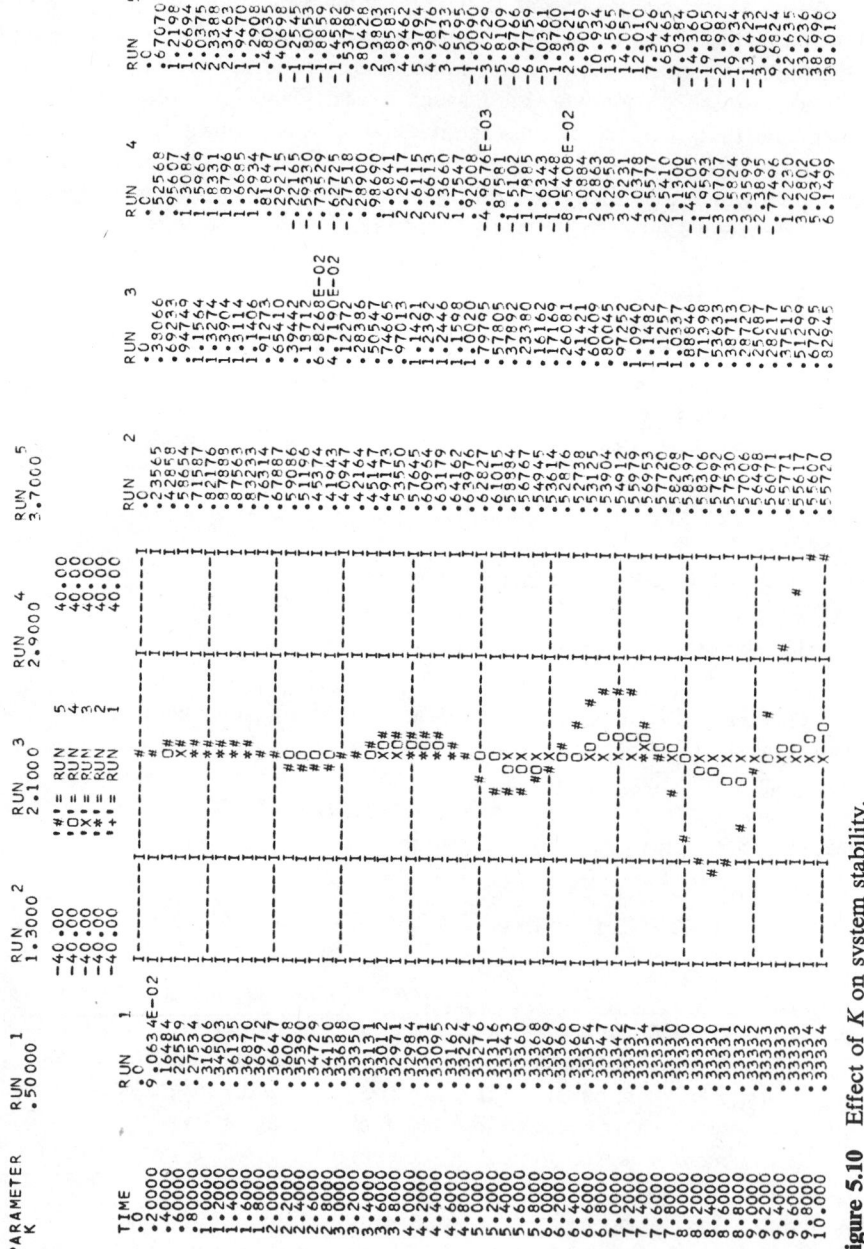

Figure 5.10 Effect of K on system stability.

196

One of the most common practical applications of dead-time models involves the transport of solid or fluid materials by conveyor or pipeline. When conveyors move at variable speed, or pipeline flow rates vary with time, the associated dead times are no longer constant and the CSMP DELAY statement is not applicable. Since time-varying dead times are analytically even more formidable than constant dead times, simulation is usually necessary in modeling studies. While the digital simulation of constant dead times is a straightforward (straightbackward?) process, variable dead times present some subtleties; we do not just let τ_{dt} be a time-varying number. Consider in Fig. 5.11 the temperature at stations 1 and 2 in a pipeline flowing at a variable velocity V. If we consider the temperature at station 1 as the input, the temperature at station 2 would be a delayed version of that at 1; however, the delay is not a fixed time interval. (Note that we are attempting to model the gross transportation effect *only*; heat and/or mass transfer *within* the fluid is neglected.) At time $= 0$ we assume the entire pipeline at some uniform initial temperature; thereafter T_1 and V vary in any manner and we wish to model the behavior of T_2. A specific

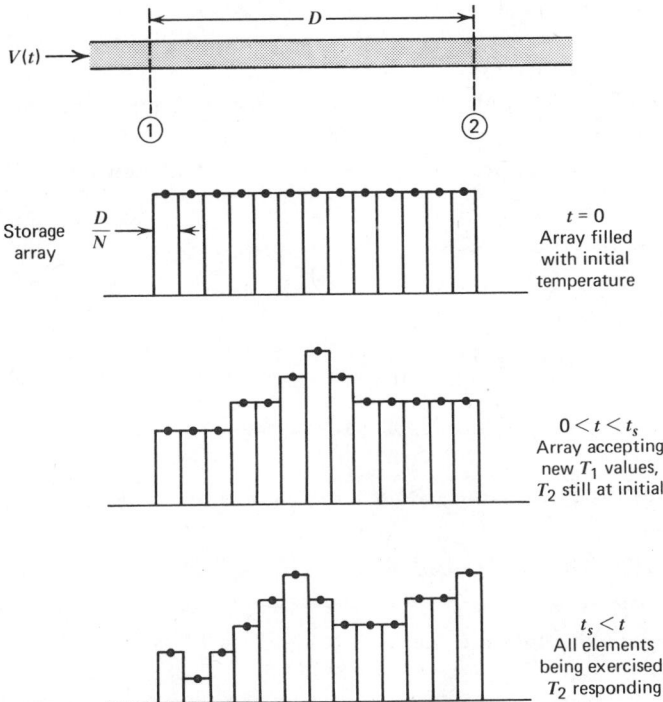

Figure 5.11   Behavior of variable dead time.

temperature which existed at 1 at $t=0$ will reach 2 at time t_s when $\int_0^t V\,dt = D$; before this time, T_2 has remained constant at the given initial value.

In the CSMP variable-dead-time model called PIPE, the distance D is sliced into N segments ($3 < N < 16378$) to give a discrete representation of the temperature profile of the pipe at any instant. The program user selects N; a typical value might be 100. At time zero the N-element storage array holding this temperature profile contains the initial uniform profile mentioned earlier. During the first DELT, if $\int V\,dt$ exceeds D/N, the left-most element of the array is updated with the new value of T_1; if not, it remains at its initial value. For subsequent DELT's, assuming V is not zero, each element of the array will be eventually filled, in sequence, until at t_s (where $\int V\,dt = D$) the array will be full of "new" values for the first time and T_2 will start to change. For each DELT after t_s, if the change in $\int V\,dt$ exceeds D/N a new temperature is entered into the left-most array element, pushing each value one element to the right and giving a new value for T_2. This description is intended only to give a general idea of the nature of a variable dead time; it does not explain the details of the actual computing algorithm.

As an example of a system including a variable dead time, consider the temperature control scheme of Fig. 5.12. A linearized version of this system, with a constant dead time, was analytically studied in the author's earlier work.[9] We here wish to use CSMP to study the start-up transient associated with bringing the pump up to speed. While the pump is accelerating, the mass flow rate G will be varying, causing a variable dead time between process temperature T_C and downstream measured temperature $T_{C,D}$. An energy balance on the tank gives:

$$Mc_p\frac{dT_C}{dt} = Q_M + Gc_p(T_I - T_C) - UA(T_C - T_U) \qquad (5.11)$$

$M \triangleq$ liquid mass in tank, 36 kg

$c_p \triangleq$ liquid specific heat, 2100 watt-sec/kg-C°

$Q_M \triangleq$ heater heating rate, watts

$$= 4600 + 10^7 v_E + 10^5 \int v_E\,dt \qquad (5.12)$$

The 4600 watts is the base heating rate needed to maintain T_C at the desired

[9]Doebelin, E. O., *Dynamic Analysis and Feedback Control*, McGraw-Hill, N. Y., 1962, p. 229.

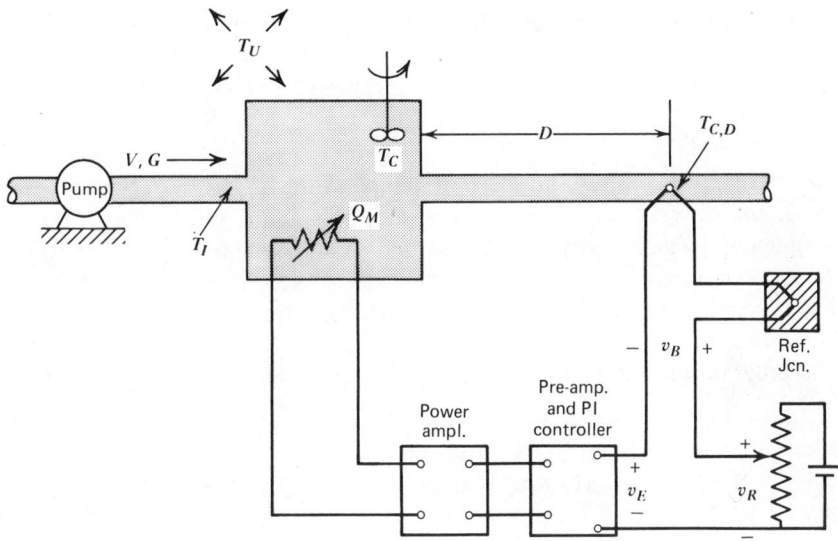

Figure 5.12   Temperature control system with variable dead time.

value for nominal design conditions, while the other two terms represent a proportional-plus-integral (PI) controller using the error voltage v_E as input signal. This heater is also subject to the limits that $Q_M \geq 0$ (no *cooling*) and $Q_M \leq 20000$ (max. power rating).

$$G \triangleq \text{liquid mass flow rate, kg/sec}$$

We wish G to be steady at 0.001 until $t = 200$ sec, at which time it begins a ramp to 0.05 at $t = 400$ sec where it levels off; this defines the pump acceleration schedule. Pipe area is such that flow velocity $V = 60G$ m/sec.

$T_I \triangleq$ liquid inlet temperature, 10 °C

$T_C \triangleq$ liquid temperature in tank, °C

$U \triangleq$ tank wall heat transfer coefficient, 5.86 watts/m²-C°

$A \triangleq$ tank wall heat transfer area, 3.62 m²

$T_U \triangleq$ ambient temperature, 20 °C

The thermocouple is modeled as a linear first-order system:

$$\frac{T_{tc}}{T_{C,D}}(s) = \frac{5 \times 10^{-5} \text{ volts/C}^{\circ}}{10s + 1} \tag{5.13}$$

We will start T_C at 45°C and desired temperature T_V at 50°C, a 5° step command, and observe the transient caused by this as the pump accelerates to operating speed. When the system has settled to steady state, we will drop T_V back to 45° to compare the behavior for the same 5° step change when the dead time has become constant at 2 sec ($V = 3$ m/sec, $D = 6$ m).

The program used to model this system follows.

```
INIT
PARAM    U=5.86,A=3.62,CP=2100.,KP=1.E7,KI=1.E5
PARAM    M=36.,TU=20.,QMO=4600.,TI=10.
PARAM    KTC=5.E-5,TAUTC=10.
         K2=M*CP
DYNAM
         TV=50.-5.*STEP(1900.)
         G=.001+.000245*(RAMP(200.)-RAMP(400.))
         V=60.*G
         VR=TV*KTC
         VE=VR-VB
         VE1=VE*KI
         QMI=INTGRL(0.0,VE1)
         QMP=VE*KP
         QMU=QMO+QMP+QMI
         QM=LIMIT(0.0,20000.,QMU)
         TCDOT=(QM+G*CP*(TI-TC)-U*A*(TC-TU))/K2
         TC=INTGRL(45.,TCDOT)
         TCD=PIPE(100,45.,6.,V,TC,1)
         TTC=REALPL(45.,TAUTC,TCD)
         VB=KTC*TTC
TIMER    FINTIM=3000.,DELT=.5,OUTDEL=20.
OUTPUT   TV,V,QM,TC,TCD
OUTPUT   TIME,TV,V,QM,TC,TCD
PAGE     XYPLOT,HEIGHT=5.5,WIDTH=8.
```

In the variable-dead-time statement TCD=PIPE(100, 45., 6., V, TC, 1), the 100 is the number of "slices" N into which distance D is segmented. Changing this to 500 increased the program cost from $3.45 to $4.14 (CPU time is about 10 sec) and showed no significant change in results, so 100 segments seems

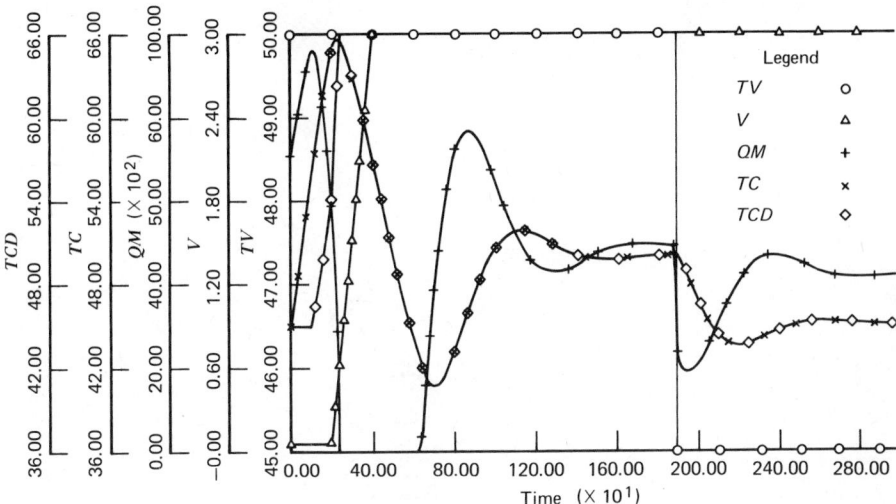

Figure 5.13 Response of temperature control system.

adequate for this problem. The next entry, 45., is the initial state of the pipeline, 6. is the transport distance D, V is the transport velocity, and TC is the input to the dead time. The last entry could be either integer 1 or 2; it is the degree of interpolation used by the algorithm in interpolating the storage arrays. Linear interpolation (integer 1) seems to be adequate in most cases.

Figure 5.13 shows the off-line graph produced by a drum plotter in response to the PAGE XYPLOT statement. We note that temperature $T_{C,D}$ is clearly delayed with respect to T_C at early times, where pump speed is slow. As the pump accelerates, the dead-time decreases and at about 240 sec the delay between T_C and $T_{C,D}$ becomes so small that their graphs appear essentially identical thereafter. The long dead times before $t = 240$ cause a large overshoot of controlled temperature to about 66°C, 16°C above the desired value. However, the system eventually recovers and returns T_C to the setpoint at about 1600 sec. At 1900 sec the 5° step change in T_V induces another transient in the system, but now, with the pump at design speed and the dead time small, T_C only undershoots the new desired value by about 1.5°C.

5.7 ARRAY-HANDLING TECHNIQUES FOR LARGE DISCRETIZED MODELS OF DISTRIBUTED SYSTEMS

When distributed-parameter systems are approximated by discretized models, using physical lumping schemes or mathematical finite difference or finite element methods, high accuracy may require a large number of lumps (CSMP

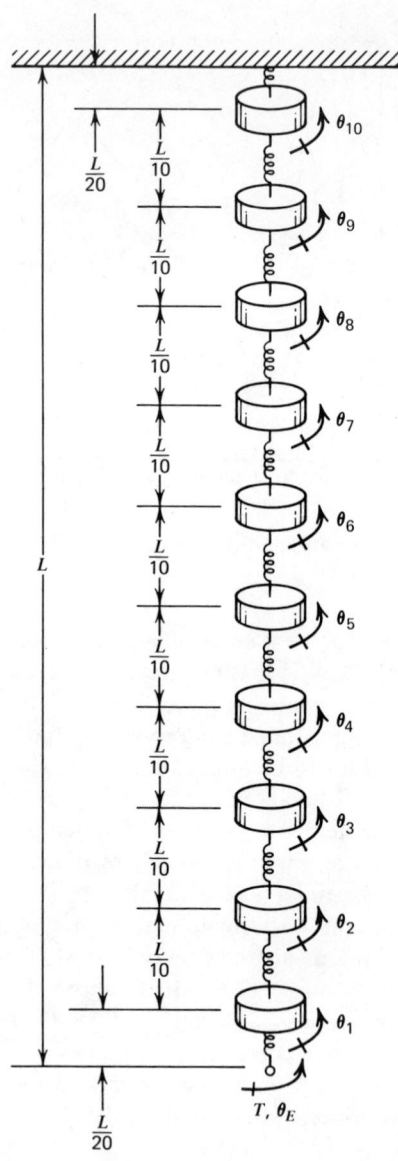

Figure 5.14 Ten-lump model of torsionally vibrating rod.

allows use of up to 300 integrators). To simplify and speed up the use of such models, CSMP provides convenient methods of handling large arrays of integrators. To illustrate this capability let us consider a step input of torque T to an approximate, lumped model of the torsional vibrating system of Section 4.1. Figure 5.14 shows such a model which uses 10 lumps; Chapter 8 will provide details on how one chooses the size and number of lumps. By applying Newton's law, Σ Torque $= J\alpha$, to each disk, one obtains 10 equations in 10 unknowns. The 10 disk inertias are each equal to that of a rod section of length $L/10$; the 9 interior springs have spring constants equal to those of rod sections of length $L/10$, while the two end springs are twice this stiff. Thus the "total inertia" and "total stiffness" of the lumped and distributed models are equal, but the *distribution* of these effects is, of course, discrete in the first case and smooth in the second.

While one could write a CSMP model of this system using only techniques already shown in this chapter, a more efficient approach, which becomes even more desirable when the number of lumps increases, will now be outlined.

```
FIXED           I
TABLE           BDIC(1 – 10) = 10*0.0
TABLE           BIC(1 – 10) = 10*0.0
PARAM           W = .3,G = 12.E6,R = .25,L = 36.
INIT
                RHO = W/386.
                KU = 5.*3.14159*G*R**4./L
                J = 3.14159*L*RHO*R**4/20.
                KSJ = KU/J
RERUN           KU,J,KSJ
DYNAM
                TI = STEP(0.0)
                BD = INTGRL(BDIC,BD2,10)
                B = INTGRL(BIC,BD,10)
PROCEDURE       BD2 = VIBRAT(KU,J,KSJ,TI,BD,B)
                BD2(1) = ((B(2) – B(1))*KU + TI)/J
                BD2(10) = ( – B(10)*2.*KU + (B(9) – B(10))*KU)/J
                DO 5 I = 2,9
            5   BD2(I) = (B(I – 1) – B(I) + B(I + 1) – B(I))*KSJ
ENDPROCEDURE
                BE = B(1) + TI/(2.*KU)
                BLO2 = (B(5) + B(6))/2.
TIMER FINTIM = .002,DELT = 5.E – 6,OUTDEL = 2.E – 5
PAGE GROUP,WIDTH = 100
OUTPUT B(1 – 5)
OUTPUT B(6 – 10)
```

```
OUTPUT BE,BLO2
OUTPUT B(1 – 10)
PAGE CONTOR
```

The statement FIXED I declares variable I to be an integer quantity, so that it can later be used as an index in the DO loops, which allows us to compactly specify our set of equations. We will use the subscripted variable B(I) to represent θ_I, where I goes from 1 to 10 for the 10 disks. The two TABLE statements each specify 10 zero initial conditions for angles B and angular velocities BD of the 10 disks. Physical parameters for a steel rod of length 36 inches and radius 0.25 inches are given in the PARAM statement, while the cards between INIT and RERUN calculate the needed spring constant KU, disk inertia J, and their ratio KSJ, which appears often in the equations. We use RERUN here in a way unintended by CSMP's authors. To get a printout of the results of INIT calculations usually requires some "separate" FORTRAN programming; if no more than three quantities (in our case KU, J, KSJ) are required, RERUN produces the desired printout more simply.

In the DYNAM section the B and BD INTGRL statements use the so-called "specification form," which permits the use of arrays for the inputs, outputs, and initial conditions. The BD statement, for example, specifies an array of 10 integrators whose initial conditions are found in the array specified by the earlier TABLE BDIC statement, whose inputs are the subscripted variables BD2(I), and whose outputs are the subscripted variables BD(I). Between the statements PROCEDURE and ENDPROCEDURE is found a small FORTRAN program that uses subscripted variables (*not allowed* in "ordinary" CSMP programming) to compactly set up the equations for the accelerations of the 10 disks. When the general capabilities of FORTRAN are needed within CSMP, several approaches, one of which is PROCEDURE, are available. Statements between PROCEDURE and ENDPROCEDURE are not sorted internally but are treated as a single functional entity by the CSMP sorting algorithm. They must thus obey all the rules of standard FORTRAN, such as being in strict order. The PROCEDURE card itself lists:

1 Names of all outputs, to left of equal sign, separated by commas (our example has only one output from the PROCEDURE, the acceleration BD2).
2 The function name. This is a dummy name and can be chosen at will; in our case we called it VIBRAT.
3 All the parameters and variables needed "inside" the PROCEDURE that must be fetched from elsewhere in the program; in our case parameters KU, J, KSJ and variables TI, BD, and B.

The "working" statements inside the PROCEDURE calculate separately the accelerations BD2(1), BD2(10) of the end disks and use a DO loop for the other 8 accelerations, whose equations are all of identical form. (Note that if we

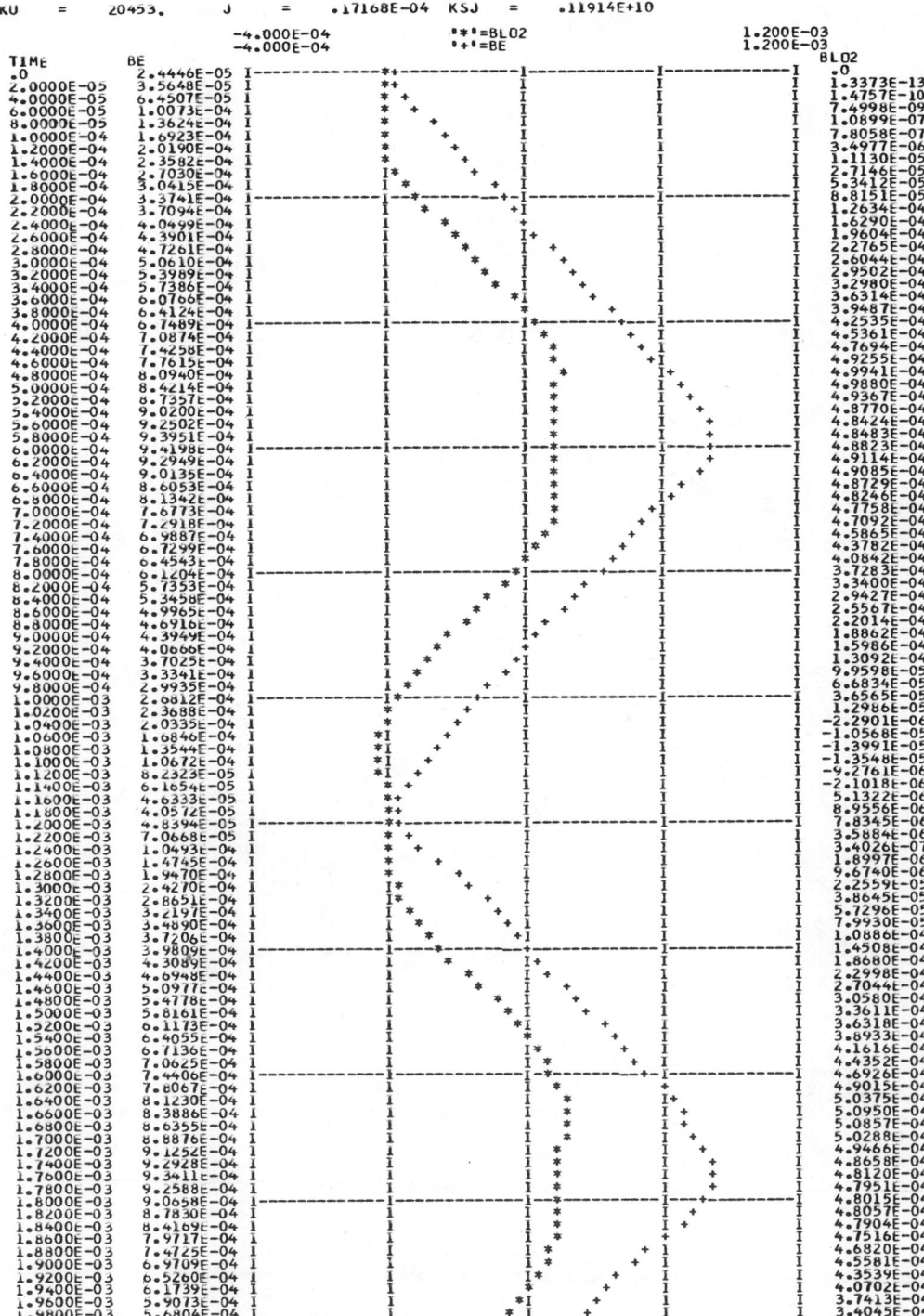

Figure 5.15 Step response of torsional vibration.

205

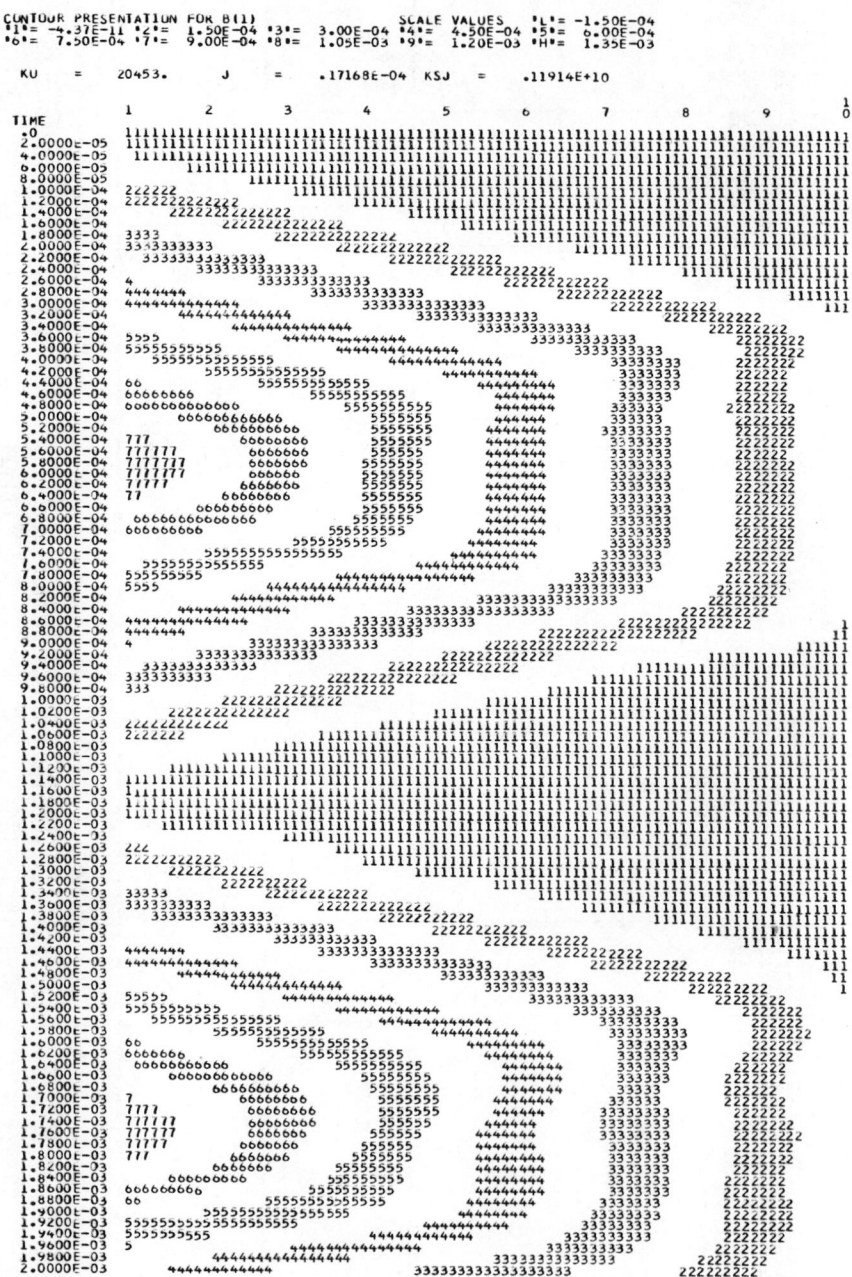

Figure 5.16 "Contour map" display for torsional rod vibration.

decided to expand the model to, say, 100 lumps, very little additional programming is necessary. This not only saves time and effort but also reduces the chances of human error since the computer, rather than the human programmer, assembles the set of 100 equations.) To compare the results of our 10-lump model with the "exact" behavior graphed in Figs. 4.4 and 4.5, we compute the end deflection BE and the mid-point (L/2) deflection BLO2 as shown; OUTPUT BE, BLO2 produces the graph of Fig. 5.15, which shows good agreement with the distributed-parameter results. A "contour map" type of output is often revealing for systems with spacewise variations; the statements OUTPUT B(1 − 10) and PAGE CONTOR produce the display of Fig. 5.16.

5.8   USE OF SORT/NOSORT FOR FORTRAN PROGRAMMING WITHIN CSMP

One of the features of CSMP is that it allows use, when necessary, of the very general programming capabilities of FORTRAN. This facility can be implemented in a number of ways; the previous section showed the use of PROCEDURE, we now give an example using the SORT/NOSORT capability. In a study related to the operation of vibrating conveyors it was desired to investigate the motion of a "bouncing particle," modeled as a mass/spring/damper system in Fig. 5.17.

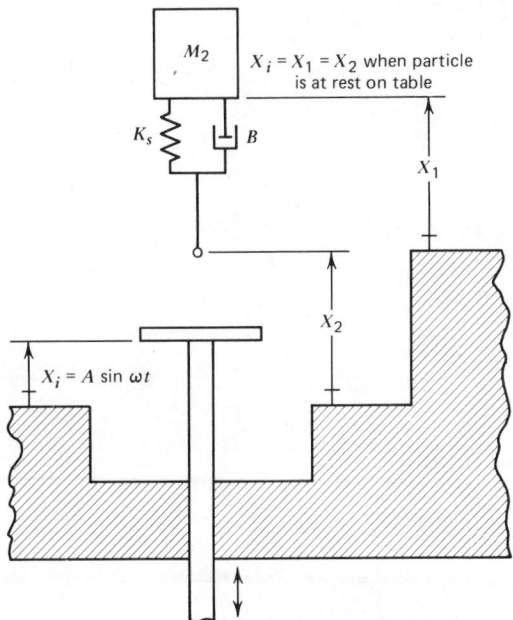

Figure 5.17   Bouncing particle system.

If the conveyor table is moved vertically with displacement $X_i = A \sin \omega t$, when A and/or ω are sufficiently large, the particle will intermittently make and leave contact with the table, requiring us to change the equation of particle motion each time this happens. While CSMP has considerable capability for dealing with intermittent effects, no direct way of handling this particular phenomenon was apparent, so a FORTRAN approach was tried, leading ultimately to the following program.

```
METHOD       RKSFX
INIT
PARAM        A = 10.,W = 10.
NOSORT
PARAM        M2 = 1.,KS = 400.,B = 20.,W2 = 386.
INCON        X10 = 0.0,X20 = 0.0,X2D0 = 0.0
             X1DOT = W*A*COS(W*TIME)
             FKB = B*(X1DOT − X2DOT) + KS*(X1 − X2) + W2
             IF(FKB)20,20,10
DYNAM
NOSORT
      5      IF(SW − 1.0)7,7,6
      6      IF(FKB)20,20,10
      7      IF(X1 − XI)10,10,20
*                PARTICLE IN CONTACT
      10     XI = A*SIN(W*TIME)
             X1DOT = W*A*COS(W*TIME)
             FKB = B*(X1DOT − X2DOT) + KS*(X1 − X2) + W2
             X2DOT2 = (FKB − W2)/M2
             SW = 2.0
             GO TO 30
*                PARTICLE FLOATING
      20     XI = A*SIN(W*TIME)
             X2DOT2 = − W2/M2
             X1DOT = (− KS*(X1 − X2) − W2 + B*X2DOT)/B
             FKB = 0.0
             SW = 0.0
             GO TO 30
*                INTEGRATING SECTION
      30     X2DOT = INTGRL(X2D0,X2D0T2)
             X2 = INTGRL(X20,X2DOT)
             X1 = INTGRL(X10,X1DOT)
TIMER FINTIM = 3.,DELT = .001,OUTDEL = .012
PRINT SW,XI,X1,X2,FKB,X2DOT2,X2DOT,X1DOT
OUTPUT XI,X1,X2
```

```
PAGE GROUP
OUTPUT TIME,XI,X1,X2
PAGE GROUP,XYPLOT,HEIGHT = 5.5,WIDTH = 6.5
LABEL BOUNCING PARTICLE
```

When this program was initially run using the usual (default) Runge–Kutta variable-step-size integration algorithm, the computation self-aborted because DELT was made (automatically) excessively small. This occasionally happens when the physical processes in a model exhibit discontinuities; it can usually be handled by using one of the fixed-step-size integration algorithms such as the Runge–Kutta called by METHOD RKSFX. In the INIT section, parameter numerical values and initial conditions $X_1(0)$, $X_2(0)$, and $\dot{X}_2(0)$ are set and the value of the force FKB in the spring/damper combination at time equals zero is calculated. If FKB is positive, particle/table contact is insured and the first IF statement branches the calculation into the PARTICLE IN CONTACT section. (The first NOSORT statement insures that all steps between it and DYNAM will be performed exactly in the sequence given.) If FKB is negative or zero, the calculation is branched into the PARTICLE FLOATING section. Actually, for the *physical* model given, FKB could never be negative; however, computationally, FKB could *start* to go negative and our program will note this and, within one DELT, switch the program to the PARTICLE FLOATING section, which has a statement setting FKB exactly equal to zero.

For the numerical values given, FKB = 2386 at $t = 0$, so the PARTICLE IN CONTACT branch is initially entered. Here FKB is again calculated (since the FKB statement between INIT and DYNAM is unavailable after $t = 0$), $\dot{X}_2$ is calculated and a "flag" SW is set equal to 2.0. The GO TO 30 statement now branches to the integrating section where $\dot{X}_2$, X_2, and X_1 are computed, whereupon the program returns to the beginning of the DYNAM section to start the next computing increment. Here, IF(SW-1.0)7,7,6 checks whether we have just come from the CONTACT or FLOATING section. If we have been in CONTACT we branch to 6, which checks FKB to see if we are *still* in contact and then branches to either CONTACT or FLOATING. If we had been FLOATING, 5 branches us to 7, which checks whether X1 and XI have *regained* contact, branching to the proper path in either case. Note that when we are in the FLOATING section, $\ddot{X}_2$ is just the gravitational acceleration, but $\dot{X}_1$ is determined by the "relaxation" of the spring/damper combination.

In a model with branching of this sort, a tabular printout such as that produced by the PRINT statement is quite helpful in debugging and making sure that the program is sequencing properly. Figure 5.18 shows the off-line drum plot produced by the XYPLOT statement, clearly showing the "table-following," "free-flight," and "impact" features of the particle's motion. This particular program does not have any SORT sections but, in general, SORT and NOSORT sections can be freely intermixed in CSMP if necessary. In a SORT

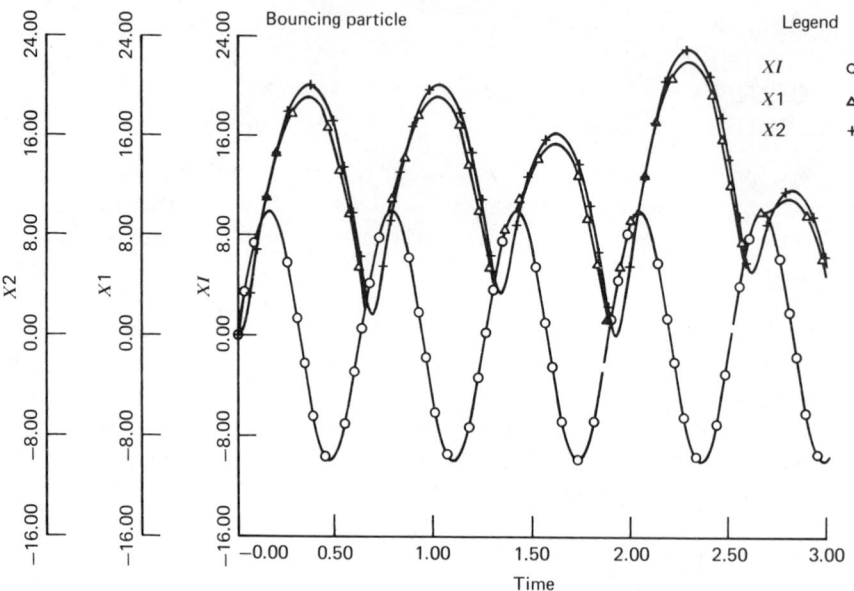

Figure 5.18 Displacements in bouncing particle system.

section, statement sequence is unimportant since the sorting algorithm will put the statements in the order needed for computation. When no PROCEDURE, SORT, or NOSORT statements appear (as in our examples of Sections 5.4–5.6), the entire program is automatically sorted.

5.9 RANDOM SIGNALS IN CSMP

Due to possible large running times (and thus costs) associated with obtaining statistically reliable results, random signal studies may be more efficiently carried out, in many cases, using analog or hybrid methods rather than pure digital. However, digital methods such as CSMP are quite practical in some problems and if no analog facility is available, they might be used in any case. While CSMP provides two random number generators, one with Gaussian distribution and one with uniform, these are not usually directly usable in system dynamics studies and the user must augment them with some additional processing, This can be accomplished in various ways, one of which will now be outlined.

In analog simulation and physical laboratory testing with random signals, the usual approach is to use a signal generator of Gaussian distribution and flat (out to some cutoff frequency) spectral density, and to then filter the output of this device to obtain the spectrum desired for the particular test, using Eq.

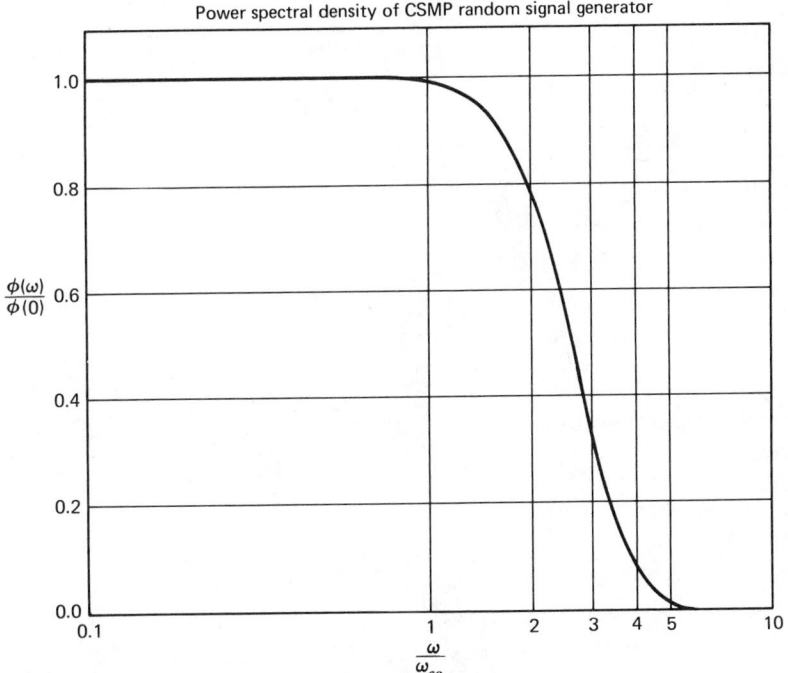

Power spectral density of CSMP random signal generator

Figure 5.19   Power-spectral density of CSMP random signal generator.

(3.107) as a basis for design of the filter. We will follow the same approach in CSMP. It can be shown[10] that a random signal generator with Gaussian distribution and spectral density as in Fig. 5.19 can be constructed with the following CSMP statements.

```
METHOD   RKSFX
         YP = IMPULS(0.0,HLDT)
         YG = GAUSS(ODDI,MN,SD)
         YH = ZHOLD(YP,YG)
         YF1 = CMPXPL(0.0,0.0,.67,OMFIL,YH)
         YRAND = CMPXPL(0.0,0.0,.67,OMFIL,YF1)
```

Figure 5.20 shows how these statements produce the desired random signal YRAND. A fixed-step-size integration routine, such as RKSFX, is used to insure consistent timing of IMPULS, which controls the frequency content of

---

[10]Schaefer, W. S., "Applied Digital Simulation of Random Signals and Step Response of Heat Exchanger Optimized for Random Signals," M.Sc. Thesis, Dept. of Mech. Eng., The Ohio State University, Columbus, 1972.

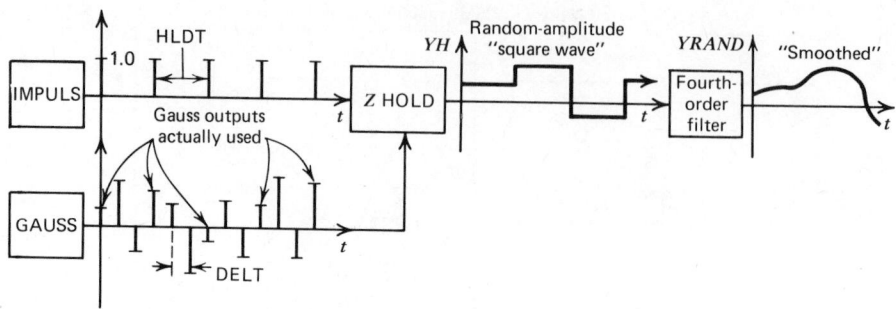

Figure 5.20   Scheme for generating random signal.

the final signal through the zero-order hold action of ZHOLD. Numerical values to be inserted are:

HLDT $\triangleq$ hold time $= 0.5505/\omega_{co}$

ODDI $\triangleq$ any odd integer, $1, 3, 5$, etc.

MN $\triangleq$ mean value of GAUSS $=$ mean value of YRAND desired

OMFIL $\triangleq$ filter corner frequency $= 3.18\omega_{co}$

SD $\triangleq$ standard deviation of GAUSS $= (\pi\phi_{WN}/\text{HLDT})^{1/2}(\text{OMFIL})^4$

$\phi_{WN} \triangleq$ power spectral density of YRAND in its flat range

(ϕ_{WN} is defined according to

$$\overline{\text{YRAND}^2} = \int_0^\infty \phi_{\text{YRAND}}\, d\omega$$

and has the dimensions (physical quantity)^2/(rad/sec)).

Numbers for ϕ_{WN} and ω_{co} (see Fig. 5.19) are chosen to meet the needs of the particular application. The total mean-square value of YRAND (including the non-flat portion) will be 3.89 $\phi_{WN}\omega_{co}$. If one wants to generate *several* random signals that are uncorrelated, simply use a *different* odd integer in each YG statement; this changes the "seed" of the random number generator to produce independent random signals. In choosing a proper DELT on the TIMER card, in addition to the usual considerations one must here be sure that HLDT is an integer multiple of DELT.

To show how this random signal generator is used, consider the system of Fig. 5.21. Here two mechanical vibrating systems, separated at rest by a clearance space CLEAR, are driven by the same random force. We wish to calculate the probability that the two masses "bump" and will do this by computing the percentage of the time that $(X_A - X_B - \text{CLEAR})$ is greater than zero. Our model will ignore the dynamic effects on the motions of the two masses caused by any collisions. This approximation can be justified if collisions

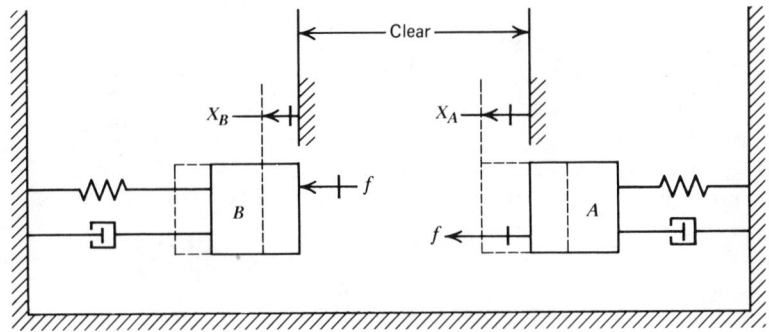

Figure 5.21   Mechanical system with random driving force.

are rare. Let us assume that the driving force has a "flat" spectrum as in Fig. 5.19 so that YRAND need not be processed through a spectrum-shaping filter but can be applied directly as the input force. Suppose that the random force can be described by $\omega_{co} = 2753$ rad/sec and $\phi_{WN} = 0.04$ N^2/(rad/sec), giving HLDT = 0.0002 sec and SD = 1.46×10^{17}, while the mechanical systems are taken as second order with gain = 1.0 and $\zeta = 0.1$ for both, while $\omega_{nA} = 2000$ rad/sec and $\omega_{nB} = 500$ rad/sec

A CSMP program might go as follows.

```
TITLE   MOVING PART INTERFERENCE STUDY
METHOD   RKSFX
PARAM      NBUMP1 = 0.0,NBUMP2 = 0.0,CLEAR = 50.0
          YP1 = IMPULS(0.0,.0002)
          YG1 = GAUSS(1,0.0,1.46E17)
          YH1 = ZHOLD(YP1,YG1)
          YF11 = CMPXPL(0.0,0.0,.67,8750.,YH1)
          YRAND1 = CMPXPL(0.0,0.0,.67,8750.,YF11)
          YG2 = GAUSS(3,0.0,1.46E17)
          YH2 = ZHOLD(YP1,YG2)
          YF12 = CMPXPL(0.0,0.0,.67,8750.,YH2)
          YRAND2 = CMPXPL(0.0,0.0,.67,8750.,YF12)
          XA1 = CMPXPL(0.0,0.0,.1,2000.,YRAND1)
          XA = 4.E6*XA1
          XBC = CMPXPL(0.0,0.0,.1,500.,YRAND1)
          XB1 = 2.5E5*XBC
          XBD = CMPXPL(0.0,0.0,.1,500.,YRAND2)
          XB2 = 2.5E5*XBD
          XAC = XA + CLEAR
          BUMP1 = XA − XB1−CLEAR
          BUMP2 = XA − XB2 − CLEAR
```

```
NOSORT
        IF(KEEP.NE.1)GO TO 40
        IF(BUMP1)20,20,10
  10    NBUMP1=NBUMP1+1.
  20    IF(BUMP2)40,40,30
  30    NBUMP2=NBUMP2+1.
  40    CONTINUE
TIMER FINTIM=.1,DELT=.00005,OUTDEL=.0002,PRDEL=.001
TERMINAL
        PBUMP1=NBUMP1/2000.
        PBUMP2=NBUMP2/2000.
        WRITE(6,50)PBUMP1,PBUMP2
  50    FORMAT('0','PBUMP1=',E10.5,5X,'PBUMP2=',E10.5)
OUTPUT YRAND1,YRAND2
PAGE    GROUP
OUTPUT YRAND1,XA
OUTPUT XA,XB1,XB2
PAGE    GROUP
OUTPUT XAC,XB1,XB2
PAGE    GROUP
OUTPUT TIME,XAC,XB1,XB2
LABEL   RANDOM IMPACTS
PAGE    XYPLOT,GROUP,HEIGHT=5.5,WIDTH=6.5
PRINT   BUMP1,NBUMP1,BUMP2,NBUMP2
END
```

In the PARAM statement, the number of collisions, NBUMP1 and NBUMP2, are initialized at zero and the clearance space is specified. Then the random force YRAND1 is computed, following the rules given earlier. It was desired to explore also the effect of the random force on system B having identical ω_{co} and ϕ_{WN} to that acting on A but being statistically independent of it, so a *second* random force YRAND2 is generated using a different odd integer (3) as the seed for the random number generator. The program is set up to solve both these problems simultaneously; this is why *two* NBUMP parameters were needed earlier. We thus have one variable XA representing the motion of mass A due to force YRAND1, but two variables XB1 and XB2 giving the motions of mass B for the two different random forces described above. The variables BUMP1 and BUMP2 indicate a collision whenever they are greater than zero.

Since some FORTRAN logic is necessary to "count" the collisions, we enter a NOSORT section. The first statement here employs the variable KEEP, which we have not encountered earlier, but which is present in every CSMP program. It is in the nature of a "flag" that CSMP automatically assigns the values 0 or 1, depending on whether calculations for the current integration step are incomplete or complete. Since we wish to check for collisions only once for each complete integration step DELT, the first IF statement bypasses the collision calculation unless KEEP = 1. If KEEP = 1, then the collision-counting scheme of the NOSORT section is entered and NBUMP1 and/or NBUMP2 are updated by adding 1 to the accumulated number of collisions. Note that the definition of "one collision" is that there be interference for the current DELT. Thus if, say three successive DELT's all have interference, this is counted as three "collisions," not one. This definition is not as unreasonable as it might first appear, since our real goal is to compute the "probability" of collision over the total time interval FINTIM. This probability is calculated in the TERMINAL section, *after* the complete FINTIM has been run through, by taking the ratio of NBUMP1 (or NBUMP2) to 2000, which is the number of DELT's contained in FINTIM. This definition of probability (PBUMP1, PBUMP2) as a ratio of the interference *time* to the total time is conventional and reasonable and justifies the definition of "collision" used earlier.

Constraints of space do not allow presentation here of all the graphic and tabular results produced by the above program, so only a few selected items will be discussed. The table requested shows the gradual accumulation of impacts, NBUMP1 ending at 93 while NBUMP2 goes to 42; the respective probabilities

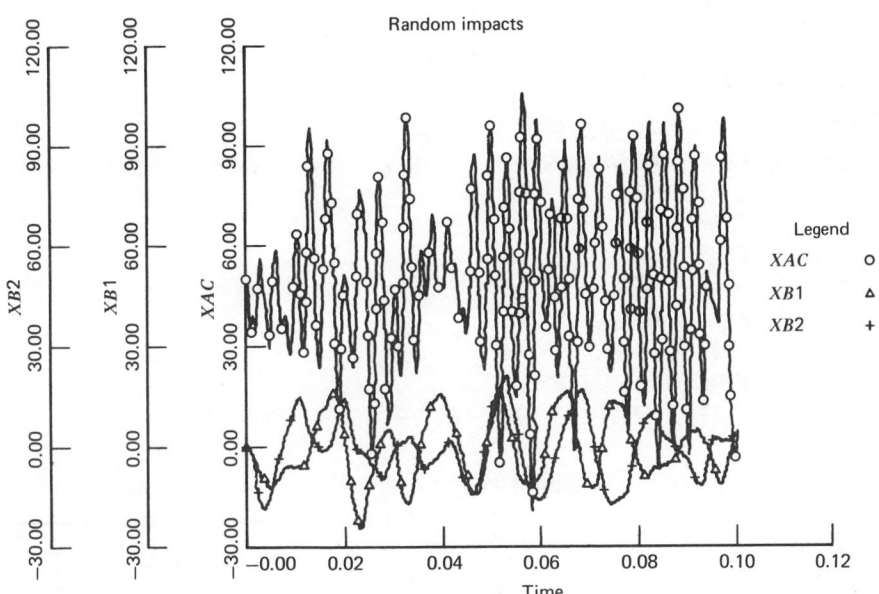

Figure 5.22   Response of mechanical system with random force.

are thus 0.0465 and 0.021. It appears that when the random force on B is statistically independent of that on A, fewer impacts result; this seems reasonable. Figure 5.22, the XY plot called for, clearly shows the random nature of the motions and the occurrence of interference. Finally we should point out that while some of the numerical results in this problem could be obtained analytically (without simulation), one needs only to introduce nonlinearities in the system models (often present in realistic problems) to make simulation a necessity.

5.10 SAMPLED-DATA SYSTEMS

In Fig. 4.11 we showed the results of a CSMP study of a sampled-data feedback control system as a means of checking theoretical results and providing response details difficult to obtain from theory. Details of this program follow below.

```
METHOD   RKSFX
INIT
            C1 = KD/T + KP + KI*T
            C2 = 2.*KD/T + KP
            C3 = KD/T
PARAM    T = .1,KD = 1.8,KP = 2.6,KI = 2.
PARAM    KPL = 1.,TAUPL = 1.
DYNAM
            R = STEP(0.0)
            RI = INTGRL(0.0,R)
            RII = INTGRL(0.0,RI)
            CDOT = 1.8*R + 2.6*RI + 2.*RII − 2.8*C − 2.6*CI − 2.*CII
            C = INTGRL(0.0,CDOT)
            CI = INTGRL(0.0,C)
            CII = INTGRL(0.0,CI)
            E1 = R − CS
            SW = IMPULS(0.0,T)
            ESH = ZHOLD(SW,E1)
            ED1 = DELAY(20,.1,ESH)
            ED2 = DELAY(40.,.2,ESH)
NOSORT
            IF(TIME.GT.0.0)GO TO 10
            CMPO = 0.0
     10     IF(SW.EQ.1)GO TO 15
            GO TO 20
     15     CMPO = CMPO + C1*ESH − C2*ED1 + C3*ED2
     20     CONTINUE
```

```
SORT
          M=INTGRL(0.0,CMPO)
          CA=REALPL(0.0,TAUPL,M)
          CS=KPL*CA
TIMER     FINTIM=3.,DELT=.005,OUTDEL=.02
OUTPUT    CMPO,C,CS
```

Because of the precise timing required by the sampling switch, a fixed-step integration algorithm RKSFX is used. Numerical values are entered with PARAM statements and initial calculations carried out under INIT. Statements between R and CII model the continuous feedback system used for comparison with the sampled-data system. This continuous system has open loop

$$\frac{C}{E}(s) = \frac{1.8s^2 + 2.6s + 2}{s^2(s+1)} \tag{5.14}$$

giving closed loop

$$\frac{C}{R}(s) = \frac{1.8s^2 + 2.6s + 2}{s^3 + 2.8s^2 + 2.6s + 2} \tag{5.15}$$

$$(s^3 + 2.8s^2 + 2.6s + 2)C = (1.8s^2 + 2.6s + 2)R \tag{5.16}$$

Equations, such as (5.16), involving differentiation of the *input* (*R* in this case) can be handled several ways in CSMP. Integration of both sides of (5.16) twice (division by s^2) gives

$$\dot{C} = 1.8R + 2.6\int R\,dt + 2\int\int R\,dt - 2.8C - 2.6\int C\,dt - 2\int\int C\,dt \tag{5.17}$$

which is what was implanted in the present program. Actually, this is not the best method since the "open-loop" integrators used will gradually "drift." A better approach represents (5.15) as

$$\frac{L}{R} = \frac{1}{s^3 + 2.8s^2 + 2.6s + 2}, \qquad C = (1.8s^2 + 2.6s + 2)L \tag{5.18}$$

which can be implemented without open-loop integrators as

```
LD3=R-2.8*LD2-2.6*LD-2.*L
LD2=INTGRL(0.0,LD3)
LD=INTGRL(0.0,LD2)
L=INTGRL(0.0,LD)
C=1.8*LD2+2.6*LD+2.*L
```

Finally, CSMP III provides a module TRANSF that models the general form

$$\frac{Y}{X}(s) = \frac{a_m s^m + a_{m-1} s^{m-1} + \cdots + a_1 s + a_0}{b_n s^n + b_{n-1} s^{n-1} + \cdots + b_1 s + b_0}, \qquad n \geqslant m$$

and requires only that the a's and b's be entered properly and that all initial conditions be zero.

Modeling of the sampled-data system begins with the E1 statement which forms the error signal from the input R and the output CS of the sampled system. Statements SW and ESH model the sample and hold operation; ESH samples E1 every T seconds and holds it till the next sampling instant. Delayed versions (ED1, ED2) of ESH, needed in the difference equation of the digital computer controller, are easily obtained with DELAY statements. We next go into a NOSORT section to carry out some FORTRAN logic. The first two statements here allow us to set the output CMPO of the computer controller at some initial value (here taken 0.0) at time = zero. The remaining statements insure that CMPO is updated only when the sampling switch is activated (*not* at every DELT) and actually compute CMPO according to its difference equation (Eq. (4.45)). We now leave the NOSORT section to model the remaining (continuous) parts of the system $1/s(s+1)$ with INTGRL and REALPL statements. It should be clear that with a program of this sort, guided by the modest theoretical background of Section 4.2, one can evaluate the performance of various proposed combinations of numerical values for controller and controlled system quite effectively and thereby accomplish a practical trial-and-error sort of design.

5.11   STIFF-SYSTEM PROBLEMS

In recent years, considerable attention has been focused on numerical methods that are capable of efficiently dealing with equations which represent so-called "stiff" systems. This brief section is intended to make the reader aware of the nature of the stiff-system problem and techniques for handling it. In the realm of linear ordinary differential equations with constant coefficients, a stiff system is defined as one for which the roots of the characteristic equation exhibit a wide spread (say 1000 to 1 or more) in numerical value. Taking the mechanical system of Fig. 2.11 as example, this would happen if, say, M_1 and M_2 were about equal, but K_{s2} were 10^6 times as stiff as K_{s1}. Taking, for convenience, $B_1 = B_2 = 1.0$, $M_1 = M_2 = 1.0$, $K_{s1} = 1.0$, and $K_{s2} = 10^6$, a computer root finding program gives the roots as $-0.5 \pm 0.866i$ and $-0.5 \pm 1000i$. This example reveals that the terminology "stiff system" is somewhat of a misnomer since the origin of the behavior lies not in the stiffness of the entire system but rather in the stiffness of a *component* embedded in an "otherwise-soft" system. For lack of a more descriptive and equally concise name, the stiff-system terminology has become widely accepted and presents no difficulty once one understands the basic definition. It should also be clear that "stiffness" is not limited to mechanical systems (where its relation to *spring* stiffness is obvious) but is basically a mathematical phenomenon.

The difficulties related to the computer analysis of stiff systems become apparent when one tries to apply the standard numerical integration algorithms to the solution of the differential equations. These difficulties can in fact be anticipated if one considers the following points. In a stiff system, transient responses will contain components of widely disparate time scale. To maintain accuracy and stability in the numerical integration, the time step DELT must be chosen small enough to suit the rapidly varying components of the solution, but the total problem time FINTIM must be chosen large enough to show the decay of the slower components, leading to intolerably large computing times and costs. The "standard" (not specifically designed for stiff problems) variable-step algorithms (such as the RKS we usually use in CSMP) are sometimes of some help in such a situation since they continuously adjust DELT. However, in really difficult stiff systems these methods also fail to reduce computing times to acceptably short values to meet the requirements of cost in batch processing or response time in on-line interactive systems using graphics terminals.

Before going further we should make the point that stiff-system numerical difficulties are sometimes *needlessly* created by insufficient emphasis on the use of "engineering judgment" at the *modeling* stage of the analysis. Remember that the diagram of Fig. 2.11 is *not* the real physical system being considered but rather a *model*, chosen by the analyst to represent some aspects of the real system's behavior. Thus when K_{s2} becomes one million times as stiff as K_{s1}, it may be appropriate to *reconsider* the model and replace the "moving part" M_2 by an "immovable wall." Certainly the immovable wall presently shown at the right of Fig. 2.2 is not a *real* feature of the system but a *model* representing a very stiff real object. When K_{s2} approaches the stiffness of *that* object, might it not make sense to treat M_2 as immovable? We raise these questions here, not to imply that stiff-system studies are merely mathematical exercises with no practical significance, but rather to try to maintain some balance in emphasis between intuitive engineering and mathematical abstraction. Special efforts to maintain such a balance may be increasingly necessary because of the growth in availability of "canned programs" for creating sophisticated mathematical models for a wide range of engineering systems. While the author would agree that such programs, properly used, are a real boon to engineering design, the tendency to replace refined judgment (as a means of attaining frugal models) with raw computer power untempered with critical thought, is a trend engineers might embrace with some caution.

Returning to our technical discussion, nonlinear and/or time-variable-coefficient models present situations where we may not be able to *avoid* stiffness problems by redefining the model as suggested above. Thus these are perhaps the main application areas where stiff-system methods have a contribution to make. For a mechanical system with a nonlinear spring whose force is given by KX^3, when displacement X goes from 0.01 to 1.0, the local "stiffness" of the spring changes by 10^6 to 1, giving rise to numerical problems resembling those in a linear system with two springs differing by 10^6 to 1 in stiffness. When a number of such nonlinearities are embedded within a large set of simultaneous differential equations, stiff-system difficulties may arise that are resolvable only by the employment of special integrating algorithms tailored to these problems. A variety of such algorithms[11,12] have been available for some time. These were at first employed in special purpose programs designed to deal with specific physical problems. Later, certain of the algorithms were interfaced with general purpose simulation languages.

In CSMP III, an algorithm[13] for stiff systems may be invoked with the statement METHOD STIFF. Some examples[13] of its use follow:

$$\frac{dy_1}{dt} = -2000y_1 + 1000y_2 + 1, \qquad y_1(0) = 0$$

$$\frac{dy_2}{dt} = y_1 - y_2, \qquad y_2(0) = 0 \tag{5.19}$$

These linear simultaneous equations exhibit a ratio of roots of about 4000 to 1 and required 12350 passes through the standard RKS algorithm to obtain the solution between $t = 0$ and $t = 4$. The STIFF algorithm required only 480 passes, an improvement of about 25 times.

$$\frac{dz_1}{dt} = 0.04(1 - z_1) - (1 - z_2)z_1 + 0.0001(1 - z_2)^2, \qquad z_1(0) = 0$$

$$\frac{dz_2}{dt} = -10000\frac{dz_1}{dt} + 3000(1 - z_2)^2, \qquad z_2(0) = 0 \tag{5.20}$$

These nonlinear equations were solved 40 to 50 times faster using STIFF as compared to RKS. Further examples and details are found in the reference, which also puts forth (based on operating experience at that time) the following

[11] Gear, C. W., *Numerical Initial Value Problems in Ordinary Differential Equations*, Prentice-Hall, Englewood Cliffs, N.J., 1971.

[12] Calahan, D. A., *Computer-Aided Network Design*, 2nd ed., McGraw-Hill, N.Y., 1972.

[13] Fowler, M. E. and Warten, R. M., "A Numerical Integration Technique for Ordinary Differential Equations with Widely Separated Eigenvalues," *IBM J. of Res. and Dev.*, Vol. II, No. 5 (1967), p. 537.

observations:

1   For linear systems with constant coefficients in which the largest root is real, the speed advantage over RKS increases when the ratio of largest to smallest roots increases.

2   If the largest root is complex, the RKS algorithm is preferred to STIFF.

3   For nonlinear and/or variable-coefficient equations that can be approximated by piecewise linear systems with constant coefficients, rules 1 and 2 hold for each piece.

A more recent reference[14,15] documents some further improvements in the stiff-system capabilities of CSMP III; however, these have not yet been made part of the standard package provided CSMP users. The new algorithm, STIFFX, has been tried on a wide variety of "real life" models, including a chemical process of order 41 (41 first-order equations), electronic circuits of order 49 and 84, and a public utility problem of order 139. The improvement in speed, compared to STIFF, ranged from 4.3 to 1 for the lowest-order problem to 19 to 1 for the highest.

5.12   DIGITAL FREQUENCY-RESPONSE CALCULATION

This chapter has emphasized time-domain response calculations using simulation languages. Since Chapter 3 showed the great utility of frequency-response methods, we wish in this section to address briefly the computer aids available to the engineer in this area, even though they are not usually considered to be "simulation." Recall first that frequency response methods apply rigorously only to linear systems with constant coefficients. For such systems, special purpose programs for evaluating and plotting $W(i\omega)$ for a selected range of frequencies are widely available in computer libraries and, in fact, one can rather easily generate such a program himself if a "canned" one is not available. While the most efficient programs will use a language such as FORTRAN, CSMP and SPEAKEASY can also be employed. To use CSMP for linear system frequency response,[16] one merely lets the independent variable TIME play the role of frequency; the integrating capabilities of CSMP are not used at all. SPEAKEASY actually computes frequency response quite easily and efficiently; the program associated with Fig. 3.35 shows the technique.

Occasionally, frequency-response techniques are applied, in approximation, to nonlinear systems, using digital simulation such as CSMP in the time domain. This is accomplished by applying a sinusoidal input to the nonlinear CSMP

---

[14] Gourlay, A. R. and Watson, H. D. D., "An Implementation for Gear's Algorithm for CSMP III," in *Stiff Differential Systems*, ed. by R. A. Willoughby, Plenum, N.Y., 1974.

[15] Gourlay, A. R. and Watson, H. D. D., "Implicit Integration for CSMP III and the Problem of Stiffness," *Simulation*, Feb. (1976), p. 57–61.

[16] Speckhart, F. H. and Green, W. L., *A Guide to Using CSMP*, Prentice Hall, Engelwood Cliffs, N.J., 1976, p. 194

model and letting TIME run until the response variable of interest goes into a periodic steady state. If the nonlinearity is not extreme and/or the sinusoidal input amplitude is not large, the periodic steady state will be "close to" sinusoidal and one can estimate its amplitude and phase (relative to the input), thereby getting one point on the system "frequency-response" curve. Repetition with different values of frequency produces the desired curves. Since nonlinear system response is sensitive to input *amplitude*, this may also have to be explored, leading to *families* of frequency-response curves, one pair for each amplitude. Clearly, this procedure can quickly lead to excessive computer costs; but sometimes it can be justified.

Having above quickly surveyed and reviewed frequency-response calculations (some of which were met earlier in this text), we now turn to the main purpose of this section, the introduction of a frequency-response calculation method based on *matrix techniques*. This approach has certain features and advantages that make it of considerable practical interest. To implement it, one needs to have available in his computer library, subroutines for manipulating matrices with complex-number elements. The SPEAKEASY language we have employed earlier is particularly easy to use in this case so we will explain the method in SPEAKEASY terms. It will not be difficult, however, to adapt this method to whatever matrix subroutines are available to the reader. The method is based on the fact that, when initially derived from the appropriate physical laws, system models consist of sets of simultaneous differential equations. If one desires a transfer function relating a chosen output/input pair, it is necessary to use substitution and elimination, determinants, and so on to combine the n equations in n unknowns into a single equation relating the chosen input and output. These "manual" manipulations become increasingly tedious and error-prone as n gets larger. If the desired end result of such manipulations is only to obtain output/input *frequency-response graphs* for known numerical values of parameters, and if the models are linear with constant coefficients, matrix computer operations on the *original* set of n equations allow us to achieve this result with *no* "manual" manipulations whatever.

To see how the method is used, consider a general system with inputs q_{ia}, q_{ib}, and so on and outputs $q_{o1}, q_{o2}, \ldots, q_{on}$ for which a model consisting of n linear differential equations with constant coefficients has been written. If these equations are Laplace transformed with zero initial conditions and then s is replaced by $i\omega$, we obtain a set of n *algebraic* equations in the n unknowns $Q_{o1}(i\omega)$, $Q_{o2}(i\omega)$, and so forth:

$$A_{11}Q_{o1}(i\omega) + A_{12}Q_{o2}(i\omega) + \cdots + A_{1n}Q_{on}(i\omega) = C_{1a}Q_{ia}(i\omega) + C_{1b}Q_{ib}(i\omega) + \cdots$$

$$A_{21}Q_{o1}(i\omega) + A_{22}Q_{o2}(i\omega) + \cdots + A_{2n}Q_{on}(i\omega) = C_{2a}Q_{ia}(i\omega) + C_{2b}Q_{ib}(i\omega) + \cdots$$

$$\vdots \qquad\qquad\qquad\qquad\qquad\qquad\qquad\qquad\qquad\qquad (5.21)$$

$$A_{n1}Q_{o1}(i\omega) + A_{n2}Q_{o2}(i\omega) + \cdots + A_{nn}Q_{on}(i\omega) = C_{na}Q_{ia}(i\omega) + C_{nb}Q_{ib}(i\omega) + \cdots$$

Here the A's and C's are *known* functions of frequency ω. We now choose which input, say Q_{ik}, we wish to exercise, set it at $1.0\underline{/0°}$ (complex number representation of $1.0\sin(\omega t + 0°)$), and set all other inputs to zero. If we now let ω take on various numerical values covering the range of our interest, the A's and C's become numbers (complex) and we have a set of simultaneous algebraic equations to solve. From linear algebra we know that the unknowns can be found by first obtaining the *inverse* of the matrix of A coefficients and then matrix multiplying this by the column vector of the C coefficients that remain on the right-hand side when all Q_i's other than Q_{ik} are set to zero:

$$
\begin{vmatrix} A_{11} & A_{12} \dots A_{1n} \\ A_{21} & A_{22} \dots A_{2n} \\ \vdots \\ A_{n1} & A_{n2} \dots A_{nn} \end{vmatrix}^{-1} \begin{vmatrix} C_{1k} \\ C_{2k} \\ \vdots \\ C_{nk} \end{vmatrix} = \begin{vmatrix} Q_{o1}(i\omega) \\ Q_{o2}(i\omega) \\ \vdots \\ Q_{on}(i\omega) \end{vmatrix}
\tag{5.22}
$$

Inverse of matrix of A coefficients · Column vector of C coefficients · Column vector of unknowns

Since Q_{ik} was set to $1.0\underline{/0°}$, the unknowns $Q_{o1}(i\omega), Q_{o2}(i\omega)$, and so on are actually the sinusoidal transfer functions $(Q_{o1}/Q_{ik})(i\omega), (Q_{o2}/Q_{ik})(i\omega)$, and so on. With "one pass" through the matrix calculations, we thus find the frequency responses between *all* the outputs and the chosen single input Q_{ik}. By choosing a *different* input to be $1.0\underline{/0°}$ and making the rest zero, *all* the possible sinusoidal transfer functions between any selected input and output can be found.

As an example, in the system of Fig. 2.11, rather than manually working out the determinants (2.117) and (2.118) to finally arrive at transfer functions (2.121) and (2.124), if we are satisfied with numerical results for frequency response, we can work directly with the original equations:

$$
\begin{aligned}
\left[M_1(i\omega)^2 + B_1(i\omega) + K_{s1} \right] X_1(i\omega) + \left[-K_{s1} \right] X_2(i\omega) = F_1(i\omega) \\
\left[-K_{s1} \right] X_1(i\omega) + \left[M_2(i\omega)^2 + B_2(i\omega) + K_{s1} + K_{s2} \right] X_2(i\omega) = F_2(i\omega)
\end{aligned}
\tag{5.23}
$$

If our interest is in $(X_1/F_1)(i\omega)$ and $(X_2/F_1)(i\omega)$, we set $F_1(i\omega) = 1.0\underline{/0°}$ and $F_2(i\omega) = 0.0$. To show the details of an actual program we now proceed using SPEAKEASY.

```
PROGRAM MFR
DOMAIN COMPLEX
ANGLES DEGREES
FREQ = A1D(1:);MAGX1 = FREQ;MAGX2 = FREQ;PHASX1 =
    FREQ;PHASX2 = FREQ
```

```
FOR L = 1, 101,1;S = (L − 1)*1I;J = IMAG(S)
M = MATRIX(2,2)
M(1) = 2.*S*S + .1*S + 100, − 100.
M(2) = − 100.,4.*S*S + .05*S + 300.
INVM = INVERSE(M);RITSID = VECTOR(2:1.0,0.0)
PROD = INVM*RITSID
X1 = PROD(1);X2 = PROD(2);MX1 = ABS(X1);MX2 = ABS(X2)
MAGX1(L) = MX1;MAGX2(L) = MX2
PHX1 = PHASE(X1);PHX2 = PHASE(X2)
PHASX1(L) = PHX1;PHASX2(L) = PHX2;FREQ(L) = J
ENDLOOP L
TABULATE FREQ,MAGX1,PHASX1,MAGX2,PHASX2
END
EXECUTE MFR
```

The first statement, PROGRAM MFR, alerts SPEAKEASY to the fact that we wish to use the program mode; the next two statements allow us to use complex numbers and treat trigonometric functions in degrees rather than radians. FREQ = A1D(1:) sets up an empty one-dimensional array (A1D) for a set of frequency values that will be generated later. MAGX1 = FREQ and so on do the same for the amplitude ratios and phase angles of the two transfer functions we are computing. The statement FOR L = 1, 101, 1 starts a computing loop (similar to FORTRAN DO LOOP) that ends with ENDLOOP L. A set of $s = i\omega$ values for $\omega = 0, 1, 2, \ldots, 100$ is generated by S = (L − 1)*1I, while J = IMAG(S) gets us the real ω's needed as FREQ. Our coefficient matrix M is defined as 2 by 2 and then read in, one row at a time, by M(1) = ... and M(2) = ..., using numerical values $M_1 = 2$, $B_1 = 0.1$, $K_{s1} = 100$, $M_2 = 4$, $B_2 = 0.05$, and $K_{s2} = 200$. The inverse of M is computed with the single statement INVM = INVERSE(M) while RITSID = VECTOR(2:1.0,0.0) sets up the column vector for the C coefficients with a 1.0 for $F_1(i\omega)$ and a zero for $F_2(i\omega)$. (If we wish to repeat this problem to get $(X_1/F_2)(i\omega)$ and $(X_2/F_2)(i\omega)$, we need only change this to RITSID = VECTOR(2: 0.0, 1.0).) The matrix multiplication of INVM by RITSID is accomplished in SPEAKEASY with the same symbol (*) as "ordinary" multiplication, a further SPEAKEASY convenience. PROD is then the column vector of Q_0's, in our case, PROD(1) = $X_1(i\omega)$ and PROD(2) = $X_2(i\omega)$. The magnitudes and phases of complex numbers X1 and X2 are obtained using ABS and PHASE. The values to be tabulated are subscripted variables to be placed in their respective arrays; MAGX1(L) = ... and so on fill these arrays as L runs through its range. When the loop has run through all L values, a TABULATE statement produces the desired table.

While the above example was a relatively simple one, the general pattern of the program stays exactly the same as the size of the matrix increases, thus no new programming "tricks" are needed to deal with complicated problems. We will be using this approach on several more involved models in later chapters. In concluding this chapter we wish to point out that both the digital simulation techniques (such as CSMP) for obtaining time responses and also the matrix method just shown for linear-system frequency responses deal *directly* with the basic simultaneous equations and also produce information on *all* the system variables, not just a single output/input pair. They thus share some of the advantages of the state-variable methods mentioned in Chapter 4.

BIBLIOGRAPHY

1 Chu, Y., *Digital Simulation of Continuous Systems*, McGraw-Hill, N.Y., 1969.

2 CSMP Users Manual, SH19-7001-2, IBM Corp.

3 Korn, G. and J. Wait, *Continuous System Simulation*, Prentice-Hall, Englewood Cliffs, N.J., 1977.

4 Norling, R. L., Continuous Time Simulation of Forces and Motion Within an Automotive Engine, SAE Paper 780665, 1978 (CSMP Application).

5 Ören, T. J. (ed.), *Annotated Bibliographies of Simulation*, Simulation Councils Inc., La Jolla, Cal., 1976.

6 Rosko, J. S., *Digital Simulation of Physical Systems*, Addison-Wesley, Reading, Mass., 1972.

7 *Simulation*, The Soc. for Computer Simulation, 1010 Pearl St., La Jolla, Cal., 92037.

8 Smith, J. M., *Mathematical Modeling and Digital Simulation for Engineers and Scientists*, John Wiley and Sons, N.Y., 1977.

9 The Speakeasy-3 Reference Manual, ANL-8000, Argonne National Lab.

10 Speckhart, F. H. and W. L. Green, *A Guide to Using CSMP*, Prentice-Hall, Englewood Cliffs, N.J., 1976.

PROBLEMS

5.1 Modify the CSMP program for the air bearing of Section 5.4 to include:
  (a) The change of Fig. P2.8.
  (b) The change of Fig. P2.9.
  (c) Both the above changes.
  (d) The use of an experimentally-measured pressure/flow relation for $\dot{m}_i$ and an experimentally-measured pressure, gap height/flow relation for $\dot{m}_o$.

5.2 The mass Fig. P5.1 is to be moved, starting from rest, from $x = 0$ to $x = x_f$, arriving at x_f with zero velocity. The applied force magnitude f_0 is fixed and known; however the time T is not. Write a CSMP program, using the TERMINAL techniques of Section 5.5, to find the required time T.

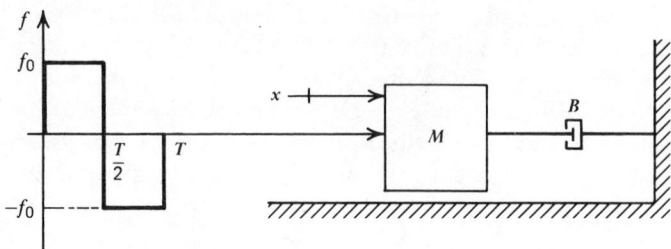

Figure P5.1

5.3 In Fig. P2.1, let each sliding block have mass M_i and viscous damping B_i. Letting F_1 represent a time-varying applied force with waveform as in Fig. P2.4, write a CSMP program to find the motions of all the blocks, assuming initial conditions are all known.

5.4 Repeat Problem 5.3 if block number one has, in addition to viscous friction, a dry friction force. Be careful to note that, before slippage occurs, a dry friction force adjusts itself to just match the applied force and that the friction force becomes a fixed constant only after slippage starts.

5.5 Write a CSMP program to compare the behavior of the two damper models of Fig. P2.2 for any given input force f_i.

5.6 Write a CSMP program to find q_o in Fig. P2.3.

5.7 Write a CSMP program for the system of Fig. P2.5.

5.8 Write a CSMP program for the system of Fig. P2.6.

5.9 Write a CSMP program for the system of Fig. P2.10.

5.10 Use CSMP's arbitrary function generator to solve for $x(t)$ in Problem 3.14.

5.11 Two approximations sometimes used for analysis of systems with dead times are:

$$e^{-\tau_{DT}s} \approx -\tau_{DT}s + 1$$

$$e^{-\tau_{DT}s} \approx \frac{2 - \tau_{DT}s}{2 + \tau_{DT}s}$$

Modify the CSMP program for the system of Fig. 5.9 to allow comparison of the exact dead-time model with each of the two approximations.

5.12 Modify the CSMP program of Section 5.7 to accomodate a 20-lump model.

5.13 Write a CSMP program to find the step response of the lumped pipeline model of Eq. (9.93).

5.14   Compare CSMP models based on:
  (a)   Equation (9.118).
  (b)   Equations (9.116) and (9.117).
  for the hydrostatic transmission of Section 9.5.1.

5.15   Compare CSMP models based on:
  (a)   Equation (10.5).
  (b)   Equations (10.2, (10.3, and (10.4).
  for the electrodynamic vibration shaker of Section 10.1.

PART II

Generalized Modeling Methods

CHAPTER 6

Model Development by Experimental Testing ("System Identification")

In Part II we discuss some aspects and techniques of physical system modeling that are applicable to all kinds of systems, irrespective of their detailed physical nature. Of major importance will be Chapter 6, where we describe the most widely used methods of defining mathematical models by means of physical testing of the actual systems (or scale models). Such physical testing serves two major functions. First, if a theoretical model has been developed, at some point its validity must be checked by actual experimental tests if a high level of confidence in its predictive capabilities is desired. At the early stages of model development such testing is particularly useful in revealing gaps and flaws in the theoretical treatment, which can then be corrected in subsequent refined models. Second, some systems defy refined theoretical treatment and thus experimentation is the *only* effective means of obtaining models of useful accuracy. The dynamic models of human operator response in steering or piloting tasks (discussed in a later chapter), which are used in man/machine system studies, are a good example of this class.

While the main application of models (whether derived by theory or experiment) is as *design tools* to allow efficient study of the effects of parameter variation on system performance in an attempt to design "optimal" systems, we should point out that experimentally derived models sometimes serve other important functions. All real systems actually *change* their behavior as time goes by; components wear, heat exchange surfaces become gradually fouled, stressed parts fatigue, chemical catalysts are depleted, and so on. If such a system is under automatic control, and if the controller was "tuned" to the process (for optimal response) at startup, this optimum behavior is gradually lost as time goes by. Various *"adaptive" control*[1] schemes are in use to deal with this problem. An obvious method, germane to our present discussion, uses an on-line model-measuring capability to continuously or intermittently evaluate the process behavior from measurements of system variables, and thus supply the

---

[1]Davies, W. D. T., *System Identification for Self-Adaptive Control*, Wiley-Interscience, N. Y., 1970.

controller with up-dated information so that it can tailor its control actions to the current process model. The increasing use of computer control is making such schemes more practical since the data processing needed to go from raw measurements to process models is often easily accommodated in a computer already justified for basic control needs. A different, but related, class of applications is found in the sophisticated *preventive maintenance*[2] techniques being introduced in some industries. Here, measurements made on newly installed and properly operating machinery or process equipment define a "benchmark" model of system behavior. At suitably timed intervals thereafter, these same signals are measured and evaluated against the benchmark model to detect incipient deterioration or malfunction and pinpoint system components that need to be repaired or replaced. In large and/or critical systems, such applications of experimental modeling can make significant contributions to system reliability.

6.1 FREQUENCY-RESPONSE (SINUSOIDAL) METHODS AND APPARATUS

Of the various test methods we discuss, the "classical" frequency-response method has been most widely used in the past, and despite recent technical advances that have made some of its competitors attractive, there is little doubt that this basic method will continue to be important in the future. While mathematically based (like many other methods) on linear, constant-coefficient models, it is successfully employed with real-world systems, including some that have significant nonlinear behavior. For multiple-input/multiple-output systems, a single sinusoidal input will usually excite *all* the outputs, allowing measurement simultaneously of all these sinusoidal transfer functions, if that is desired. For transfer functions having a *different* input, one must of course apply the sinusoidal signal to *that* input, again allowing measurement of *all* the output/input pairs related to that input. Note the similarity to the matrix method of *calculating* frequency response from a theoretical model, as explained in Section 5.12.

In principle, one needs only a source of sinusoidal excitation, suitable measurement transducers, and a two-channel recorder (multi-channel if *several* output/input pairs are measured at once) to gather frequency-response data. Although measurement transducers[3] to change most physical variables into proportional voltages are readily available, devices for producing the needed sinusoidal *input* signals (motions, temperatures, flow rates, etc.) must often be

---

[2]"Detection, Diagnosis, and Prognosis," *Proc. of Mech. Failure Prevention Group*, Nat'l. Bur. of Stds. Rept. NBSIR 73-252, Sept. 1973.
[3]Doebelin, E. O., *Measurement Systems* (rev. ed.), McGraw-Hill, N. Y., 1976.

designed and built by the experimenter. Since versatile electronic signal generators are widely available to provide voltages of different frequencies, amplitudes,
and waveforms, a common approach is to design a voltage-to-physical variable
"transducer" (often a feedback control system[4]) of suitable power level to drive
the system being tested and then use a commercial voltage source as input to
this transducer. For mechanical motions and forces, electrodynamic or electrohydraulic shakers are commercially available in a wide range of capabilities, so
one may prefer to buy or rent one of these rather than designing from scratch.
Also, low pressure (<15 psig) and low-frequency (<20 Hz) air pressure signals
can be generated from commercially available voltage-to-pressure transducers.
Most other sinusoidal input generating equipment (temperature, flow rate,
chemical composition, etc.) will, however, need to be designed.

If the physical systems to be tested were nearly linear and devoid of noise
effects, the "two-channel recorder" scheme would be a feasible, although tedious, method of frequency-response testing, and of course if more sophisticated
equipment were not available, this approach would have to be used. Because of
the widespread use of frequency-response methods, it is, however, economically
feasible for electronic instrument firms to design and market special frequency-
response analyzers, and any laboratory that uses this technique regularly can
justify an investment in such equipment since it greatly speeds the measurements
while also significantly improving accuracy. Such analyzers can be constructed
on various basic principles and are also available with a wide range of options,
from "barebones" systems to completely automated facilities that require little
operator attention, and interface with computers for further sophisticated data
processing. We now discuss the operation of two different basic types of
analyzers.

6.1.1   BAND-PASS TRACKING FILTER ANALYZERS OF HETERODYNE TYPE

All types of frequency-response analyzers are faced with the same fundamental
problem. When we apply a perfect sinusoidal input to a *real* system and wait for
transient effects to disappear, the system output will *not* be the perfect sine wave
predicted by linear mathematics. The distortions present in real systems are of
two main types; harmonics at integer multiples of the excitation frequency, and
random noise. Harmonics are present because real systems are always somewhat
nonlinear and when one excites a nonlinear system with a sinusoidal input, the
steady-state output is a periodic (but nonsinusoidal) wave for which a Fourier
series analysis reveals a discrete spectrum of harmonics. Random appearing
components in the output signal can arise from various sources; random noise

---

[4]Doebelin, E. O., *Dynamic Analysis and Feedback Control*, McGraw-Hill, N. Y., 1962.

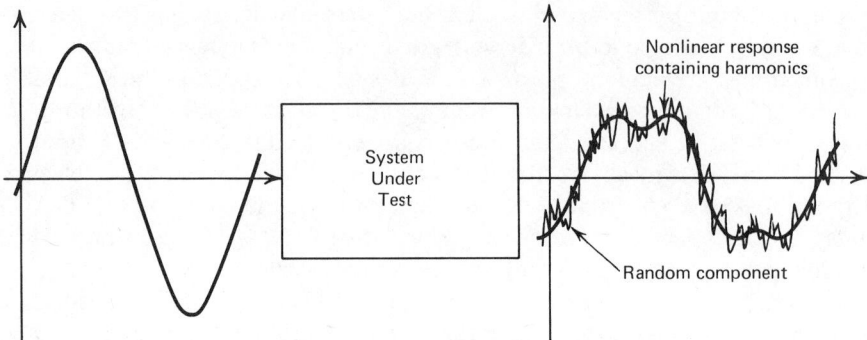

Figure 6.1 Response of real-world system to sinusoidal input.

contributed by measurement amplifiers and transducers, "nonrepeatable" nonlinearities such as dry friction and deadspace in mechanisms, time-varying
environmental effects such as ambient temperature, and so on. We should also
note that while great pains are taken to make the *input* signal sinusoidal, it also
can have harmonic and random content, producing system response to these
components. It should be clear from Fig. 6.1 that when the above described
effects become large relative to the basic sine-wave signal, the two-channel
recorder method of measuring frequency response becomes inaccurate due to
poorly defined amplitude and phase at the output, and more sophisticated data
processing is needed.

When, as in real-world measurements, the output (and sometimes also the
input) signals are nonsinusoidal, we define amplitude ratio and phase angle not
for the total signals but for the fundamental-frequency *components* of these
signals. A main function of frequency-response analyzers is to somehow extract
these components from the distorted signals measured; this is clearly some sort
of band-pass filtering operation, but it can be accomplished in several different
ways. In a heterodyne-type tracking filter[5] (see Fig. 6.2) a sweep oscillator (in
our example set to a test frequency of 157 Hz) provides both an input drive
signal to the system under test and also a tuning signal to the tracking filter.
Modulation techniques in the carrier converter translate the 157 Hz tuning
signal to $157 + 100,000 = 100,157$ Hz. It is then applied, in combination with the
signal to be measured (Fig. 6.2 shows the physical system *input* signal being
measured; the "switch" shown indicates we could as easily measure the system
output) to balanced modulator 1, which then produces an output of frequency
$100,157 - 157 = 100,000$ Hz, plus the noise and harmonic signals present in the
measured signal. Thus, no matter what test frequency we might choose, balanced modulator 1 produces an output representing the measured signal's

[5]Spectral Dynamics Corp., San Diego, Calif.

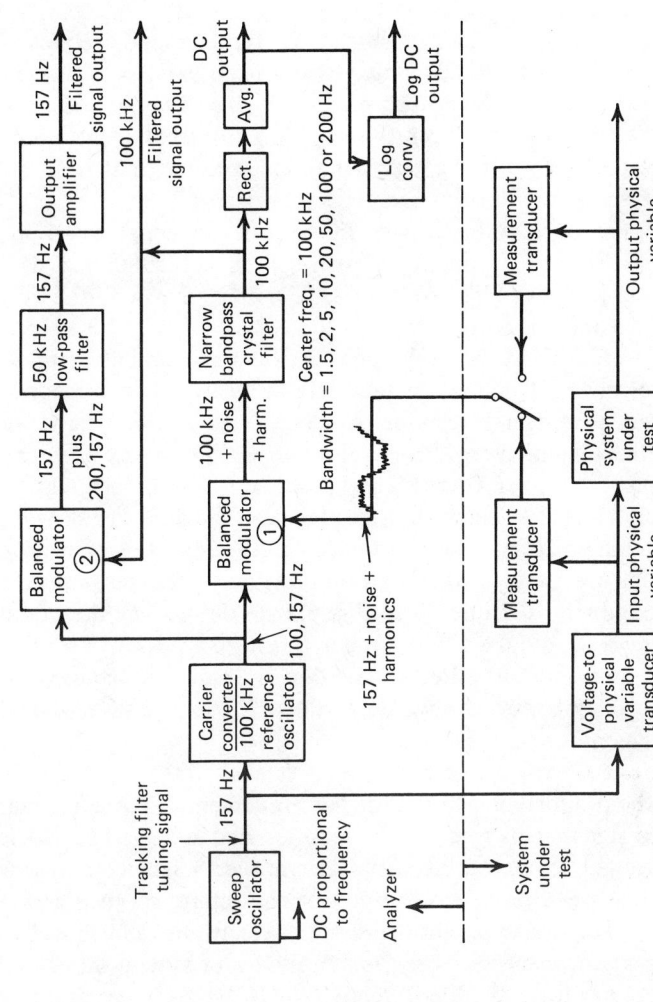

Figure 6.2 Heterodyne-type tracking filter.

characteristics, but *frequency shifted* so that its fundamental frequency is now 100 kHz. The reason for doing this is that now we can band-pass filter our signal (to remove noise and harmonics) using a single *fixed* filter tuned to 100 kHz. Rather than changing the filter center frequency to suit the test frequency (as some analyzers do), we shift the test signal spectrum so that its fundamental frequency always aligns with our fixed 100 kHz filter.

Sharply tuned narrow band-pass filters can be constructed in various ways. In the referenced instrument, piezoelectric crystal filters are used. These tend to work best at rather high frequencies. This, together with the desired 2 to 50,000 Hz test frequency range, leads to the choice of the 100 kHz operating frequency. Figure 6.2 shows that the crystal filters are available in a range of bandwidths; narrow bandwidths are necessary to resolve closely spaced peaks in the tested system's frequency response but require slow frequency sweeps and thus long test times. Since the width of resonant peaks in many physical systems tends to increase at higher frequencies, broader filters (which can be swept faster) can often be used at higher test frequencies. The analyzer shown provides automatic switching among three pre-selected filters as the sweep progresses from low to high frequency, allowing optimization of the resolution/test-time tradeoff. The bandwidths B quoted in Fig. 6.2 are defined by the points on the filter's frequency response that are 3 db down from the peak. As we move away from these -3 db points the filter attenuation is 38 db per bandwidth. Thus a filter with, say, 10 Hz bandwidth would be down 79 db at 25 Hz, giving a sharp rejection of harmonics and noise.

The output of the crystal filter will be a "clean" 100 kHz sine wave whose amplitude will be proportional to that of the fundamental frequency component of the measured signal. This sine wave is rectified and averaged to provide a DC voltage proportional to measured signal amplitude. Since we really want the amplitude *ratio* between tested system output and input, we need a *two-channel* tracking filter (two complete sets of the analyzer electronics shown in Fig. 6.2) so that we can simultaneously measure the amplitudes of system input and output (see Fig. 6.3). Rather than dividing directly the DC voltages representing output and input, logarithmic converters are used to produce voltages proportional to their logarithms. These log voltages are then *subtracted* to get the logarithm of amplitude ratio, which can then be conveniently plotted as decibels of amplitude ratio versus test frequency as we sweep from low to high frequency.

We also want available for plotting a DC voltage proportional to phase angle between output and input. The 100 kHz filtered signal outputs (Fig. 6.2) for our two signals are "phase coherent" with the fundamental components of the physical system input and output signals. That is, the phase shifts of the 100 kHz signals match those of the 157 Hz signals, thus the 100 kHz signals may be used for the phase measurements, which is convenient since they are of high *fixed* frequency. Figure 6.3 shows in simplified fashion the phase measuring

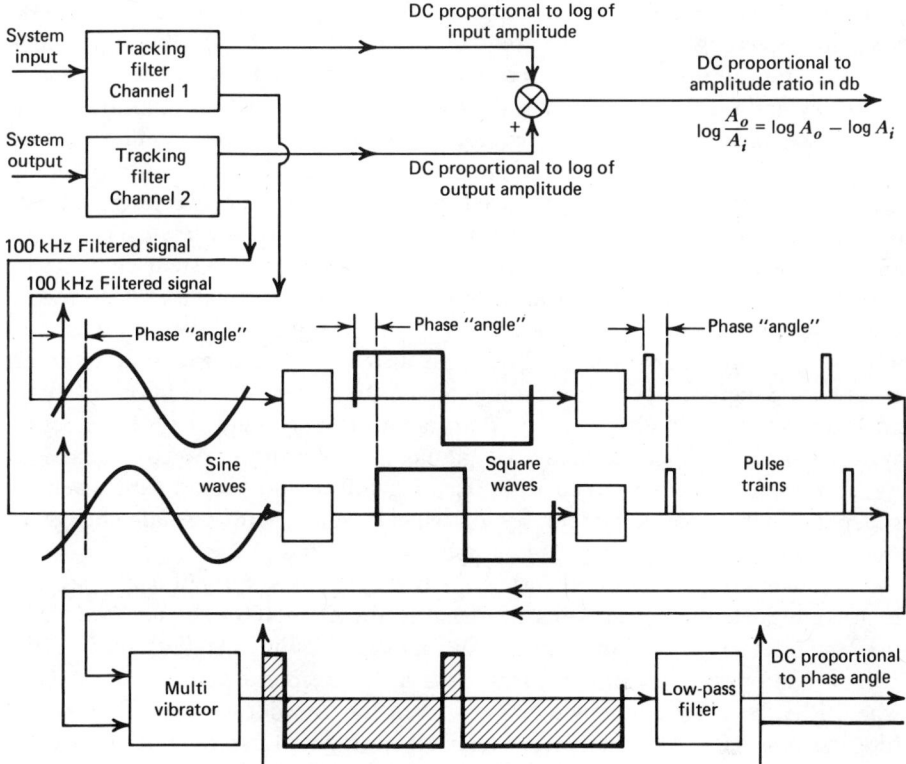

Figure 6.3   Phase angle measuring scheme.

scheme. The sine waves are first converted to square waves to make their zero crossing points (which define the relative phase) more distinct, and then the square waves are converted to pulse trains that are fed to the set and reset inputs of a bistable multivibrator. The pulse width of the multivibrator output depends on the phase relation between the two pulse train inputs; a 180° phase difference results in a symmetrical square wave with zero average value. Phase shifts more, or less, than 180° cause, respectively, waveforms with positive or negative average values that are read as DC voltages proportional to phase angle by a low-pass filter at the multivibrator output. Since the sweep oscillator provides a DC voltage proportional to test frequency, we apply this to the X axis of a two-pen XY plotter, driving the two Y axes with the decibel amplitude ratio and phase angle voltages to get continuous frequency-response graphs as we sweep the test frequency through its desired range.

Having explained in simplified terms the basic operating principles of heterodyne tracking filter analyzers, it is now appropriate to discuss certain points pertinent to their application in actual testing. We have referred several times to "sweeping" through the frequency range; to agree with the mathematical definition of frequency response we should actually *pause* at each discrete frequency and *wait* for the sinusoidal steady state before making amplitude ratio and phase measurements. The equipment described certainly allows us to operate in such a point-by-point fashion if we wish. However, we generally prefer the continuous graphs produced more quickly by the sweeping method; certain precautions must however be observed to preserve accuracy. The fastest allowable sweep rate depends on both analyzer and tested-system characteristics. If the tested system were known to be in perfect sinusoidal steady state and we "suddenly" connected the analyzer to its output signal, we would still have to *wait* for both the narrow-band filter and also the averaging circuit to reach steady state before accurate readings could be taken. In addition to these analyzer delays, it is also clear that when the driving frequency of the tested system changes (either suddenly, or gradually as during a sweep), there is a *system* response lag to contend with. This lag will be most significant when we sweep through a sharp system response peak, since here a small change in frequency causes a large change in amplitude ratio and phase.

In the analyzer referenced earlier, an averaging time of about 0.0001 sec can be used because the signal being averaged is always at 100 kHz, irrespective of the test frequency. The main delay in analyzer response is thus that of the narrow band-pass filter; these filters[6] have a 98% response time of about $4/B$, where B is the -3 db bandwidth. Thus, even for the widest filter (200 Hz), the filter delay is 0.02 sec, much larger than the averaging time, permitting calculation of the maximum allowable sweep rate to be based on filter lag only. This rate is (somewhat arbitrarily) chosen so that the frequency error in a measured peak will not exceed one filter bandwidth. Making the sweep rate equal to $B/(4/B) = B^2/4$ accomplishes this.

As to the effect of the tested system's response lag, approximate analysis[7,8] shows that to keep errors in the measured system's peak amplitude less than about 5%, and errors in peak frequency less than 0.4 B_s, the maximum allowable sweep rate should be no more than $B_s^2/8$, where B_s is the -3 db bandwidth of the resonant peak being measured. It should now be noted that the choice of the analyzer filter bandwidth B is based on the desired *frequency resolution* of the analysis, which in turn depends on the sharpness of expected peaks in the frequency response. Since the level of accuracy required in the measurement of a

[6]Burrow, R. L., Some Analog Methods of Power Spectral Density Analysis, Spectral Dynamics Corp. Tech. Publ. PS-1, 1968.
[7]Reed, W. H., Hall, A. W., Barker, L. E., Analog Techniques for Measuring the Frequency Response of Linear Physical Systems Excited by Frequency-Sweep Inputs, NASA TND-508, 1976.
[8]Drain, Bruton, Paulovich, Airbreathing Propulsion System Testing Using Sweep Frequency Techniques, NASA TND-5485, 1969.

system's frequency response varies with the particular application, no precise recommendation can be given; however, a reasonable estimate[9] is to make $B \leqslant B_{s,\min}/4$, where $B_{s,\min}$ is the -3 db bandwidth of the narrowest expected peak.

While the above guidelines are helpful in estimating allowable sweep rates (and thus total test time to cover a given frequency range), experience with the particular analyzer and class of physical system is a necessity in developing the engineering judgment necessary to run tests of adequate accuracy in minimum time. It should be noted that any errors associated with sweep rates too fast for either the analyzer and/or the tested system can be easily evaluated by letting the analyzer *dwell* at specific frequencies until steady state is assured, and then comparing swept values of amplitude ratio and phase with the "correct" steady-state values. Filter bandwidths and/or sweep rates are often changed (either manually, or automatically by the analyzer) to take advantage of conditions noted as the test progresses. Generally, wider bandwidth filters and faster sweep rates may be used as one progresses toward the high-frequency end of the sweep since system peaks often tend to have a bandwidth proportional to peak frequency. Most analyzers also allow choice of linear or logarithmic sweep rates; a linear rate is more appropriate when the sharpness of system peaks increases linearly with frequency, a logarithmic when it increases with the square root of frequency.[10] The sharpness of peaks is often specified numerically in terms of their "Q" value, where Q is defined in terms of a second-order system's damping ratio ζ by

$$Q \triangleq 1/2\zeta \tag{6.1}$$

or sometimes by

$$Q \triangleq f_c/B_s \tag{6.2}$$

where f_c is the peak frequency. The peaks in the frequency-response curves of vibrating structures, for example, are often of approximately constant Q; ideally this would require a hyperbolic[11] sweep rate; however, since most analyzers do not provide this, logarithmic may be used as an approximation.

Since peaks and valleys in the frequency response are often of prime interest, analyzers such as that referenced earlier may provide special features to enhance ease of use and accuracy at such points. At peaks and valleys,

[9]Bendat, J. S. and A. G. Piersol, Analog Power Spectral Density Analyzers, Honeywell Corp. Rept. D-2149, 1964, pp. 5–7.
[10]Broch, J. T., On the Measurement of Frequency-Response Functions, B & K Tech. Rev., No. 4, 1975, B & K Instruments, Cleveland, Ohio, p. 11.
[11]Broch, *op. cit.*

amplitude ratio and/or phase may change very rapidly and the sweep rate should be reduced to preserve measurement accuracy and also not overtax the pen speed capabilities of plotters. An automatic *sweep rate override* control measures the rates of change of amplitude and phase signals, and when they exceed a preset value, reduces the sweep rate appropriately, and returns it to the "normal" value when the peak has passed. Another useful feature is a *resonant dwell control system*. In physical systems where *spacewise* variation of system variables is of interest, measurement of "mode shapes" is often desirable. A mode shape is a curve or surface showing the spacewise variation of a system variable (structural deflection, fluid pressure, etc.) when the system is in steady sinusoidal operation at one of its resonant frequencies. To measure a mode shape, the system must be kept *precisely* at the resonant peak, not an easy task for lightly damped resonances. A resonant dwell control system is a servo system that controls the drive oscillator's frequency so that it is "locked" to the resonant condition. At resonance, the phase angle is some definite value, such as 90° (recall, say, a second-order system). The servo system compares the actual phase angle (from the analyzer's phasemeter) with the desired value (say 90°) and, whenever they are not equal, changes the drive frequency until 90° is again achieved, thus "locking" the oscillator to the system's resonance point even if it is changing. At the same time, a separate *amplitude* servo system compares a desired amplitude with the actual system response and adjusts oscillator *amplitude* to keep system response constant. In this way, a resonant condition of fixed output amplitude is automatically maintained, allowing convenient measurement of mode shapes.

6.1.2 "FOURIER FILTER" ANALYZERS

Another type of analyzer, which employs rather different principles, has been variously called a *Fourier filter-, wattmeter-, or correlation-type* instrument. We will base our discussion on a recent commercial model.[12] Figure 6.4 shows the basic principal, where the system under test is as in Fig. 6.2 and the signal coming from it to the analyzer could be either the system input or output (or both, in a two-channel analyzer). The drive oscillator in the referenced instrument changes frequency in discrete steps and dwells at each frequency because the averaging is done by integrating over an integer number of cycles. (Analyzers of this class *can* use sweep oscillators and low-pass-filter averaging if desired.) In addition to the system drive signal, the oscillator also provides $\sin \omega t$ and $\cos \omega t$ signals to the multipliers. If the tested system were perfectly linear and noise free, the analyzed signal (whether system input or output) would be of form $R \sin(\omega t + \phi)$ and the indicated multiplying and averaging operations lead

---

[12]EMR Telemetry, Weston Instruments, Schlumberger, Model 1170, Sarasota, Florida, 1975.

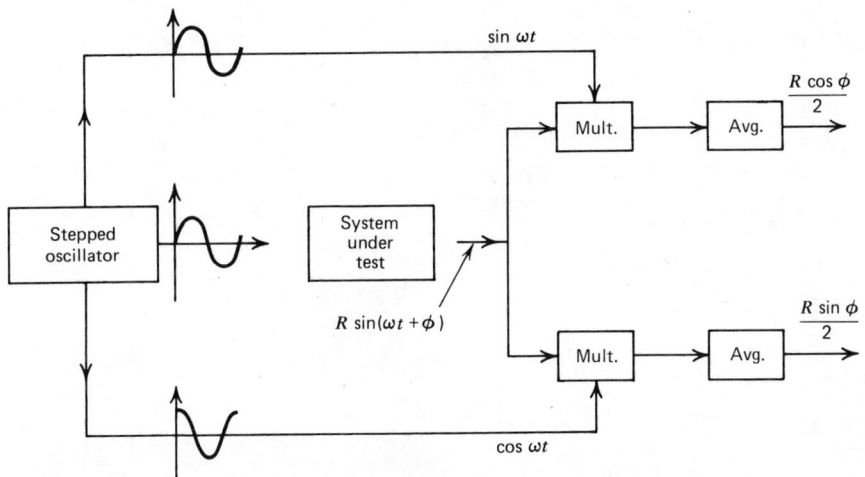

Figure 6.4 Principle of Fourier filter analyzer.

to:

$$\frac{1}{NT}\int_0^{NT} R\sin(\omega t + \phi)\sin\omega t\, dt = \frac{R\cos\phi}{2} \qquad (6.3)$$

$$\frac{1}{NT}\int_0^{NT} R\sin(\omega t + \phi)\cos\omega t\, dt = \frac{R\sin\phi}{2} \qquad (6.4)$$

since the averaging is done by integration over an integer (N) number of cycles of the waveform of period T. The desired amplitude R and phase ϕ are easily calculated from the measured $R\cos\phi/2$ and $R\sin\phi/2$, the referenced instrument using digital methods. Similarly, amplitude *ratio* and phase *shift* of the measured system are easily calculated in a two-channel analyzer.

If the system under test is nonlinear, harmonics $R_n\sin(n\omega t + \phi_n)$ will be present in the analyzed signal; however, their effect is *completely* cancelled out (at least theoretically) because:

$$\frac{1}{NT}\int_0^{NT} R_n\sin(n\omega t + \phi_n)\sin\omega t\, dt = 0, \quad n = 2,3,\ldots \qquad (6.5)$$

$$\frac{1}{NT}\int_0^{NT} R_n\sin(n\omega t + \phi_n)\cos\omega t\, dt = 0 \qquad (6.6)$$

(Equations (6.3) to (6.6) are closely related to the Fourier series equations and lead to the "Fourier filter" terminology sometimes applied to these analyzers. The "correlation" and "wattmeter" terminology relates to the multiplying and averaging operations performed.) With this type of analyzer, the averaging time must be some integer number of cycles, the absolute minimum being one. While,

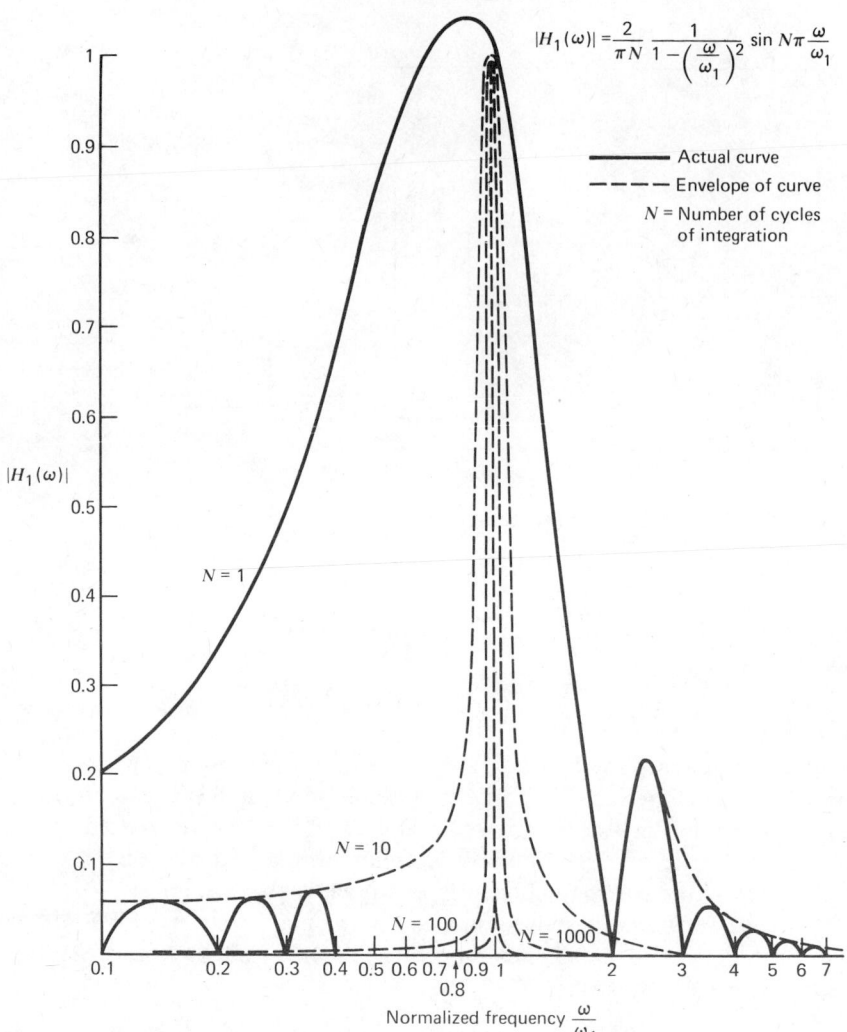

$$|H_1(\omega)| = \frac{2}{\pi N} \frac{1}{1 - \left(\frac{\omega}{\omega_1}\right)^2} \sin N\pi \frac{\omega}{\omega_1}$$

Actual curve
Envelope of curve
N = Number of cycles of integration

$|H_1(\omega)|$

Normalized frequency $\dfrac{\omega}{\omega_1}$

Figure 6.5   Frequency response of Fourier filter.

theoretically, harmonics are perfectly rejected even for $N=1$, in a practical instrument, greater accuracy is obtained for $N > 1$. For random noise or non-harmonic sine waves, a detailed analysis[13] gives the "frequency response" of this analyzer as shown in Fig. 6.5. We see that for N large, *all* frequency components

[13]Elsden, C. S. and A. J. Ley, "A Digital Transfer Function Analyzer Based on Pulse Rate Techniques," *Automatica,* **5** (1969), pp. 51–60.

other than the test frequency are strongly rejected. Although no physical band-pass filter hardware (such as the crystal filters of Section 6.1.1) is used, we see that the *effect* is the same. While this Fourier filter has no physical response lag (as does a crystal filter), the need for large N to get good rejection results in similar restrictions on testing speed. Physical band-pass filters are difficult to implement at low frequencies; the analyzer of Section 6.1.1 is designed for the range 2–50,000 Hz. Fourier filters do not suffer from this limitation; the referenced instrument covers the range 0.0001 to 9999.0 Hz, though of course operation at very low frequencies will be tedious.

When the drive oscillator is stepped rather than swept, one must control the time between application of a new frequency and initiation of the measurement cycle. Time delays of 0.1, 1, 10, or 100 sec may be selected. Also, output graphs will now be produced in a "point plotting" rather than "line drawing" mode. Again, linear or logarithmic stepping rates are selectable and can be intermixed. For those analyzers that use swept frequency and low-pass filter averaging, to keep errors in resonant peak amplitude less than 5% and errors in peak frequency less than 0.13 B_s, one must use $\tau < 0.15 t_b$, where τ is the low-pass filter's time constant and t_b is the time required to sweep through B_s.[14,15] Remember that these errors are due to the analyzer *only* and are in *addition* to those due to system response lags quoted in Section 6.1.1.

6.1.3   USE OF DIGITAL FAST FOURIER TRANSFORM (FFT) ANALYZERS FOR SINUSOIDAL TESTING[16]

The digital FFT analyzers discussed in more detail later in this chapter, while particularly suited for pulse and random testing, are also usable with sinusoidal excitation. In such instruments the "filtering" needed to extract the fundamental frequency component of the analyzed signals is accomplished by digital computer software (programs/algorithms) rather than analog hardware (crystal filters, etc.). Since FFT analyzers are capable of dealing with "fast" transient test signals, the choice of the actual sinusoidal sweep rate is now determined by considerations *other* than analyzer limitations; power limits of the physical system input driving device being perhaps the most common. Assuming a desired sweep rate has been selected, one puts the analyzer into a "redundant averaging" mode and selects the number of spectra to be averaged such that the time to accumulate all of them equals the total time of the frequency sweep.

Usually, FFT analyzers average spectra that are nonredundant. That is, the sample of signal time history stored in the instrument's input buffer (which will be Fourier transformed to get the frequency spectrum) contains no sample points in common with a prior sample or a subsequent sample. For slow-sweep

[14]Reed, Hall, Barker, *op. cit.*
[15]Drain, Bruton, Parlovich, *op. cit.*
[16]Personal communication from G. F. Lang, Nicolet Scientific Corp.

sinusoidal signals, however, highly redundant averaging is employed and the instrument averages every spectrum which it computes. For example, suppose we sweep from 5 to 500 Hz in 100 seconds. We set the analyzer for the 0–500 Hz range; this might correspond in a typical analyzer to a time window (sample time) of 0.8 sec. Thus a complete sample is gathered in 0.8 sec, and since this sample can be transformed in about 0.03 sec, when the input buffer is loaded with the *next* sample, about 0.77 sec of this sample is "old" (redundant) data and only about 0.03 is "refreshed." If we observe an oscilloscope display of the spectrum as the sweep progresses, instead of the usually instantaneous and complete spectrum, we see the display slowly appear from left to right (low frequency to high) and the total spectrum is not visible until the sweep has completed. At this point the averaging is manually stopped, the final and complete spectrum is frozen in the memory and we can observe it on a fast oscilloscope display or read it out slowly for hardcopy on a plotter.

6.1.4   FITTING ANALYTICAL TRANSFER FUNCTIONS TO EXPERIMENTALLY MEASURED FREQUENCY RESPONSE DATA

While the measured curves of amplitude ratio and phase angle found by frequency-response testing are often useful in themselves, it is sometimes necessary or desirable to find *analytical* transfer functions that closely fit these experimental data. Since pulse and random signal testing *also* lead to amplitude ratio and phase angle curves, a similar requirement for curve fitting may exist there. While methods for fitting curves to empirical data are a well-known part of classical mathematics, the need to fit *two* curves (amplitude ratio and phase) simultaneously presents some unique problems and thus requires a treatment somewhat different from that familiar to most students.

Since sinusoidal, pulse, and random test methods are all based on linear mathematics, the types of functions to be employed in curve fitting will be limited to this class. Furthermore, most measured data can be fitted adequately by lumped (rather than distributed) models, so we generally consider transfer functions in the form of a ratio of polynomials in s, sometimes augmented with dead times to take care of situations where the polynomial ratios fit the amplitude ratio well but are inaccurate in phase. For systems of moderate complexity and/or in situations where the need for curve fitting is occasional rather than frequent, sophisticated computer techniques may be inappropriate and a cut-and-try graphical approach based on the logarithmic amplitude ratio curve may be most effective. Recall from Eq. (3.40) that the measured decibel versus log frequency curve can be synthesized from a properly chosen combination of steady-state gain, integrators or differentiators, and first- and second-order systems. Sharp resonant peaks not only give away the presence of second-order terms but also allow easy numerical estimates of ω_n and ζ. Slopes of "straight line" portions of the curve (-20 db/decade, $+40$ db/decade, etc.)

may give clues as to the type of term needed and also numerical values. A phase angle approaching $-90°$ at low frequency indicates an integrator, asymptotic phase angle of $-270°$ at high frequency means that the highest power of s in denominator is three more than in numerator, and so on. If the decibel curve is well fitted but more phase lag is needed at high frequency, try a dead time. Also keep in mind that any function, no matter *how* it was arrived at (including guesswork or clairvoyance), is a satisfactory solution if it *fits the data*.

When a more "scientific" approach to curve fitting is desired, several methods documented in the literature are available. Levy's[17] method, one of the earliest proposed, will be briefly presented. It requires that the measured frequency response be given in terms of its real and imaginary parts:

$$\frac{Q_o}{Q_i}(i\omega) = F(i\omega) = R(\omega) + iI(\omega) \tag{6.7}$$

Let the analytical function we wish to fit to this data be given by

$$G(i\omega) = \frac{A_0 + A_1(i\omega) + A_2(i\omega)^2 + A_3(i\omega)^3 + \cdots}{B_0 + B_1(i\omega) + B_2(i\omega)^2 + B_3(i\omega)^3 + \cdots} \tag{6.8}$$

The analyst must choose which and how many A's and B's to use, drawing on all available information and past experience to make a best and simplest first choice. The method will then compute the numerical values of the chosen A's and B's. Most curve-fitting methods will share this feature; the analyst must choose the *form* of the function and then the computer produces numerical coefficient values. While Levy's method can be set up for *any* number of A's and B's, it is *not* good practice to blindly use large numbers of coefficients in the hope that the computer will grind out "zeros" for those coefficients not really needed. This is because matrix inversion is involved and excessive numbers of coefficients can lead to ill-conditioned numerical problems.

Returning to the method, we write

$$G(i\omega) = \frac{(A_0 - A_2\omega^2 + A_4\omega^4 - \cdots) + i\omega(A_1 - A_3\omega^2 + A_5\omega^4 - \cdots)}{(B_0 - B_2\omega^2 + B_4\omega^4 - \cdots) + i\omega(B_1 - B_3\omega^2 + B_5\omega^4 - \cdots)} \tag{6.9}$$

$$\triangleq \frac{\alpha + i\omega\beta}{\sigma + i\omega\tau} \triangleq \frac{N(\omega)}{D(\omega)} \tag{6.10}$$

We now define the error of fitting, $\epsilon(\omega)$, as

$$\epsilon(\omega) \triangleq F(i\omega) - G(i\omega) = F(i\omega) - \frac{N(\omega)}{D(\omega)} \tag{6.11}$$

[17]Levy, E. C., "Complex-Curve Fitting," *IRE Trans. Auto. Cont.*, May (1959).

If one tries to minimize the square of the error $\epsilon(\omega)$ (conventional least-squares method), he is led to a set of simultaneous *nonlinear* algebraic equations whose solution is difficult even by computer methods. Levy saw that if he multiplied $\epsilon(\omega)$ by $D(\omega)$, this difficulty would be overcome (however, the method is then *not* a true least-squares analysis on $\epsilon(\omega)$, which occasionally causes trouble, which we comment on later). We now have

$$D(\omega)\epsilon(\omega) = D(\omega)F(i\omega) - N(\omega) \triangleq a(\omega) + ib(\omega) \tag{6.12}$$

$$|D(\omega)\epsilon(\omega)| = |a(\omega) + ib(\omega)| = \sqrt{a^2(\omega) + b^2(\omega)} \tag{6.13}$$

At any typical frequency ω_k:

$$|D(\omega_k)\epsilon(\omega_k)|^2 = a^2(\omega_k) + b^2(\omega_k) \tag{6.14}$$

If we pick, say, $m+1$ points (frequencies) from the measured curves of $F(i\omega)$, we can define overall error E as

$$E \triangleq \sum_{k=0}^{m} \left[a^2(\omega_k) + B^2(\omega_k) \right] \tag{6.15}$$

Using standard calculus minimization procedures, the A's and B's can now be found so as to minimize E. Levy worked out the details for the general case and we will shortly give his results only since the intermediate steps are tedious. To clearly show the essence of what is involved, however, let us do a specific example.

Suppose we have experimental data on $F(i\omega)$ at five frequencies as given by

ω, rad/sec	$R(\omega)$	$I(\omega)$
0	1.000	0.000
1	0.500	−0.500
2	0.200	−0.399
3	0.099	−0.298
4	0.059	−0.236

and we wish to try to fit this with

$$G(i\omega) = \frac{A_0}{1 + B_1(i\omega)} \tag{6.16}$$

Then

$$\epsilon(\omega) = F(i\omega) - \frac{A_0}{1 + B_1(i\omega)} \tag{6.17}$$

$$D(\omega)\epsilon(\omega) = \left[1 + B_1(i\omega) \right] F(i\omega) - A_0 \tag{6.18}$$

Combining Equations (6.7), (6.10), and (6.15) gives the general result

$$E = \sum_{k=0}^{m} \left[(R_k\sigma_k - \omega_k\tau_k I_k - \alpha_k)^2 + (\omega_k\tau_k R_k + \sigma_k I_k - \omega_k\beta_k)^2 \right] \tag{6.19}$$

In our present example

$$G(i\omega) = \frac{A_0}{1 + B_1(i\omega)} = \frac{\alpha + i\omega\beta}{\sigma + i\omega\tau} \tag{6.20}$$

so we see that

$$\left. \begin{array}{ll} \beta_k = 0, & \alpha_k = A_0 \\ \sigma_k = 1.0, & \tau_k = B_1 \end{array} \right\} \text{ for any } k \tag{6.21}$$

For our example, Eq. (6.19) thus becomes

$$\begin{aligned} E = & \left[(1 - A_0)^2 + (0)^2 \right] + \left[(0.5 + 0.5B_1 - A_0)^2 + (0.5B_1 - 0.5)^2 \right] \\ & + \left[(0.2 + 0.798B_1 - A_0)^2 + (0.4B_1 - 0.399)^2 \right] \\ & + \left[(0.099 + 0.894B_1 - A_0)^2 + (0.297B_1 - 0.298)^2 \right] \\ & + \left[(0.059 + 0.944B_1 - A_0)^2 + (0.236B_1 - 0.236)^2 \right] \end{aligned} \tag{6.22}$$

To minimize E with respect to A_0 and B_1 we set $\partial E/\partial B_1 = \partial E/\partial A_0 = 0$ to get, after some manipulation,

$$\begin{aligned} -5A_0 + 3.136B_1 &= -1.858 \\ -3.136A_0 + 1.132B_1 &= 0 \end{aligned} \tag{6.23}$$

These two equations in two unknowns are easily solved to get $A_0 = B_1 = 1.0$, which is correct since the table of "experimental" data was actually "manufactured" from $1/(i\omega + 1)$.

We see from this example that there are essentially two steps in the computation procedure. We first have to generate "n equations in n unknowns" and then solve these for the A's and B's. Levy has worked out a general procedure for the first step; the second step is a standard matrix operation available in most computer libraries. Levy's calculation scheme goes as follows. First, we *always* take $B_0 = 1.0$. Then get the n equations in n unknowns from the matrix in Fig. 6.6. (While Levy's method can be set up for any number of A's and B's, one rarely needs more than 10 of each, so Fig. 6.6 was designed to handle problems up to this size.) The numerical coefficients needed in the matrix

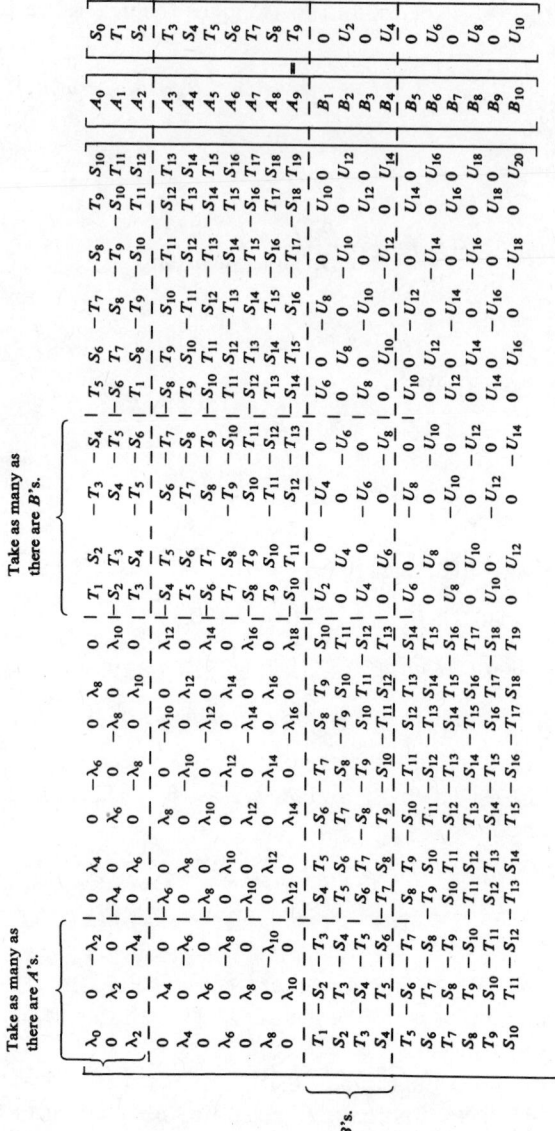

Figure 6.6 Matrix scheme for Levy's method.

are computed from

$$\lambda_h \triangleq \sum_{k=0}^{m} \omega_k^h, \qquad T_h \triangleq \sum_{k=0}^{m} \omega_k^h I_k$$

$$S_h \triangleq \sum_{k=0}^{m} \omega_k^h R_k, \qquad U_h \triangleq \sum_{k=0}^{m} \omega_k^h (R_k^2 + I_k^2) \qquad (6.24)$$

(A regular user would write a computer program to evaluate these.) If less than 10 A's and B's are wanted, the matrix collapses into a smaller one, as indicated in Fig. 6.6. For example, if we choose

$$G(i\omega) = \frac{A_0 + A_1(i\omega) + A_2(i\omega)^2}{1.0 + B_1(i\omega) + B_2(i\omega)^2 + B_3(i\omega)^3 + B_4(i\omega)^4} \qquad (6.25)$$

the matrix becomes

$$\begin{bmatrix} \lambda_0 & 0 & -\lambda_2 & T_1 & S_2 & -T_3 & -S_4 \\ 0 & \lambda_2 & 0 & -S_2 & T_3 & S_4 & -T_5 \\ \lambda_2 & 0 & -\lambda_4 & T_3 & S_4 & -T_5 & -S_6 \\ T_1 & -S_2 & -T_3 & U_2 & 0 & -U_4 & 0 \\ S_2 & T_3 & -S_4 & 0 & U_4 & 0 & -U_6 \\ T_3 & -S_4 & -T_5 & U_4 & 0 & -U_6 & 0 \\ S_4 & T_5 & -S_6 & 0 & U_6 & 0 & -U_8 \end{bmatrix} \begin{bmatrix} A_0 \\ A_1 \\ A_2 \\ B_1 \\ B_2 \\ B_3 \\ B_4 \end{bmatrix} = \begin{bmatrix} S_0 \\ T_1 \\ S_2 \\ 0 \\ U_2 \\ 0 \\ U_4 \end{bmatrix}$$

Once a set of A's and B's is found, one compares the frequency response of this model with the measured data. If the fit is judged inadequate, Levy's method is rerun, using more A's and B's, as guided by results of the first run, hopefully achieving satisfactory results after a few tries.

After Levy's method had been used for several years, it was found that it was sometimes unsatisfactory when very wide ranges of frequencies were used or when the denominator function $D(\omega)$ had a wide variation. An iterative procedure[18] that starts with Levy's values of the A's and B's and minimizes $\Sigma[\epsilon(\omega_k)]^2$ (*true* least-squares method) rather than $\Sigma|D(\omega_k)\epsilon(\omega_k)|^2$ was developed to overcome these problems. The reference clearly explains the method and shows an example (nuclear reactor) where Levy's scheme fails, but the modified approach works well in a few iterations.

---

[18]Sanathanan, C. K. and J. Koerner, "Transfer Function Synthesis as a Ratio of Two Complex Polynomials," *IEEE Trans. Auto. Cont.*, Jan. (1963), p. 56–58.

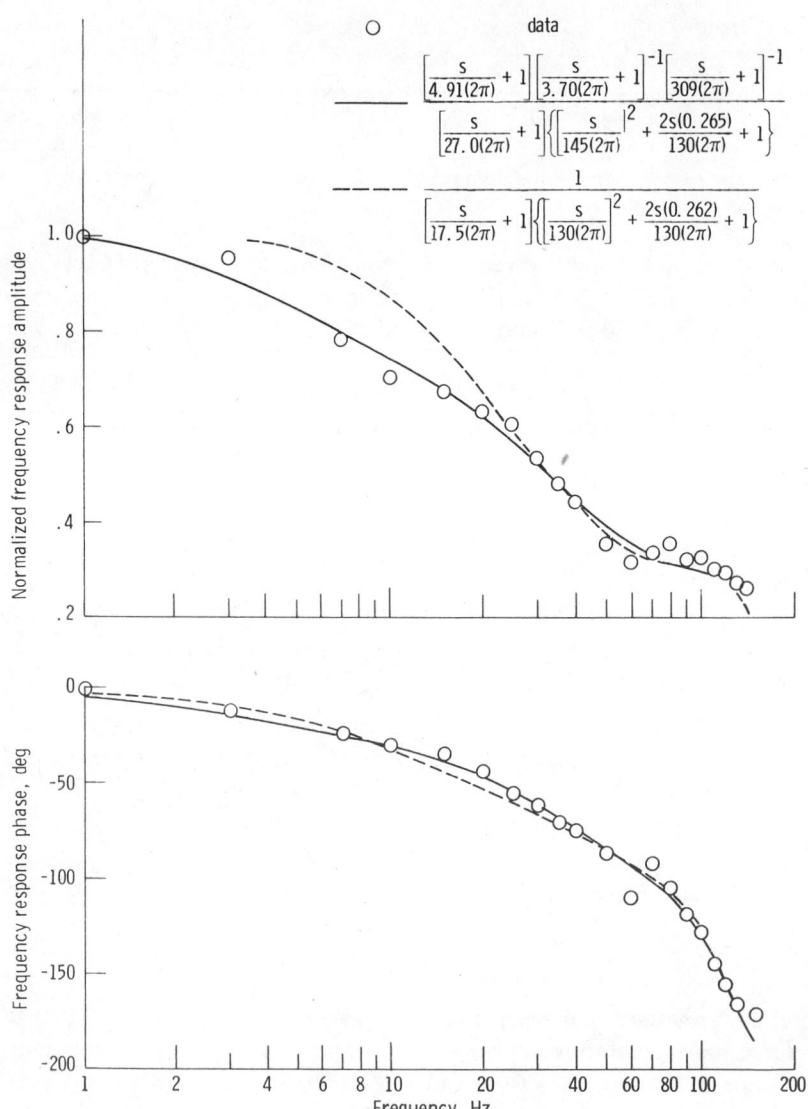

Figure 6.7 Quality of curve fit obtained with conjugate gradient technique.

A more recent effort[19] in this area uses a conjugate gradient search technique to minimize the integral of the absolute value of the error squared between model and measured data. Allowed forms of model include all the terms in Eq. (3.40), even the dead time, which Levy's method does not allow. The program user enters a "best guess" of model form and numerical parameters whereupon the program iterates these in a systematic manner until an optimum combination is found. Figure 6.7 shows some results for data measured on a jet engine supersonic inlet, where system input was the area of a bypass door opening and output was inlet pressure. Both a three-parameter and a six-parameter model were tried; the more complex model exhibits a better fit, as expected. The results shown were obtained in 19 iterations (2.3 sec CPU time) for the simpler model and 93 iterations (20 sec CPU time) for the more complex. Another use for such programs (Levy's included) is to find simplified versions of complex transfer functions known analytically. One merely enters the calculated (rather than measured) frequency response of the complex model as if it were measured data and asks for the best-fit model of whatever simpler form is of interest. If the simpler model reasonably matches (over the frequency range of interest) the frequency response of the more complex, the two systems *must* be nearly dynamically equivalent and use of the simpler model is justified.

6.2   PULSE TESTING METHODS AND APPARATUS

As in the case of sinusoidal testing, pulse testing methods are based on the representation of the physical system in terms of a linear, constant coefficient mathematical model. In Section 3.7 we saw that for a transient (pulse) input, the system sinusoidal transfer function $W(i\omega)$ was given by the ratio of the Fourier transforms of the output and input signals:

$$W(i\omega) = \frac{Q_o(i\omega)}{Q_i(i\omega)} \tag{6.26}$$

In pulse testing, a suitably chosen input pulse $q_i(t)$ is applied to the system input, causing a transient output response $q_o(t)$. The recorded $q_i(t)$ and $q_o(t)$ are then each transformed into the frequency domain using Eq. (3.79); division of $Q_o(i\omega)$ by $Q_i(i\omega)$ then gives the system model as the usual pair of amplitude ratio and phase angle graphs of the sinusoidal transfer function. If an analytical function for $W(i\omega)$ is desired, the curve-fitting methods of Section 6.1.4 can be applied.

The mathematically ideal input pulse would be a perfect impulse $\delta(t)$ since its frequency spectrum is perfectly flat out to infinite frequency and thus the

[19]Seidel, R. C., Transfer-Function-Parameter Estimation from Frequency-Response Data—a Fortran Program, NASA TM X-3286, 1975.

frequency response of the system under test would be uniformly exercised, revealing all its details. Also, for a perfect impulse input we would need only to transform $q_o(t)$ since $Q_i(i\omega)$ would be *known* to be a simple constant. The infinite height and infinitesimal duration of the perfect impulse are of course not achievable in the real world, and in fact we should not even strive too hard to achieve the infinite height since excessively large values of physical variables will drive the system into nonlinear operation, violating our assumptions of essentially linear behavior.

If one tentatively chooses a particular input pulse shape and duration, Fourier transformation reveals the frequency spectrum of this test pulse. If past experience and/or rough calculations have provided an estimate of the highest frequency at which the system to be tested can respond with significant amplitude, we can require that the test pulse's frequency spectrum extend "with significant magnitude" out to this frequency. If, however, no information about the tested system's dynamics is available, we might take the approach of applying the *fastest* pulse we can (with the available hardware), computing $W(i\omega)$ out to the highest frequencies thus exercised, and then simply stating that the system model so derived must not be *used* above this frequency range since the test did not *exercise* the system in this range and thus its response there is simply unknown. To aid in the design of suitable test pulses, calculated pulse frequency spectra such as those of Fig. 6.8 are helpful. Note first that pulses of certain symmetrical shapes have *zeros* in their spectra (such as at $\omega = 2\pi/T$ for curve A), thus they provide no excitation at such frequencies, and we must stay below the lowest such frequency in making calculations of $W(i\omega)$. Even for pulses (such as C) that do not exhibit zeros, reliable results require that we limit our calculations to frequencies below the point where $|F(i\omega)|/|F(i0)|$ becomes "too small," say about 0.2. In actual practice, calculations are usually carried to higher frequencies than really valid and the erratic behavior of the plotted $W(i\omega)$ used to discard the unreliable results.

When the frequency spectrum of single-pulse transient inputs cannot be adjusted to meet test needs, consideration of frequency-swept transients or pulse train inputs may be in order. A frequency-swept transient is related to the "slow" frequency sweeps discussed under sinusoidal testing in Section 6.1; however, the sweep rate is so fast that the system never gets into "sinusoidal steady state" and the input is thus properly considered transient (see Fig. 6.9 for some examples). While the exact frequency spectrum of such transient frequency-modulated (FM) signals is difficult to calculate, useful approximate results are available. The reference[20] shows that for a swept-sine transient, to obtain a reasonably flat spectrum from the initial frequency ω_i to the final frequency ω_f, the variation of frequency with time should be linear; fortunately

[20]Reed, Hall, Barker, *op. cit.*

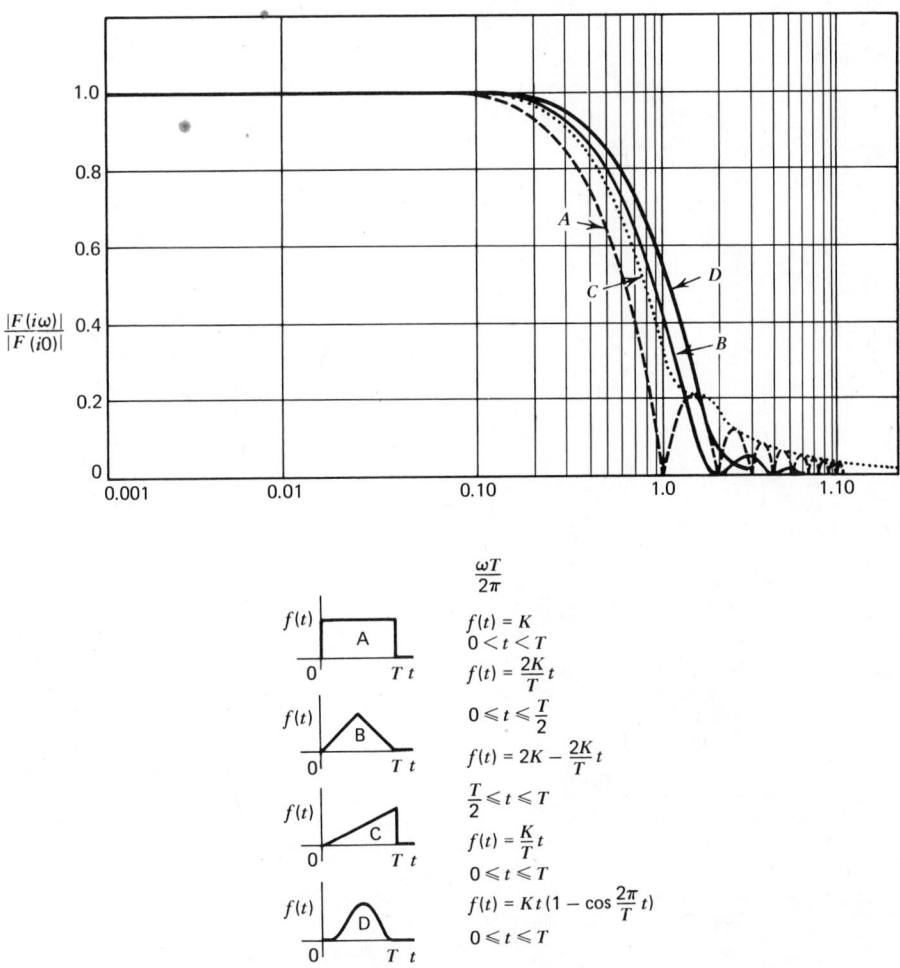

Figure 6.8   Frequency spectra of common test pulses.

most electronic oscillators provide such a sweep. The frequency function for such a transient will have a magnitude approximated by $1.25A_0/\sqrt{\dot\omega}$, where $\dot\omega$ is the sweep rate in (rad/sec)/sec. Figure 6.10[20] compares this magnitude with that obtained from a single transient half-sine pulse and also a step input of identical amplitude A_0 for a specific numerical case, clearly showing that stronger system excitation can be obtained with the same peak value of input time function. Magnitude estimates for triangular and square wave swept signals were $1.02A_0/\sqrt{\dot\omega}$ and $1.77\,A_0/\sqrt{\dot\omega}$, respectively. Measurements to define spectrum shape for these two signals were not made; however, one would expect stronger

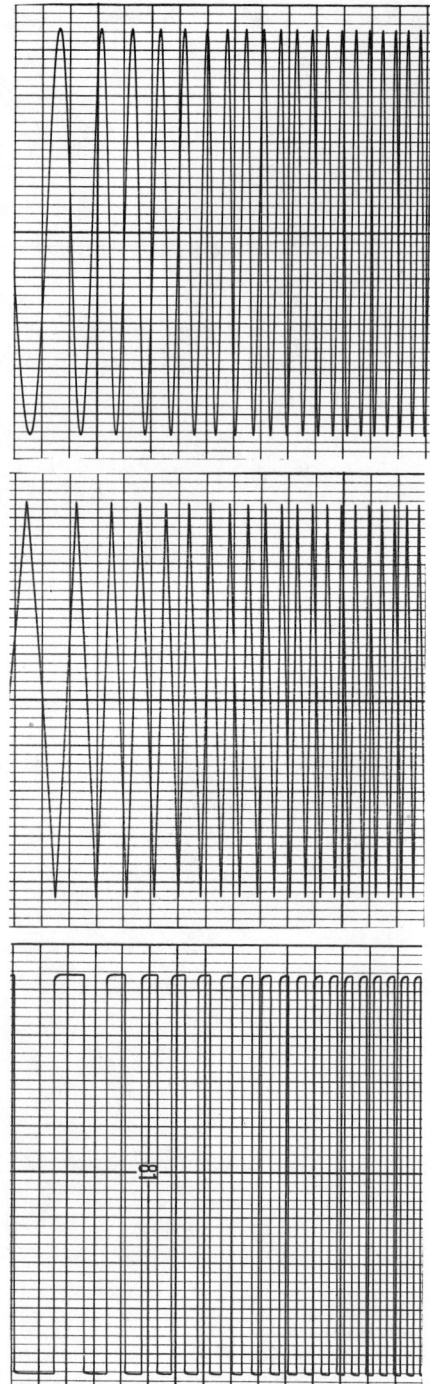

Figure 6.9 Three frequency-swept transients.

254

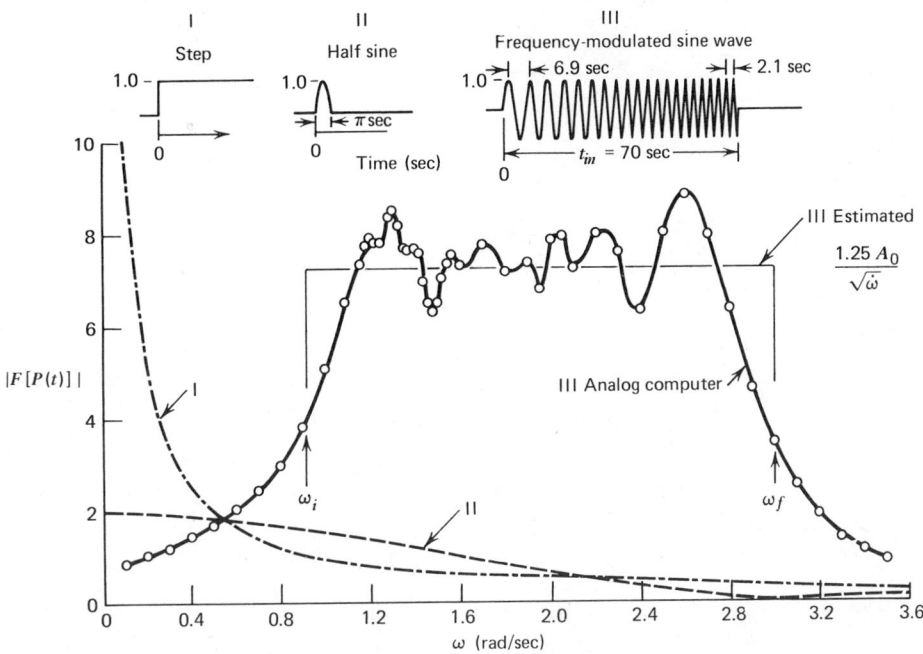

Figure 6.10 Spectrum comparison of three test signals.

frequency content beyond ω_f (the *steady*-frequency waves are rich in harmonics) and perhaps a flatter spectrum between ω_i and ω_f, for the same reason.

Another approach to obtaining large spectrum levels with modest peak values of the time function utilizes repetitive identical pulses (pulse trains). A commercially available structural impact exciter[21] of this type produces the force/time function shown in Fig. 6.11; the repetition rate is variable from 10 to 40 pulses per second while T_p is about 70 μ sec and F_0 can be as large as $13500N$ (3000 lb$_f$). For 10 pulses per second ($T_r = 0.1$) we get a force frequency spectrum (Fourier series) that is fairly flat out to about 10 kHz. While a *single* force pulse of the given shape would exhibit a Fourier transform of similar flatness with frequency, it would require a much higher *peak* value to achieve the same system response as produced by a train of smaller pulses. The repetitive pulse technique[22] can thus avoid system nonlinear response due to excessive input size and also reduce the requirements for wide dynamic range on signal conditioners and analyzers.

[21]Wilcoxon Research Model F20, Bethesda, Md.
[22]Noiseux, D. U. and B. G. Watters, A Simple Source of Intense Vibrations, ASME Paper 65-WA/MD-11, 1965; Schloss, F., *Impedance Measurements Using Impact Techniques*, 82nd Meeting, Acoustical Soc. of Amer., Denver, October 19, 1971; Multiple Impact Test Method, Appl. Note 6, Hewlett–Packard Corp., June 1977.

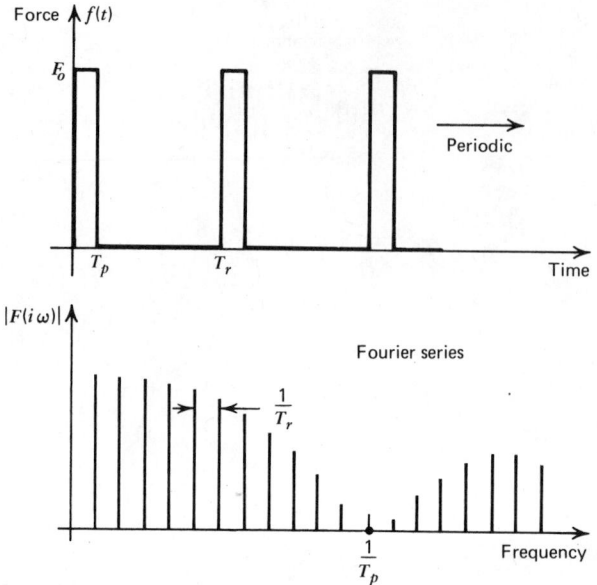

Figure 6.11   Pulse train and its spectrum.

Most of the earlier applications of pulse testing were for slow-acting systems for which sinusoidal testing could be (and perhaps had been) done, but at considerable expense and delay, because of the very low frequencies involved. Pulse testing offered the possibility of much shorter test times and lower costs. A good example of this type is found in the marine propulsion system tests of Banham.[23] The systems tested included components of steam-generating and propulsion equipment (boilers, heat exchangers, turbine-driven blowers, etc.) for naval vessels, in addition to various pneumatic devices used to control the overall system. The dynamics of these systems had previously been routinely studied with sinusoidal testing; Banham examined pulse testing as an alternative test method and concluded that savings of 60% ($18,000 per set of steam generator response tests at three load levels) could be achieved with only a slight decrease in accuracy. A triangular form of input pulse was chosen; Fig. 6.12 shows some typical input/response pairs, indicating the degree to which the goal of triangular input was actually achieved in each case (the *actual* transient, not the perfect triangle, was of course analyzed). Results of the Fourier transformation for a typical output/input pair, together with an analytical curve fit (labeled "theoretical"), are shown in Fig. 6.13.

[23]Banham, J. W., Jr., "Obtain Process Dynamics by Pulse Testing," *Cont. Eng.*, April (1965), pp. 83–88; Banham, J. W., Jr., Development of Experimental Technique for Frequency Response Analysis by the Pulse Test Method, Naval Boiler and Turbine Lab, Rept. for Project B-622, Philadelphia, PA, 1964.

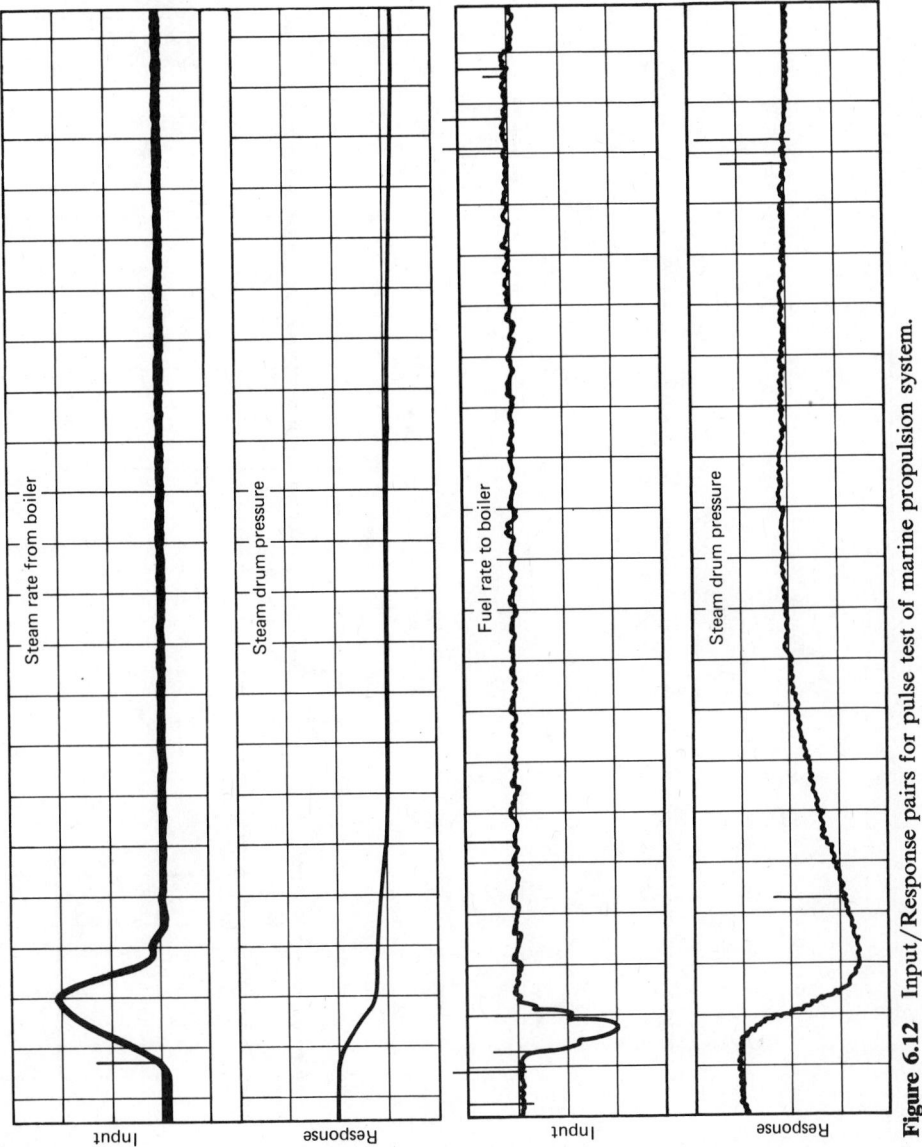

Figure 6.12 Input/Response pairs for pulse test of marine propulsion system.

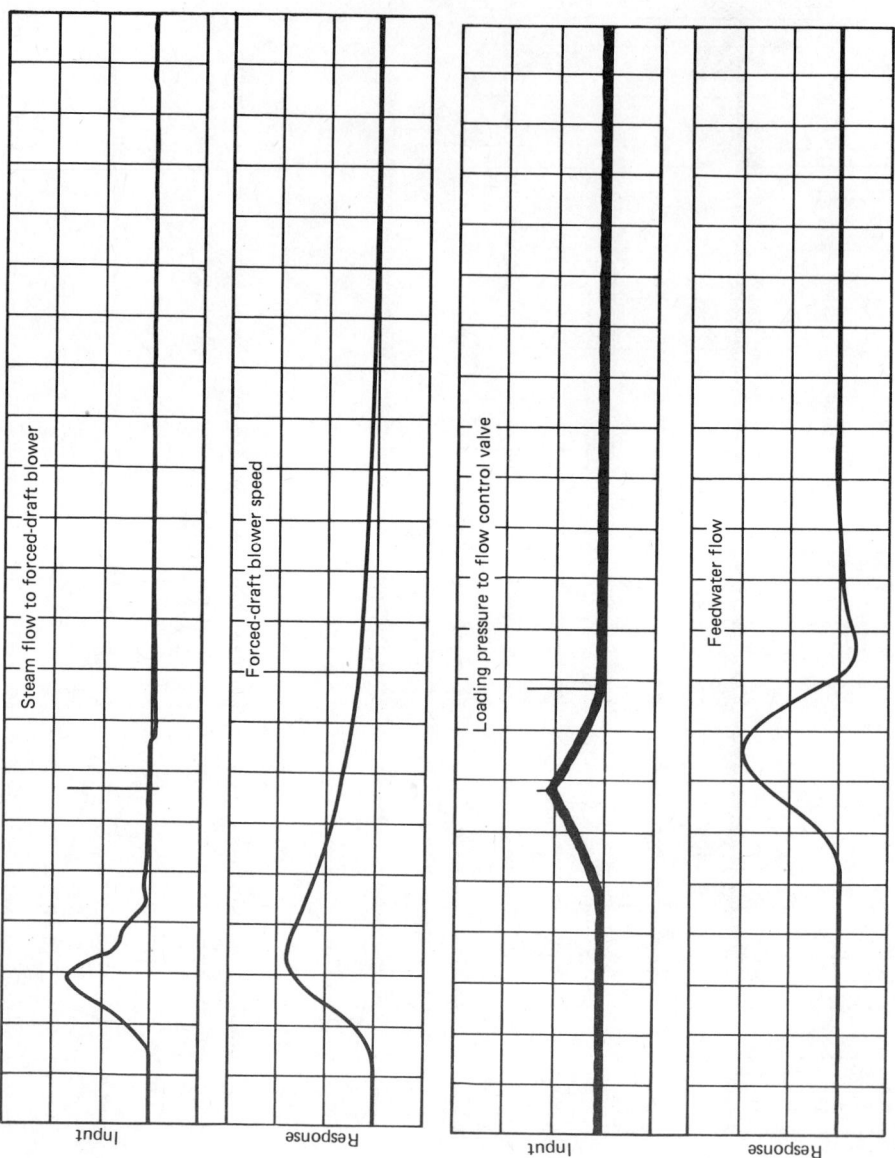

Steam flow to forced-draft blower

Forced-draft blower speed

Loading pressure to flow control valve

Feedwater flow

Input

Response

Input

Response

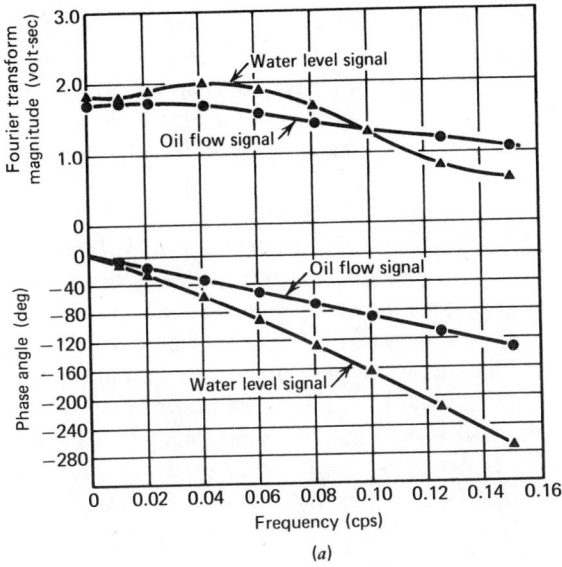

(a)

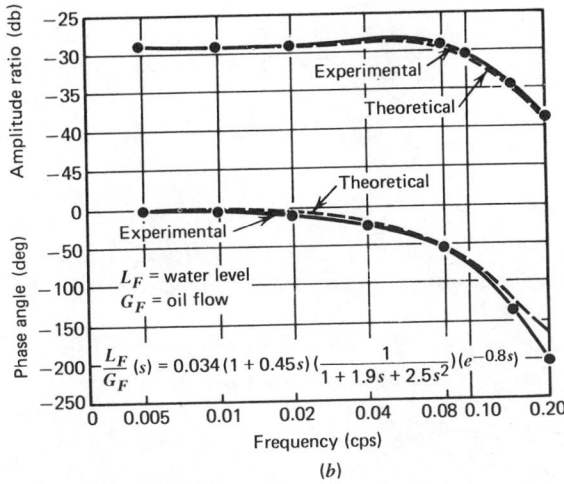

(b)

Figure 6.13   System frequency response obtained from pulse test data.

When the system being tested has one or more *integrating* effects in its transfer function, the response to a transient input whose area is not zero (the usual case) is not itself a "transient," since it will *not* return to zero as time goes by. (In Fig. 6.12, two of the four responses shown there exhibit this feature.) This behavior creates a mathematical problem since the Fourier transform of such response signals will not go to zero at high frequency. Various data processing methods may be employed to deal with this situation. If sufficiently noise-free (and accurate) approximate differentiating hardware or software is available, one can differentiate the output time history, once for each integration present in the tested system, thus giving a time signal that *does* return to zero. Fourier transformation is now possible; however, the "$Q_o(i\omega)$" calculated in this way must be divided by $i\omega$ at each frequency to recover the true $Q_o(i\omega)$.

Banham (Fig. 6.12), who used an analog computer mechanization of the Fourier transform calculations, employed two matched band-pass filters (tuned to the frequency being analyzed) to filter $q_i(t)$ and $q_o(t)$ (thus ensuring that both signals went to zero) before they were Fourier analyzed. The $Q_i(i\omega)$ and $Q_o(i\omega)$ obtained in this way are not the "true" functions, but their *ratio* will be the true $W(i\omega)$ since each was operated on by the *same* filter transfer function. Still another approach[24] modifies the output transient by subtracting from it a step function whose size is $q_o(t_f)$, where t_f is the time at which q_o levels off at its nonzero steady state. This new transient $q_{om}(t)$ *does* level off at zero and thus can be Fourier transformed; however, to get $Q_o(i\omega)$ we must add the transform of the step function:

$$Q_o(i\omega) = Q_{om}(i\omega) + \frac{q_o(t_f)}{i\omega} \tag{6.27}$$

(We are here using the Fourier transform of a step function formally as its Laplace transform with $s = i\omega$.) This approach prevents us from calculating $Q_o(i0)$ and thus $W(i0)$; however, the system steady-state gain can usually be easily obtained by "static calibration."

Turning now to the methods of actually calculating the needed Fourier transforms, early workers such as Banham used special purpose analog computers marketed specifically for pulse testing, general purpose analog computers programmed to generate the sine and cosine signals and carry out the multiplication and integration required in Eq. (3.79), or used classical numerical methods employing hand or machine computation. In using the analog methods, $q_i(t)$ and $q_o(t)$ would usually be tape-recorded since $Q_i(i\omega)$ and $Q_o(i\omega)$ are calculated "one ω at a time" by multiplying, say, $\sin\omega t$ by $q_i(t)$ and then integrating. This must be repeated for every ω desired, thus we must be able to

[24]LaVerne, M. E. and A. S. Boksenbom, Frequency Response of Linear Systems from Transient Data, NACA Rept. 1977, 1950.

reproduce $q_i(t)$ electrically whenever we need it. The band-pass filtering mentioned earlier as a means of dealing with signals that do not return to zero was also routinely used (even for signals that *did* return to zero) to improve the "signal-to-noise ratio" of the calculation. Since we are computing one point (one ω) of, say, $Q_o(i\omega)$, at a time, all the frequency content *away* from this ω is really irrelevant; however, the computing hardware "feels" the *total* signal and must be scaled so as to not over-range for this total signal. If we band-pass filter *before* entering the computing hardware, the extraneous frequency content has been reduced, leaving a smaller signal (but now mainly "information" rather than "noise") that can be scaled up for greater accuracy of computation.

While these analog methods can be made to work, the hardware maintenance requirements, set-up and execution times, and lack of versatility have led to their being largely superceded by digital techniques, performed "off-line" on a general purpose digital computer or in "real time" by a special purpose computer or "instrument," often using FFT (Fast Fourier Transform) algorithms. For off-line use on a general purpose computer one can write his own program or procure one ready-made from various sources,[25] the more recent programs usually using some form of FFT algorithm. To use these programs one must first digitize the time records of $q_i(t)$ and $q_o(t)$, either manually or by electronic sampling and analog/digital conversion, taking care to anti-alias (low-pass) filter any frequency content above the highest analysis frequency and then sample at a rate of at least 2.5 points per cycle of the highest analysis frequency.

Performance improvements and cost reductions in the electronics and computer technology associated with these spectrum analysis techniques has led to the availability of a number of commercially available instruments and minicomputer-based systems that are very fast and easy to use and of sufficiently low cost to encourage wide industrial application. One such system[26] was used to pulse test the support structure for an experimental 100-kw wind turbine,[27] to aid in the diagnosis of vibration problems (see Fig. 6.14). The input force pulse was produced by swinging a 272 kg cylindrical mass from a crane and allowing it to impact the tower at the 28 m level. Force measurements from a load cell attached to the swinging mass, and tower motion measurements from attached accelerometers were fed to a two-channel analyzer, producing the compliance (displacement/force) transfer function shown. By moving the accelerometer to different locations on the tower and repeating the pulse test, mode shapes were also found.

[25]Frequency Response from Pulse Test Data, Instrument Society of America, Publ. Dept., 400 Stanwix Ave., Pittsburgh, PA 15222.

[26]Hewlett–Packard 5451 B Fourier Analyzer.

[27]Linscott, B. S., Shapton, W. R., and Brown, D., Tower and Rotor Blade Vibration Test Results for a 100-Kilowatt Wind Turbine, NASA TM X-3426, 1976.

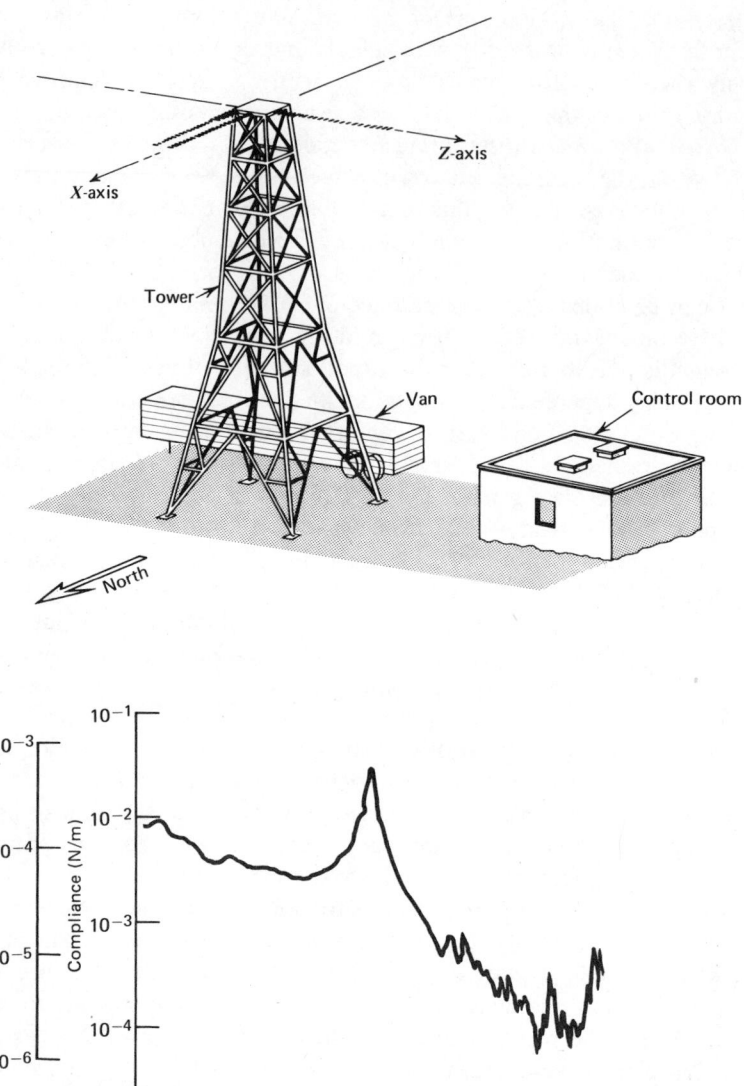

Figure 6.14 Pulse test of wind-turbine tower.

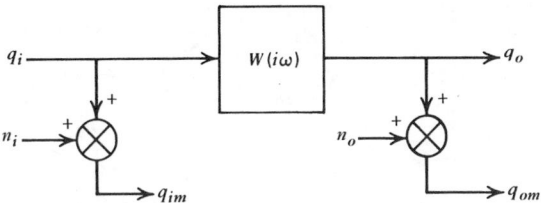

Figure 6.15 System with measurement noise.

While computation of the system transfer function as the ratio of output Fourier transform to input Fourier transform is generally successful if the system is relatively noise free, more reliable results are obtained in practice by using the ratio of cross-spectral density between output and input divided by input power spectral density.[28] These spectral densities, which we introduced in Sections 3.8 and 3.9 to describe random signals, can be computed for *any* signals using their Fourier transforms as follows:

$$W(i\omega) = \frac{\phi_{io}(i\omega)}{\phi_i(i\omega)} \triangleq \frac{Q_o(i\omega)Q_i^*(i\omega)}{Q_i(i\omega)Q_i^*(i\omega)} \tag{6.28}$$

where the starred quantities denote complex conjugates. The advantage in using the spectral densities (rather than the transforms themselves) is not apparent from Eq. (6.28) since Q_i^* clearly cancels out.

It is necessary to consider a system with measurement noise n_i and n_o at the input and output, respectively (see Fig. 6.15), and use *averaging* of several pulse tests to illustrate the benefits of this approach. If we Fourier transform the measured signals q_{im} and q_{om}, we can get an estimate $W_m(i\omega)$ of the transfer function from

$$W_m(i\omega) \triangleq \frac{Q_{om}(i\omega)}{Q_{im}(i\omega)} = \frac{Q_o(i\omega) + N_o(i\omega)}{Q_i(i\omega) + N_i(i\omega)} \tag{6.29}$$

However, this does not give $W(i\omega)$ because of the presence of N_o and N_i. Furthermore, averaging of several W_m's from repeated tests does not improve W_m since the spectra N_i and N_o are essentially the same each time and thus no "noise cancellation" occurs. If we now employ spectral densities:

$$W_m(i\omega) = \frac{Q_o(i\omega) + N_o(i\omega)}{Q_i(i\omega) + N_i(i\omega)} \frac{\left[Q_i(i\omega) + N_i(i\omega)\right]^*}{\left[Q_i(i\omega) + N_i(i\omega)\right]^*}$$

$$= \frac{\phi_{io} + \phi_{ino} + \phi_{oni} + \phi_{nino}}{\phi_i + \phi_{ini} + \phi_{nii} + \phi_{ni}} \tag{6.30}$$

[28]Halvorsen, W. G. and D. L. Brown, "Impulse Technique for Structural Frequency Response Testing," *Sound and Vibration*, Nov. (1977).

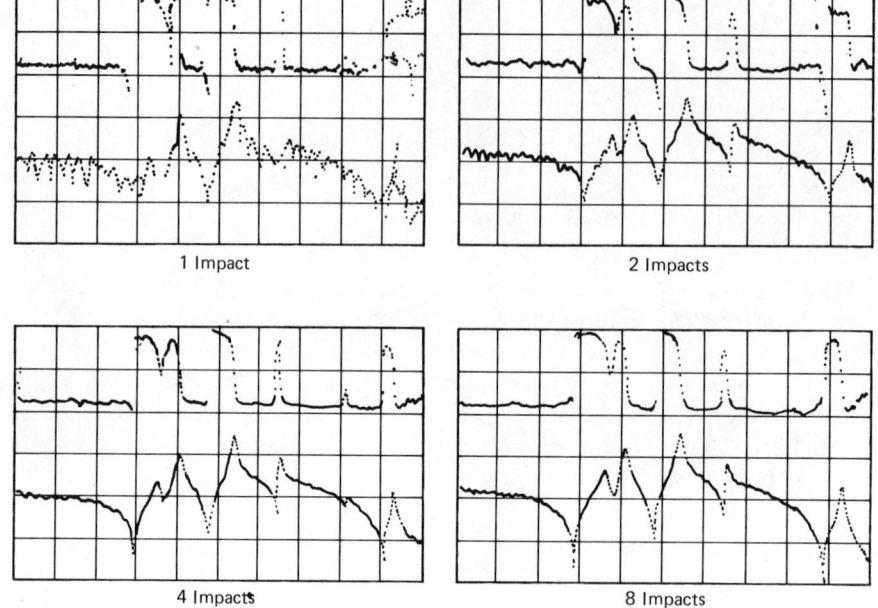

1 Impact   2 Impacts

4 Impacts   8 Impacts

Figure 6.16   Improvement in spectrum quality by averaging.

If the noise signals n_i and n_o are unrelated to each other and to q_i, all the cross spectra involving them would be zero for samples of infinite length, giving

$$W_m(i\omega) = \frac{\phi_{io}}{\phi_i + \phi_{ni}} = \frac{W(i\omega)}{1 + \phi_{ni}/\phi_i} \tag{6.31}$$

which shows no error due to n_o. To make W_m approach W we must thus make ϕ_{ni}/ϕ_i as small as possible and *average* enough tests to make the cross spectra involving n_i and n_o approach their infinite-sample zero values. These requirements and the assumptions underlying the analysis are often approximated sufficiently well in practice that the calculation of $W(i\omega)$ from the cross and power spectral densities is quite common.

Figure 6.16[29] shows some actual results of this averaging technique as applied to pulse testing of a beam structure using an impact hammer and one of the commercially available analyzers.[30] The acceleration/force transfer function phase and amplitude ratio shown exhibit a definite reduction in uncertainty as

[29] Appl. Manual DSP-003 9/75, Spectral Dynamics Corp., San Diego, CA 1975.
[30] Spectral Dynamics Model SD 360.

more tests are averaged; however, little is gained by going beyond about eight. As a further aid in judging the validity of test results, most analyzers provide a measurement of the *coherence function* γ^2. In Fig. 6.15, we define

$$\gamma_{io}^2(\omega) \triangleq \frac{|\phi_{io}(i\omega)|^2}{\phi_i(\omega)\phi_o(\omega)} \tag{6.32}$$

and note that for this ideal situation (perfect linear system, no noise) $\gamma_{io} \equiv 1.0$ since $|\phi_{io}|/\phi_i = |W|$ and $\phi_o = |W|^2\phi_i$. For a real situation with measurement noise as in Fig. 6.15, the measured coherence function γ_{iom}^2 would be[31]:

$$\gamma_{iom}^2(\omega) = \frac{|\phi_{iom}|^2}{\phi_{im}\phi_{om}} = \frac{|\phi_{io}|^2}{(\phi_i + \phi_{ni})(\phi_o + \phi_{no})} \tag{6.33}$$

$$= \frac{\gamma_{io}^2}{(1 + \phi_{ni}/\phi_i)(1 + \phi_{no}/\phi_o)} \tag{6.34}$$

We see that if ϕ_{ni} and ϕ_{no} are not zero, then the measured coherence function will be less than one. In practice, measured values less than one are taken as evidence of noise and/or nonlinear behavior which reduce our confidence in the validity of the linear model $W(i\omega)$ that we have measured. Coherence values of 0.9 or greater are desirable but lesser values may be acceptable to the extent that an approximate model is acceptable. Note that coherence varies with frequency, thus the model may enjoy high confidence in some frequency ranges but not in others. When the amplitude ratio has high peaks and low valleys, one can expect poor coherence (almost zero) near the valleys since very little system response is present here and noise dominates the measured signal. Under such conditions, poor coherence does not really invalidate the model.

When studying spacewise variation of system response (mode shapes) with the pulse test method, a practical application of the reciprocity theorem may be utilized in some instances, structural dynamics being a good example. Rather than applying a pulse input force at the desired driving point and measuring motion response at different spatial locations with an array of motion sensors, we instead use one fixed motion sensor but sequentially apply the input force (using a special instrumented hand-held hammer) at the spatial locations of interest. This creates a family of transfer functions with a space-varying input location and a fixed output location. However, the reciprocity theorem tells us

---

[31]Halvorsen, W. G. and J. S. Bendat, "Noise Source Identification Using Coherent Output Power Spectra," *Sound and Vibration*, Aug. (1975).

that this is equivalent to the situation where the input location is fixed and the output locations vary spatially, thus we obtain the mode shape data we originally wanted.

We conclude this section by giving a brief description of one of the commercial FFT analyzers[32] to indicate the general capabilities of this class of instrument as currently available from several manufacturers. For *system* (as opposed to *signal*) analysis, one requires a two-channel instrument so that system excitation and response data can be *simultaneously* gathered and processed. Each channel of analog input data is first passed through carefully matched anti-aliasing (low-pass) filters to remove frequency content above the desired analysis range. The desired frequency range may be selected from 16 available, starting with a 0–1 Hz range and going in the 1-2-5 sequence up to 0–100 kHz. Associated with each range is a time window that defines the length of the data sample gathered and processed. The 0–1 Hz range requires 400 sec of data while the 0–100 kHz range needs only 0.004 sec. For any selected range, the time functions are sampled and digitized at 1024 points and the frequency spectra computed at 400 points, giving a frequency resolution ranging from 0.0025 Hz for the 0–1 Hz range to 250 Hz for the 0–100 kHz range, and using a sampling rate of 2.56 points per cycle of the highest frequency. Low-frequency data requires a long time sample (400 sec for the 0–1 Hz range), but once the sample is gathered, the transform calculation takes only about 30 milliseconds, this processing time being the same no matter what frequency range is selected.

The most recent instruments of this type use a microprocessor to control many aspects of the data processing, giving very versatile and convenient operation. A completely annotated built-in oscilloscope display presents the results of a wide variety of useful calculations that may be selected from a keyboard. These include excitation and response time histories (anti-alias filtered), Fourier transforms, power-spectral density, cross-spectral density, transfer function, coherence function, impulse response, probability density function, auto-correlation, and cross-correlation. Many results can be stored in memory and later recalled for comparison with current data. Several different forms of averaging can be requested. Any information displayed on the oscilloscope can be read out slowly to produce hardcopy on a digital plotter. Up to nine complete patterns of control settings may be stored in memory and recalled by entering a single ID number, allowing quick setup of standard conditions. A frequency expansion option ("zoom") allows greatly increased resolution by locating the 400 frequency points in any selected 50, 100, 500, 1000 or 2000 Hz band in the 0 to 100 kHz range. For example, the band 80,000–80,050 could be analyzed with resolution of 0.125 Hz.

[32]Nicolet Scientific Model 660, Northvale, N. J.

6.3   RANDOM-SIGNAL TESTING

The use of naturally occurring or intentionally applied random excitation signals as a means of determining a linear model for system response is based on the relation

$$W(i\omega) = \frac{\phi_{io}(i\omega)}{\phi_i(\omega)} \qquad (6.35)$$

Methods for computing the required cross- and power-spectral densities were discussed in Sections 3.8 and 3.9. Today the calculation is most often done by way of Fourier transforms as in Eq. (6.28). In fact, the computation procedure for pulse testing and random signal testing is essentially the same. A curious feature of random signal testing, however, is that transformation of "short" records gives "poor" estimates for the cross and power spectral densities themselves but "good" estimates for the tested system's frequency-response function,[33] which is computed directly from these "inaccurate" spectra. This apparent contradiction can be explained qualitatively as follows. From our discussion of random signals in Section 3.8 it should be clear that accurate estimates of frequency spectra require a certain minimum record length; records of shorter length simply have inadequate information content to be statistically reliable. When we consider *system response*, however, we are processing *simultaneous* records of input and output and the fact that the signals are "random" does not affect the output/input relation. In fact, our definition of "pulses" or "transients" really puts no restriction on the waveform. Thus a "deterministic" triangular pulse is not really in a different class from a "random" waveform in so far as system response calculations based on time histories that have already occurred are concerned. Even the apparent difference in time duration (pulse is "transient," random signal is "ongoing") melts away in practice, where *every* signal processed *must* be of finite duration.

6.3.1   RANDOM ERRORS IN MEASURED POWER SPECTRA

Since the spectra themselves are sometimes of interest, we now present several useful results related to their statistical reliability. If a numerical value $\phi(f)$ of the power-spectral density of a random signal has been measured at a given cyclic frequency f Hz by an analog measuring scheme using a filter of effective bandwidth B Hz and a record length of T sec, it is possible to compute "confidence intervals" for this estimate. Effective bandwidth[34] refers to the fact that real filters do not have the perfect "flat-topped and vertical-sided"

[33]Personal communication from L. Enochson, Gen. Rad. Corp., Time/Data Div., Santa Clara, Calif.
[34]Bendat, J. S. and A. G. Piersol, *Random Data: Analysis and Measurement Procedures*, Wiley-Interscience, N.Y., 1971, p. 277.

frequency response of the mathematically ideal narrow band-pass filter, thus one must *define* bandwidth in such a way that the theoretically derived formulas for ideal filters can be used with good accuracy for real filters. This effective bandwidth is usually defined by experimental procedures and its numerical value is supplied by the analyzer manufacturer. In statistics, a 90% confidence interval refers to a tolerance band around an estimated value. The limits of this band are such that we are 90% sure that the true value of our estimated quantity lies somewhere within this band. By 90% sure we mean that if we apply this procedure routinely to various practical problems we will be right 9 times out of 10.

It can be shown[35] that such confidence intervals for power-spectral density estimates can be calculated using the standard statistical tables for the chi-square distribution:

$$\frac{n}{\chi^2_{n,\alpha/2}}\phi(f) \leqslant \text{true value of } \phi(f) \leqslant \frac{n}{\chi^2_{n,1-\alpha/2}}\phi(f) \qquad (6.36)$$

$$n \triangleq \text{degrees of freedom} = 2BT$$

$$1-\alpha \triangleq \text{confidence level}$$

$$\chi^2 \triangleq \text{numerical value from } \chi^2 \text{ table}$$

For example, suppose $n=100$ and we decide that we wish to be 95% sure that our confidence interval will include the true value; this makes $\alpha=5\%=0.05$, and we need to look up $\chi^2_{100,0.025}$ and $\chi^2_{100,0.975}$. These values are, respectively, 130 and 74.2, so our confidence interval extends from $0.77\phi(f)$ to $1.35\phi(f)$, a -23%, $+35\%$ tolerance band around $\phi(f)$. The effect of n and α on these confidence limits is displayed in Fig. 6.17.[36] Note that sharper filters (smaller B) require longer records to maintain a given confidence level. If B is chosen to resolve the most closely spaced peaks in $\phi(f)$ and if confidence level and $\pm$ tolerance band are also chosen, Fig. 6.17 can be used to decide the necessary record length and also to examine cost/accuracy trade-offs involving α, B, T and tolerance band.

When digital FFT techniques are used to compute the power-spectral density, statistical accuracy considerations take a different form.[37] It is found that a spectrum computed directly from a single record, *no matter what the record length T might be*, has an unacceptably large random error (the standard deviation is as large as the quantity being estimated) and thus averaging of some sort is always necessary. *Increasing T increases the frequency resolution but does not reduce the random error.* The most common averaging method is ensemble

[35]Bendat and Pierson, *op. cit.*, p. 192.
[36]Weston Instruments, Sarasota, Fla., Brochure TM711-C-1.
[37]Bendat and Piersol, *op. cit.*, p. 189.

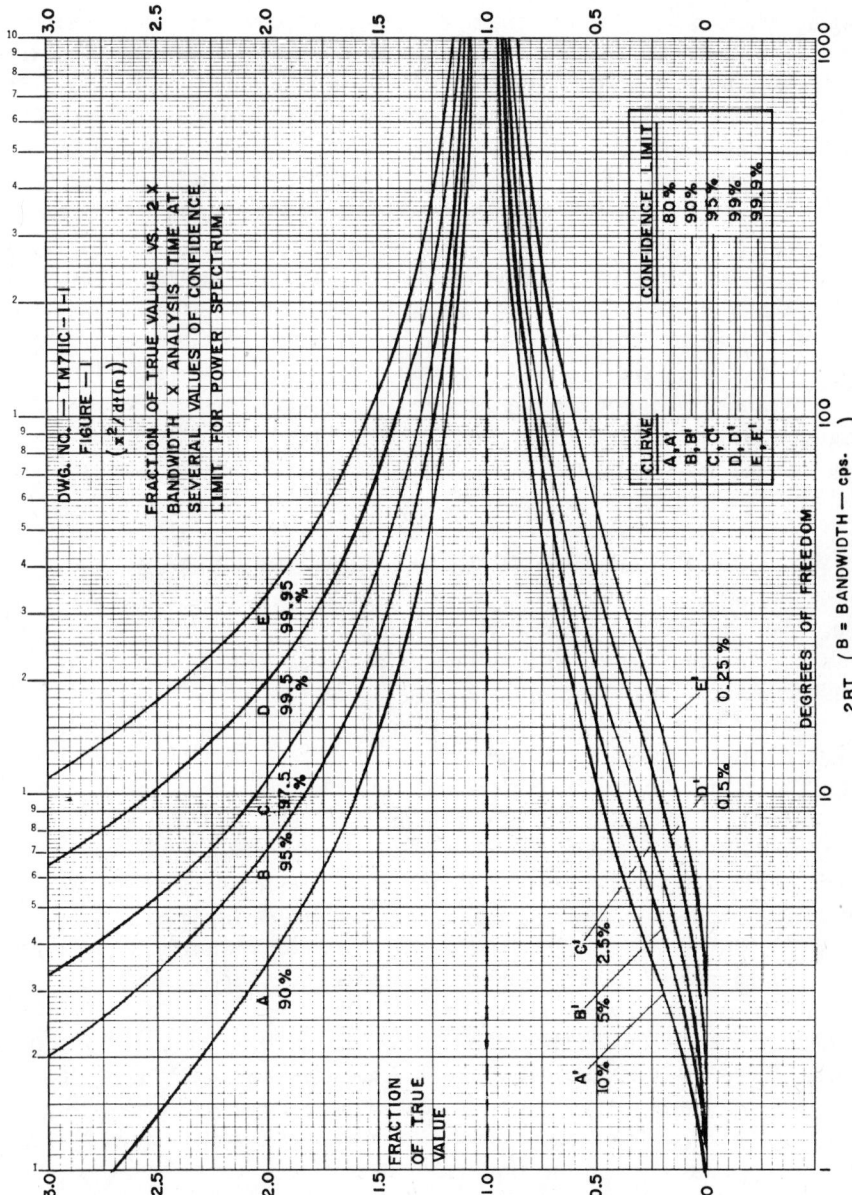

Figure 6.17 Confidence intervals for power-spectral density measurements.

averaging, where spectra are computed for M successive records, each of length T_i, and then the final spectrum is computed by simple averaging of these individual spectra at each frequency. Clearly it will take time $(M)(T_i)$ to gather these M records; however, a spectrum computed from a *single* record of this length will not have the statistical reliability of a spectrum computed by *averaging* M spectra of length T_i. It can be shown[37] that Eq. (6.36) may be used for such digital calculation if we interpret T to be $(M)(T_i)$ and if B is taken equal to $1/T_i$, the frequency resolution of the spectrum, thus giving $n = 2M$ as the degrees of freedom. An alternative procedure ("smoothing over frequency") gathers a single record of length $(M)(T_i)$, computes a single spectrum, averages groups of M adjacent frequency points, and plots a final spectrum of averaged points with frequency spacing $1/T_i$. Theoretically the above two procedures are essentially equivalent; however, most commercial analyzers utilize ensemble averaging. (The use of "cyclic frequency" f rather than "radian frequency" ω in this section has no significance other than to remind the reader that both are used in the literature.)

6.3.2 RANDOM ERRORS IN MEASURED TRANSFER FUNCTIONS

Based on Eq. (6.35), to obtain a system's sinusoidal transfer function we must measure the input-output cross-spectral density and the input power-spectral density, for an input signal with sufficient frequency content to exercise the system's dynamics. This *could* be done by gathering time histories for q_i and q_o over a period T_a to T_b and for q_i over T_c to T_d. That is, the records used to compute ϕ_{io} and ϕ_i need not be coincident in time. Such a procedure is, of course, foolish since both the power- and the cross-spectral estimates so obtained are subject to the (large) errors of the type discussed in Section 6.3.1. These component errors produce an even larger error in a $W(i\omega)$ computed from ϕ_{io} and ϕ_i. The preferred procedure clearly is to use *simultaneous* records to compute the spectra and $W(i\omega)$. Then, in fact, the errors discussed in 6.3.1 are theoretically zero; however, uncertainties arise from other sources. These other sources include measurement noise, system nonlinearities, and response components in the output caused by inputs other than the one measured. The coherence function discussed in Section 6.2 is a measure of the deviation from noisefree, perfectly linear, single-input/single-output behavior. For linear, single-input/single-output systems with measurement noise, a theoretical analysis[38] is available for calculating uncertainties in measured system transfer functions. The probability P that the amplitude ratio $|W(i\omega)|$ is in error by less than

[38]Goodman, N. R., On the Joint Estimation of the Spectra, Cospectrum and Quadrature Spectrum of a Two-Dimensional Gaussian Process, Scientific Paper No. 10, Eng. Statistics Lab, New York University, New York, N.Y., 1957.

$\pm\,\epsilon\%$ and that the phase angle is in error by less than $\pm 0.01\epsilon$ radians is given by:

$$P \approx 1 - \left[\frac{1 - \gamma_{io}^2(f)}{1 - \gamma_{io}^2(f)\cos^2(0.01\epsilon)} \right]^n \tag{6.37}$$

where n is degrees of freedom as in Eq. (6.36) (use $2M$ when M spectra are averaged). Note that this result is in a form such that one cannot independently choose error levels for both amplitude ratio and phase angle; if we want 10% on amplitude ratio we *must* choose 0.1 radians on phase angle. However, the allowed values are reasonable ones, thus the formula is practically useful, particularly in the graphical form of Fig. 6.18.[39] Note that as $\gamma^2 \rightarrow 1.0$, the uncertainty in the transfer function approaches zero even for short record lengths (n small). It also appears that uncertainty due to *any* low value of coherence can be overcome by sufficiently large sample size. However, if significant nonlinearity is present in the tested system, large sample size cannot make the computed linear model an accurate predictor of system nonlinear behavior. Due to the convenience of Fig. 6.18, the above error analysis is widely used, even though it is approximate. A more correct (but less convenient) theory is available[40] if greater accuracy is felt necessary.

If Eq. (6.37) is used to help in choosing the required record length (or number of spectra to be averaged) necessary to obtain a desired accuracy at a chosen confidence level, a numerical value of γ^2 is needed. In most cases γ^2 is not known until *after* the tests have been run, and even then we have only an *estimate*,[41] not the true value. Thus one must usually employ tentative values of γ^2 based on judgment and/or past experience and refine these as measured values become available. When a fast FFT analyzer is being used on a system where it is easy to quickly repeat the test and accumulate averages based on several (or many) spectra, one can often forego the uncertainty calculations presented in this section and simply "watch" the graphic displays of $W(f)$ and $\gamma^2(f)$ change as more tests are run and averaged, terminating the testing when a "visual judgment" indicates that little further reduction in uncertainty is likely (see Fig. 6.16). When direct Fourier transforms are used to compute the spectra, a *single* spectrum always gives a calculated γ^2 of exactly 1.0, irrespective of the true value of coherence, so averaging must *always* be used to get a valid estimate of coherence.

[39]Enochson, L. D., Frequency-Response Functions and Coherence Functions for Multiple-Input Linear Systems, NASA CR-32, 1964.
[40]Bendat and Piersol, *op. cit.*, p. 202
[41]Bendat and Piersol, *op. cit.*, p. 193.

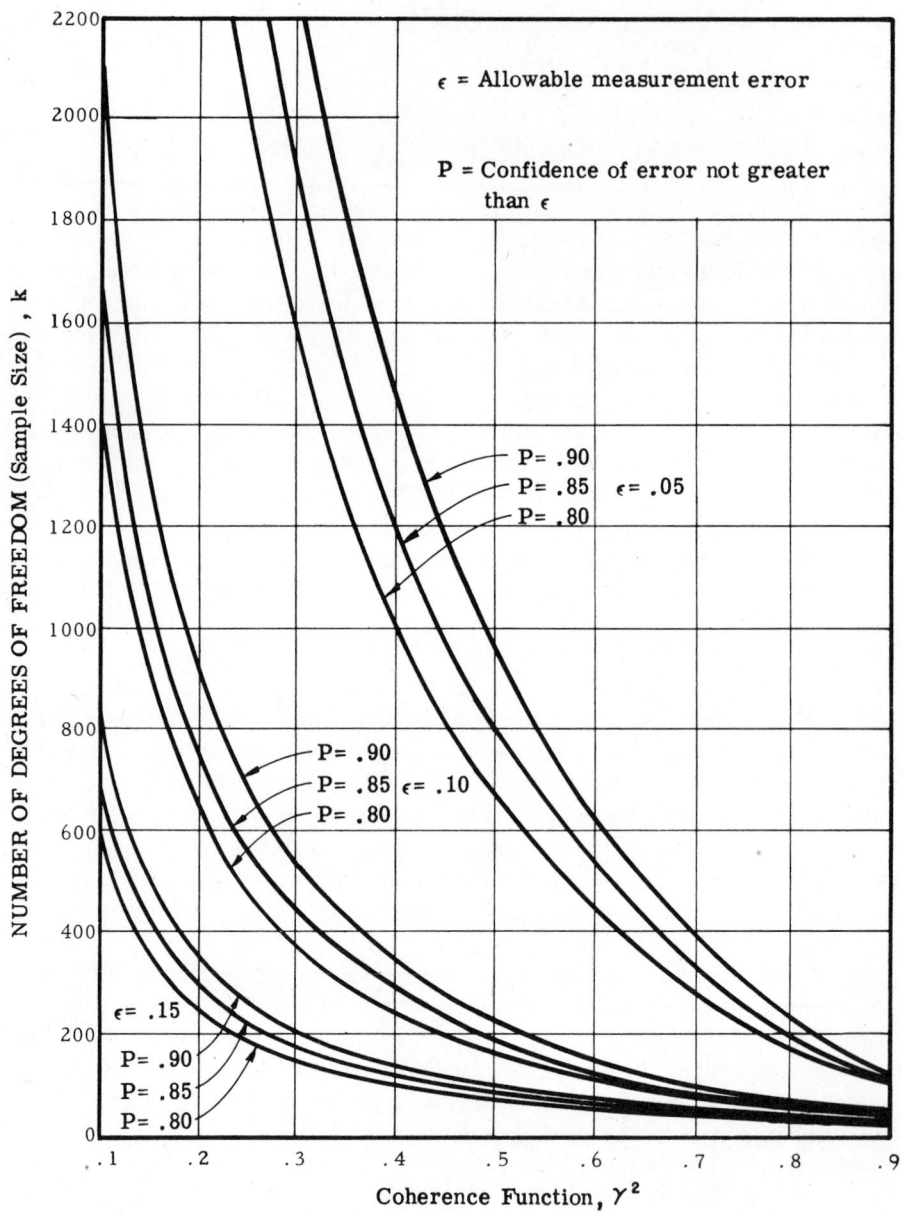

DATA FOR FREQUENCY RESPONSE FUNCTION
MEASUREMENT CONFIDENCE

Figure 6.18   Confidence intervals for measured transfer functions.

6.3.3   INPUT SIGNALS: CONTINUOUS, BINARY, AND PSEUDO-RANDOM

When random-signal testing makes use of the "naturally occurring" random disturbances present in many physical processes, these system input signals are generally of a continuous nature and, in fact, often exhibit an amplitude distribution close to the Gaussian ideal. The aircraft dynamics testing displayed in Fig. 3.52, where aircraft motions were stimulated by naturally occurring atmospheric turbulence, is a good example of this class. Whether we employ naturally occurring random inputs or intentionally apply "artificial" ones, it is of course necessary that such inputs have power spectral densities whose magnitudes are large enough to allow accurate measurement of input and response, and whose frequency content covers the range of frequencies for which we need a system model. If it is possible to run tests for inputs of several different levels (total-signal RMS value) and identical frequency content, one can explore system nonlinearity in a manner similar to running sinusoidal tests at different amplitudes. As in the case of sinusoidal and pulse testing, one usually generates the desired form of random signal as a voltage/time waveform and then transduces this signal to the physical form appropriate to the input of the system under test. In the past, electronic random-signal generators relied mainly on naturally-occurring random processes in certain vacuum tube or semiconductor devices as a source of wide band continuous Gaussian random voltage signals that could then be suitably filtered and shaped for test purposes. More recently, as part of the general trend toward digital electronics, random voltage waveforms have been "manufactured" by starting out with suitably chosen binary digital waveforms and filtering these to obtain good approximations to continuous Gaussian signals. In our upcoming discussion of these devices, we shall see that the binary signals *themselves* can sometimes serve directly as the input for our system testing.

We base our explanation of this class of signal generators on a specific commercial instrument.[42] While various binary waveforms might be employed, the most common appears to be the *maximum length pseudo-random binary sequence* (PRBS). Such sequences are easily generated with a shift register using appropriate feedback through an exclusive OR gate as in Fig. 6.19, where the device is set up to produce a 15-bit sequence. By changing the feedback configuration, sequences of length $N = 15$ to 1,048,575 bits may be obtained, where $N = 2^n - 1$ and $n = 4, 5, 6, \ldots, 20$. For a 15-bit sequence length, only the first four shift register stages are needed. Suppose we start the system up with $1, 0, 0, 0$ respectively in the four stages. Taking the output signal as the contents of stage 1, we begin the sequence with a logical one, which produces, say, a -5 volt output (let logical zero produce $+5$ volts). The shift register is driven by periodic

---

[42]Model 3722A Noise Generator, Hewlett–Packard Journal, Sept. (1967).

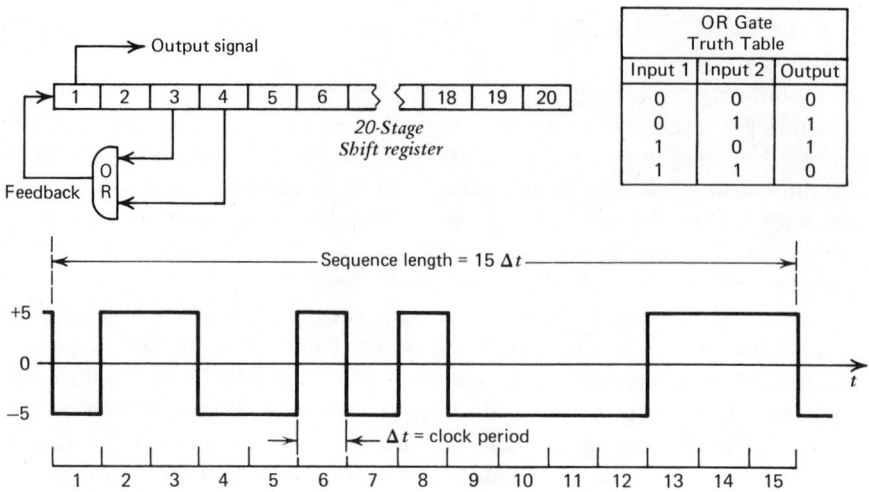

Figure 6.19 Generation of pseudo-random binary sequence.

clock pulses at intervals adjustable over the range 1 μsec, 3.33 μsec, 10 μsec,...,333 sec. (This adjustable clock rate gives control over the frequency content of our random signals, the binary signals having bandwidth adjustable from 0.00135 Hz to 450 kHz while the Gaussian signals derived from them by filtering go from 0.00015 Hz to 50 kHz.) When a clock pulse occurs, the shift register acts like a conveyor belt and shifts all data one stage to the right, giving us the pattern 0100 since the OR gate had two zero inputs and thus a zero output. The next clock pulse gives 0010, the next 1001, the next 1100, and so on until the 14th pulse gives 0001 and the 15th gives 1000 (the initial state) whereupon the whole sequence is repeated cyclically.

The properties of maximum length PRBS signals have been worked out in some detail[43]; we here summarize only the major points. If the particular sequence chosen (such as the 15-bit described above) is allowed to repeat cyclically, we have a periodic function, for which the frequency spectrum is discrete and can be computed using Fourier series or indirectly from the autocorrelation function,[44] giving the result:

$$\text{average value} = E/N$$

$$\text{amplitude of } n\text{th harmonic} = E\frac{\sqrt{N+1}}{N}\left|\frac{\sin(\pi n/N)}{\pi n/N}\right| \tag{6.38}$$

[43]Uhrig, R. E., *Random Noise Techniques in Nuclear Reactor Systems*, Ronald Press, N.Y., 1970, p. 286; Davies, W. D. T., *System Identification for Self-Adaptive Control*, Wiley-Interscience, N.Y., 1970, p. 44.
[44]Lawrence, C. B. and A. Pearson, Measurement Techniques Using a Pseudo Random Binary Sequence, Atomic Energy of Canada Ltd., Rept. AECL-3601, Chalk River, Ontario, 1970.

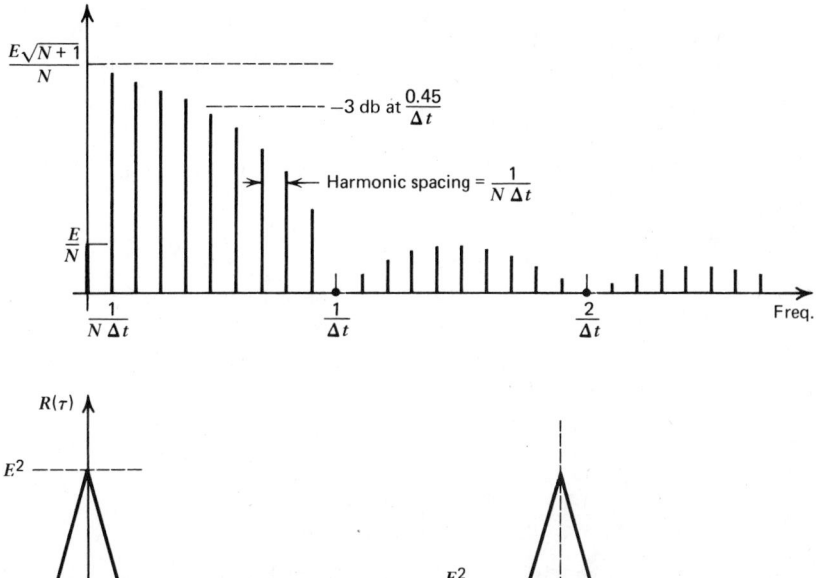

Figure 6.20   Frequency spectrum and autocorrelation function for pseudo-random binary sequence.

Here N is the sequence length, $n = 1, 2, 3, \ldots, \infty$, E is the $\pm$ output level, and the fundamental frequency is at $1/(N\Delta t)$ Hz, where Δt is the clock period in seconds. Figure 6.20 displays this spectrum and also the autocorrelation function. Ideal white noise would exhibit a single impulse function at the origin for its autocorrelation function and a continuous and perfectly flat frequency spectrum, thus the PRBS is an approximation to ideal random noise to the extent that these time and frequency domain characteristics are approximated. For practical system testing the PRBS spectrum must be designed so that:

1   The spectrum is reasonably flat out to the highest frequency of interest in the system transfer function. From Fig. 6.20 this will be somewhere between $0.45/\Delta t$ and $1.0/\Delta t$, thus we adjust Δt to meet this need.
2   Since the spectrum is discrete, there is *no* excitation between the harmonics, thus they must be spaced closely enough that no system peaks or valleys can hide between them. This requires that sequence length N be sufficiently large.

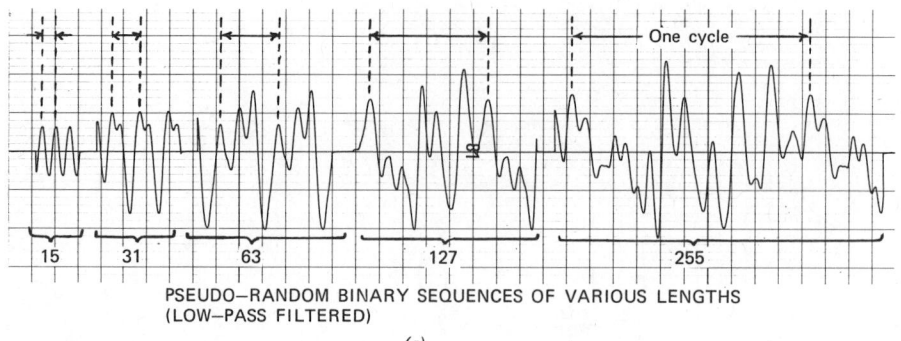

PSEUDO–RANDOM BINARY SEQUENCES OF VARIOUS LENGTHS
(LOW–PASS FILTERED)

(a)

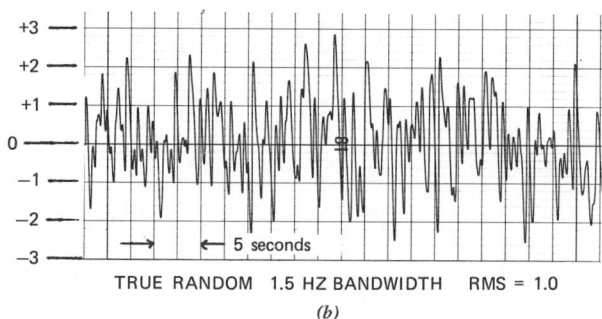

TRUE RANDOM   1.5 HZ BANDWIDTH   RMS = 1.0

(b)

Figure 6.21   Low-pass filtered pseudo-random and random signals.

3 Harmonic magnitudes must be large enough to allow accurate measurement of system input and output without causing excessive nonlinear response; this is adjusted by the output level E.

It is also necessary[45] that the system under test be subjected to at least one complete sequence (sometimes called the "settling sequence") before gathering of input and output data commences.

Should a truly random binary signal with continuous (rather than discrete) spectrum be required, the referenced instrument provides a mode of operation in which the shift register feedback and OR gate are disconnected and the first stage of the register is controlled by a semiconductor noise source. Just before each clock pulse, the noise source is sampled by a level detector that decides, on arrival of the shift pulse, whether a one or a zero is to be placed in the first stage of the register. Since the random signal is not periodic, there is no repeated

[45]Davies, W. D. T., *System Identification for Self-Adaptive Control*, Wiley-Interscience, N.Y., 1970, p. 137.

pattern in the resulting series of ones and zeros and we get a binary signal whose spectrum is continuous with the same shape as the envelope of the discrete spectrum of a PRBS with the same clock period.

If we send a binary signal (either a periodic PRBS or a truly random one) through a low-pass filter with cutoff frequency about $\frac{1}{20}$ of the clock frequency, we "round off the corners" and get a smoothly varying signal that approaches a Gaussian amplitude distribution for large N, quite acceptable accuracy being attained for $N \geqslant 8191$. A filtered periodic PRBS is of course still periodic, giving a discrete spectrum with harmonics at the same frequencies as the PRBS itself; however, the flat (-3 db) range now extends to only $1/(20\Delta t)$ Hz. The truly random binary signal gives a truly random continuous signal when filtered and leads to a continuous spectrum that is flat (-3 db) to $1/(20\Delta t)$ Hz. In Fig. 6.21a, waveforms of filtered PRBS signals of several sequence lengths, produced by the Hewlett–Packard instrument referenced earlier, are shown. The truly random signal produced when the shift register is triggered by the random noise source is displayed in Fig. 6.21b.

6.3.4 APPLICATION EXAMPLES

Further details on the aircraft dynamic testing of Fig. 3.52 are given in Figs. 6.22 and 6.23. The transfer function measured had as input the vertical gust velocity (ft/sec) of atmospheric turbulence impinging on the aircraft, while the output was the vertical acceleration (ft/sec^2) of the vehicle center of mass. Using the methods of Fig. 6.18 and measured values of γ^2, the 90% confidence bands shown were calculated. In addition to direct measurement of coherence, an attempt was made to predict it from estimates of "noise" sources that were felt to contribute to reduction of coherence. Figure 6.23 enumerates these sources and compares predicted with measured coherence values over the frequency range of interest. The referenced report concludes "...with appropriate precautions, flight tests in rough air of a few minutes duration may be used to obtain reliable estimates of airplane frequency-response functions".

An application[46] of PRBS signals to structural dynamics testing is shown in Figs. 6.24 and 6.25. Using a sequence length $N = 511$ and clock period $\Delta t \approx$ 0.001 sec, our excitation signal will be flat within -3 db from 0 to 450 Hz with a harmonic spacing of about 2 Hz. The FFT analyzer is set up to acquire exactly one cycle of the PRBS signal, using a sampling rate of 1 millisecond. At this sample rate, aliasing will occur for frequencies above about 400 Hz, so anti-aliasing measures are necessary because the PRBS spectrum extends beyond 400 Hz. Rather than using the FFT analyzer's anti-aliasing filters on the force and acceleration signals (the usual approach), the experimenter chose to low-pass

---

[46]Roth, P., Measurement of Transfer Functions with Wide Dynamic Range, Appl. Note 140-5, Hewlett–Packard Corp., 1973.

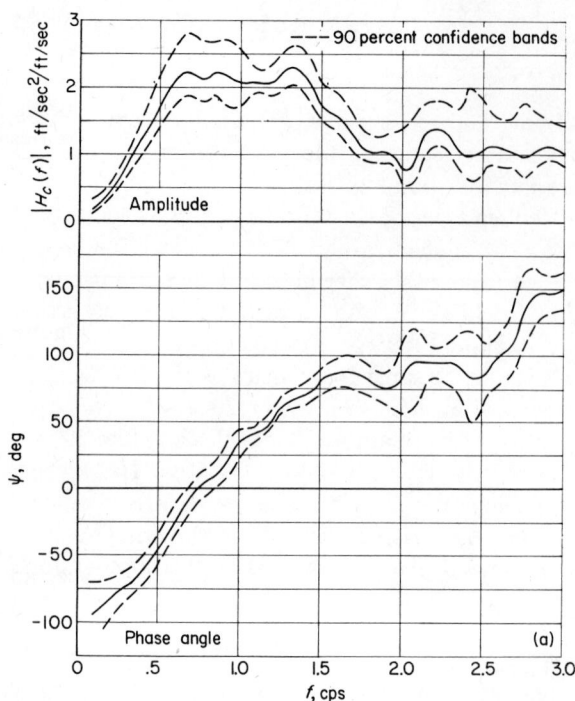

Figure 6.22 Aircraft transfer function measured with random signals.

Noise source	Estimated percent reductions in the coherency functions by noise source for frequencies of—			
	< 0.3 cps	0.3 to 2 cps	2 to 3 cps	> 3 cps
Instrument errors			?	?
Reading errors	0	0 to 10	10 to 25	25
Side gusts	20			
Head-on gusts	10	10	10	10
Spanwise gust variations	10	5 to 20	20 to 30	30
Pilot control motions	10			
Total	50	15 to 40	40 to 65	65

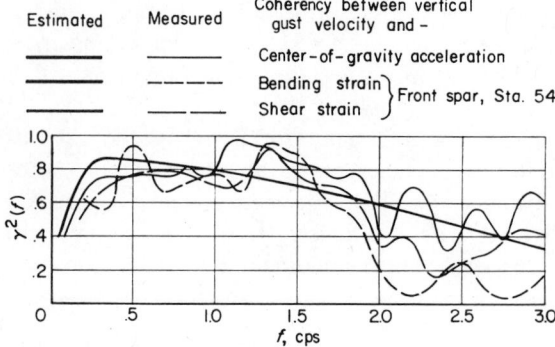

Figure 6.23 Prediction of coherence function.

278

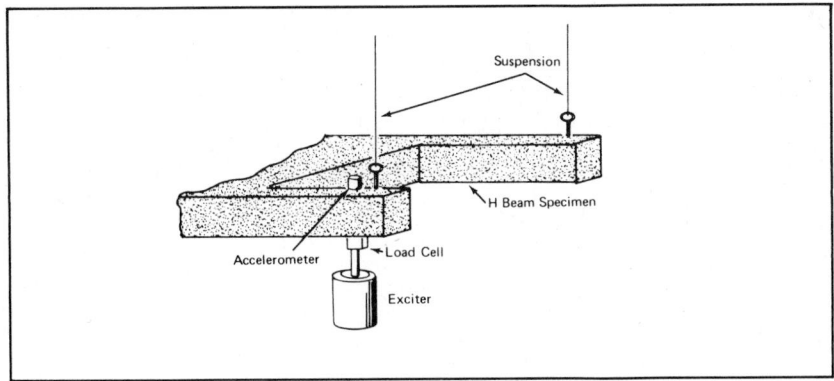

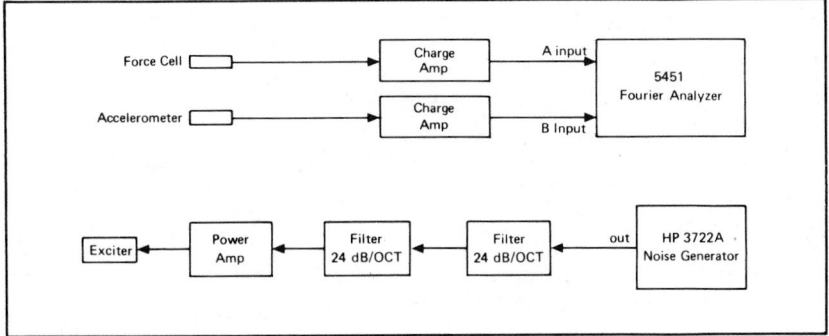

Figure 6.24   Setup for binary random structural dynamics test.

filter the PRBS signal at about 400 Hz *before* applying it to the vibration shaker. While this analyzer computes spectra out to half the sampling frequency (500 Hz in this case), results above 390 Hz are suppressed in the graphs of Fig. 6.25 since aliasing is beginning to make these computations unreliable. These graphs are ensemble averages of four complete PRB sequences and the author cautions that excessive numbers of averages can lead to propagation of truncation errors. He also notes that the anti-resonant "valley" at about 220 Hz (which should be a true null and thus go as "deep" as the one at 150 Hz) could be more correctly measured by redesigning the experiment for frequency resolution finer than the 2 Hz here used.

Our final example[47] of random-signal testing involves the dynamic response of a nuclear power reactor's neutron flux to input perturbations of reactivity caused by control rod motion. The input signal (see Fig. 6.26) was a 16-bit

[47]Serdula, K. J., *et al.*, Frequency-Response Measurements of the Gentilly Nuclear Reactor Dynamics, Atomic Energy of Canada Ltd., Repr. AECL-4370, Chalk River, Ontario, 1973.

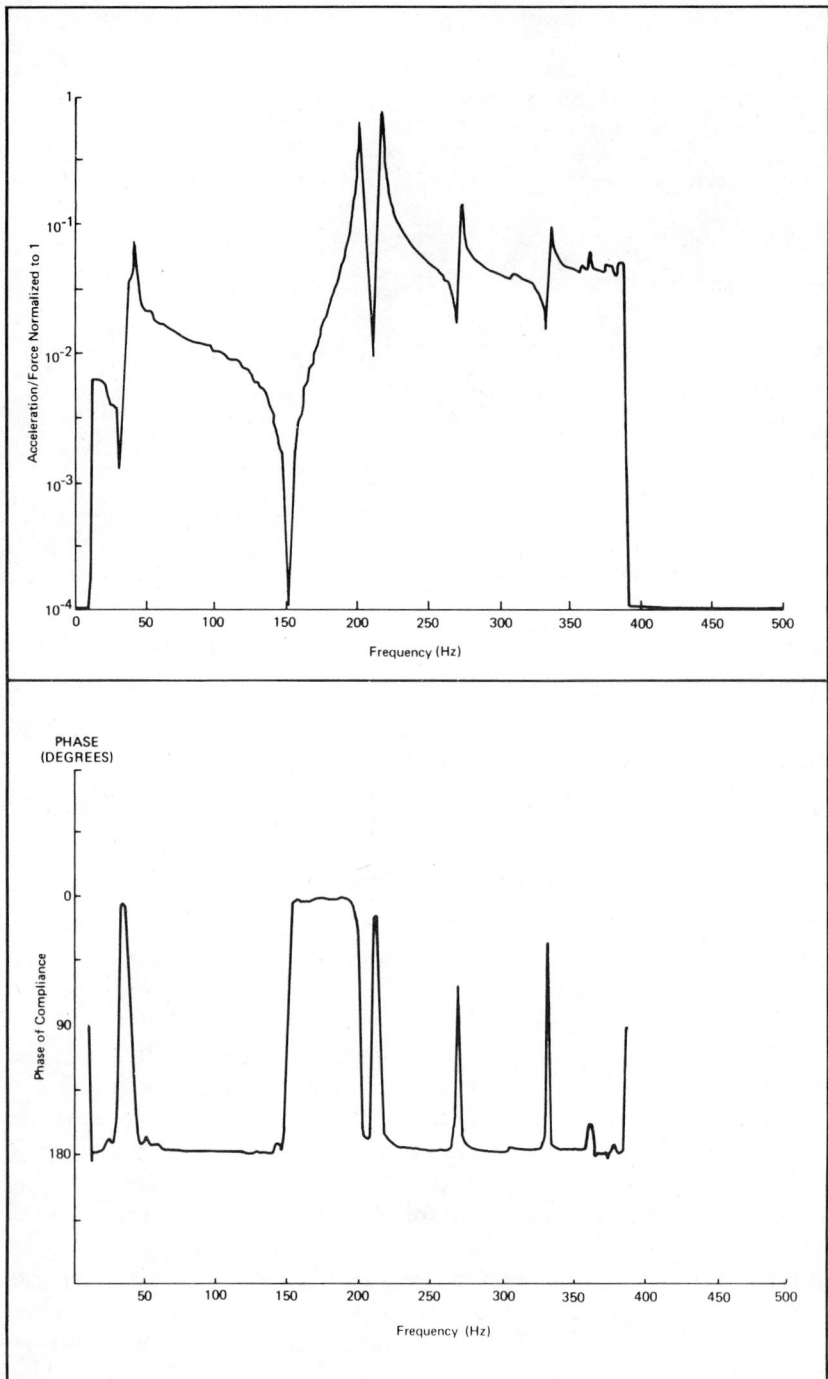

Figure 6.25 Results of binary random structural dynamics test.

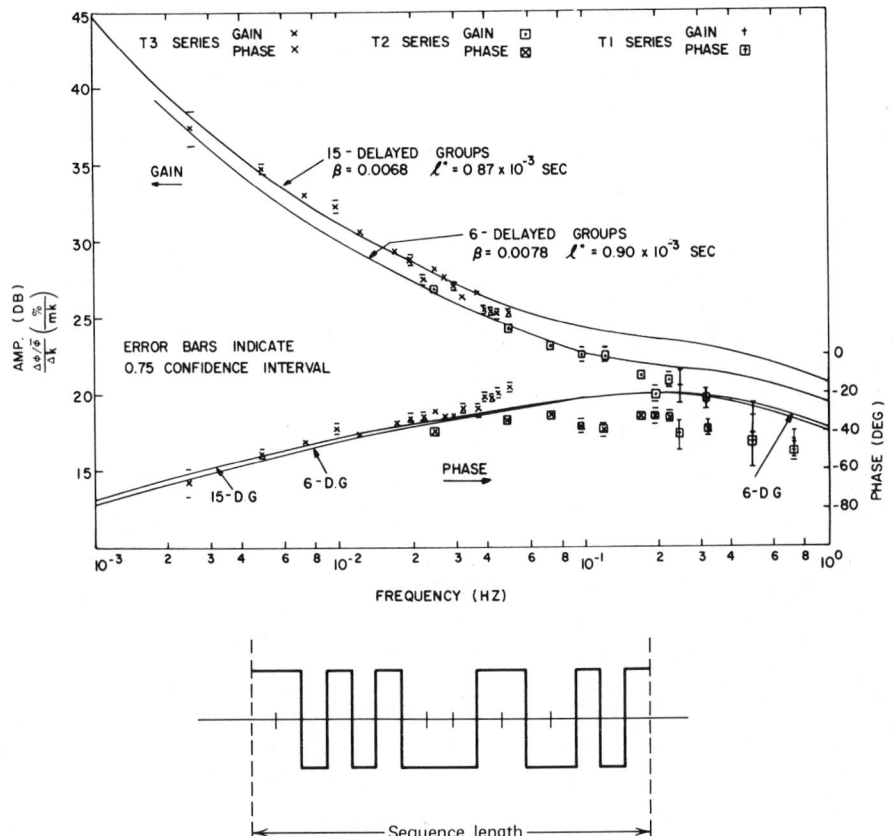

Figure 6.26   Binary random tests of nuclear reactor.

pseudo-random binary sequence, but not the maximum length type discussed in 6.3.3. A signal with *zero* mean value was desired since an integrating effect in the tested system would cause a gradual drifting away from the desired operating point for any input signal with nonzero average value. Also, several tests (T1, T2, T3 of Fig. 6.26) with different clock periods were run to optimize test signal frequency content for particular frequency ranges. The periods of the complete 16-bit sequences were 4.096, 40.96, and 409.6 sec, respectively, for tests T1, T2, and T3. Each test consisted of 11 complete sequences, the first used to allow settling of transients and the remaining 10 used to actually record data, 640 total points being gathered in each case. The data was FFT processed[48] on a general purpose computer as 10 blocks of 64 points each and then averaged,

[48]MAC/RAN Program, University Software Systems, El Segundo, Calif.

leading to the transfer function results of Fig. 6.26. Actually, each test produced 32 discrete frequency points ($\frac{64}{2}$) but any of these that had coherence less than 0.75 were discarded as being unreliable (*most* of the plotted points had $\gamma^2 >$ 0.95). Discarded points generally corresponded to frequencies at which the input signal had small amplitude. The solid lines on Fig. 6.26 refer to two different theoretically calculated models, included for comparison with the measured behavior.

6.4 PARAMETER-TRACKING (STEEP DESCENT) METHODS

The test methods discussed to this point in Chapter 6 have shared the following features:

1 The system model is a linear, constant-coefficient one.
2 A particular form of input signal is stipulated.

We now discuss a class of methods that require neither of these restrictions. While this desirable generality would seem to make such methods of wide applicability, the literature actually reveals less use than the sine, pulse, and random techniques already explored. While this modest level of application must be indicative of certain practical defects, we nevertheless judge the methods of sufficient interest to present a brief discussion.

While this class of methods goes by several names, perhaps the most common is *parameter tracking*, and the *equation-error*[49] technique is considered one of the better specific approaches. We will use a number of specific examples to illustrate it. Consider a linear constant-coefficient system as in Fig. 6.27. Its equation could be written as

$$0 = a_1 \dot{q}_o + q_o - q_i \tag{6.39}$$

Let $a_1^*(t)$ be a computed estimate of a_1, define the *equation error* $E(t)$ by

$$E(t) \triangleq a_1^*(t)\dot{q}_o + q_o - q_i \tag{6.40}$$

and subtract to get

$$E(t) = \left[a_1^*(t) - a_1 \right] \dot{q}_o \tag{6.40a}$$

The scheme for computing $a_1^*(t)$ uses "steep descent" concepts and defines

$$\dot{a}_1^* \triangleq -G \frac{\partial E^2}{\partial a_1^*} = -2GE\dot{q}_o \tag{6.41}$$

---

[49]Hofmann, L. G., P. M. Lion, and J. J. Best, Theoretical and Experimental Research on Parameter Tracking Systems, NASA CR-452, 1966.

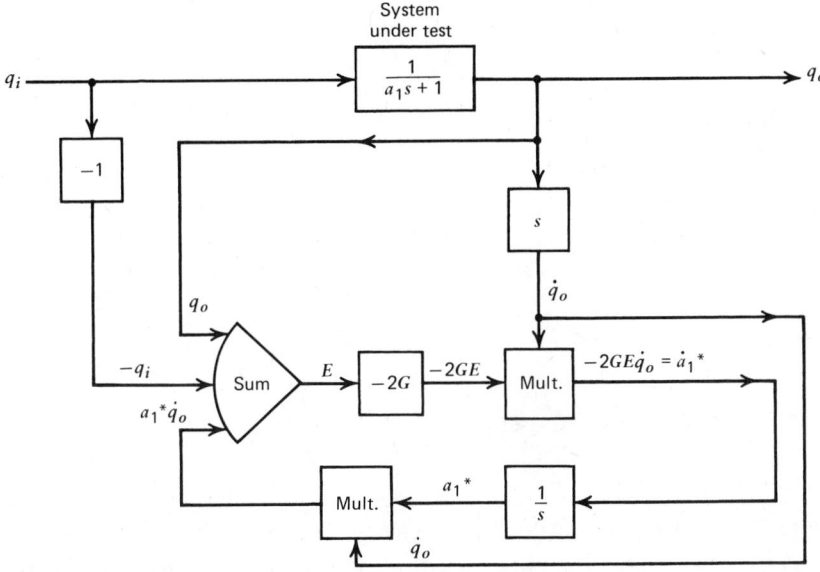

Figure 6.27 Parameter tracking of simple linear system.

This is called the "parameter adjustment law" and G is an unspecified positive constant. At a given instant of time, $\dot{q}_o$ would be some specific value and at that instant, Eq. (6.40a) shows a parabolic relation between E^2 and a_1^*, as in Fig. 6.28. At a time instant t_1, where a_1^* is below the correct value a_1, note that the slope $\partial E^2/\partial a_1^*$ is such that Eq. (6.41) will cause a_1^* to drive toward a_1 and that it drives faster if at this instant $\dot{q}_o$ is large. If at some other time t_2 we find a_1^* above a_1, we again see that a_1^* is driven toward the correct value a_1. In fact, a_1^* will remain stationary only if $\dot{q}_o \equiv 0$ (system is not being exercised) and/or $a_1^* \equiv a_1$. Thus it appears that a_1^* is always driven toward the correct value a_1 and the rate of convergence depends on G, E, and $\dot{q}_o$. Large values of G seem to be desirable except that the feedback nature of Fig. 6.27 warns of possible instability for excessively large G. Actually, detailed analysis[50] shows that these schemes are rarely unstable.

For the simple example above only, an analytical solution[51] is possible:

$$a_1^*(t) = (a_{10}^* - a_1)e^{-2G\int_0^t \dot{q}_o^2(\tau)d\tau} + a_1 \qquad (6.42)$$

where a_{10}^* is the initial value assumed for a_1^* at $t=0$ (taken as zero if no better

[50]Hofmann, *et al.*, *op. cit.*, p. 5.
[51]Miller, B. J., A Theoretical and Analog Study of a Steep Descent Coefficient Computer, PH.D. Dissertation, Mech. Eng. Dept., The Ohio State Univ., 1962, p. 41.

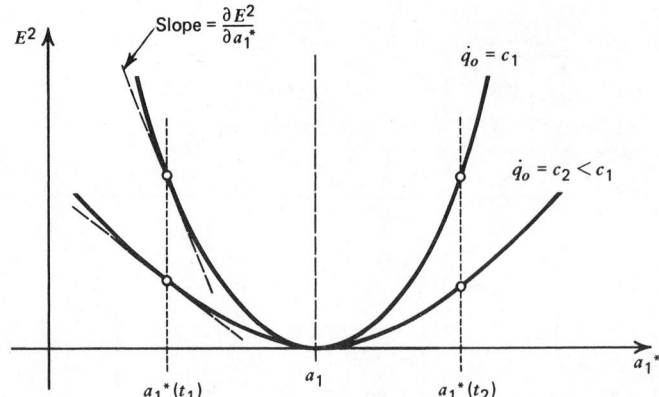

Figure 6.28   Geometric interpretation of parameter adjustment law.

information is available). From (6.42) it is clear that, irrespective of the nature (sine, pulse, random, etc.) of the input signal, as long as $\dot{q}_o$ is not identically zero, a_1^* will eventually converge to a_1. While not rigorously proven, intuition indicates that if a_1 varied with time (time-varying linear system), a_1^* should "track" a_1 so long as a_1 does not change too fast for the convergence rate of the computing scheme. Analog and digital simulation verify this conjecture and also have shown that the method works for multi-parameter tracking of both linear and nonlinear systems. Since analytical solutions are generally not possible, simulation is used to study the behavior of specific tracking systems and the remainder of our presentation consists mainly of CSMP studies of several examples.

When studying parameter tracking schemes on CSMP, one first needs to simulate the system under test, using conventional CSMP methods. When we then set up the various computing loops to track the parameters of interest, we require "measured" values of q_i, q_o and certain of their derivatives. In a CSMP (or analog) simulation of a physical system, all these "measured" signals naturally appear when we simulate the system under test, so when we get ready to simulate the parameter-tracking system, all the needed inputs are already available in "perfect" form. Since a real-world (rather than simulation) application of parameter tracking would require *actual measurement* of these signals, some of which (the derivatives) might be difficult to measure accurately, simulation studies can give unrealistically favorable pictures of tracking system performance. Fortunately, once we recognize this, we can easily use simulation to include practically any type of real-world "defect" in our system studies and thus get results more characteristic of the actual implementation.

A CSMP study of the system of Fig. 6.27 might go as follows.

```
PARAM     A1 = 1.0,K = 1.0,A1S0 = 0.0
          Q1 = RAMP(0.0) − 2.*RAMP(3.0) + RAMP(6.0)
          Q2 = SIN(TIME)
          Q3 = SIN(TIME/5.)
          Q4 = SIN(TIME*5.)
          QI = B1*Q1 + B2*Q2 + B3*Q3 + B4*Q4
PARAM     B1 = 1.,B2 = 0.0,B3 = 0.0,B4 = 0.0
          QODOT = (K*QI − QO)/A1
          QO = INTGRL(0.0,QODOT)
          E = A1S*QODOT + QO − K*QI
PARAM     G = 1.0
          A1SDOT = − 2.*G*E*QODOT
          A1S = INTGRL(0.0,A1SDOT)
TIMER     FINTIM = 10.,DELT = .01,OUTDEL = .2
OUTPUT    QI,QO,QODOT,E,A1S
END
PARAM     G = 10.
END
PARAM     G = 1.0,B1 = 0.0,B2 = 1.,B3 = 0.0,B4 = 0.0
END
PARAM     G = 10.
END
PARAM     G = 1.0,B1 = 0.0,B2 = 1.,B3 = 1.,B4 = 1.0
END
PARAM     G = 10.
END
```

The multiple runs produced by this program explore the effects of the nature of q_i (triangular pulse, single sine, multiple sine) and the size of G (1.0,10.). In each case A1S converges to the correct value, the larger G value simply making the convergence more rapid. Figure 6.29 shows partial results.

Of the many application areas and potential difficulties that might be studied let us next briefly examine the "model-mismatch" question. Since real-world systems *never* conform precisely to *any* specific mathematical equation, any model we might choose will always be mismatched with the real system to some degree. What happens to a parameter-tracking system under such circumstances? Let us examine the case of a real-world second-order system we (wrongly) presume to be first order in building our parameter tracker. For the actual system let

$$\ddot{x} + C\dot{x} + x = Kf = K(\sin \omega_1 t + \sin \omega_2 t) \qquad (6.43)$$

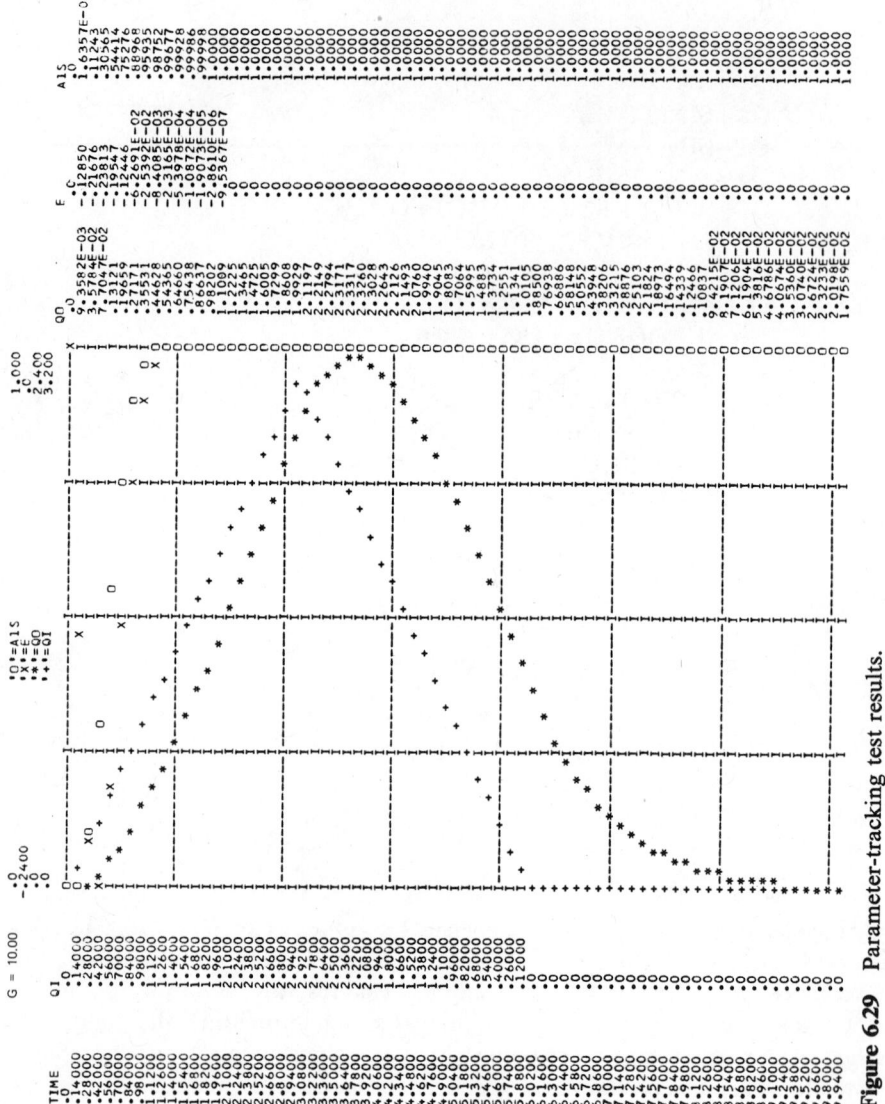

Figure 6.29 Parameter-tracking test results.

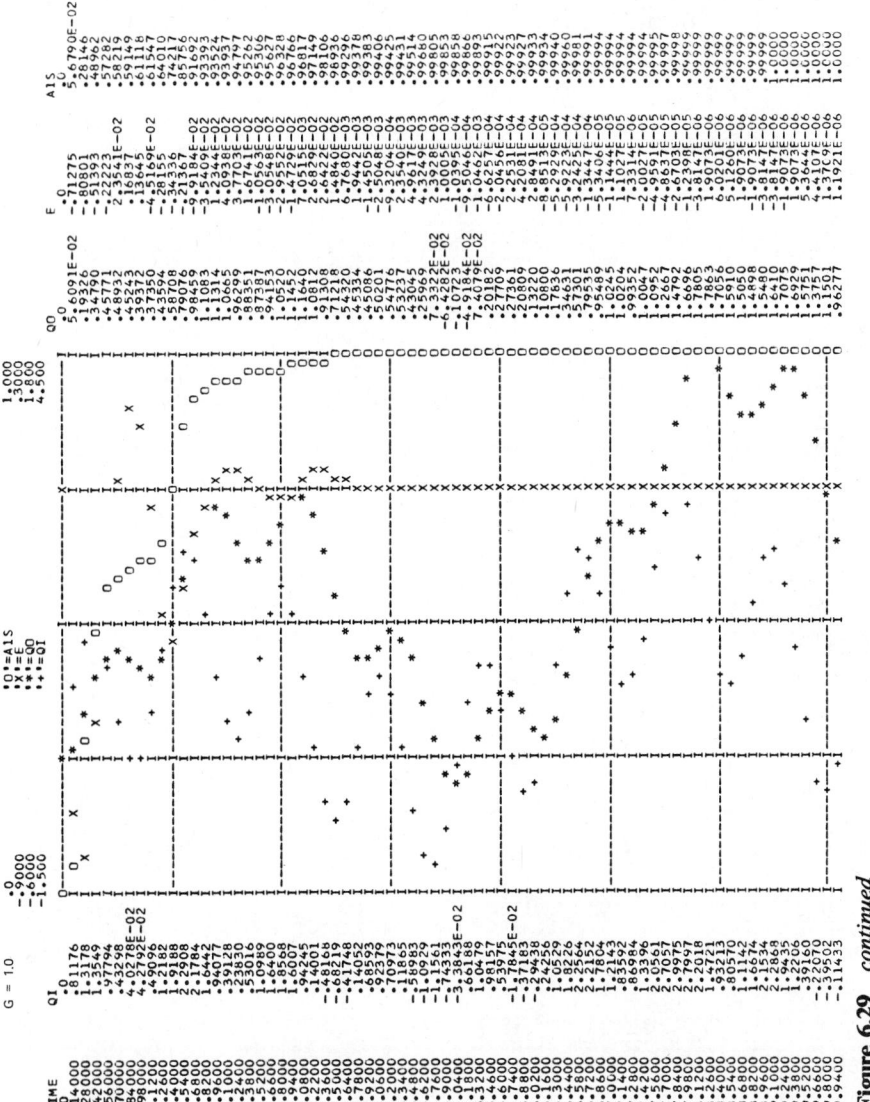

Figure 6.29 *continued*

287

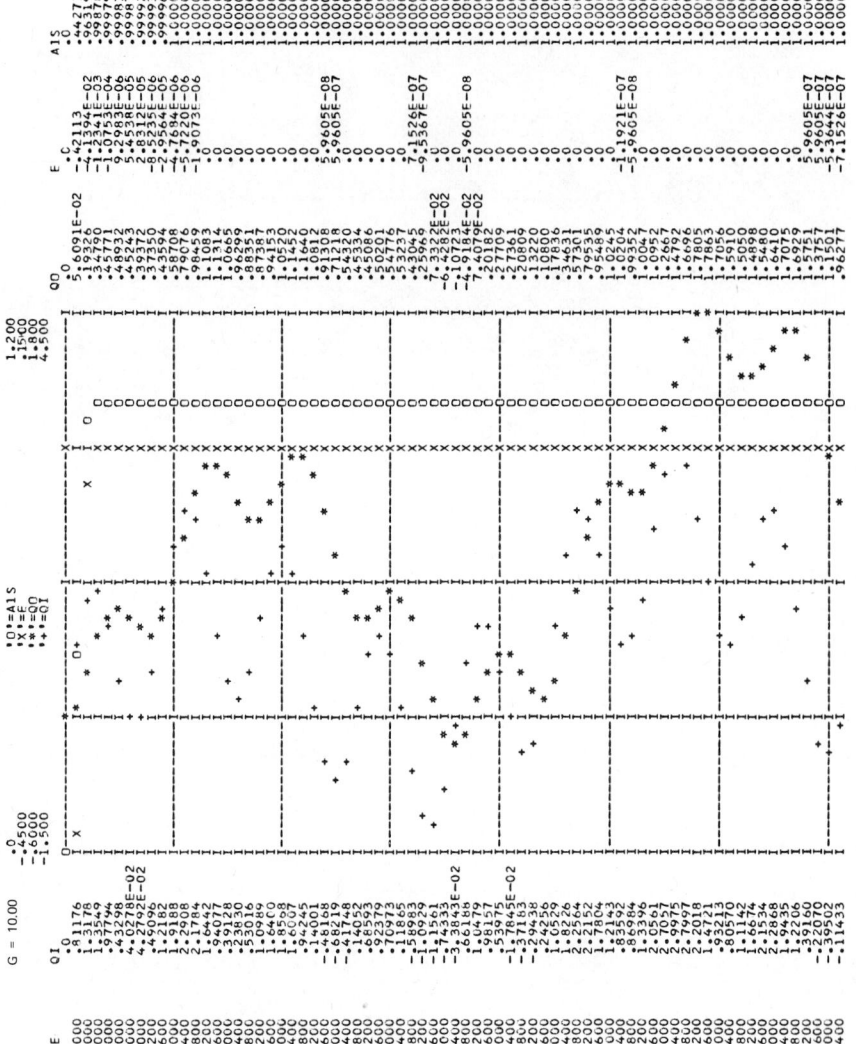

Figure 6.29 *continued*

288

while we wish to find C and K in the presumed model

$$C\dot{x} + x = Kf = K(\sin \omega_1 t + \sin \omega_2 t) \tag{6.44}$$

thus giving

$$E = C^* \dot{x} + x - K^* f \tag{6.45}$$

and

$$\dot{C}^* = -2GE\dot{x}, \qquad \dot{K}^* = 2GEf \tag{6.46}$$

Note that the definitions of E, $\dot{C}^*$, and $\dot{K}^*$ follow exactly the same recipe in this two-parameter tracker as in a single-parameter scheme. The pattern is in fact the same for *any* number of parameters, thus no new tricks need be invented for the more general cases. Since a second-order system can be approximated as first order for low frequencies, let us take $\omega_1 = 0.1$ and $\omega_2 = 0.2$ to exercise our second-order system with $\omega_n = 1.0$. Let us also take $K = 1.0$ and try two values of C (0.2 and 1.4) corresponding to light ($\zeta = 0.1$) and heavy ($\zeta = 0.7$) damping.

The CSMP results of Fig. 6.30 show that C^* and K^* now do not converge to steady values but rather fluctuate in what appears to be a periodic pattern. One interpretation of this would be (remember that in a real application we *would not know* the system was really second order) that our system is first order with *time-varying* coefficients $C(t)$ and $K(t)$. If, on the other hand, we insist on a constant-coefficient model, we could estimate average values for C and K from the CSMP time histories and charge the fluctuations to modeling errors, measurement noise, or whatever. If this is done for the case where C is actually 0.2, we get $C \approx 0.24 \pm 0.24$ and $K = 0.99 \pm 0.19$, thus the average values are good estimates but the fluctuation is excessive, especially for C. When we go to the heavily damped system a similar procedure produces $C \approx 1.45 \pm 0.25$ and $K \approx 0.98 \pm 0.18$, giving somewhat more confidence in a constant-parameter model. A CSMP run comparing x of the actual system with that predicted by the constant-parameter first-order model with $C = 1.45$, $K = 0.98$ gives the results of Fig. 6.31, showing that the model mimics the behavior of the real system quite closely for the given input. However, if we now excite the system with ($\sin t + \sin 3t$) rather than ($\sin 0.1t + \sin 0.2t$), frequencies high enough to bring out its second-order features, the tracking system completely "fails," giving the results of Fig. 6.32. This single example illustrates apparent pitfalls in this method. However, one might take the point of view that the lack of convergence *signals* a model mismatch, telling us to try another model, and thus is not really a fatal defect in the method since we are not led to use an incorrect model. While not infallible, if good convergence *is* obtained, there is a high probability that we have a good model.

MERGED OUTPUT PRESENTATION FOR KS

PARAMETER   RUN   1   RUN   2
    C   .20000   1.4000

                           .0   '*'= RUN   2   2.400
                           .0   '+'= RUN   1   2.400

    TIME   RUN   1
    .0   .0   *-------I---------I---------I---------I   RUN   2
  2.7000   .57155   I   * +I   I   I   I   .0
  5.4000   1.0809   I   I   *   + I   I   I   .41210
  8.1000   .91452   I   I   *+   I   I   I   .74581
 10.800   1.0947   I   I   *+   I   I   I   .90744
 13.500   1.0932   I-------I------+*I---------I---------I   1.0279
 16.200   1.1576   I   I   +I *   I   I   1.1592
 18.900   1.4179   I   I   I+   *   I   I   1.3542
 21.600   1.4100   I   I   I+   *   I   * I   1.8011
 24.300   1.0071   I   I   I +   *   I   I   2.2371
 27.000   1.1556   I-------I---------+*I---------I---------I   1.3588
 29.700   1.0938   I   I   *   + I   I   I   1.1123
 32.400   1.0591   I   I   *   + I   I   I   .83349
 35.100   1.0784   I   I   *I   I   I   .78298
 37.800   1.0960   I   I   *I   I   I   1.0568
 40.500   1.1424   I-------I---------*I---------I---------I   1.0802
 43.200   1.1365   I   I   *+I   I   I   1.1486
 45.900   1.0901   I   I   *I   I   I   1.1024
 48.600   1.0604   I   I   *I   I   I   1.0718
 51.300   1.0451   I   I   *I   I   I   1.0589
 54.000   1.0247   I-------I---------*---I---------I---------I   1.0468
 56.700   1.0005   I   I   *I   I   I   1.0327
 59.400   .95381   I   I   * I   I   I   1.0125
 62.100   .78955   I   I   *   I   I   I   .97221
 64.800   .90721   I   I   +* I   I   I   .81959
 67.500   .98392   I-------I---------*-I---------I---------I   .92046
 70.200   1.0122   I   I   *I   I   I   .98335
 72.900   1.0305   I   I   *I   I   I   1.0074
 75.600   1.0462   I   I   *+I   I   I   1.0232
 78.300   1.0629   I   I   *I   I   I   1.0373
 81.000   1.0885   I-------I---------*-I---------I---------I   1.0535
 83.700   1.1352   I   I   *I   I   I   1.0795
 86.400   1.0844   I   I   *I   I   I   1.1276
 89.100   1.0899   I   I   *I   I   I   1.0835
 91.800   1.1188   I   I   *I   I   I   1.0848
 94.500   1.1690   I-------I---------*---------I---------I   1.1128
 97.200   1.0873   I   I   * I   I   I   1.1700
 99.900   1.0869   I   I   * I   I   I   1.0821
102.60   1.1134   I   I   *I   I   I   1.0834
105.30   1.1565   I   I   *I   I   I   1.1067
108.00   1.0808   I-------I---------*-I---------I---------I   1.1459
110.70   1.0603   I   I   *I   I   I   1.0725
113.40   1.0451   I   I   *I   I   I   1.0603
116.10   1.0301   I   I   *I   I   I   1.0490
118.80   1.0111   I   I   *I   I   I   1.0364
121.50   .97907   I-------I---------*--I---------I---------I   1.0194
124.20   .88394   I   I   *   I   I   I   .98889
126.90   .87994   I   I   *   I   I   I   .89223
129.60   .97990   I   I   *I   I   I   .88168
132.30   1.0095   I   I   *I   I   I   .97538
135.00   1.0262   I-------I---------*--I---------I---------I   1.0032
137.70   1.0393   I   I   *+ I   I   I   1.0196
140.40   1.0520   I   I   * I   I   I   1.0332
143.10   1.0678   I   I   * I   I   I   1.0475
145.80   1.0963   I   I   +*I   I   I   1.0677
148.50   1.0796   I-------I---------*-I---------I---------I   1.1101
151.20   1.0856   I   I   *-I   I   I   1.0863
153.90   1.1121   I   I   *+I   I   I   1.0812
156.60   1.1795   I   I   I *   I   I   1.1018
159.30   1.0966   I   I   * I   I   I   1.1692
162.00   1.0857   I-------I---------*-I---------I---------I   1.0864
164.70   1.1003   I   I   * I   I   I   1.0819
167.40   1.1684   I   I   *I   I   I   1.0958
170.10   1.0881   I   I   * I   I   I   1.1687
172.80   1.0644   I   I   * I   I   I   1.0766
175.50   1.0488   I-------I---------*-I---------I---------I   1.0634
178.20   1.0342   I   I   +* I   I   I   1.0521
180.90   1.0171   I   I   * I   I   I   1.0400
183.60   .99071   I   I   * I   I   I   1.0247
186.30   .92577   I   I   * I   I   I   .99969
189.00   .81130   I-------I----*-----I---------I---------I   .93455
191.70   .96654   I   I   * I   I   I   .80956
194.40   1.0037   I   I   * I   I   I   .96231
197.10   1.0223   I   I   * I   I   I   .99749
199.80   1.0358   I   I   * I   I   I   1.0157
                                                                            1.0296

Figure 6.30   Effects of model mismatch in parameter tracking.

290

MERGED OUTPUT PRESENTATION FOR CS

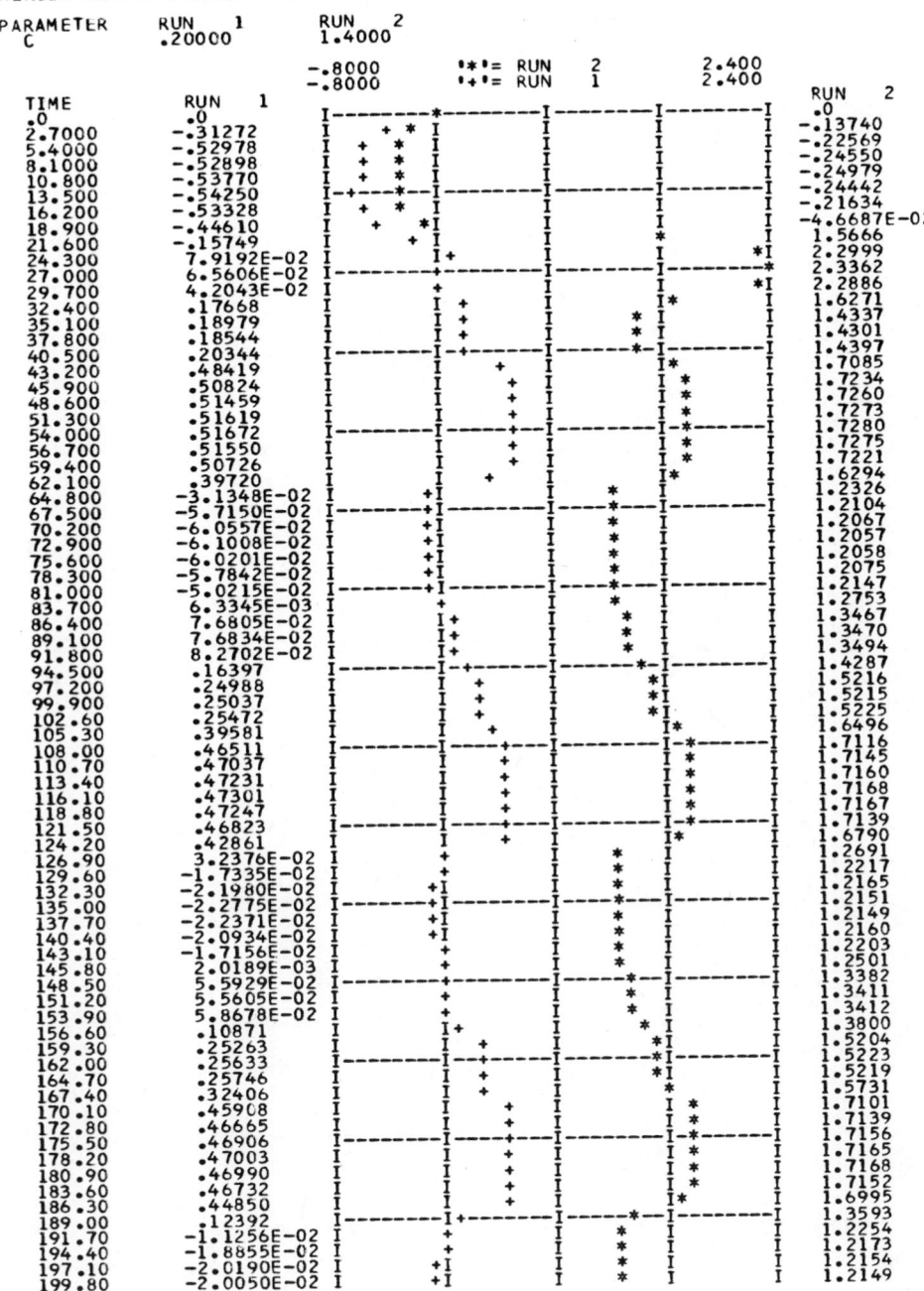

Figure 6.30 *continued*

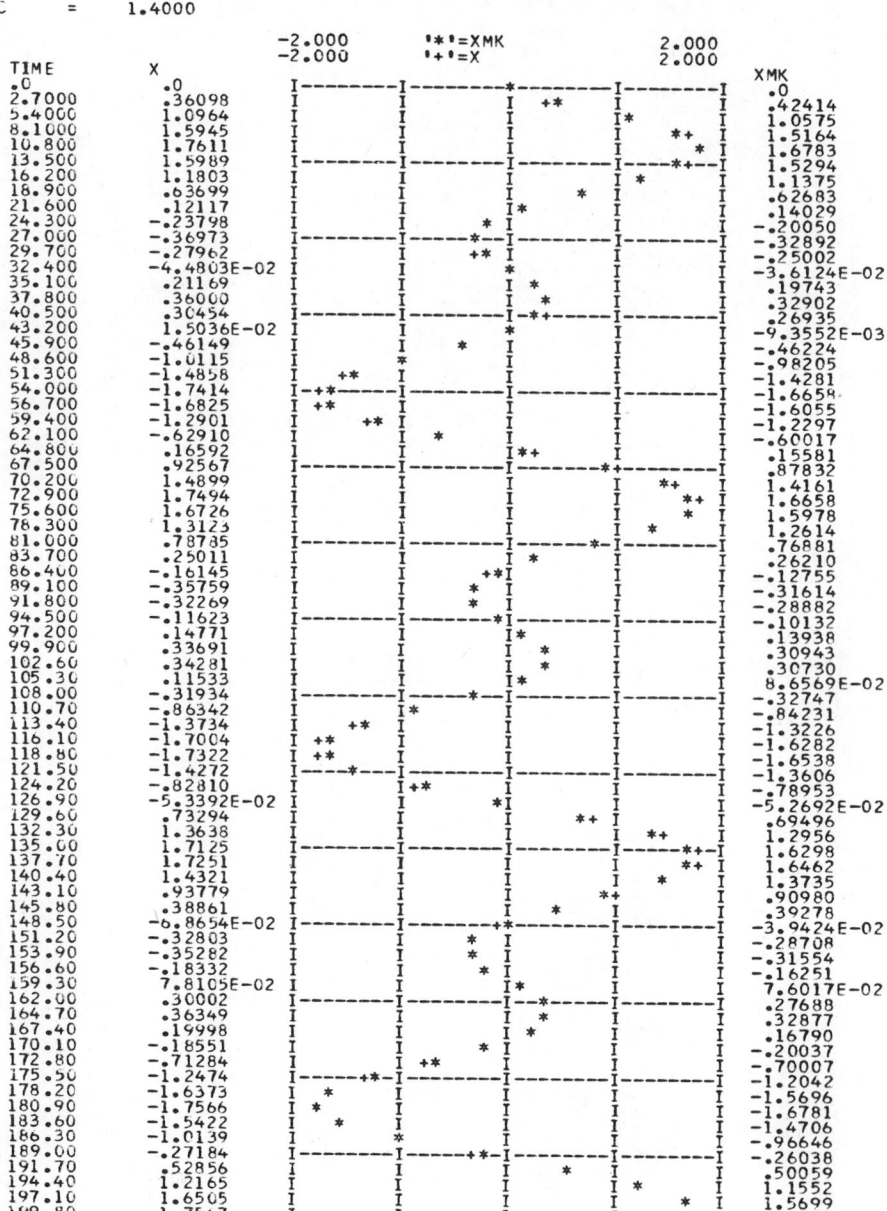

Figure 6.31 Performance of first-order model of second-order system; low-frequency input.

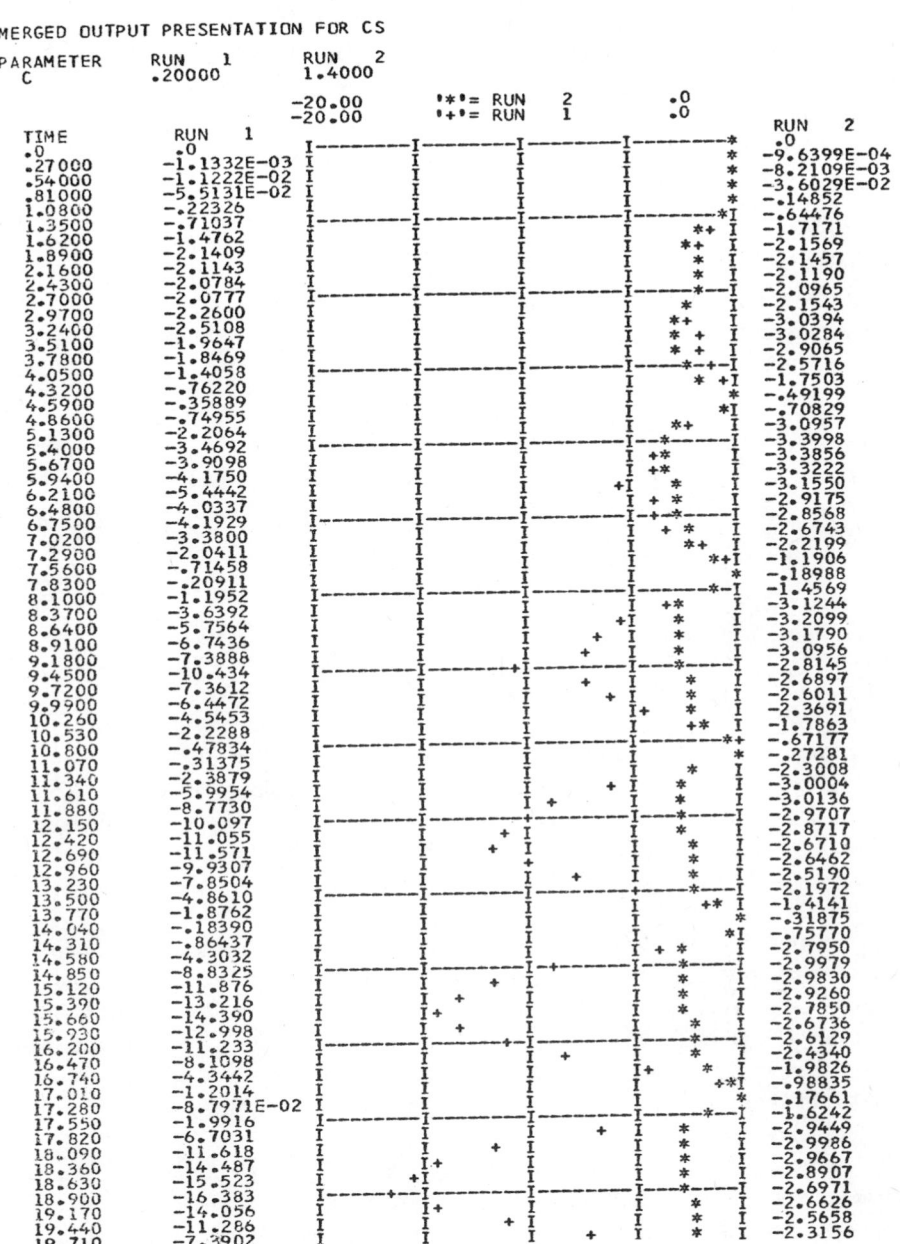

Figure 6.32 Performance of first-order model of second-order system; high-frequency input.

Since we indicated the method is directly applicable to thoroughly nonlinear systems, our next example is from this area. Consider a mass/spring/damper system with nonlinear damper and nonlinear spring given by

$$M\ddot{x} + B\frac{\dot{x}}{|\dot{x}|}|\dot{x}|^{1.8} + K_1 x - K_2 x^2 + K_3 x^3 = f_i(t) \tag{6.47}$$

Let us take $M = 2.1$, $B = 10$ and, for a typical automotive suspension air spring, $K_1 = 165$, $K_2 = 37.1$, and $K_3 = 10.3$. Suppose M is known but B, K_1, K_2, and K_3 are to be found experimentally by parameter tracking. Definition of E and the individual parameter-adjustment laws proceeds in this nonlinear example exactly as in the linear examples before:

$$E = M\ddot{x} + B^*\frac{\dot{x}}{|\dot{x}|}|\dot{x}|^{1.8} + K_1^* x - K_2^* x^2 + K_3^* x^3 - f_i(t) \tag{6.48}$$

$$\dot{B}^* = -2GE\frac{\dot{x}}{|\dot{x}|}|\dot{x}|^{1.8}, \qquad \dot{K}_1^* = -2GEx, \qquad \dot{K}_2^* = 2GEx^2, \qquad \dot{K}_3^* = -2GEx^3 \tag{6.49}$$

For excitation we use:

$$f_i = 300(\sin t + \sin 4.7t + \sin 9t + \sin 17t + \sin 31t) \tag{6.50}$$

This force, a mixture of sine waves with noninteger-multiple frequencies, while having a discrete spectrum, is not periodic and looks quite random. The frequencies were chosen to span a range both above and below the "linearized natural frequency" associated with the operating point $x = 0$, while the amplitude covers the range expected in the system's practical application. (A CSMP run for $0 < t < 12$ gave $-1373 < f_i < 1106$.)

An initial guess of $G = 10$ gave stable results except for the K_3^* computing loop, which appeared to be diverging in an oscillatory fashion, the excursions being -24 to $+53$ at $t \approx 8$. By reducing G *for this loop only* to 1.0 (all other loops kept $G = 10$), stability was attained and the results of Fig. 6.33 were obtained. This scheme of using different G's for each computing loop is of general utility in this method, and while not always necessary, such "tuning" will often give better performance.

Our final example involves tracking of time-varying parameters in a linear system. For the longitudinal (pitch axis) motion of an aircraft, the differential equation relating aircraft pitch rate $\dot{\theta}$ to horizontal tail (elevator) angle δ is

$$a_2\dddot{\theta} + a_1\ddot{\theta} + \dot{\theta} = b_1\dot{\delta} + b_0\delta \tag{6.51}$$

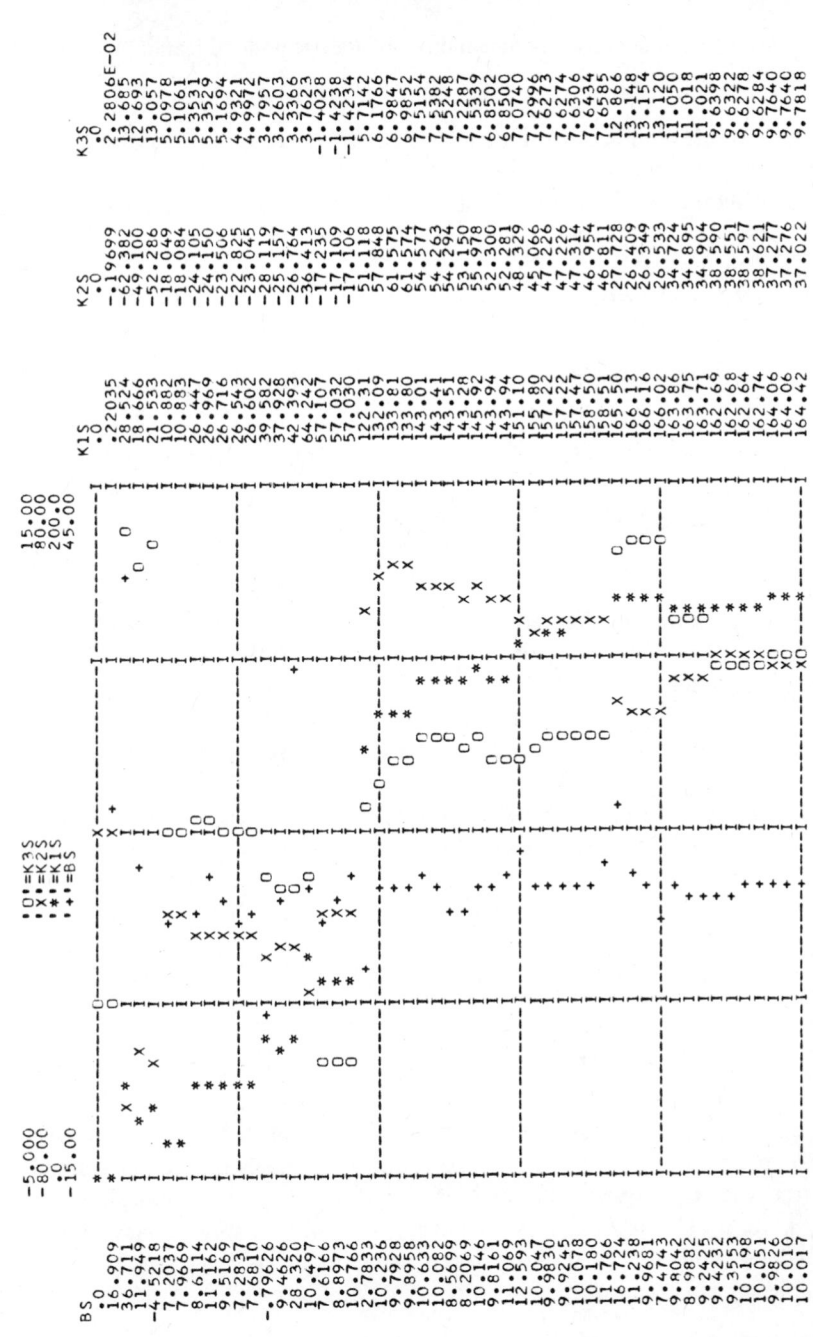

Figure 6.33 Parameter tracking of nonlinear system.

295

When the aircraft changes altitude and/or flight speed, all four coefficients vary with time. Let us assume that the aircraft, flying at constant speed, maintains constant altitude for 15 seconds, climbs for 15 seconds, and then levels off, such that

$$a_2(t) = 0.04 + 0.064(t - 15)\mu(t - 15) - 0.064(t - 30)\mu(t - 30) \qquad (6.52)$$

$$a_1(t) = 4 - 0.0667(t - 15)\mu(t - 15) + 0.0667(t - 30)\mu(t - 30) \qquad (6.53)$$

$$b_1(t) = 0.5 - 0.0133(t - 15)\mu(t - 15) + 0.0133(t - 30)\mu(t - 30) \qquad (6.54)$$

$$b_0(t) = 5 - 0.2(t - 15)\mu(t - 15) + 0.2(t - 30)\mu(t - 30) \qquad (6.55)$$

Let the excitation $\delta(t)$ be derived from a thrice repeated 15-bit PRBS with clock period 1 second by integrating it and then low-pass filtering with $\tau = 0.5$ sec to give the curve labelled DELHS in Fig. 6.34a. Note the drift of the mean value of DELHS due to integration of a PRBS, which has nonzero mean. If this drift is objectionable, one could bias it out by adding the appropriate constant to the PRBS (before it is integrated).

Suppose we wish to find and track the time-varying coefficients a_1 and b_0 over the period $0 < t < 45$, assuming all other coefficients are known and that we also have accurate starting values of a_1 and b_0 at $t = 0$. The computing scheme is set up exactly as in previous examples:

$$E = a_2\ddot{\theta} + a_1^*\ddot{\theta} + \dot{\theta} - b_1\delta - b_0^*\delta \qquad (6.56)$$

$$\dot{a}_1^* = -GE\ddot{\theta}, \qquad \dot{b}_0^* = GE\delta \qquad (6.57)$$

Our CSMP simulation assumes "perfect" values of all needed signals are available and that our assumed form of equation matches the actual system. An initial trial with $G = 5$ gave good results on b_0 but not on a_1. Increasing G to 20 in the a_1 loop (G left at 5 for b_0) gave the results of Fig. 6.34b, which are improved but still not satisfactory for a_1. Further trials were not pursued; an acceptable solution may or may not exist. One might try:

1 Various combinations of G's in the two loops.
2 Different types and amplitudes of excitation.
3 Various filters used on the $\ddot{\theta}$ and/or δ signals in Eq. (6.57).

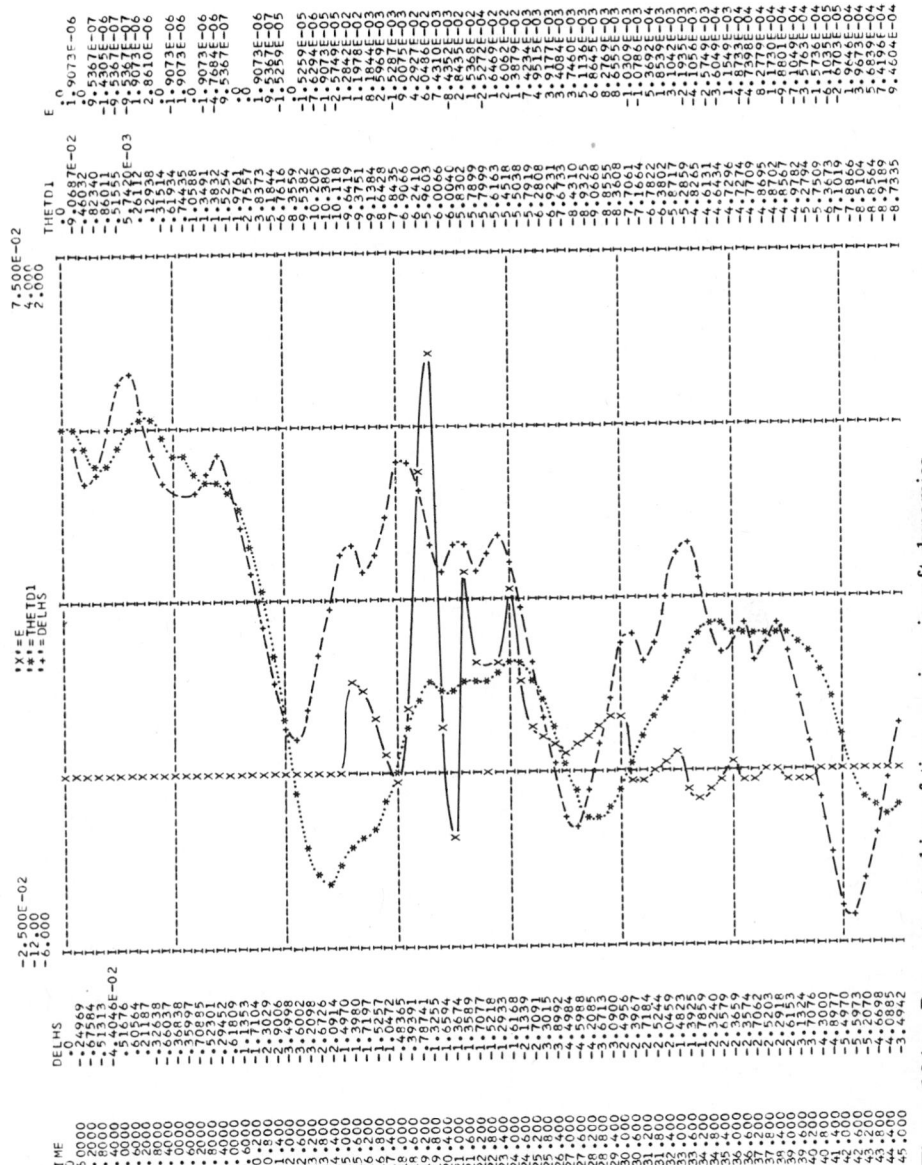

Figure 6.34a Parameter tracking of time-varying aircraft dynamics.

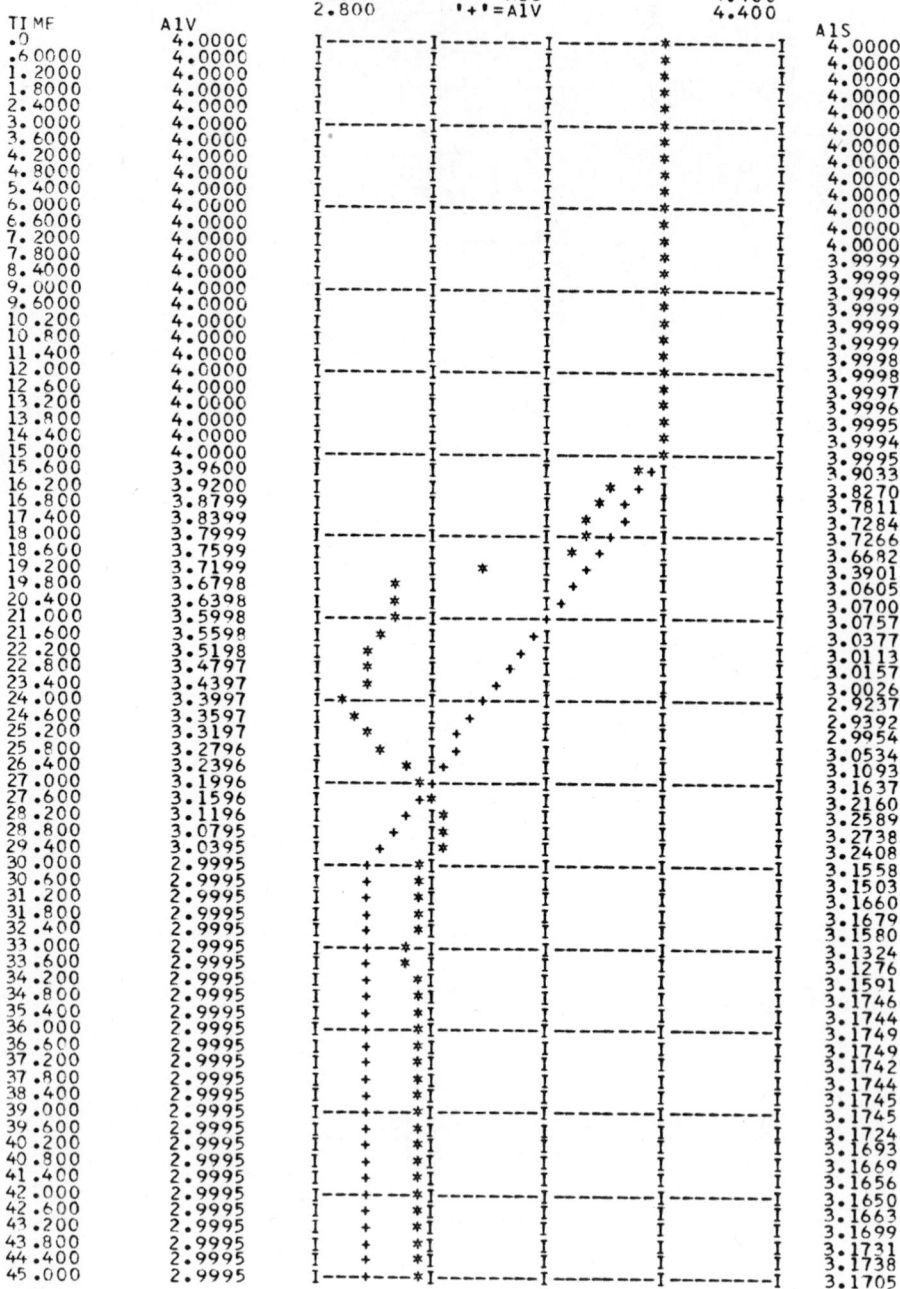

| | | | 2.800 | '*'=A1S | 4.400 | |
| | | | 2.800 | '+'=A1V | 4.400 | |

TIME	A1V	A1S
.0	4.0000	4.0000
.60000	4.0000	4.0000
1.2000	4.0000	4.0000
1.8000	4.0000	4.0000
2.4000	4.0000	4.0000
3.0000	4.0000	4.0000
3.6000	4.0000	4.0000
4.2000	4.0000	4.0000
4.8000	4.0000	4.0000
5.4000	4.0000	4.0000
6.0000	4.0000	4.0000
6.6000	4.0000	4.0000
7.2000	4.0000	4.0000
7.8000	4.0000	4.0000
8.4000	4.0000	3.9999
9.0000	4.0000	3.9999
9.6000	4.0000	3.9999
10.200	4.0000	3.9999
10.800	4.0000	3.9999
11.400	4.0000	3.9998
12.000	4.0000	3.9998
12.600	4.0000	3.9997
13.200	4.0000	3.9996
13.800	4.0000	3.9995
14.400	4.0000	3.9994
15.000	4.0000	3.9995
15.600	3.9600	3.9033
16.200	3.9200	3.8270
16.800	3.8799	3.7811
17.400	3.8399	3.7284
18.000	3.7999	3.7266
18.600	3.7599	3.6682
19.200	3.7199	3.3901
19.800	3.6798	3.0605
20.400	3.6398	3.0700
21.000	3.5998	3.0757
21.600	3.5598	3.0377
22.200	3.5198	3.0113
22.800	3.4797	3.0157
23.400	3.4397	3.0026
24.000	3.3997	2.9237
24.600	3.3597	2.9392
25.200	3.3197	2.9954
25.800	3.2796	3.0534
26.400	3.2396	3.1093
27.000	3.1996	3.1637
27.600	3.1596	3.2160
28.200	3.1196	3.2589
28.800	3.0795	3.2738
29.400	3.0395	3.2408
30.000	2.9995	3.1558
30.600	2.9995	3.1503
31.200	2.9995	3.1660
31.800	2.9995	3.1679
32.400	2.9995	3.1580
33.000	2.9995	3.1324
33.600	2.9995	3.1276
34.200	2.9995	3.1591
34.800	2.9995	3.1746
35.400	2.9995	3.1744
36.000	2.9995	3.1749
36.600	2.9995	3.1749
37.200	2.9995	3.1742
37.800	2.9995	3.1744
38.400	2.9995	3.1745
39.000	2.9995	3.1745
39.600	2.9995	3.1724
40.200	2.9995	3.1693
40.800	2.9995	3.1669
41.400	2.9995	3.1656
42.000	2.9995	3.1650
42.600	2.9995	3.1663
43.200	2.9995	3.1699
43.800	2.9995	3.1731
44.400	2.9995	3.1738
45.000	2.9995	3.1705

Figure 6. 34 b

298

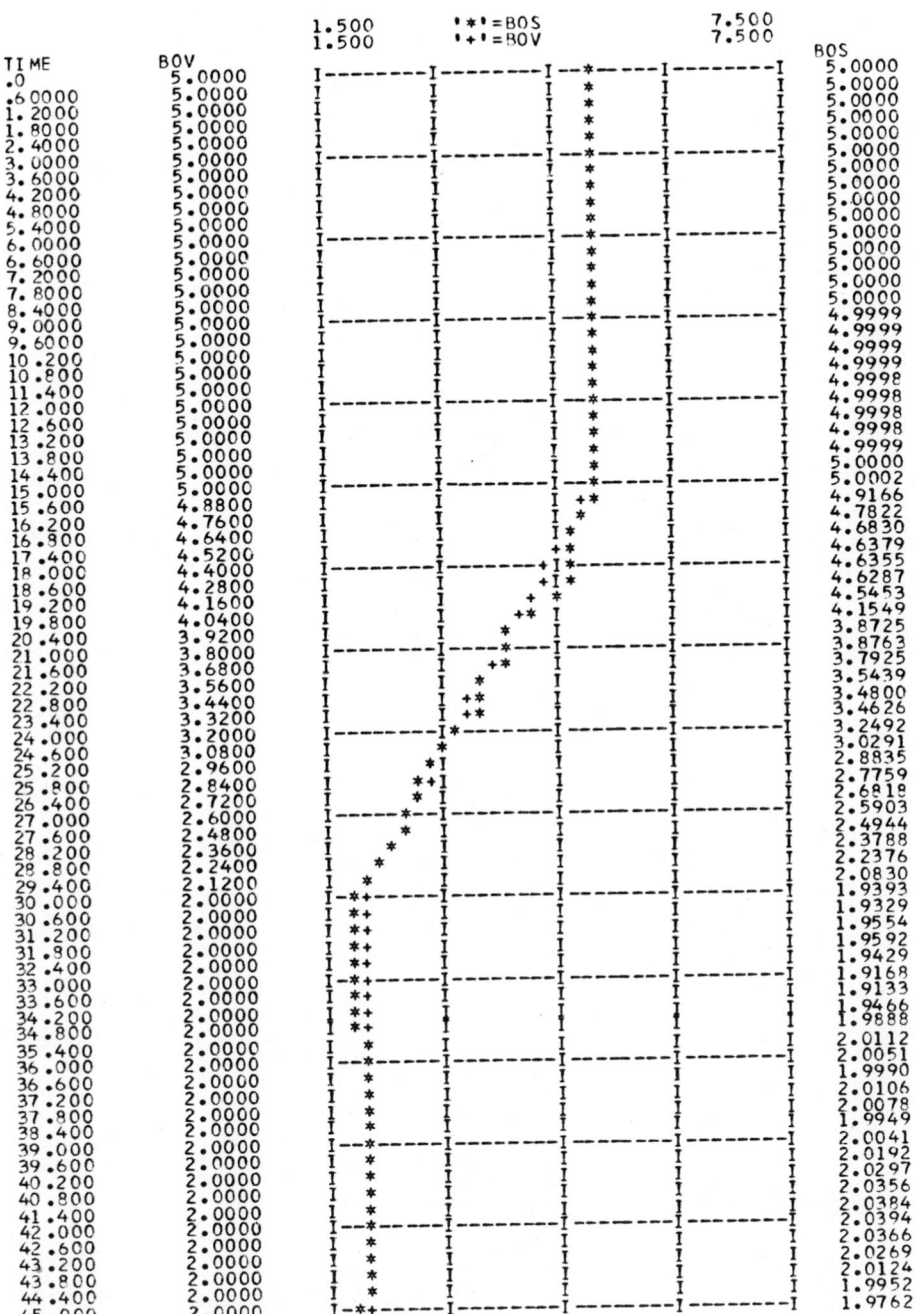

Figure 6.34b *continued*

6.5   MULTIPLE REGRESSION AND LEAST-SQUARES METHODS

In many areas of science and engineering where analysis of relationships among variables by means of basic physical laws is difficult or impossible, recourse has been had to statistical regression or least-squares methods as a means of developing the desired models Such techniques have been particularly prevalent in biology, agriculture, medicine, sociology, and economics. Many of the classical applications have been to steady-state (rather than dynamic) conditions and we begin with these. However, the methods are also applicable to the dynamic (differential equation) models that have been our emphasis in this text, and we also treat these briefly.

6.5.1   MULTIPLE REGRESSION FOR STATIC MODELS

In engineering practice we often run experiments in which an output variable depends on one or more input variables. Sometimes we understand the theory behind the physical system well enough to formulate a mathematical model that can predict the output if all the inputs are given. In the ideal case, we know both the form of the model and also numerical values for all the parameters. For example, in a cantilever beam

$$\delta = \frac{PL^3}{3EI} \tag{6.58}$$

with input variables P, L, E, and I while δ is the output. The crudest form of this relation could be expressed as

$$\delta = f(P, L, E, I) \tag{6.59}$$

where we indicate that we know on *what* δ depends but not *how*. A refinement would be

$$\delta = C \frac{PL^3}{EI} \tag{6.60}$$

in which C is an unknown constant. Here we indicate that we know specifically the *form* of the mathematical function but are not sure of the numerical value of C.

Multiple regression methods are useful modeling tools in those situations where we know (or can at least postulate on a tentative basis) the *form* of the relationship among the dependent and independent variables and wish to use experimental data to find the "best" values of the coefficients in the equation. Most engineering students encounter the simplest version of this method quite early in their training since it is nothing more than the least-squares fitting of the best straight line to a set of x,y data points. Let us review this simple application. We have run an experiment and obtained a table of x,y data. Based on theoretical knowledge of the physical process, past experience with similar phenomena, visual evidence by "eyeballing" the plotted data, and/or intuition, we decide that a straight-line relationship is reasonable and we wish to find the best possible such line. (Of course, if the data show little "scatter," complex mathematics is hardly necessary and we draw in the best line by eye.)

If we describe the desired line by

$$y = mx + b \tag{6.61}$$

then the vertical (dependent variable) deviation of an individual data point x_i, y_i from the line would be

$$\text{vertical deviation} \triangleq \text{residual} \triangleq y_i - mx_i - b \tag{6.62}$$

In the least-squares method the *definition* of the best straight line requires that the sum of the squares of the residuals for all n data points be made a minimum. That is, by proper choice of m and b we wish to minimize D, where

$$D \triangleq \text{sum of squared residuals} = \sum_{i=1}^{n} (y_i - mx_i - b)^2 \tag{6.63}$$

From calculus we know this requires

$$\frac{\partial D}{\partial m} = 0 \quad \text{and} \quad \frac{\partial D}{\partial b} = 0 \tag{6.64}$$

$$\frac{\partial D}{\partial m} = 0 = 2 \sum (y_i - mx_i - b)(-x_i) \tag{6.65}$$

$$\frac{\partial D}{\partial b} = 0 = 2 \sum (y_i - mx_i - b)(-1) \tag{6.66}$$

which lead to the so-called *normal equations*

$$\left(\sum x_i^2 \right) m + \left(\sum x_i \right) b = \sum x_i y_i$$

$$\left(\sum x_i \right) m + (n) b = \sum y_i \tag{6.67}$$

Since these are linear algebraic equations they are easily solved to get

$$m = \frac{n\sum x_i y_i - \sum x_i \sum y_i}{n\sum x_i^2 - \left(\sum x_i\right)^2} \qquad (6.68)$$

$$b = \frac{\sum y_i \sum x_i^2 - \sum x_i y_i \sum x_i}{n\sum x_i^2 - \left(\sum x_i\right)^2} \qquad (6.69)$$

These values of m and b are guaranteed to give the line with the minimum possible sum of squared residuals.

We now wish to generalize the allowable form of model to include *many* independent variables affecting *one* dependent variable. Because it adequately fits many physical relations and also leads to convenient mathematics (linear algebra), we choose the linear form

$$y = a_1 x_1 + a_2 x_2 \cdots + a_r x_r \qquad (6.70)$$

The linearity of the above model lies in the fact that the coefficients a_i appear in a linear fashion with respect to the dependent variable y. The independent variables x_i can actually take on any form so long as the coefficients appear linearly. This allows us to develop models with a wide (though *not* unlimited) variety of useful forms. A few examples will illustrate this capability

I.
$$y = a_1 + a_2 x_2 + a_3 x_2^2 + a_4 x_2^3 \qquad (6.71)$$

This would just be a polynomial curve fit of y versus x_2, where we have taken $x_1 \equiv 1.0$ to accommodate the constant term a_1. We merely define $x_2 \triangleq x_2, x_3 \triangleq x_2^2$ and $x_4 \triangleq x_2^3$. This gives the standard form

$$y = a_1 x_1 + a_2 x_2 + a_3 x_3 + a_4 x_4 \qquad (6.72)$$

so we can use standard computer programs, entering values of x_2^2 wherever x_3 is called for and x_2^3 wherever x_4 appears.

II.
$$y = a_1 x_1 x_2 + a_2 x_3^3 + a_3 \sin x_4 \qquad (6.73)$$

We here use the same trick as in I above, defining $x_a \triangleq x_1 x_2, x_b \triangleq x_3^3, x_c \triangleq \sin x_4$ to again obtain the standard form.

III.
$$y = a_0 \frac{x_1^{a_1}}{x_2^{a_2}} \qquad (6.74)$$

Here the coefficients a_1 and a_2 do *not* appear linearly; however, a transformation allows us to proceed as usual.

$$\log\frac{y}{a_0} = \log\frac{x_1^{a_1}}{x_2^{a_2}} = a_1 \log x_1 - a_2 \log x_2 \qquad (6.75)$$

$$\log y = \log a_0 + a_1 \log x_1 - a_2 \log x_2 \qquad (6.76)$$

Now define $z \triangleq \log y, a_a \triangleq \log a_0, x_a \equiv 1.0, x_b \triangleq \log x_1, x_c \triangleq -\log x_2$, giving

$$z = a_a x_a + a_1 x_b + a_2 x_c \qquad (6.77)$$

which is of course again the standard form. In this last example, however, it is important to note that, thinking in terms of the least-squares interpretation, the set of a's found by the computing process will be that set which minimizes the sum of the squares of deviations of the actual z values from the model-predicted z values. Thus it is the (log y) deviations squared, *not those of y itself*, that are minimized, and therefore the basic definition of what constitutes the best fit has been altered. As long as we recognize this change in interpretation, we can still *use* the set of a values calculated by the process in a predictive model for y. We cannot now, however, be sure that it is the *best* model (in a least-squares sense on y) of the type postulated. In practice this uncertainty is often acceptable because:

1 We can certainly check the "fit" of the model-predicted y values versus actual y values and make a judgment as to model accuracy.

2 The alternative procedure (setting up a computation scheme that *does* result in a minimization on y rather than log y) results in the need to solve a set of simultaneous *nonlinear* algebraic equations to find the coefficient values. This is considerably more difficult (even on a computer) than solving a set of linear equations.

Since the least-squares method leads to sets of linear equations, the standard computer programs generally employ matrix methods in solving for the unknown coefficients. Recall that if there are r coefficients to be found, we require $n \geqslant r$ data points to obtain a solution since each data point gives one equation. When $n = r$ a set of a's that precisely satisfy all r equations can be found *without* the use of least-squares methods. For example, m and b in $y = mx + b$ can be found from "two equations in two unknowns" if only two sets of x, y values are given since two points determine one and only one straight line. It is when *more* points that do *not* all fall on a single line are given that least-squares methods are needed. If we substitute our n sets of observed data into Eq. (6.70), we generate a set of n equations in r unknowns, which may be put in matrix form

$$y = XA \qquad (6.78)$$

where

$$y \triangleq \begin{bmatrix} y(1) \\ y(2) \\ \vdots \\ y(n) \end{bmatrix} \qquad A \triangleq \begin{bmatrix} a_1 \\ a_2 \\ \vdots \\ a_r \end{bmatrix} \tag{6.79}$$

$$X \triangleq \begin{bmatrix} x_1(1) & x_2(1) & \cdots & x_r(1) \\ x_1(2) & x_2(2) & \cdots & x_r(2) \\ \vdots & \vdots & & \vdots \\ x_1(n) & x_2(n) & \cdots & x_r(n) \end{bmatrix} \tag{6.80}$$

When $n = r$ (least-squares *not* needed), matrix inversion gives the solution for the a's as $X^{-1}y$. When $n > r$, it can be shown[52] that least-squares methods lead to the following matrix solution:

$$A = (X^T X)^{-1} X^T y \tag{6.81}$$

where $X^T \triangleq$ transpose of matrix X (interchange rows and columns). The matrix operations of Eq. (6.81) are easily performed with a general purpose language such as SPEAKEASY or special purpose least-squares routines.

The least-squares methods described above use only calculus in their derivation, no statistical concepts are necessary. Our use of the term multiple regression however implies a relation to this statistical method, so we wish to at least state this relation. If one studies the theory of statistics one finds a widely used general tool for estimation of parameters in statistical models called the "principle of maximum likelihood." When this principle is applied to the problem (called multiple regression) of finding the best values of parameters in linear models, and if one further assumes that any random effects are Gaussian and independently distributed with constant variance, then the formulas for calculating the best values of the parameters turn out to be *identical* to those given by the least-squares method. Since the least-squares procedures are easily understood and accepted by anyone with an elementary calculus background, we rely on these to establish the reasonableness of the approach. When we later make use of standard computer programs to actually carry out calculations, these programs return, in addition to the main results that are the values of the coefficients, other useful statistical data. We will at that point need to give some

[52]Hsia, T. C., *System Identification: Least-Squares Methods*, Lexington Books, D. C. Heath, Lexington, Mass., 1977, p. 19.

additional discussion on the meaning of these statistical parameters since we presume only a modest statistics background on the part of the reader. Our purpose here is to give a working tool with a minimum of time invested. As always, such an approach involves some risk of misapplying a method whose details are not completely understood; however, practical engineers must often assume and deal with such risks.

The above-discussed calculation procedures assume the availability of a set of numerical data on y, x_1, x_2, and so on. These are of course obtained by running an experiment in which some physical apparatus is "exercised" over some range of operating conditions. Sometimes the independent variables (x's) are completely under our control and we can set them at any desired specific values. Other times some of the independent variables may be measurable but *not* subject to our control. An example would be a process sensitive to outdoor temperature; we certainly can measure this but cannot pre-program specific values.

When the independent variables are completely under our control, it is always possible (and sometimes preferable) to hold all independent variables, except one, at fixed values, exercise the one selected variable (say x_1) over the desired range, record the response of the dependent variable, and from this data draw some conclusions about the dependence of y on x_1. This procedure can be repeated with each of the other independent variables. While such an approach appears reasonable from the point of view of trying to separate a complicated dependence into easily understandable component effects, it may not always be the most effective way of arriving at the final overall model. Also, in some situations, simultaneous variation of some of the "independent" variables may be physically unavoidable.

If we decide (or are forced by circumstances) to carry out our experiment in such a way that all the independent variables are exercised simultaneously, the question immediately arises as to how we shall select the *combinations* of x_1, x_2, and so on to be tested. Some of the questions involved can be adequately dealt with by common sense approaches while others require more careful mathematical study. Whole books have been written on the statistical aspects of "Design of Experiments"; our approach here must necessarily be much less ambitious. Let us consider an example in which y is some function of three independent variables x_1, x_2, and x_3. Our physical familiarity with the process will ordinarily allow us to choose the pertinent *ranges* of variation for the independent variables. For example,[53] a study of the time-to-failure of epoxy-cemented aluminum flats under steady stress at various temperatures and relative humidities used five different stress levels, three temperature levels, and three humidity

[53]Jones, W. C., *et al*. Use of Multiple Regression Analysis to Develop Predictive Models for Failure Times of Adhesive Bonds at Constant Stress, Tech. Memo. 2066, Picatinny Arsenal, Dover, N.J., March 1973.

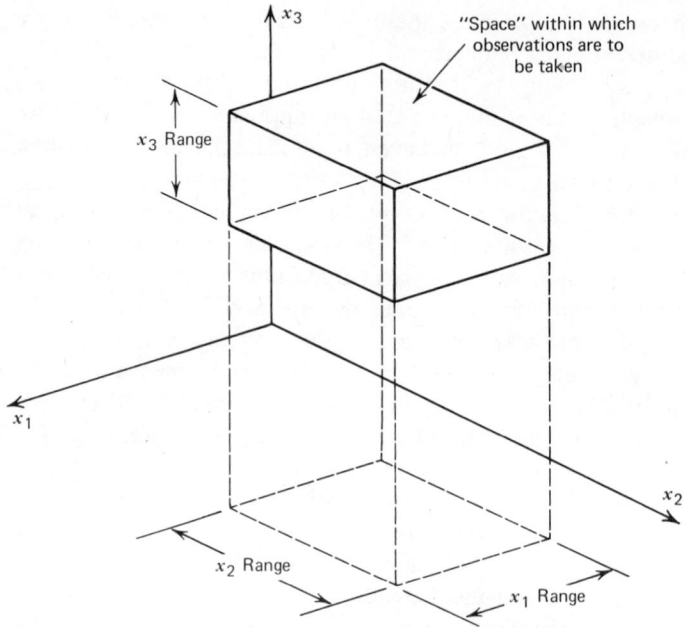

Figure 6.35   Geometrical interpretation of sample space.

levels. The *ranges* of stress, temperature, and humidity to be covered can be decided by examining the service conditions under which such epoxy-cemented parts would be used. The *number, distribution,* and *combinations* of levels are not so simply decided upon however. For experiments with three or less independent variables we may conveniently think in geometrical terms as in Fig. 6.35 Here the chosen ranges of x_1, x_2, and x_3 define a volume or *sample space* within which the desired data points will lie. It is not necessary however that sample points be *uniformly* distributed over this volume. It may be that certain portions of the volume are of greater practical interest, or exhibit a more sensitive dependence of y on the x's than other portions. Such considerations might lead us to space sample points more closely in such regions.

Determination of the *number* of sample points obviously involves tradeoffs between statistical reliability (which requires many points) and costs (which increase with more data points). Also, it is clear that coarsely spaced points will be unable to detect "fine-structure" (significant variations of y for small changes in x) in physical phenomena. *Replication* (repetition of a point in the sample space) is not necessary, but when the cost can be justified it does allow a more definitive statistical analysis of the "goodness of fit" of the model.

We are now ready to illustrate the application of the standard multiple regression computer programs available in most computer libraries. Instead of running *actual* physical experiments and then analyzing the results on the computer, we will *simulate* the physical experiment with one computer program (CSMP) and then analyze the results of this simulated experiment with another program (multiple regression). Since we can very precisely control the behavior of our computer-simulated experiment in very versatile ways, and since we will always know the *true* relation among y and the x's, we can learn about the capabilities and limitations of the multiple regression technique in a very efficient way. The CSMP program is as follows.

```
TITLE    DATA PREPARATION FOR MULTIPLE REGRESSION
METHOD   RECT needed to make time steps exactly 1.0
INIT
/        DIMENSION Y1(129), Y2(129), Y3(129), Y4(129) allocates storage
NOSORT
DYNAM
NOSORT
FIXED    I,L
         L=0
         I=0
         PUNCH=TIME/PRDEL         } sets up card punching
         L=IFIX(PUNCH)
         I=L+1
         Y1(I)=X1
         Y2(I)=X2                 } selects variables to
         Y3(I)=X3                   be punched
         Y4(I)=YMES
    8    CONTINUE
SORT
         IMP=IMPULSE(0.0, DELT) needed to generate random error
PARAM    SDY=(1.0,5.0) set std. dev. of measurement error in YMES
         YG=GAUSS(IMP,0.0,SDY) random measurement error
PARAM    A00=0.0, B00=0.0, C00=0.0, Y00=0.0 adjust means of x's, YTRU
PARAM    A1=1.0, A2=2.0, A3=4.0 true values of model coefficients
PARAM    X1A=1., X2A=1., X3A=1. sets ranges of x's
         R=RAMP(0.0)-RAMP(1.0)
         R1=DELAY(15,15.,R)      } generates 128
         R2=DELAY(31,31.,R)        values of X1
         R3=DELAY(47,47.,R)
         R4=DELAY(63,63.,R)
```

```
        A = A00 + R1 + R2 + R3 - 3.*R4        ⎫  generates 128
        AD1 = DELAY(64, 64., A)               ⎬  values of X1
        X1 = (A + AD1)*X1A                    ⎭
        B1 = DELAY(3, 3., R)
        B2 = DELAY(7, 7., R)
        B3 = DELAY(11, 11., R)
        B4 = DELAY(15, 15., R)
        B11 = B00 + B1 + B2 + B3 - 3.*B4      ⎫  generates 128
        BD1 = DELAY(16, 16., B11)             ⎬  values of X2
        BD2 = DELAY(32, 32., B11)             ⎭
        BD3 = DELAY(48, 48., B11)
        B5 = B11 + BD1 + BD2 + BD3
        BD4 = DELAY(64, 64., B5)
        X2 = (B5 + BD4)*X2A
        C1 = RAMP(0.0) - 4.*RAMP(3.0) + 3.*RAMP(4.0)    ⎫
        CD1 = DELAY(4, 4., C1)
        CD2 = DELAY(8, 8., C1)
        CD3 = DELAY(12, 12., C1)
        C2 = C00 + C1 + CD1 + CD2 + CD3           ⎬  generates 128
        CD4 = DELAY(16, 16., C2)                      values of X3
        CD5 = DELAY(32, 32., C2)
        CD6 = DELAY(48, 48., C2)
        C3 = C2 + CD4 + CD5 + CD6
        CD7 = DELAY(64, 64., C3)
        X3 = (C3 + CD7)*X3A                       ⎭
        YTRU = Y00 + A1*X1 + A2*X2 + A3*X3   true value of dependent
                                                 variable
        YMES = YTRU + YG   measured (noisy) value of dependent variable
        Y0 = INTGRL(0.0, YTRU)   dummy integration (needed to make time
                                     steps exactly 1.0)
TERMINAL
NOSORT
        DO 15 I = 2, 129                     ⎫
        WRITE(2, 10) Y1(I), Y2(I), Y3(I), Y4(I)  ⎬  punches
   10   FORMAT(4E13.5)                           cards
   15   CONTINUE                             ⎭
SORT
TIMER   FINTIM = 128., DELT = 1.0, PRDEL = 1.0
PRINT   X1, X2, X3, YTRU, YG, YMES
END
```

This program is set up to deal with three independent variables but could be easily extended to more. In terms of Fig. 6.35, the basic sample space extends from 0 to 3 for each of the independent variables and is divided into four levels: 0, 1, 2, 3. Points are *uniformly* spaced throughout the volume, thus every possible combination of x_1, x_2, and x_3 at all levels is present, for a total of 64 points. This is accomplished by making x_1, x_2, and x_3 "vary" with time as in Fig. 6.36. (We of course do not need the differential-equation-solving capabilities of CSMP to generate this data; a simple FORTRAN program could be written. However, CSMP does have a convenient random number generator we use to add "measurement noise" to our data, and also CSMP is useful when we later extend our studies to dynamic models.) By computing y at time $= 0, 1, 2, 3, \ldots, 63$, one sample of each point in the space is produced. To give one *replication* of each of these 64 points, the statement AD1 $=$ DELAY(64, 64., A), and similar statements for x_2 and x_3, repeat the pattern out to time $= 127$ to generate a total of 128 points. To allow the independent variables to take on values *other* than 0, 1, 2, 3, the parameters A00, B00, and C00 allow a shift of mean value while X1A, X2A, and X3A permit expansion of range; however, only four different *levels* will still be produced.

Once we have generated the desired set of independent variable values, we can compute the dependent variable y according to whatever relation we choose. The statement

$$YTRU = Y00 + A1*X1 + A2*X2 + A3*X3$$

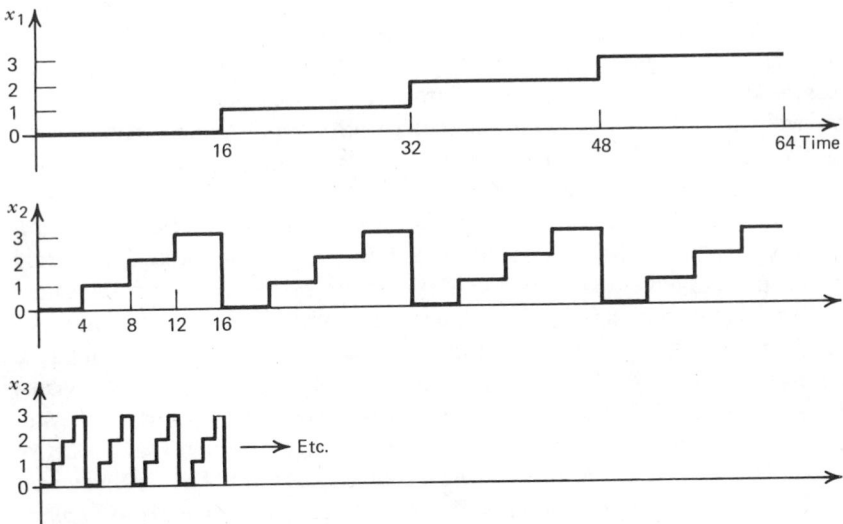

Figure 6.36   CSMP generation of x_1, x_2, x_3 combinations.

computes the "true" value of y as the linear function of x_1, x_2, and x_3 shown. Any *other* function could of course be easily inserted here to allow study of how the multiple regression program handles various models. To simulate random measurement errors in y, we wish to add to the true value a random variable with a Gaussian distribution and our choice of mean value and standard deviation. This random error is generated by the statement

$$YG = GAUSS(IMP, 0.0, SDY)$$

where 0.0 is the chosen mean value (could be set at any desired value) and SDY is the standard deviation (numerical value given on a PARAM card). In our example program, two runs are called for, one with SDY = 1.0 and a second with SDY = 5.0. The measured value of y (YMES) is obtained by adding the random error to each value of YTRU. The CSMP program then punches cards, each card carrying one "observation" of X1, X2, X3, and YMES, in a format acceptable to the multiple regression program.

The multiple regression program accepts as input the observations of the independent and dependent variables. *We* have to decide what *form* of model we will try to fit and input this information to the program. Hopefully, the program then will come back with numerical values of the coefficients in our model. Since (for our CSMP simulated experiments) we know the *true* relation among Y, X1, X2, and X3 we can readily compare the predictions of the regression analysis with the actual behavior to see whether the regression method is capable of discovering the relationship that is "buried in the data." If our assumed form of model is close to the true relation, if random measurement errors are not too large, and if enough observations are taken, we should get good results. The computer simulation approach allows efficient investigation of the meaning of the words "close," "too large," and "enough."

We will use a regression program that is more sophisticated than a basic routine—a so-called *stepwise regression analysis*. When we are carrying out *real* (rather than computer-simulated) regression studies we of course are not *sure* of the proper form of the relation among y and the x's. Thus we often try *several* different assumed forms and then choose the one that works best. Quite often this approach is implemented by postulating a model with, say, r terms and then asking the program to compute the best possible 1-term, 2-term, ..., r-term models. Since the program provides output data indicative of the predictive qualities of each model tried, one can then choose the simplest (least number of terms) model that gives adequate performance. Stepwise regression programs are available in most large computer libraries; we will employ a program that is part of SAS,[54] a widely used package of statistical analysis routines. The SAS stepwise regression program has many options we do not here need or wish to explain.

---

[54]Barr, Goodnight, Sall, and Helwig, SAS Inst. Inc., P. O. Box 10066, Raleigh, N. C. 27605.

For our purposes the program is called into action as follows:

DATA SDYEQ1; names a data set SDYEQ1 for data with SDY = 1.0
INPUT X1 X2 X3 YMEAS1; gives format of data on upcoming cards
YTRU = X1 + 2*X2 + 4*X3; computes YTRU and adds it to data set (5th
                column)
CARDS;   tells program that data cards are coming next
          } 128 cards from CSMP for SDY = 1.0
TITLE SDY = 1 Y = A1*X1 + A2*X2 + A3*X3; title for printout
PROC PRINT; prints all of data set
PROC STEPWISE; calls stepwise regression routine
MODEL YMEAS1 = X1 X2 X3 / NOINT MAXR; explained below
PROC SCATTER; calls plotting routine
PLOT YTRU*YMEAS1; plots YMEAS1 versus YTRU

In a real-world application we would of course be unable to compute YTRU and plot YMEAS1 against it; we do it here to display the amount of random noise in our data. The MODEL ... statement informs the program that YMEAS1 is the dependent variable while X1, X2, and X3 are independent. Furthermore, no intercept (constant term) is wanted in our model (NOINT) and the program is to use the stepwise regression routine called MAXR (several are available). Running the above program produces (among other things) the tabular printout of Fig. 6.37 and the graph of Fig. 6.38. From the graph we note that there is noticeable, but not excessive, random scatter around the line YMEAS1 = YTRU, thus good regression results should be expected. The tabular results bear this out; the best three-variable model found was

$$y = 0.95x_1 + 2.02x_2 + 4.04x_3 \tag{6.82}$$

quite close to the actual

$$y = 1.0x_1 + 2.0x_2 + 4.0x_3 \tag{6.83}$$

The quantity R^2, given on the printout as 0.99 for the above three-variable model, is called the *multiple correlation coefficient* and gives the percentage of the variation in dependent variable y that is "explained" by the model. Perfect correlation (100%) would have $R^2 = 1.0$, so the above model looks very good. In a real-world application we must rely heavily on the value of R^2 in choosing among alternative models since, of course, the true values of the model coefficients are unknown. Note from the printout that the best two-variable model is that which uses x_2 and x_3 in the form

$$y = 2.39x_2 + 4.41x_3 \tag{6.84}$$

and that $R^2(0.98)$ is not much smaller than for the best three-variable model,

```
                                  SDY=1        Y=A1*X1+A2*X2+A3*X3                    11:28 FRIDAY, JULY 14, 1978    4
                            MAXIMUM R-SQUARE IMPROVEMENT FOR DEPENDENT VARIABLE YMEAS1
STEP 1    VARIABLE X3 ENTERED             R SQUARE = 0.89707686
                              DF              SUM OF SQUARES        MEAN SQUARE             F        PROB>F
               REGRESSION      1              15866.93947053     15866.93947053       1106.93        0.0001
               ERROR         127               1820.44065594        14.33417839
               TOTAL         128              17687.38012647
                                  B VALUE          STD ERROR        TYPE II SS              F        PROB>F
               X3              5.95124152         0.17887407     15866.93947053       1106.93        0.0001
------------------------------------------------------------------------------------------------------------
THE ABOVE MODEL IS THE BEST  1 VARIABLE MODEL FOUND.
STEP 2    VARIABLE X2 ENTERED             R SQUARE = 0.98200396
                              DF              SUM OF SQUARES        MEAN SQUARE             F        PROB>F
               REGRESSION      2              17369.07733255      8684.53866628       3437.77        0.0001
               ERROR         126                318.30279392         2.52621265
               TOTAL         128              17687.38012647
                                  B VALUE          STD ERROR        TYPE II SS              F        PROB>F
               X2              2.39053444         0.09803362      1502.13786202        594.62        0.0001
               X3              4.41446938         0.09803362      5122.43905596       2027.71        0.0001
------------------------------------------------------------------------------------------------------------
THE ABOVE MODEL IS THE BEST  2 VARIABLE MODEL FOUND.
STEP 3    VARIABLE X1 ENTERED             R SQUARE = 0.99339814
                              DF              SUM OF SQUARES        MEAN SQUARE             F        PROB>F
               REGRESSION      3              17570.61051986      5856.87017329       6269.69        0.0001
               ERROR         125                116.76960661         0.93415685
               TOTAL         128              17687.38012647
                                  B VALUE          STD ERROR        TYPE II SS              F        PROB>F
               X1              0.95148561         0.06477969       201.53318731        215.74        0.0001
               X2              2.01821398         0.06477969       906.72699809        970.64        0.0001
               X3              4.04214892         0.06477969      3637.19633399       3893.56        0.0001
------------------------------------------------------------------------------------------------------------
THE ABOVE MODEL IS THE BEST  3 VARIABLE MODEL FOUND.
```

Figure 6.37 Tabular results of stepwise regression.

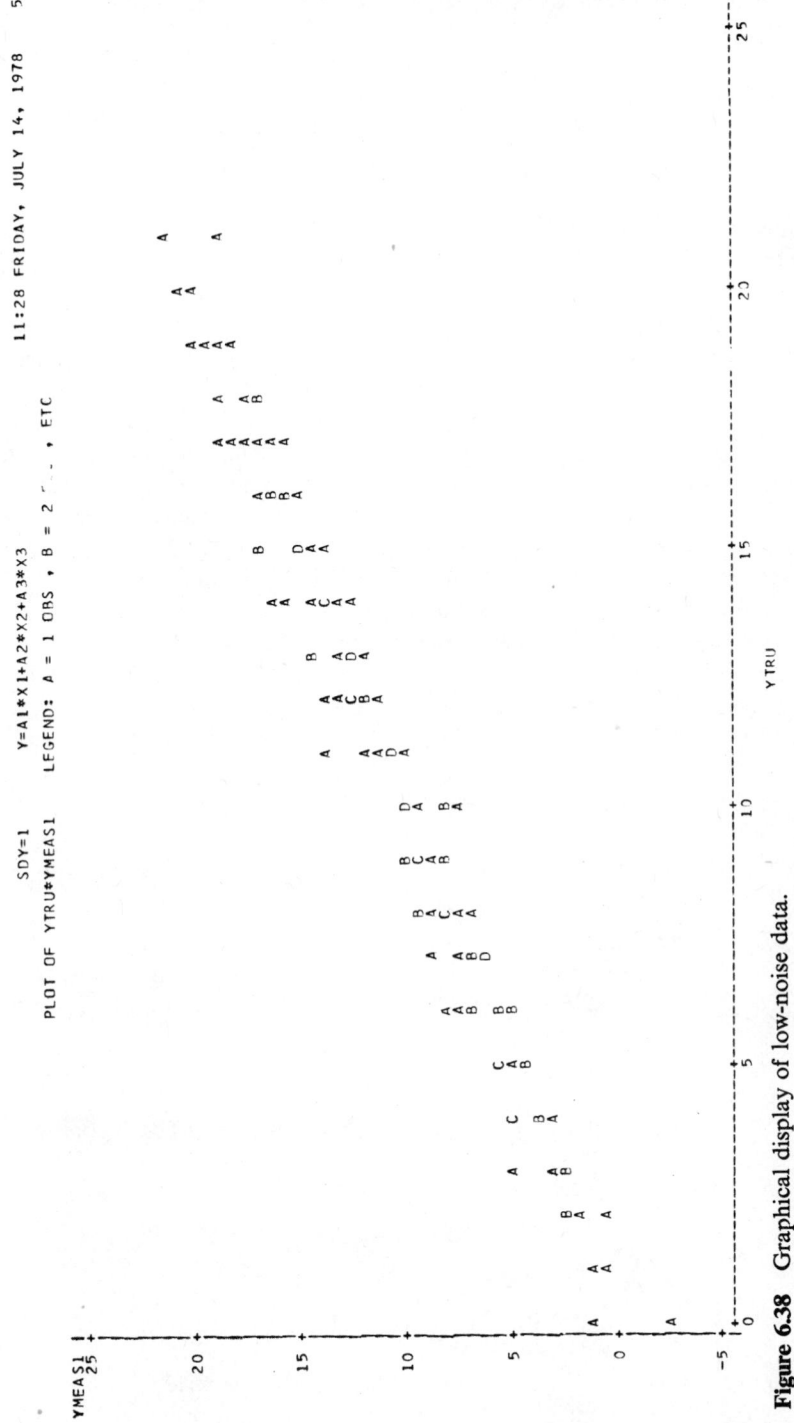

Figure 6.38 Graphical display of low-noise data.

313

thus if it were important to obtain *simple* models, one might choose this two-variable form. The best one-variable model is

$$y = 5.95 x_3 \tag{6.85}$$

and now R^2 is down to 0.90. In addition to comparison of R^2 values, graphs of model-predicted y values versus measured y values are sometimes helpful in deciding on the best model.

Of the other statistical information provided for each model on the printout, only the PROB > F value will be further explained. For each model, the computer runs a "statistical significance" test (F test) on the entire model and also on each coefficient separately. The larger the F value, the more significant is that particular result. Once the F value is computed, a "table" of the F distribution is "looked up" by the computer and it calculates the probability of obtaining an F value equal to or greater than that actually found, *assuming that the coefficient had actually been zero*. That is, even if y did *not* depend on, say, a_3, the computed F value in, say, the three-variable model would *not* be zero but it would be a smaller number than the 3893.56 value shown in Fig. 6.37. *If a_3 were really zero*, there is a certain probability of getting an F value as large as 3893.56, in our case this is $0.0001 = 0.01\%$ (PROB > F = 0.0001). This means that the coefficient $a_3 = 4.04$ is "highly significant"; there is only a chance of 1 in 10,000 (0.01%) that such a large (3893.56) F value could occur if a_3 were truly zero.

For the 128 points with SDY = 5 the graph of YMEAS5 versus YTRU (Fig. 6.39) shows very large scatter. However, the regression technique still does a quite respectable job of extracting the relation from this very noisy data, giving the following results:

Best One-Variable Model

$$y = 6.04 x_3, \qquad R^2 = 0.78, \qquad \text{PROB} > \text{F} = 0.0001$$

Best Two-Variable Model

$$y = 2.39 x_2 + 4.51 x_3, \qquad R^2 = 0.85, \qquad \text{PROB} > \text{F} = 0.0001 \text{ (both coefficients)}$$

Best Three-Variable Model

$$y = 0.76 x_1 + 2.09 x_2 + 4.21 x_3, \qquad R^2 = 0.86$$
$$\text{PROB} > \text{F} = 0.021 \, (a_1)$$
$$\text{PROB} > \text{F} = 0.0001 \, (a_2, a_3)$$

Since we have not given any details on how the stepwise regression method chooses the various models, a few final comments are in order. A "brute force"

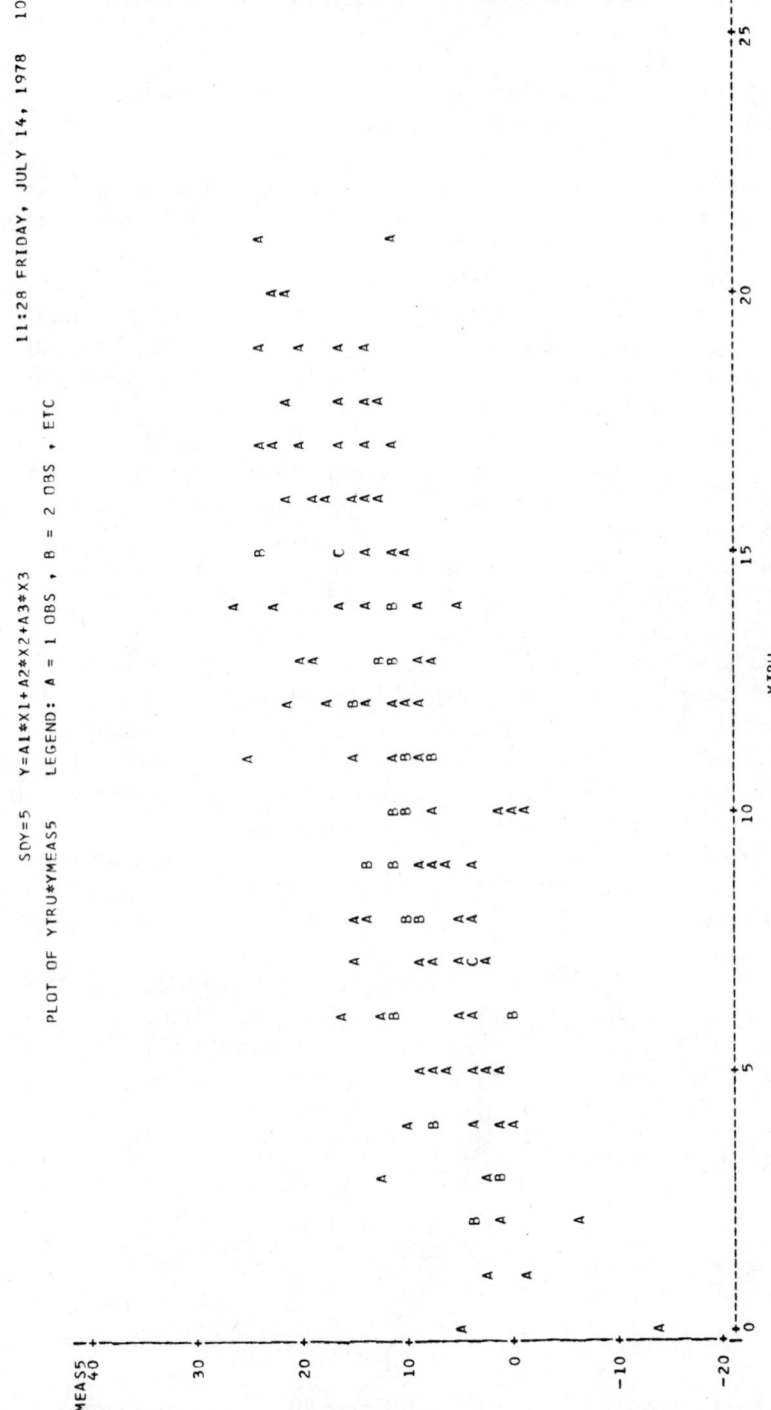

Figure 6.39 Graphical display of high-noise data.

approach to deciding on the "best possible" model would be to use an "ordinary" multiple regression program, simply rerun it for *all possible* combinations of variables, and then examine the results to select the best. This procedure assures that no possibilities are overlooked, but for large models (many independent variables) the number of combinations becomes huge, requiring excessive computer time and encouraging the invention of some more efficient (though not 100% foolproof) technique. The various *stepwise* regression programs (SAS allows a choice from five alternatives and even more can be found in the literature) represent attempts to implement these more efficient concepts. They do *not* examine every possible model but fortunately rarely overlook the best model and are significantly faster running. We have here used the SAS routine called MAXR, based mainly on its simplicity of use and the feature that it always produces a complete family of models, starting with one-variable and going to r-variable, where r is the number of independent variables. A regular user of multiple regression might wish to experiment with several of the available methods to see whether any were particularly suitable for the applications at hand. However, the MAXR method would probably be about as good as any in most cases.

It is hoped that the above brief introduction will apprise the reader of the utility of the multiple regression method in developing empirical models. Using the computer simulation method, many other illuminating examples may be contrived to demonstrate its capabilities and limitations; some of these are included as chapter problems. While the results of multiple regression are rarely clearcut and thus may be occasionally misinterpreted by the statistically-naive user (or even a statistics "expert"!), the combination of practical engineering judgment and an easy-to-use regression program can be a powerful tool in model building.

6.5.2 MULTIPLE REGRESSION FOR DYNAMIC MODELS

For a dynamic system with input q_i and output q_o, as long as the differential equation form is known or assumed and can be put in the form of Eq. (6.70), a direct application of multiple regression is possible if we are able to measure y and all the x's. For example, Eq. (6.47) could be written as

$$x = \frac{f_i}{K_1} + \frac{K_2}{K_1}x^2 - \frac{K_3}{K_1}x^3 - \frac{B}{K_1}\frac{\dot{x}}{|\dot{x}|}|\dot{x}|^{1.8} - \frac{M}{K_1}\ddot{x} \qquad (6.86)$$

and if we define

$$x_1 \triangleq f_i, \quad x_2 \triangleq x^2, \quad x_3 \triangleq x^3, \quad x_4 \triangleq \frac{\dot{x}}{|\dot{x}|}|\dot{x}|^{1.8}, \quad x_5 \triangleq \ddot{x}, \quad y \triangleq x$$

we are back to the standard regression Eq. (6.70). We must now exercise the system dynamically some way to generate time histories of y and all the x's. If the equation modeled the system perfectly and we could make perfect, noise-free measurements of y and all the x's, only five sets of readings at five different time points would be necessary to generate five equations in the five unknown coefficients, and these could be solved without recourse to least-squares procedures. To deal with the real world of imperfect models and noisy measurements we, as usual, take readings at many more then five time points and then use least-squares methods to get the best-fit model. (The time points also cannot be too closely spaced or else the matrices become "ill-conditioned.")

A CSMP-simulated experiment on the system

$$(5D+1)q_o = 1.0[\sin 0.1t + \sin 0.4t + \sin 2t] = Kq_i$$

which gathered 100 time points for $0 < t < 100$ and then used SAS to find K and T in the model $q_o = Kq_i - T\dot{q}_o$ led to the essentially perfect results $R^2 = 1.0$, $K = 1.0, T = 5.0$. For such perfect data, 100 points and least squares are of course hardly necessary; *two* points allows accurate solution for K and T. However, if we now "contaminate" q_i, q_o, and $\dot{q}_o$ with Gaussian random noise (σ, respectively of $0.1, 0.1, 0.01$), two-point solutions are now very inaccurate, whereas SAS returns $R^2 = 0.96, K = 0.97, T = 4.8$. A rerun of SAS using only the first 16 of the hundred points gave nearly as good results; $R^2 = 0.96, K = 0.98, T = 4.7$.

The direct approach just illustrated is often not workable since it requires measurement of all the derivatives of q_i and q_o that are present in the model differential equation. A number of least-squares methods that require measurement of q_i and q_o only are available.[55] We here briefly consider one[56] that produces as result the impulse response $h(t)$ of the system under test. Since the concept of impulse response is limited to linear, constant-coefficient models, this method applies to the same class of systems as the sine, pulse, and random tests of Sections 6.1–6.3. Now, however, instead of finding a graph of $W(i\omega)$, we find one for $h(t)$. Since in Section 3.4 we showed how to numerically interconvert these two system descriptions, we can follow our least-squares calculation of $h(t)$ with a conversion to a graph of $W(i\omega)$, and can then even use Levy's or other curve-fitting methods to get an analytical transfer function.

We use convolution integral to develop our least-squares technique.

$$q_o(t) = \int_0^t h(t - \tau) q_i(\tau) \, d\tau \tag{3.22}$$

We sample q_i and q_o every T seconds and then approximate them in stepwise

---

[55]Hsia, *op. cit.*
[56]Hsia, *op. cit.* p. 38.

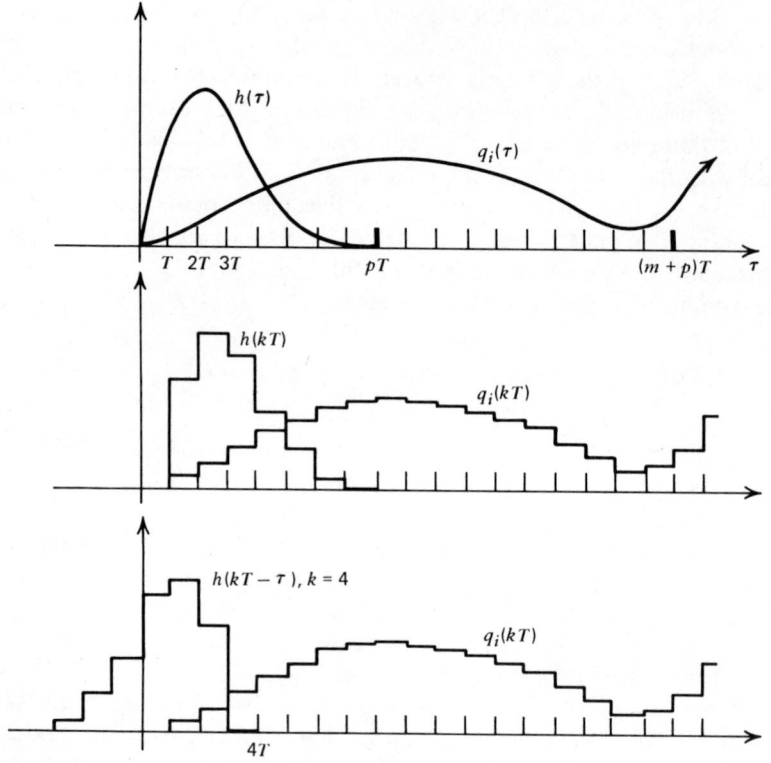

Figure 6.40 Graphical interpretation of convolution integral.

fashion:

$$q_i(t) \approx q_i(kT), \qquad q_o(t) \approx q_o(kT), \qquad kT \leqslant t < (k+1)T \qquad (6.87)$$

Equation (3.22) then can be interpreted graphically as in Fig. 6.40, leading to the result

$$q_o(kT) \approx \sum_{i=0}^{k} Th(kT - iT)q_i(iT) \qquad (6.88)$$

Taking $k = 4$ as an example, we would have

$$q_o(4T) \approx \left[q_i(4T)h(0) \right.$$
$$\left. + q_i(3T)h(T) + q_i(2T)h(2T) + q_i(T)h(3T) + q_i(0)h(4T) \right] T$$

$$(6.89)$$

Note that this is an equation of form

$$y_j = a_1 x_{1j} + a_2 x_{2j} + \cdots + a_r x_{rj} \tag{6.90}$$

where $h(0), h(T), \ldots$ are the a's. By using the data available for various values of k, we can generate a set of such equations that fit exactly the form of our earlier least-squares calculations. In Fig. 6.40 note that pT represents a time at which the tested system's impulse response has died out. This would be unknown but must be estimated before calculations can proceed. However, if we guess pT too short we simply do not find the complete course of $h(\tau)$ and must rerun the calculation with a longer pT until we get $h(\tau)$ to die out. Note also that we get p discrete points on $h(\tau)$, not a continuous curve, thus T must be chosen small enough (p large enough) to resolve any fine details in $h(\tau)$.

If we use $k < p$ in Eq. (6.88), since q_i is zero for negative time, we get some zero coefficients on the h terms rather than numerical values of q_i, thus we use the output record $q_o(kT)$ only for $p \leqslant k \leqslant (m+p)$, where $m > p$. If we took $m = p$, we would get $p+1$ equations in $p+1$ unknowns $(h(0), h(T), h(2T), \ldots, h(pT))$, while $m > p$ gives us the usual least-squares situation of more equations than unknowns. We now have in matrix form

$$y = XA \tag{6.91}$$

$$y \triangleq \begin{bmatrix} q_o(pT) \\ q_o((p+1)T) \\ \vdots \\ q_o((p+m)T) \end{bmatrix}, \qquad A \triangleq \begin{bmatrix} h(0) \\ h(T) \\ \vdots \\ h(pT) \end{bmatrix} \tag{6.92}$$

$$X \triangleq \begin{bmatrix} Tq_i(pT) & Tq_i((p-1)T) & \cdots & Tq_i(0) \\ Tq_i((p+1)T) & Tq_i(pT) & \cdots & Tq_i(T) \\ \vdots & & & \\ Tq_i((p+m)T) & Tq_i((p+m-1)T) & \cdots & Tq_i(mT) \end{bmatrix} \tag{6.93}$$

This matches exactly Eq. (6.78) so we can use standard least-squares (multiple regression) programs such as the SAS routine described earlier to solve this type of problem.

We here conclude our treatment of least-squares methods, however many useful extensions are available in the literature.[57] These include:

1 Definition of optimum input signals. (Maximum length PRBS signals are found to often give significant computational benefits and produce accurate results.)

---

[57]Hsia, *op. cit.*

2 Sequential on-line methods that continuously update the model as fresh data appears and that do not require repetitive time-consuming matrix inversion.

3 Methods for multiple-input multiple-output systems.

4 Techniques that produce difference equation, Z transfer function, and state-variable models, rather than the impulse response model.

5 Relation to the Kalman filter approach.

6.6 COMPARISON OF MODELING METHODS

While we have certainly not discussed all known techniques of model determination by experimental testing, we have included a representative sample and it would seem appropriate now to give some guidelines for *selecting* the particular approach for a certain practical application. This turns out to be a relatively difficult and space-consuming task, so we will content ourselves with some brief summary comments plus reference to other sources[58] where lengthy discussion is more appropriate. The main difficulties in presenting clearcut rules lie in the diversity of the applications, constraints of particular industrial conditions, available test equipment and computer programs, historical prejudices, economic factors, and so on. Each section of this chapter did include some comments on the applicability of the particular method under discussion, thus we now briefly gather these together and augment them with some additional discussion.

Sinusoidal testing, when process constraints allow its use, generally gives the most accurate results and provides flexibility in defining amplitude-dependent nonlinearities by repeating experiments at different excitation amplitudes. It allows concentration of the excitation at discrete frequencies, giving larger responses and better signal-to-noise ratios. Test time may be quite long and this prevents identification of system time-varying parameters, if such are present. Also, sine-wave testing is "unrealistic" in that it does not directly simulate the operating conditions of most systems. Pulse testing's main advantages are reduced test time and usually simpler test hardware. Due to poor signal-to-noise ratio, accuracy suffers somewhat. Since input magnitude versus frequency is not independently controllable as in sine testing, amplitude-dependent nonlinearities are harder to define. Random signal testing offers the possibility of identifying system parameters without the need to introduce test signals, thus the process normal operation need not be interrupted, offering convenience and economy. It

[58]Gustavsson, I., "Survey of Applications of Identification in Chemical and Physical Processes," *Automatica*, **11** (1975), pp. 3–24; Gustavsson, I., "Comparison of Different Methods for Identification of Industrial Processes," *Automatica*, **8** (1972), pp. 127–142; Lang, G. F., Shake, Rattle, or Rap (How to Conduct Vibration Tests), Appl. note 10, Nicolet Scientific Corp., Northvale, N.J., 1975; Sisson, T., *et al.*, Determination of Modal Properties of Automotive Bodies and Frames Using Transient Testing Techniques, Structural Dynamics Research Corp., SAE Paper 730502, 1973.

appears, however, that relatively few such applications have been totally successful since the naturally occurring process perturbations often have insufficient amplitude and frequency content to allow reliable calculations. Use of intentional test signals, particularly PRBS types, allows tailoring of the frequency content to suit particular test needs, provides signal-to-noise ratios better than pulse testing but worse than sine-wave, and (if analyzer and input signal are synchronized) allows fast computation. Parameter tracking offers the potential of identifying all types of systems—not just linear and constant-coefficient— using a single approach. While simulation results, such as the CSMP examples given, are impressive, few real-world applications were found in the author's literature search. Neither, however, was detailed documentation found of the *failure* of the method in specific practical instances. Least-squares, multiple regression, and other related statistical approaches for steady-state modeling are certainly well established, widely used, and often successful. Their extension to dynamic modeling is somewhat less widely appreciated in industry, not because they have been inadequately explored theoretically (*thousands* of papers have appeared in recent years), but perhaps because much of the work has been presented in mathematical terms difficult for the practicing engineer to understand and apply. As this field gets sorted out and reduced to common practice, a number of techniques will emerge that usefully augment the established methods such as sinusoidal testing.

BIBLIOGRAPHY

1 Bendat, J. S. and A. G. Piersol, *Random Data*: *Analysis and Measurement Procedures*, Wiley-Interscience, N.Y., 1971.

2 Blaser, D. A. and J-Y Chung, A Transfer Function Technique for Determining the Acoustic Characteristics of Duct Systems with Flow, Gen. Motors Res. Lab. Rept. GMR-2640, 1978.

3 Brown, D., Carbon, G. and K. Ramsey, Survey of Excitation Techniques Applicable to the Testing of Automotive Structures, SAE Paper 770029, 1977.

4 Davies, W. D. T., *System Identification for Self-Adaptive Control*, Wiley-Interscience, N.Y., 1970.

5 Eykhoff, P., *System Identification*, John Wiley and Sons, N.Y., 1974.

6 Graupe, D., *Identification of Systems*, Van Nostrand Reinhold, N.Y., 1972.

7 Hsia, T. C., *System Identification*: *Least-Squares Methods*, Lexington Books, D. C. Heath, Lexington, Mass., 1977.

8 *Identification and System Parameter Estimation, Parts 1 and 2, Proc. of 3rd IFAC Symp.*, American Elsevier, N.Y., 1973, Vols. 1 and 2.

9 Kelley, R. W., Transient Response of Linear Damped Lumped Spring-Mass Systems by Experimentally Derived Transfer Functions, Sandia Labs. Rept. SC-TM-64-1772, Albuquerque, N.M., 1965.

10 Lyster, H. N. C., A Critical Review of Techniques for the Flight Determination of Aircraft Response Characteristics, Nat'l. Res. Council of Canada Rept. 6825, Ottawa, 1962.

11 Plummer, M. C., Nuclear Power Plant Diagnostics Using Fourier Analysis Techniques, Hewlett–Packard Appl. Note 140-7, 1974.

12 Roth, P. R. (Hewlett–Packard Corp.), Noise Source for Transfer Function Testing; Noise Burst Source for Transfer Function Testing, U.S. Patents 3,988,667 October 26, 1976 and 4,023,098 May 10, 1977.

13 Uhrig, R. E., *Random Noise Techniques in Nuclear Reactor Systems*, Ronald Press, N.Y., 1970.

PROBLEMS

6.1 For the transfer function of Eq. (6.25):
  (a) Assume some numerical values for the A's and B's, compute the frequency response, and then use Levy's method on this "manufactured" data to "recover" the A and B values.
  (b) Repeat part a but "contaminate" the manufactured data with 5% errors to see the effect on the predicted A and B values.
  (c) Repeat part b with 10% errors.

6.2 Write out the Levy matrix for a model with A's through A_3 and B's through B_6.

6.3 Using the numbers from part a of Problem 6.1, try to fit the manufactured data with an "incorrect" model, one with A's through A_1, and B's through B_3. Comment on the results.

6.4 Repeat Problem 6.3 using an "incorrect" model with A's through A_3 and B's through B_5.

6.5 Calculate the Fourier series for the pulse train of Fig. 6.11.

6.6 If a bandwidth $B = 10$ Hz is adequate to resolve the closest pair of expected peaks in an analog power-spectral density measurement, what record length T is needed to obtain a $\pm 10\%$ confidence interval with 90% confidence? How much is the 90% confidence interval reduced if we double this T value?

6.7 If the measurement of Problem 6.6 is implemented by digital methods, how many spectra must be averaged to realize the desired confidence using ensemble averaging? What record length is used? If we double the number of spectra averaged, how much is the confidence interval reduced? Does record length change?

6.8 For an error of $\pm 10\%$ in amplitude ratio and ± 0.1 rad in phase, and for a confidence level of 90%, how many digital spectra must be averaged if the coherence γ^2 is estimated as 0.6 in a random-signal transfer function measurement? What record length should be used?

6.9 Choose clock rate Δt and sequence length N for a pseudo-random binary sequence used to test a system out to frequency 500 Hz if no system response peaks with spacing closer than 5 Hz are expected.

6.10 Write a CSMP program for the system of Eq. (6.43) and run the simulation if CSMP is available to you. In addition to the various runs in the text, include one where *both* low- and high-frequency inputs are present; that is, let the input be $K(\sin 0.1t + \sin 0.2t + \sin t + \sin 3t)$.

6.11 For the system of Fig. P6.1, use digital simulation (such as CSMP) to study the following aspects of parameter tracking:

(a) Set up a program to track parameter B if the damper has force $= -(\dot{x}_0/|\dot{x}_0|)B\dot{x}_0^2$. Take $K_s = 1.0$, $M = 1.0$, $f_i = 1\sin t$, let the true value of B be 1.0 but let B^* start at zero for $t=0$.

(b) What is the effect of f_i's size and frequency? Let waveform be sinusoidal.

(c) What is the effect of "gain" G? Will system go unstable for large G's?

(d) How well does the system track a *varying* B?

(e) Set up a system to track K_s only.

(f) Set up a system to track K_s and B simultaneously.

(g) What is the effect of sinusoidal noise added to the measured system signals?

(h) What is the effect of dynamics in the measuring transducers?

(i) How well does the method work if f_i is a transient pulse?

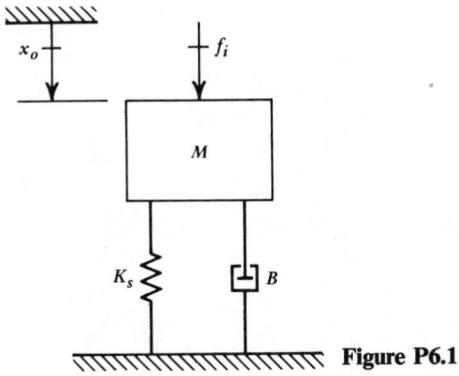

Figure P6.1

6.12 In the CSMP/SAS study of multiple regression in Section 6.5, change the CSMP model to:

$$YTRU = YOO + A1*X1*X2 + A3*X3**2$$

(You will also have to change the CSMP statements that perform the card punching.) Run the data produced by CSMP in the SAS program to see whether it is able to find "good" values for the coefficients in this model. Comment on the results.

6.13 Using the original model for the CSMP/SAS study in Section 6.5, examine the effect of sample size as follows:

(a) Rerun the problem using 64 (rather than 128) points and comment on the results.

(b) Rerun the problem using only 16 points. In part a (64 points), *all* possible combinations of values of X1, X2, and X3 are "exercised" once. When we go to *less* than 64 points we no longer can exercise the "sample space" (Fig. 6.35) completely. Decide on a scheme for choosing the 16 points and explain why you chose it. Then run the SAS program with these 16 points and comment on the results.

CHAPTER 7

Subsystem Coupling Methods

7.1 UTILITY OF SUBSYSTEM COUPLING METHODS

To make clear our intended definitions of "element," "subsystem," and "system," consider the structural dynamics modeling of an automobile frame. If we consider the entire frame as the system of interest, individual beams, plates, and shells could be considered subsystems, while the masses and springs used to define lumped models of these subsystems are the mechanical elements. While the definitions of elements are essentially fixed, those for subsystems and systems can of course change with the interest of the analyst. For example, once we understand the automobile frame's behavior we may then want to treat it as a *subsystem*, together with engine, drive train, body, and suspension subsystems, to study the entire vehicle as a system.

Having defined the overall system of interest, there of course is no *necessity* of defining subsystems, one can define the complete system entirely in terms of elements; particularly with computerized analysis techniques this is often a proper approach. There are, however, situations where the subsystem viewpoint has definite advantages and we wish to develop the concept so that the reader is aware of this alternative. One such class of situations is where a variety of systems is assembled from more-or-less standard building blocks (subsystems). Machinery hydraulic systems, for example, utilize the same pumps, valves, motors, piping, and so on arranged in different configurations to create myriad useful systems. Rather than each time breaking down, say, the pump into fluid and mechanical elements, it may be more efficient to develop only once a subsystem model for it that can be easily coupled with similar models for piping, cylinders, and so on to create system models. Familiarity with such an approach may be particularly helpful in the initial creative *design* of new systems, as opposed to the analysis of given configurations, since creative design is still essentially the province of the human mind rather than computers, and the mind is better able to bridge the gap from subsystem to system than from element to system. Also, for very large systems, computer capacity may be exceeded unless the total problem is broken into manageable parts, the results of which must then be properly combined.

Another important application area of these methods involves experimental testing such as discussed in Chapter 6. When the lack of accurate theoretical models requires experimental testing, such testing is generally least expensive and most accurate in predicting overall system behavior if it is at a subsystem level rather than an element level. Definition of the detailed nature of such experiments requires understanding of the methods by which the results will be combined with similar experimental or theoretical data for adjacent subsystems. Finally, our discussion of coupling methods will clarify the approximations inherent in the transfer function method of coupling introduced in most elementary treatments of system dynamics and probably already familiar to the reader.

7.2 TWO-PORT DEVICES, EFFORT AND FLOW VARIABLES

A fundamental limitation in the familiar transfer function description of physical hardware lies in its characterization of input and output conditions in terms of single variables at each location. When a second device is coupled to a first at its output, it will draw some power from the first. Definition of this power is impossible in terms of a single variable; two are required. The various coupling methods (impedance/mobility,[1] four-pole,[2] bond graph,[3] etc.) begin by defining *pairs* of input and output variables and then proceed to develop subsystem descriptions that correctly predict coupled behavior without the no-loading assumption implicit in the transfer function approach. For example, using impedance methods, we will see that a *complete* description of a subsystem, adequate to predict coupled-system behavior when loading is not insignificant, requires *three* pieces of information, only one of which is the familiar "unloaded" transfer function (the other two are the subsystem *input impedance* and *output impedance*).

While extension to multiport situations is possible, we here limit ourselves to the simpler two-port configuration. We define a *two-port device* as one that exchanges energy with others at only two locations (ports). Figure 7.1 shows the conventional graphic representation of this concept, which resembles an electrical network diagram but is of course more general in application than this. At each port we define two variables, one of which is selected as being of primary practical interest to us and the other, such that the product of the two variables gives the instantaneous power flowing through the port. We will use the names *effort variable* and *flow variable* here; other common usages are "across vari-

[1]Church, A. H., *Mechanical Vibrations*, John Wiley and Sons, N.Y., 1963.
[2]Vlach, J., *Computerized Approximations and Synthesis of Linear Networks*, John Wiley and Sons, N.Y., 1969.
[3]Karnopp, D. and R. Rosenberg, *System Dynamics: A Unified Approach*, John Wiley and Sons, N.Y., 1975.

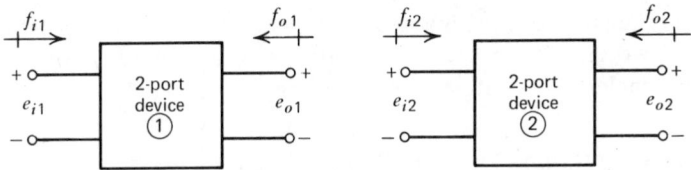

Figure 7.1 Symbology for coupling of two-port devices.

able" and "through variable." The most common applications involve:

Class of system	Effort variable (e)	Flow variable (f)
Electric circuits	Voltage	Current
Mechanical systems	Force	Velocity
	(Torque)	(Ang. velocity)
Hydraulic systems	Pressure	Volume flow rate

One can often apply two-port methods to situations that at first glance appear not to be two-port by simple redefinition of the subsystem boundaries and restriction of generality of the resulting model. The DC generator system of Fig. 7.2 would have two electrical ports and one mechanical if we insist on a model general enough to couple with *any* prime mover. If, however, we commit ourselves to one specific prime mover and *include* it within the defined subsystem, then only the two electrical ports remain. This type of reasoning also allows

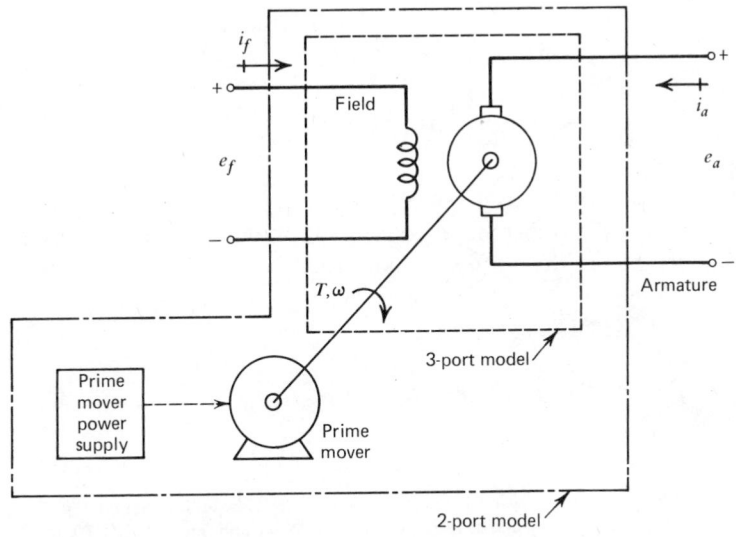

Figure 7.2 Reinterpretation of three-port system as a two-port.

us to "ignore" the thermal energy port involving the heat loss of the machinery to its surroundings; we simply restrict the model to one specific thermal environment and then include that within our boundaries.

7.3   ELEMENTARY FOUR-POLE COUPLING METHODS

It can be shown[4,5] that for systems which obey the reciprocity theorem and are comprised of linear mechanical, electrical, and/or fluid elements without internal energy sources, the relations among the "terminal" variables of a two-port device can be put in the form

$$e_i = a_{11}e_o + a_{12}f_o \tag{7.1}$$

$$f_i = a_{21}e_o + a_{22}f_o \tag{7.2}$$

The a's are called the *four-pole parameters*, and since they are related by[4]

$$a_{11}a_{22} - a_{12}a_{21} = 1 \tag{7.3}$$

we see that it takes only *three* of these quantities to completely describe the terminal behavior of any two-port, no matter how complex it might be internally. It is possible to calculate[6] the a's from the set of equations that describes the two-port device. However, we are also interested in noting how they can be *measured* directly for an existing device.

Let us use the example of Fig. 7.3 to develop these concepts. For subsystem 1, direct application of Newton's law leads to

$$f_{i1} = (1)f_{o1} + (0)v_{o1} \tag{7.4}$$

$$v_{i1} = \left(\frac{Bs + K_{s1}}{BK_{s1}}\right)f_{o1} + (1)v_{o1} \tag{7.5}$$

where force f is the effort variable and velocity v the flow (the use of f for both the general flow symbol and the specific force symbol should cause no confusion if we are careful to remember the context). Likewise:

$$f_{i2} = (1)f_{o2} + (Ms)v_{o2} \tag{7.6}$$

$$v_{i2} = \left(\frac{s}{K_{s2}}\right)f_{o2} + \left(\frac{Ms^2}{K_{s2}} + 1\right)v_{o2} \tag{7.7}$$

[4]Harman, W. W. and D. W. Lytle, *Electrical and Mechanical Networks*, McGraw-Hill, N.Y., 1962, p. 313.
[5]Vlach, J., *Computerized Approximation and Synthesis of Networks*, John Wiley and Sons, N.Y., 1969, p. 29.
[6]Vlach, *op. cit.*, p. 32.

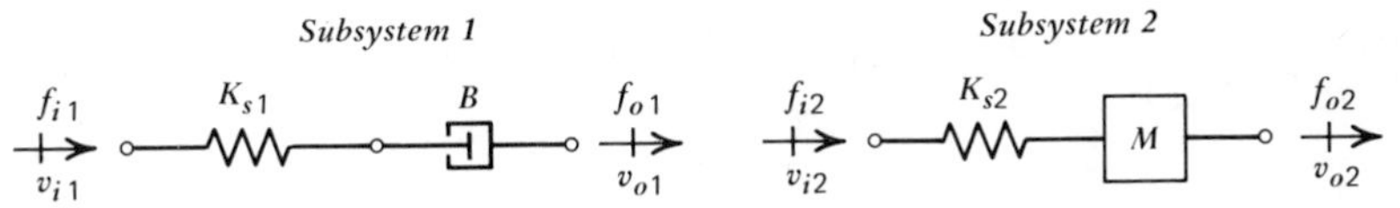

$f_{i1} \triangleq$ Force *of* previous subsystem *on* subsystem 1

$f_{o1} \triangleq$ Force *of* subsystem 1 *on* subsystem 2

Figure 7.3   Example for development of four-pole concepts.

We clearly can write the four-pole equations in matrix form:

$$\begin{bmatrix} e_i \\ f_i \end{bmatrix} = \begin{bmatrix} a_{11} & a_{12} \\ a_{21} & a_{22} \end{bmatrix} \begin{bmatrix} e_o \\ f_o \end{bmatrix} \tag{7.8}$$

The square matrix of coefficients is called the *transfer matrix,* and if we couple a chain of n two-ports such that the output of one is the input of the next ("cascade connection"), we can easily generate the equations relating the effort and flow variables of the first and last devices as

$$\begin{bmatrix} e_{i1} \\ f_{i1} \end{bmatrix} = [M_1][M_2] \cdots [M_n] \begin{bmatrix} e_{on} \\ f_{on} \end{bmatrix} = [M_s] \begin{bmatrix} e_{on} \\ f_{on} \end{bmatrix} \tag{7.9}$$

where the M's are the 2-by-2 subsystem transfer matrices of each two-port, and M_s is a 2-by-2 system transfer matrix that is their left-to-right product. For our example

$$\begin{bmatrix} f_{i1} \\ v_{i1} \end{bmatrix} = \begin{bmatrix} 1 & Ms \\ \dfrac{K_{s1} + K_{s2}}{K_{s1} K_{s2}} s + \dfrac{1}{B} & M \dfrac{K_{s1} + K_{s2}}{K_{s1} K_{s2}} s^2 + \dfrac{M}{B} s + 1 \end{bmatrix} \begin{bmatrix} f_{o2} \\ v_{o2} \end{bmatrix} \tag{7.10}$$

As a simple application of this result, suppose no third device is attached at the output of subsystem 2 and that no external force f_{o2} is applied. If we were then given either $f_{i1}(t)$ or $v_{i1}(t)$, we could find v_{o2} from

$$f_{i1} = (1)f_{o2} + (Ms)v_{o2} = Msv_{o2} \tag{7.11}$$

or

$$v_{i1} = \left(\dfrac{K_{s1} + K_{s2}}{K_{s1} K_{s2}} s + \dfrac{1}{B} \right) f_{o2} + \left(M \dfrac{K_{s1} + K_{s2}}{K_{s1} K_{s2}} s^2 + \dfrac{M}{B} s + 1 \right) v_{o2} \tag{7.12}$$

$$v_{i1} = \left(M \dfrac{K_{s1} + K_{s2}}{K_{s1} K_{s2}} s^2 + \dfrac{M}{B} s + 1 \right) v_{o2} \tag{7.13}$$

If f_{o2} were not zero but given as $f_{o2}(t)$, we again can solve using (7.11) or (7.12).

Or, suppose M were attached to a rigid wall ($v_{o2} \equiv 0$); then (7.11) gives $f_{o2} = f_{i1}$ if f_{i1} is given, while (7.12) gives

$$f_{o2} = \frac{v_{i1}}{\dfrac{K_{s1} + K_{s2}}{K_{s1}K_{s2}} s + \dfrac{1}{B}} \qquad (7.14)$$

if v_{i1} is given. Note that $f_{o2} \equiv 0$ is one way to make the power drain ("loading") at the output of device 2 equal to zero, while $v_{o2} \equiv 0$ is the other way, since in either case the power fv is zero.

While we have emphasized getting from e_{i1}, f_{i1} to e_{on}, f_{on} in one jump (Eq. (7.9)), if "interior" variables are of interest, they may also be obtained with this method. Thus in Fig. 7.4 if our interest is in e_b and/or f_b, we could proceed as follows:

1   Get the overall ("a to c") matrix. At c either e or f will be zero, while at "a" either e or f will be a given input. There are then just 2 unknowns at "a" and c and these can both be found from the two equations provided by the overall matrix.

2   Form either the "a to b" matrix or the "b to c" matrix; either provides two equations. If the "a to b" matrix is used, both e and f at "a" are known, so both e and f at b can be found. Similarly, if the "b to c" matrix is used, both e and f are known at c, again allowing solution for e and f at b.

Turning now to the implementation of these methods using experimental measurements, Equations (7.1) and (7.2) suggest a number of possibilities. We could arrange our apparatus so that, say, $e_o \equiv 0$, apply a sinusoidal sweep, pulse, or random drive signal at, say, e_i and, using methods of Chapter 6, compute

$$a_{12}(i\omega) = \frac{e_i(i\omega)}{f_o(i\omega)}, \qquad a_{22}(i\omega) = \frac{f_i(i\omega)}{f_o(i\omega)} \qquad (7.15)$$

Note, as is clear from our example Equations (7.4)–(7.7), that the four-pole parameters are in general sinusoidal transfer functions and when measured experimentally we have only *graphs* of amplitude ratio and phase versus frequency, not analytical functions. However, these numerical values are sufficient to carry out the operations of Eq. (7.9) as long as we have this data for all the a's of each matrix over the frequency range of interest. That is, the a's of M_s are computed as curves of amplitude ratio and phase versus frequency, "one frequency at a time," from curves for the a's of each M_1, M_2, and so on. Returning to Equations (7.1) and (7.2), repeating the above testing with $f_o \equiv 0$ would give

$$a_{11}(i\omega) = \frac{e_i(i\omega)}{e_o(i\omega)}, \qquad a_{21}(i\omega) = \frac{f_i(i\omega)}{e_o(i\omega)} \qquad (7.16)$$

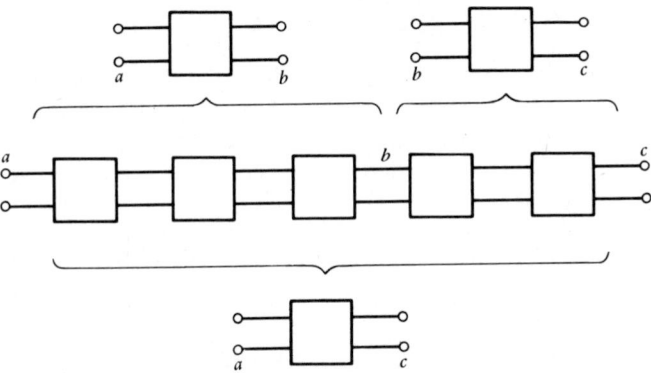

Figure 7.4   Treatment of "interior" variables.

Only three of the four measurements in (7.15) and (7.16) are really necessary because of the relation in Eq. (7.3). However, if all four are made, Eq. (7.3) serves as a good check on the numerical results. We see thus that, compared to the transfer function description of a device, which requires measurement of only one frequency response function, the more complete four-pole model requires measurement of *three* such functions.

The measurements described above are not always easy to make with the necessary accuracy. For example, making the motion v_{o1} exactly zero when measuring a_{11} and a_{21} for subsystem 1 in Fig. 7.3 is impossible since no "perfectly rigid wall" is available in a real-world laboratory. Of course the motion need not be *exactly* zero; however, at high frequency *all* the motions become small and it becomes increasingly difficult to keep the motion of the "fixed point" small *relative* to that of intended moving points. Of several possible alternative measuring schemes, we now describe one.[7] Connect a "load" (second device) at the output of the four-pole of interest, drive the four-pole input with a suitable test signal, measure both e's and both f's of the four-pole and express them as frequency functions in the usual way. Now attach a *different* load to the four-pole and repeat the procedure. At each frequency, Equations (7.1) and (7.2) can now be written twice, giving four equations in the four unknown a's and allowing for their solution, "one frequency at a time." (Actually, only *three* measurements are again needed due to the availability of Eq. (7.3).)

[7]Molloy, C. T., "Application of Four-Pole Parameters to Torsional Vibration Problems," *ASME J. Eng. Ind.*, Feb. (1962), p. 30.

7.4   ELEMENTARY IMPEDANCE COUPLING METHODS

An alternative but closely related coupling method utilizes impedance concepts. Equations (7.1) and (7.2) are actually a particular pair of a set of 12 possible equations one could write relating the e's and f's of a given two-port. That is, since *any two* of the variables may be considered as independent variables, one could write

$$
\begin{array}{lll}
e_i = e_i(e_o, f_o), & e_o = e_o(e_i, f_i), & f_o = f_o(e_i, e_o) \\
f_i = f_i(e_o, f_o), & f_o = f_o(e_i, f_i), & f_i = f_i(e_i, e_o)
\end{array}
\tag{7.17}
$$

$$
\begin{array}{lll}
e_o = e_o(e_i, f_o), & f_o = f_o(e_o, f_i), & e_o = e_o(f_i, f_o) \\
f_i = f_i(e_i, f_o), & e_i = e_i(e_o, f_i), & e_i = e_i(f_i, f_o)
\end{array}
$$

The upper left pair of equations is of course our four-pole model, but all the other equations are also potentially useful. As an example of their utility, suppose we have an "unloaded" ($f_{o1} \equiv 0$) transfer function relating e_{o1} to e_{i1} for device 1 in Fig. 7.1 and we wish to determine the effect on the e_{o1}/e_{i1} relation of connecting device 2. For device 1 we can write

$$
e_{o1}(s) = W_u(s)e_{i1}(s) + Z_{go1}(s)f_{o1}(s)
\tag{7.18}
$$

$$
\left.\frac{e_{o1}}{e_{i1}}(s)\right|_{f_{o1}=0} \triangleq W_u(s) \triangleq \text{unloaded transfer function}
\tag{7.19}
$$

$$
\left.\frac{e_{o1}}{f_{o1}}(s)\right|_{e_{i1}=0} \triangleq Z_{go1}(s) \triangleq \text{generalized output impedance}
\tag{7.20}
$$

Note that (7.18) is of exactly the same general form as (7.1) or (7.2) except that we have given the parameters names and explicitly shown the dynamic nature of the relation by the use of the Laplace variable s. Also, W_u and Z_{go} can be found theoretically or by experiment, just as the four-pole a's. For device 2 we use the equation

$$
e_{i2}(s) = Z_{gi2}(s)f_{i2}(s) + Z_{gt2}(s)f_{o2}(s)
\tag{7.21}
$$

$$
\left.\frac{e_{i2}}{f_{i2}}(s)\right|_{f_{o2}=0} \triangleq Z_{gi2}(s) \triangleq \text{generalized input impedance}
\tag{7.22}
$$

$$
\left.\frac{e_{i2}}{f_{o2}}(s)\right|_{f_{i2}=0} \triangleq Z_{gt2}(s) \triangleq \text{generalized transfer impedance}
\tag{7.23}
$$

If device 2 has no third device connected at its output and, say, $f_{o2} \equiv 0$, since $e_{o1} \equiv e_{i2}$ and $f_{o1} \equiv -f_{i2}$ when the two devices are connected, we can combine our equations to get

$$W_l(s) \triangleq \text{loaded transfer function} \triangleq \frac{e_{o1}}{e_{i1}}(s)$$

$$= \left[\frac{1}{\dfrac{Z_{go1}(s)}{Z_{gi2}(s)} + 1} \right] W_u(s) \qquad (7.24)$$

$$\underbrace{\phantom{\left[\frac{1}{\dfrac{Z_{go1}(s)}{Z_{gi2}(s)} + 1} \right]}}_{\text{loading effect}}$$

While this equation applies directly to the coupling of only two devices, it is of considerable practical utility since it clearly shows under what conditions we may couple subsystems accurately using the familiar transfer function (here called *unloaded* transfer function) method, and what additional subsystem information (the two impedances) is needed to get accurate coupling when loading is *not* negligible. Note that just as in the four-pole method, *three* pieces of information are needed for any subsystem to allow its accurate coupling within a cascaded system. Furthermore, if $Z_{gi2} \gg Z_{go1}$, then the unloaded transfer function is a good approximation to the loaded. When we employ the sinusoidal version of (7.24), the degree of approximation can actually be expressed numerically, and it may be that the approximation is good over certain ranges of frequency but not others.

Equation (7.24) is just one example of many useful results that may be obtained by interpreting Equations (7.17) using impedances and transfer functions in a fashion similar to (7.18). For instance, if subsystem 2 of Fig. 7.1 actually had e_{o2} (rather than f_{o2}) equal to zero, corresponding in a mechanical system to a freely moving (rather than clamped) output point, we need only select the equation

$$e_{i2}(s) = Z_{gi2}^*(s)f_{i2}(s) + W_u^*(s)e_{o2}(s) \qquad (7.25)$$

$$\left. \frac{e_{i2}}{f_{i2}}(s) \right|_{e_{o2}=0} \triangleq Z_{gi2}^*(s) \triangleq \text{generalized input impedance} \qquad (7.26)$$

$$\left. \frac{e_{i2}}{e_{o2}}(s) \right|_{f_{i2}=0} \triangleq W_u^*(s) \triangleq \text{unloaded transfer function} \qquad (7.27)$$

Since $e_{o2} \equiv 0$, $W_u^*(s)$ plays no role in this example and, proceeding exactly as for Eq. (7.24), we arrive at the same result, except that Z_{gi2}^* replaces Z_{gi2}. These two impedances are of course different, and we could give them identifying names such as "blocked-output input impedance" and "free-output input impedance." However, this is not really necessary so long as we calculate or measure them according to their defining equations. Also note that the a's of the four-pole model are relatable to the W's and Z's of the various impedance models (compare Equations (7.18) and (7.1) to see that $W_u = 1/a_{11}$, for example).

Let us conclude this section by describing a class of applications of considerable practical significance in mechanical systems design. In large equipment involving diverse technologies, major subsystems may be manufactured by different contractors at remote locations, with subsystems being brought together at a single final assembly point only after each has been individually completed. For example, aircraft frames and aircraft engines are produced by companies thousands of miles apart. While a *theoretical* vibration analysis of the complete airframe/engine system can be performed at either or both manufacturing locations at any time after drawings are completed, *experimental* studies of the complete system cannot be performed until final assembly. Discovery of design faults at this late stage can cause severe economic and scheduling problems. Thus a capability for experimental testing of each subsystem at the respective manufacturer's facility and a proper coupling of these results to predict behavior of the assembled system can be a valuable tool.

As an example of such a situation consider the ship hull structure of Fig. 7.5 wherein location A is the attachment point of some rotating machinery that produces vibration-exciting dynamic forces f_{i1} and location B is the attachment point of an electronics package that can withstand only limited vibration. It is desired to determine the force f_{i2} that will be applied to the electronics package when it has been connected. However, it is preferable to run separate vibration

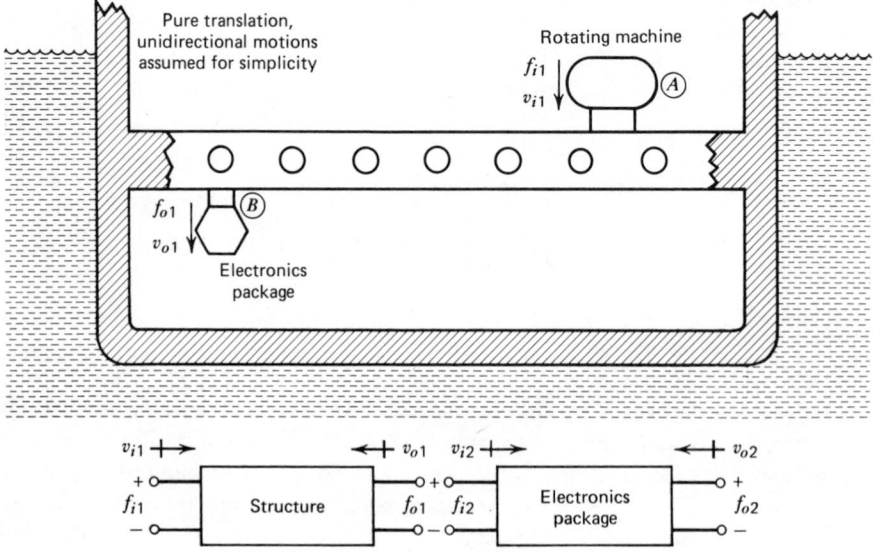

Figure 7.5 Impedance methods applied to structural dynamics coupling problem.

tests (one at the shipyard, the other at the electronics plant) on each subsystem and then *calculate* from these measurements what the force will be.

Let us assume for simplicity that the frequency spectrum of the input force f_{i1} is known from theory or experiment. Our problem then becomes one of finding $(f_{i2}/f_{i1})(s)$ for the "loaded" condition. This application fits exactly the conditions of Eq. (7.24) since $f_{o1} \equiv f_{i2}$ when the subsystems are joined. It appears then that we must carry out three measurements:

$$\left. \frac{f_{o1}}{f_{i1}}(i\omega) \right|_{v_{o1}=0} \qquad \left. \frac{f_{o1}}{v_{o1}}(i\omega) \right|_{f_{i1}=0} \qquad \left. \frac{f_{i2}}{v_{i2}}(i\omega) \right|_{f_{o2}=0} \tag{7.28}$$

In the block diagram of Fig. 7.5 the interpretation of the "output" variables v_{o2}, f_{o2} may not be clear. The electronics package actually contains many individual components, each of which will have its own vibratory motion that we might select to call v_{o2}. Whichever we might choose, the external force f_{o2} on this component is by definition zero (the parts are allowed to freely vibrate), thus our selection of the definition of input impedance as in (7.28) (rather than the one requiring $v_{o2} \equiv 0$) is required simply to match the actual physical situation. Also, definition of a *specific* electronic component's motion as v_{o2} is unnecessary since we are not asked to calculate this motion. Finally, the block labeled "structure" in Fig. 7.5 includes the attached machine in all our measurements. However, the machine is *not* rotating and producing the exciting forces since we provide a sine, pulse, or random test force from a vibration shaker.

Figure 7.6 shows the three test setups required to gather the needed data. We cannot here go into practical instrumentation details; however, certain common practices and the existence of potential problem areas should at least be mentioned. In Fig. 7.6a the f_{o1} load cell and "wall" cannot be perfectly rigid, violating $v_{o1} \equiv 0$. Should these flexibilities be sufficient to cause significant error, calculated corrections may be possible if the flexibilities are known. The vibration shaker should apply its force at the same point and in the same manner (in so far as possible) as the force produced by the rotating machine in actual use, while the f_{i1} load cell should faithfully measure this force. In Figs. 7.6b and 7.6c where velocity measurements are indicated, accelerometers will often be used, their signals being integrated to obtain velocity. When impedance data is used only in *ratios*, such as in Eq. (7.24), this integration is not really necessary since ratios of velocities and accelerations would be the same. Assuming all the measurements can be made with sufficient accuracy, we can calculate our desired results one frequency at a time from

$$\frac{f_{o1}}{f_{i1}}(i\omega) = \left. \frac{f_{o1}}{f_{i1}}(i\omega) \right|_{v_{o1}=0} \left[\frac{1}{\dfrac{Z_{go1}(i\omega)}{Z_{gi2}(i\omega)} + 1} \right] \tag{7.29}$$

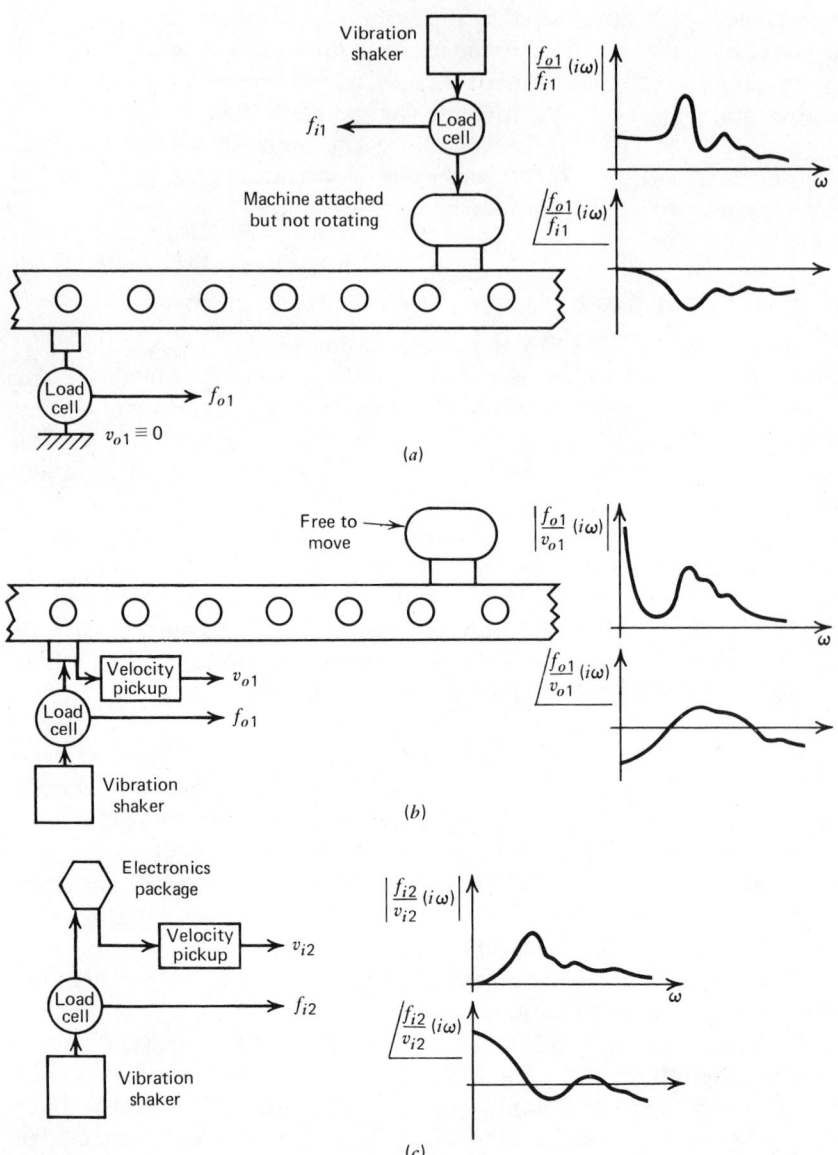

Figure 7.6 Experimental measurements needed to predict coupled-system response.

and display them as the usual amplitude ratio and phase curves versus frequency. Given the spectrum of f_{i1} produced by the rotating machine, we can then compute that of f_{o1} applied to the electronics package.

7.5   ADVANCED COUPLING METHODS

The intended scope of this text prevents us from extending the simple concepts introduced above to their ultimate generality; however, many results of this sort are available in the literature. In fact the main problem is often one of *selecting* the best method for a particular application from the many available. An area on which particular attention has recently been focused is that of structural dynamics, the field chosen for the examples of this chapter. Let us briefly discuss a few references from this field to indicate the nature of recent work. The space shuttle program employs a reuseable orbiter vehicle that will be coupled with many different payloads to accomplish the various intended missions. Vibration test programs could be considerably simplified if the subsystem coupling technique of testing the orbiter and various payloads separately could be shown to be valid. A preliminary study[8] ran sinusoidal tests on simplified mechanical models of orbiter and various payloads to compare the accuracy of several coupling methods. Figures 7.7 and 7.8 show the configurations of these models. The orbiter and payload are to be attached at points 1, 2, 3, and 4, point 5 is the desired output location for the payload, and points 6, 7, 8, and 9 are possible driving input locations.

Note that this application requires extension of our earlier results to *multiport* configurations (the subsystems interface at more than one point) and also to *multi-directional* motions (both horizontal and vertical forces and motions are present). These extensions are accomplished by use of appropriate matrix formulations. Even these relatively simple mechanical models require extensive test programs. The payload was excited, in turn, at f_1, f_2, f_3, f_4, and f_5 with acceleration measurements taken at a_1, a_2, a_3, a_4, and a_5 in each case (total of 25 frequency response functions). For the orbiter, forces 1,2,3,4,6,7,8,9 and the same accelerations (64 total functions) were measured. For the combined system, force was applied at f_8 and accelerations 5, 6, 7, 8, and 9 were measured.

When the subsystem data were coupled using a method called "transmission matrix," satisfactory results were not obtained. Another technique called "admittance matrix" (admittance is essentially the reciprocal of impedance) led to good results, as did a combination of impedance and admittance methods which gave the results of Fig. 7.9. Admittance techniques generally utilize free rather than clamped boundary points, which can be an advantage. However, in

[8]Kana, D. D. and L. M. Vargas, Prediction of Payload Vibration Environments by Mechanical Admittance Test Techniques, NASA CR-2591, 1975.

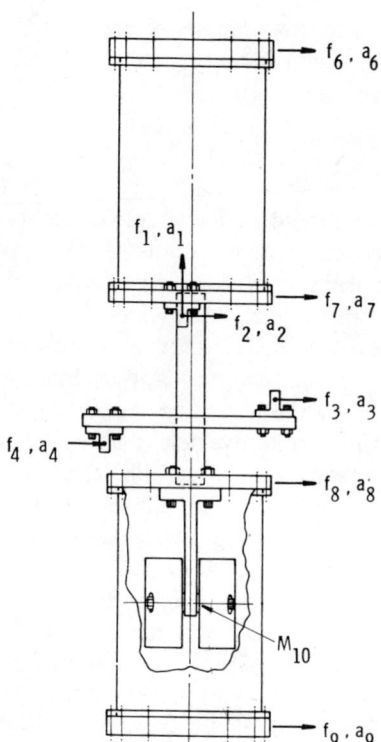

Figure 7.7 Orbiter model.

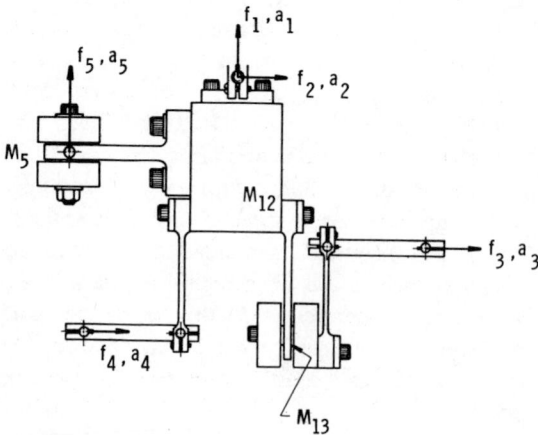

Figure 7.8 Flexible payload
model.

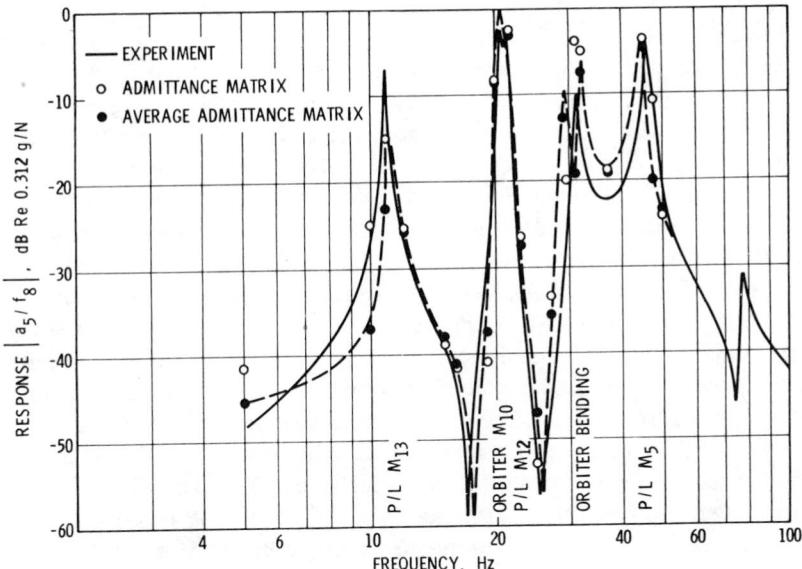

Figure 7.9 Comparison of test results with subsystem coupling calculations.

this case, the blocked impedance measurements actually gave slightly better accuracy. It was also noted that a faster and more automated test method was needed to make the method practical. An extension[9] of this study used transient excitation and Fourier transform methods. Various input transients were tried; the form finally used was a rapidly swept sine of increasing amplitude given by

$$f(t) = \left(F_0 + F_2 t^2 \right) \sin\left[\left(\omega_0 + A_2 t^2 \right) t \right] \tag{7.30}$$

where $0 \leqslant t \leqslant 4$ sec and frequency went from 5 to 100 Hz. While some questions were still unresolved, the report concludes that the transient method is not only much faster (seconds versus hours) but also, at least for these tests, gave better correlation with theory than did the sinusoidal methods. To achieve these results, *averaging* of about five repeated runs of each transient test was found necessary.

Motivated by difficult dynamic problems involved in coupling helicopter subsystems (airframe, engine, rotor, ordnance, etc.), a very comprehensive analytical study[10] of these questions was carried out. Methods were developed for handling quite general situations, including multiple attachment points and

[9]Kana, D. D. and L. M. Vargas, NASA CR-2787, 1977.
[10]Flannelly, W. G., A. Berman, and N. Giansante, Research on Structural Dynamic Testing by Impedance Methods, USAAMRDL Report 72-63A, Vols. I–IV, Kaman Aerospace Corp., Bloomfield, Conn., 1972.

up to six degrees of freedom (three translations, three rotations) at any point in the model. Techniques for dealing with tests using free boundary conditions or arbitrarily constrained conditions such as flexible (rather than perfectly rigid) mounting points were also detailed. Many example numerical cases, using "manufactured" data contaminated with random and bias errors to simulate real measurement data were run to test the "robustness" of the methods. Very favorable results were obtained and the methods were considered to show much promise for actual experimental application. However, no experimental work was carried out in the above reference.

A later reference[11] from the helicopter field describes very extensive experimental sinusoidal impedance tests on helicopter airframes and engines to study coupling methods. Completely analytical coupling methods based on mobility and modal synthesis (NASTRAN) were also compared. These two methods both gave good results and no definite preference was stated; however, the mobility method is anticipated to save computer time. The results of the coupling calculations based on *experimental* engine, airframe, and complete system measurements were however very disappointing. (It should be noted that this system is quite complex, engine measurements[12] being taken at 16 excitation and 20 response points, including both two-dimensional and three-dimensional translations (but no rotations) at various locations.) The report concludes that the test program was overly ambitious for the state of the art and suggests that a program of simpler experiments be undertaken with the hope of gradually expanding system complexity as experience is gained.

Our final references [13] describe methods that overcome some of the difficulties encountered in using measured dynamic characteristics for subsystem coupling purposes. The second item in footnote 13, page 138, indicates that past difficulties with the use of measured data may have been due not so much to inaccuracies in the data as measured, but rather to overlooking the need to maintain accuracy in certain *relations* among parameters. An example is shown where 2% errors in basic data are quite tolerable when the necessary relations are enforced precisely, but these same errors cause very large errors when the relationships are ignored. The necessary relationships are maintained by fitting (using improved Levy or other methods) the measured frequency response functions with analytical forms. By basing all further calculations on this

[11]White, J. A., OH-58A Propulsion System Vibration Investigation, USAAMRDL-TR-74-47, Bell Helicopter Co., 1974.

[12]Parker, W. H., T63 Engine Vibratory Characteristics Analysis, USAAMRDL-TR-74-87, Detroit Diesel Allison, 1974.

[13]Klosterman, A. L. (Structural Dynamics Res. Corp., Cincinnati), A Combined Experimental and Analytical Procedure for Improving Automotive System Dynamics, SAE Paper 720093, 1972; Klosterman, A. L., Ph.D. Dissertation, Dept. of Mech. Eng., Univ. of Cin., 1971; McClelland, W. A., and A. L. Klosterman, Using NASTRAN for Dynamic Analysis of Vehicle Systems, SAE Paper 740326, 1974.

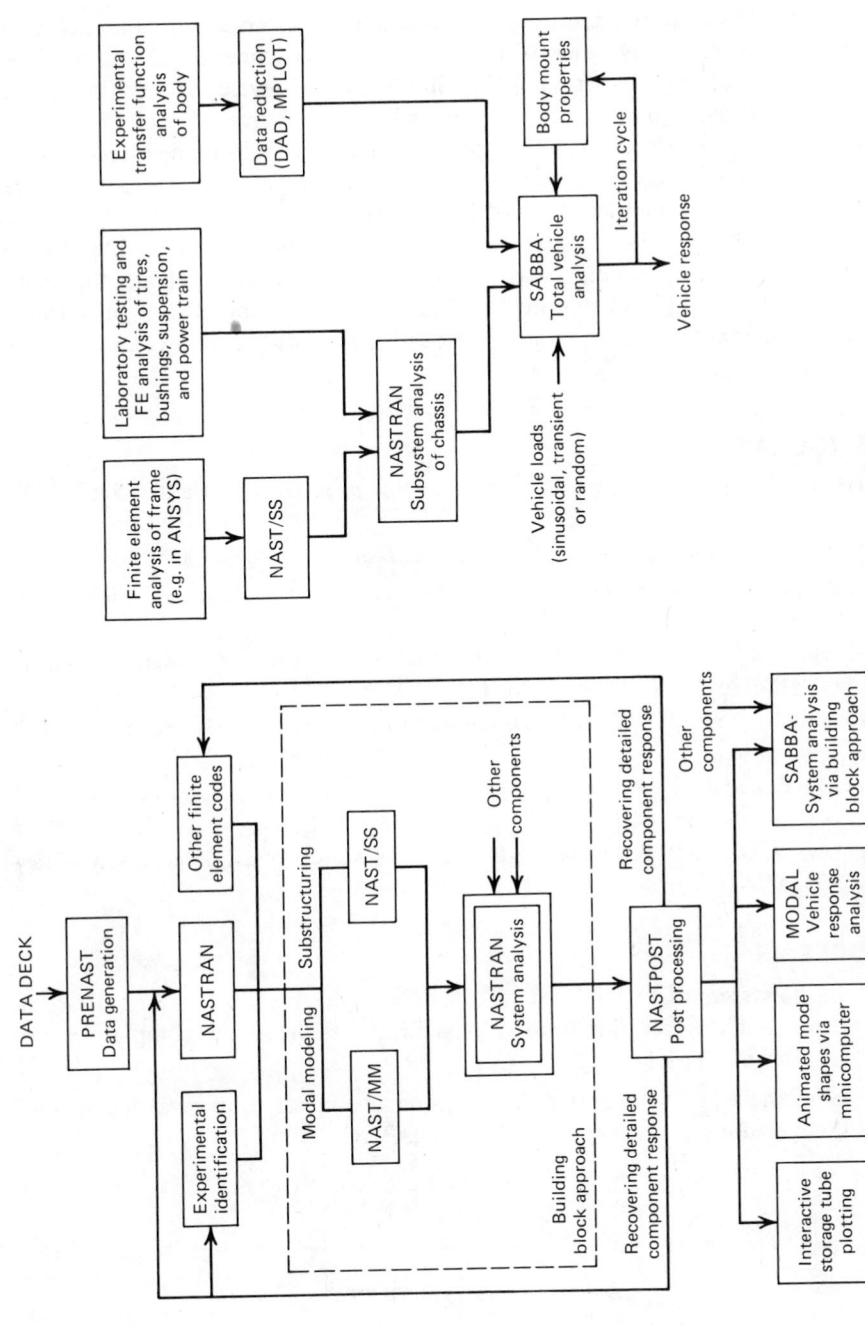

Figure 7.10 Organization of modeling approach for structural dynamics problems.

341

analytical representation, the necessary internal consistency is maintained and the results are relatively insensitive to the usual experimental errors. Various other techniques such as "residual flexibility" were developed to allow more accurate representation of complex systems by a relatively small number of modes. This emphasis on obtaining the *simplest* workable models results in a greater chance of success when coupling several subsystems. The methods are set up (footnote 13, McClelland and Klosterman) so that it is relatively easy to combine analytical finite element models, for the subsystems amenable to such treatment, with the models based on experimental measurements for those subsystems that require this approach. The necessary computer programs are all linked together with interactive graphics display devices to give a comprehensive design analysis tool (see Fig. 7.10).

BIBLIOGRAPHY

1 Hatter, D. J., *Matrix Computer Methods of Vibration Analysis*, Halsted Press, N.Y., 1973.

2 Hopkins, R. B., *Design Analysis of Shafts and Beams*, McGraw-Hill, N.Y., 1970.

3 Karnopp, D. and R. Rosenberg, *System Dynamics: A Unified Approach*, John Wiley and Sons, N.Y., 1975.

4 Klosterman, A. L., A Combined Experimental and Analytical Procedure for Improving Automotive System Dynamics, SAE Paper 720093, 1972.

5 Koenig, H. E. and W. A. Blackwell, *Electromechanical Systems Theory*, McGraw-Hill, N.Y., 1961.

6 Pestel, E. and F. Leckie, *Matrix Methods in Elastomechanics*, McGraw-Hill, N.Y., 1963.

7 Pilkey, W. D. and P. Y. Chang, *Modern Formulas for Statics and Dynamics*, McGraw-Hill, N.Y., 1978.

PROBLEMS

7.1 In the system of Fig. 7.3, attach a third subsystem as in Fig. P7.1. Find the transfer matrix for this third subsystem and then get the complete system transfer matrix relating $f_{i1}, v_{i1}, f_{o3}, v_{o3}$.

7.2 Find the transfer matrix of a system made up of four cascaded spring/ damper units of the type K_{s3}, B_3 in Fig. P7.1.

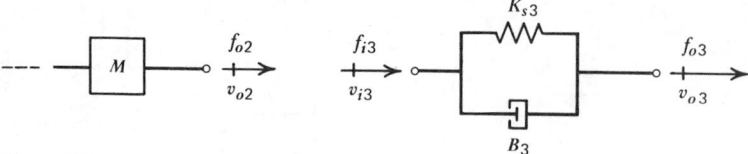

Figure P7.1

7.3 In problems of form Eq. (7.9), of the four quantities $e_{i1}, f_{i1}, e_{on}, f_{on}$, two will be known and two will be unknown. If we are interested in calculating frequency response for known numerical values, the matrix multiplication $[M_1][M_2]...[M_n]$ need not be done "manually" (as in Eq. (7.10)), but can be carried out numerically by a computer for $s = i\omega$. Once the elements of $[M_s]$ are known as numbers (at a given ω), we can solve for the two unknown variables as soon as the two knowns are given. As an example, for the system of Fig. 7.3, let $f_{o2} \equiv 0.0$ and let f_{i1} be a sinusoidal force of unit amplitude and frequency ω. Using SPEAKEASY's matrix multiplication capability to compute $[M_s]$ from $[M_1][M_2]...[M_n]$ in Eq. (7.9), write a SPEAKEASY program to compute v_{i1} and v_{o2} over a selected range of frequencies.

7.4 For the systems of Fig. P7.2:

(a) Find the transfer function $(e_{if}/e_i)(s)$ when the two circuits are not connected.

(b) Find $(e_{if}/e_i)(s)$ when the circuits are connected, using impedance coupling methods of Section 7.4.

(c) Repeat part b using "ordinary" circuit analysis of the complete, joined system.

(d) From results of part b, find $(e_o/e_i)(s)$.

(e) Find $(e_o/e_i)(s)$ from "ordinary" circuit analysis of the complete system.

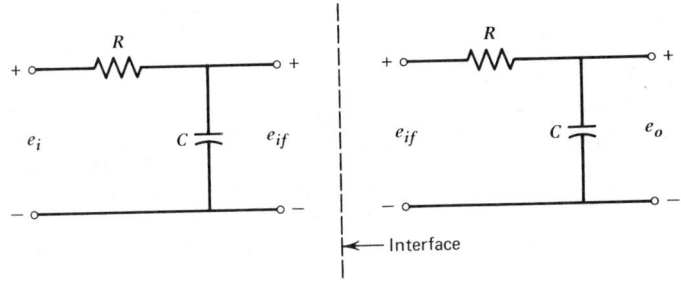

Figure P7.2

7.5 For the text example of Fig. 7.5, some of the necessary experimental measurements require certain points in the system to be "clamped" (zero motion). Since this may sometimes be difficult to carry out in actual practice, an approach using "free" rather than "clamped" measurements might be desirable. To develop this method for this example, suppose it is desired to find the velocity at B caused by a known force at A. Using impedance methods, show that:

$$\frac{v_{o1}}{f_{i1}}(s)\bigg|_{\text{loaded}} = \frac{v_{o1}}{f_{i1}}(s)\bigg|_{\substack{\text{unloaded} \\ (f_{o1}=0)}} \left[\frac{1}{1 + Z_{gi}/Z_{go}}\right]$$

where

$$\frac{v_{o1}}{f_{i1}}(s)\bigg|_{\substack{\text{unloaded} \\ (f_{o1}=0)}} \triangleq \frac{1}{Z_{\text{tr}}}, \qquad Z_{\text{tr}} \triangleq \text{transfer impedance}$$

(*Hint:* Try the relations $f_{o1} = f_{o1}(e_{i1}, e_{o1})$ and $f_{i2} = f_{i2}(e_{i2}, e_{o2})$.)

7.6 In Fig. P7.3 the mass M_2 is to be rigidly fastened to M_1 at the interface.

(a) Using impedance methods, find $(f_{i2}/f_{i1})(s)$, expressing it in the form:

$$\frac{f_{i2}}{f_{i1}}(s)\bigg|_{\text{loaded}} = \left[\frac{1}{Z_{go}(s)/Z_{gi}(s)+1} \right] \left[\frac{f_{o1}}{f_{i1}}(s)\bigg|_{\text{unloaded}} \right]$$

$$= \frac{f_{o1}}{f_{i1}}(s)\bigg|_{\text{loaded}}$$

Put Z_{gi} and Z_{go} in standard form (define K's, ζ's, ω_n's).

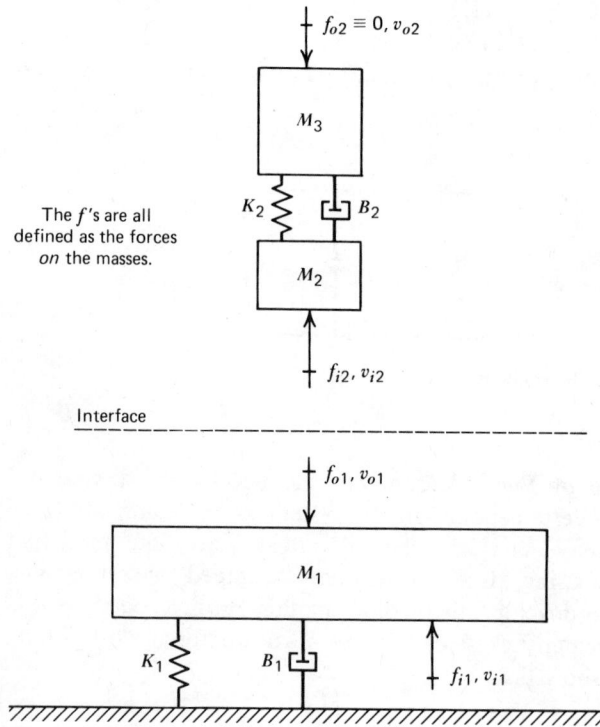

Figure P7.3

(b)   Using the results of part a, put $(f_{i2}/f_{i1})(s)$ in the form of a ratio of polynomials:

$$\frac{f_{i2}}{f_{i1}}(s) = \frac{Ks^2\left[s^2/\omega_n^2 + 2\zeta s/\omega_n + 1\right]}{As^4 + Bs^3 + Cs^2 + Ds + 1}$$

(c)   Check the results of part b by getting $(f_{i2}/f_{i1})(s)$ by conventional methods.

(d)   What does $(f_{i2}/f_{i1})(i\omega)$ approach as ω goes to zero and infinity? Show that these mathematical results agree with physical reasoning.

(e)   Take numerical values:

$$M_1 = 0.1, \qquad K_1 = 1000, \qquad B_1 = 2.0$$
$$M_2 = 0.25, \qquad K_2 = 187{,}500, \qquad B_2 = 75, \qquad M_3 = 0.75$$

You should now have:

$$\frac{f_{i2}}{f_{i1}}(s) = \frac{0.714(s^4 + 400s^3 + 1{,}000{,}000s^2)}{s^4 + 320s^3 + 788{,}000s^2 + 1{,}713{,}000s + 714{,}000{,}000}$$

Using available computer programs for frequency response, compute and graph $(f_{i2}/f_{i1})(i\omega)$ for $10 \leqslant \omega \leqslant 10{,}000$.

CHAPTER 8

Discretization of Distributed Systems

8.1 PHYSICAL LUMPING

Since real-world physical systems are, at the macroscopic level, of a spatially distributed nature, and since the partial differential equations that describe such situations are often unsolvable, we frequently need to discretize ("lump") the physical configuration to proceed with practical solutions. Such lumping can be done on a somewhat intuitive physical basis or by using more mathematical approaches of finite differences or finite element methods. When physical lumping is used, there is no need to even write the governing partial differential equations, let alone solve them. One simply divides the space into sub-regions using judgment, past experience, and available rules-of-thumb. Within a given sub-region (lump), we assume the unknowns either do not vary at all with location, or vary in a known way. The only remaining unknown variation is then with respect to time, leading to ordinary differential equations.

The torsional vibrating system of Fig. 4.1 is to be modeled in approximate discrete form using physical lumping. We need to decide how the lumps will be defined and also how many lumps are needed. When the object being discretized has uniformly distributed mass and elasticity, as in this case, it is difficult to make an argument for anything other than equal-size lumps. Should the object have obvious changes in cross-section and/or material properties over its volume, then one might argue for finer lumping in regions of rapid change. Figure 8.1 thus shows equal size lumps, three in number. We next need to define rules for locating the lumped inertia and compliance of each lump and computing their numerical values. From our background in elementary mechanics, choosing the location of the lumped inertia at the center of mass of the respective lump seems most reasonable. Once the J's are thus located, the K's are most reasonably defined as those of lengths of rod equal to the distances between the J's, while the J's themselves are the inertias of rod sections of length equal to the individual lump length. For the three-lump model shown,

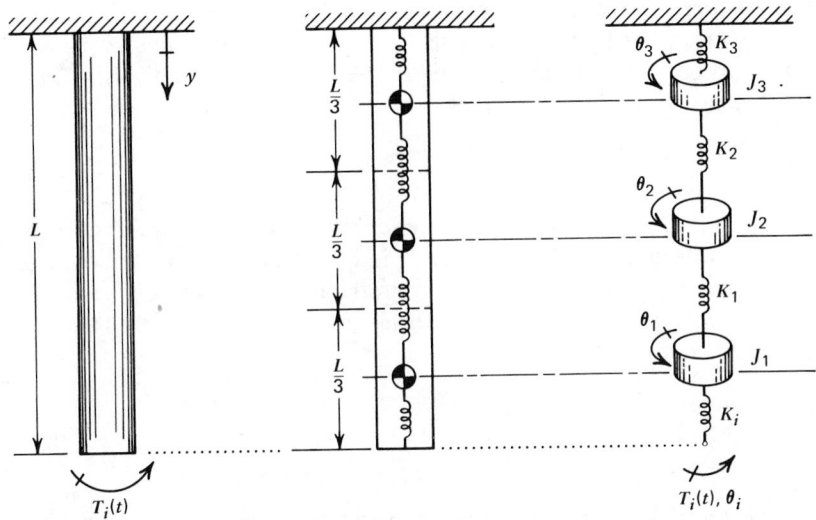

Figure 8.1 Lumped model for rod torsional vibration.

this results in

$$K_i = K_3 = 2K_1 = 2K_2 = \frac{3\pi r^4 G}{L}, \qquad J_1 = J_2 = J_3 = \frac{\pi \rho L r^4}{6} \qquad (8.1)$$

While these lumping rules seem logically defensible, they are *not* the only possible ones. Different lumping rules will give somewhat different results for small numbers of lumps, but all valid methods should agree more and more closely as the number of lumps is increased.

In choosing the *number* of lumps, rules of thumb based on general distributed-parameter results are sometimes available, whereas in other cases one must rely on past experience. Of course one can always rerun the problem with more lumps, and if results do not change greatly this is generally a good indication that the lumping is fine enough. For distributed systems governed by the wave equation, useful lumping guides are generally available. Here the spatial variation of the unknowns follows a sinusoidal function of wavelength λ, where

$$\lambda = c/f \qquad (8.2)$$

$c \triangleq$ velocity of propagation

$f \triangleq$ frequency of oscillation $\qquad (8.3)$

When we lump we replace the smooth sinusoidal spacewise variation by a set of discrete points joined by straight lines, and experience shows that if we take about 10 lumps (10 points) per wavelength the sine wave is well approximated. The velocity of propagation (recall, in fluids, the "speed of sound") is a fixed and known number, given in our example by (see Eq. (4.37))

$$c = \sqrt{G/\rho} \qquad (8.4)$$

Finally, in computing λ we must use for f the *highest* frequency of interest in the model since this will give the shortest wavelength and thus the finest lumping. Recall that a model must only be accurate up to the highest frequency present in the input driving function, in our case the torque T_i. If we can estimate this f_{max}, then the lump size l is given by

$$l = \frac{\sqrt{G/\rho}}{10 f_{max}} \qquad (8.5)$$

To compare lumped and distributed models with respect to mathematical effort involved versus results obtained, and also to indicate quantitatively the nature of the improvement achieved with finer lumping, we now carry through analysis of the system of Fig. 8.1 for several numbers of lumps. For the simplest (one-lump) model we get

$$T_i - \frac{\pi r^4 G}{L} \theta = \frac{\pi \rho L r^4}{2} \ddot{\theta} \qquad (8.6)$$

giving a single natural frequency

$$\omega_n = \frac{1.414}{L} \sqrt{\frac{G}{\rho}} \qquad (8.7)$$

which may be compared with the first natural frequency predicted by the distributed analysis (Eq. (4.29))

$$\omega_{n1} = \frac{1.571}{L} \sqrt{\frac{G}{\rho}} \qquad (8.8)$$

To get the mode shape, remove T_i and allow the system to freely vibrate at ω_n. Our one-lump model solves directly *only* for the motion at location $L/2$; however, we can infer the motions of all other locations from the known properties of the lumped elements. For the region $L/2$ to L there can be no change in dynamic deflection since this spring is massless and carries no torque. For the region 0 to $L/2$ the deflection must change *linearly* with distance since

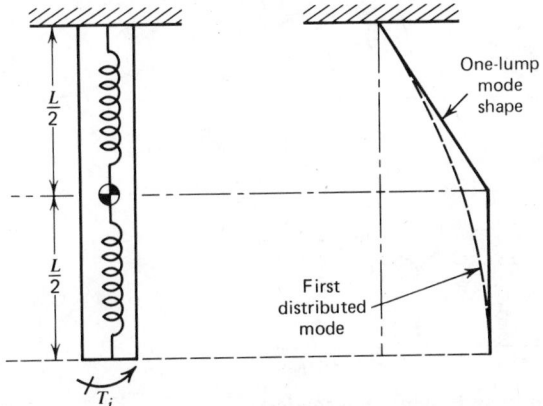

Figure 8.2 Mode shapes for one-lump and distributed models.

this spring is massless and linear. Figure 8.2 compares the one-lump mode shape with the "correct" lowest mode shape ($\frac{1}{4}$ of a sine wave) of the distributed model. We see that a one-lump model gives fair accuracy for both the natural frequency and mode shape of the first mode of vibration, but gives no clue whatsoever as to the existence of an infinite number of higher natural frequencies, a potentially catastrophic error if excitation is present at these higher frequencies.

For a two-lump model

$$T_i - \frac{\pi r^4 G}{L}(\theta_1 - \theta_2) = \frac{\pi \rho L r^4}{4}\ddot{\theta}_1 \tag{8.9}$$

$$\frac{\pi r^4 G}{L}(\theta_1 - \theta_2) - \frac{2\pi r^4 G}{L}\theta_2 = \frac{\pi \rho L r^4}{4}\ddot{\theta}_2 \tag{8.10}$$

leading to two natural frequencies

$$\omega_{n1} = \frac{1.53}{L}\sqrt{\frac{G}{\rho}}, \qquad \omega_{n2} = \frac{3.70}{L}\sqrt{\frac{G}{\rho}} \tag{8.11}$$

Note that our first mode has improved from 1.41 to 1.53 (correct value 1.57) and we now also predict a second mode at 3.70 (correct value 4.71). Substitution of

$$\theta_1 = A_1 \sin \omega t, \qquad \theta_2 = A_2 \sin \omega t \tag{8.12}$$

into either (8.9) or (8.10) ($T_i = 0$), and then letting $\omega = \omega_{n1}$ and ω_{n2} leads to

$$A_2 = 0.42A_1 \text{ at } \omega = \omega_{n1}, \qquad A_2 = -2.4A_1 \text{ at } \omega = \omega_{n2} \tag{8.13}$$

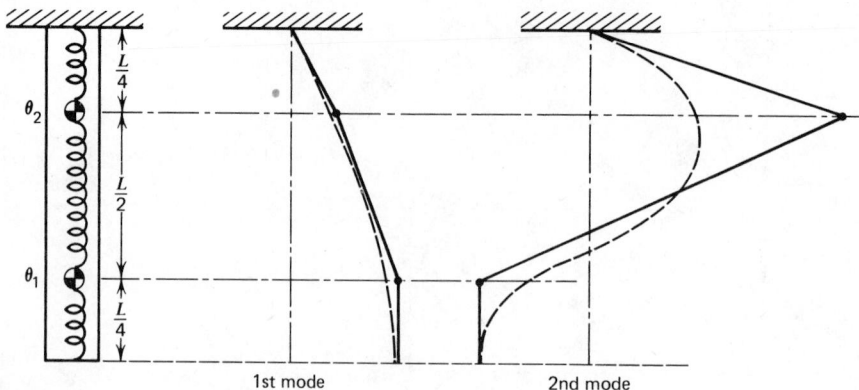

1st mode   2nd mode

Figure 8.3 Mode shapes for two-lump and distributed models.

giving the mode shapes of Fig. 8.3. Mode shape one is now quite well fitted by the three straight line segments, explaining the rather accurate value for ω_{n1}. It should now be clear that as we add more lumps, more natural frequencies are predicted, their numerical values become more accurate (the lower ones generally being the best), and since additional lumps add more discrete points to the mode shapes, these also become more correct. It can also be shown that sinusoidal transfer functions, such as $(\theta_1/T_i)(i\omega)$ or $(\theta_2/T_i)(i\omega)$ will become accurate to higher frequencies, the rule of Eq. (8.5) being a good guideline here.

Further example comparisons of lumped and distributed model results are of course possible for any partial differential equations models that are solvable. We will encounter some of these in later chapters devoted to specific physical problem areas; others may be found in references.[1]

8.2 FINITE-DIFFERENCE METHODS

When a mathematical, rather than physical method of discretization is desired, the classical approach is that of finite differences. Here one must first derive correctly the partial differential equations describing the physical situation. In our example of Section 8.1 this is Eq. (4.4):

$$\frac{\partial^2\theta}{\partial y^2} = \frac{\rho}{G}\frac{\partial^2\theta}{\partial t^2} \tag{8.14}$$

It is then necessary to replace the partial differential equations with approximate ordinary differential equations (if space is discretized but time left continuous),

[1]Doebelin, E. O., *System Dynamics*, C. E. Merrill, Columbus, Ohio, 1972, p. 458.

or approximate algebraic equations (if both space and time are discretized). To form these approximate equations one needs to use so-called "differences" and "difference quotients."

Consider Fig. 8.4 where the unknown angular displacement θ and its spacewise partial derivatives are displayed. Think of these figures as a "snapshot" at some instant of time, showing how θ (and its rates of change with respect to y) vary with y at that instant. In the finite-difference method we must work with discrete values of y and θ rather than smoothly changing ones. In most cases we consider a grid of y values separated by a constant spacing h. Since we will have information about θ only at these discrete points, we must

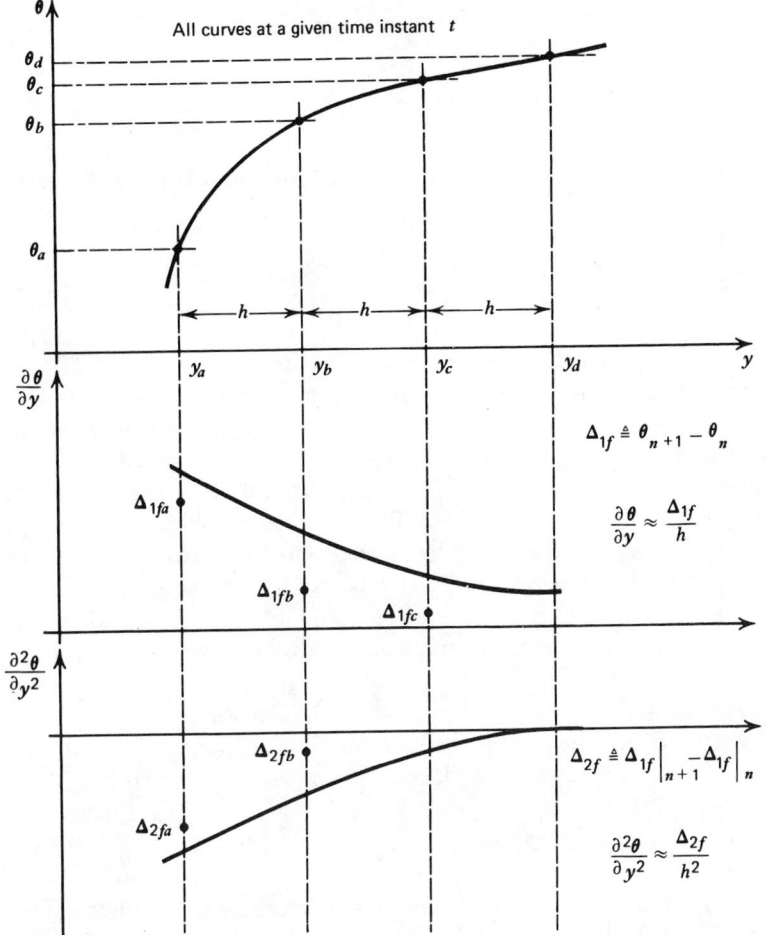

Figure 8.4 Definition of finite differences for rod torsional vibration.

develop approximations for all the derivatives that appear in our original partial differential equation in terms of these point values of θ.

In numerical analysis the process of representing a function given by a set of discrete pairs of values is called *interpolation*. This involves fitting a given form of function to the discrete points; polynomial functions are usually used unless the function is known to be periodic, then trigonometric functions may be used. If only two points such as $(y_a, \theta_a), (y_b, \theta_b)$ in Fig. 8.4 are used, we can fit only a unique first-degree polynomial (straight line) to these two points since if we assume $\theta = b_0 + b_1 y$ we can write only two equations to find b_0 and b_1:

$$\theta_a = b_0 + b_1 y_a, \qquad \theta_b = b_0 + b_1 y_b \tag{8.15}$$

$$b_0 = \frac{\theta_a y_b - \theta_b y_a}{y_b - y_a} = \frac{\theta_a y_b - \theta_b y_a}{h} \tag{8.16}$$

$$b_1 = \frac{\theta_b - \theta_a}{y_b - y_a} = \frac{\theta_b - \theta_a}{h} \tag{8.17}$$

Having found the constants b_0 and b_1 we can now differentiate $\theta \approx b_0 + b_1 y$ to get an approximation to $\partial \theta / \partial y$:

$$\frac{\partial \theta}{\partial y} = \frac{d\theta}{dy}\bigg|_{t=\text{const}} \approx b_1 = \frac{\theta_b - \theta_a}{h} \tag{8.18}$$

Note that this "slope" is exactly what we would have intuitively guessed from the geometry of Fig. 8.4 without thinking about "interpolation" at all. It is also clear that the formula $\theta \approx b_0 + b_1 y$ will give *zero* if we try to use it to approximate any *higher* derivatives than the first. By using *three* points we could go to a polynomial $\theta \approx b_0 + b_1 y + b_2 y^2$, find b_0, b_1, and b_2, and differentiate to get expressions for both $\partial \theta / \partial y$ and $\partial^2 \theta / \partial y^2$. By using more points we can get polynomials of higher degree, which can be differentiated a sufficient number of times to get approximate expressions for derivatives of any order we need in our differential equation. (This is not the *only* method of generating difference equations. One can approximate the derivatives by use of power series without explicitly employing interpolation at all.)

It is conventional to express the interpolation formulas in terms of the so-called *differences*, of various orders, of the function being approximated. In Fig. 8.4 the *first forward difference* of θ, at location n, would be given by

$$\Delta_{1f} \triangleq \theta_{n+1} - \theta_n \tag{8.19}$$

Note that if we have tabulated values of θ and y we can similarly tabulate values of Δ_{1f}; however, it takes *two* values of θ to get *one* value of Δ_{1f}. Also, when we reach the rightmost value of y we will not be able to compute a Δ_{1f} for it. From Eq. (8.18) we see that Δ_{1f} is the same as the numerator of the expression giving the approximate first derivative. By dividing the difference Δ_{1f} by the grid spacing h we define the *difference quotient*:

$$\frac{\Delta_{1f}}{h} \triangleq \text{difference quotient} = \frac{\theta_{n+1} - \theta_n}{h} \approx \frac{\partial \theta}{\partial y} \tag{8.20}$$

which is our approximation to the first partial derivative.

From Fig. 8.4 it is clear that, having formed the *first* difference values (to approximate the first partial derivative) one can easily define at location n a second forward difference by

$$\Delta_{2f} \triangleq \Delta_{1f}\big|_{n+1} - \Delta_{1f}\big|_n \tag{8.21}$$

and use this to approximate the second partial derivative. Note that a *single* value of Δ_{2f} requires *three* point values of θ, in agreement with our earlier result that a second-degree polynomial (requiring solution for three "fitting constants") was necessary to get nonzero values for the second derivative. In defining an approximation for the second derivative one need not actually fit a polynomial and then differentiate it; one can work directly from the differences and the basic definition of the second derivative as the rate of change of the first derivative:

$$\frac{\partial^2 \theta}{\partial y^2} \approx \frac{\dfrac{\partial \theta}{\partial y}\Big|_{n+1} - \dfrac{\partial \theta}{\partial y}\Big|_n}{h} \approx \frac{\dfrac{\theta_{n+2} - \theta_{n+1}}{h} - \dfrac{\theta_{n+1} - \theta_n}{h}}{h} = \frac{\Delta_{2f}}{h^2} \tag{8.22}$$

The *forward* differences used up to this point could be used at the leftmost and the interior points of a range of y, but cannot directly be used at the rightmost points since they require values to the right of the point under consideration. Actually, all our results up to this point could be expressed as well in terms of *backward* differences Δ_{1b}, for example:

$$\frac{\partial \theta}{\partial y}\Big|_n \approx \frac{\theta_n - \theta_{n-1}}{h} \triangleq \frac{\Delta_{1b}}{h} \tag{8.23}$$

These backward differences could be used at the rightmost and interior points but not directly at the leftmost points. In actual practice, better accuracy is usually achieved if *central* differences are used for all interior points, while forward and backward differences are employed only for, respectively, the leftmost and rightmost points ("boundary conditions"), where central differences cannot directly be used. (Actually, by use of fictitious points outside the boundary, one can derive methods that allow use of central differences every-where. However, these techniques are beyond the scope of this brief presentation.) A commonly used central difference formula to approximate the second partial derivative is:

$$\frac{\partial^2\theta}{\partial y^2}\bigg|_n \approx \frac{\theta_{n+1} - 2\theta_n + \theta_{n-1}}{h^2} \tag{8.24}$$

Note that this has the same form as (8.22) except that now we use points at, behind, and ahead of location n, whereas (8.22) uses only points at and ahead of n.

Let us now apply the methods briefly sketched above to the torsionally vibrating rod of Fig. 8.1. The first decision involves choice of the number of grid points to be used and where they will be located. Since Eq. (8.14) involves second derivatives we must use at least three points, and since the boundary conditions involve values at the ends, we require points at $y = 0$ and $y = L$. With these restrictions the simplest model would be as in Fig. 8.5, where we have taken $T_i \equiv 0$ to study the free vibration problem. Note that in the finite difference method, no "mechanical schematic" diagram showing specific lumped inertias and springs is defined. Rather than applying Newton's law to such a mechanical model, we go directly from the partial differential equation (which *did* use Newton's law) to the approximate ordinary differential equations by purely mathematical means. For the model of Fig. 8.5 at $y = L/2$ we can write

$$\frac{\partial^2\theta}{\partial y^2} \approx \frac{\theta_3 - 2\theta_2 + \theta_1}{(L/2)^2} \approx \frac{\rho}{G}\frac{d^2\theta_2}{dt^2} \tag{8.25}$$

Note that we *could* approximate $\partial^2\theta_2/\partial t^2$ by finite differences, resulting in

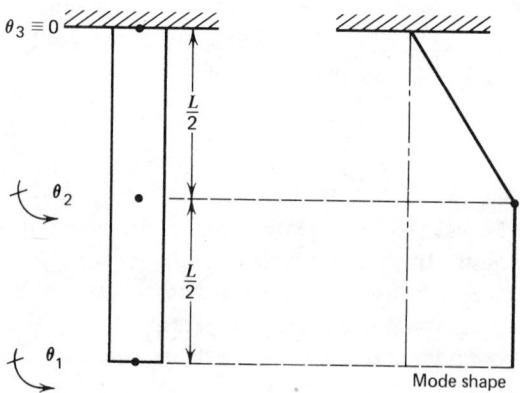

Figure 8.5   Simplest finite-difference rod model.

algebraic equations rather than the ordinary differential equation (8.25). However, if we wish to use an analog computer or digital simulation (such as CSMP) to study our system, this additional discretization of time is neither necessary nor desirable.

Formula (8.24) cannot be applied at $y = 0$ and $y = L$ since it would require points *outside* the system boundaries. However, at $y = L$ we have a known boundary condition of zero torque, and since torque is proportional to torsional strain $\partial\theta / \partial y$ we can write:

$$\left.\frac{\partial\theta}{\partial y}\right|_{y=L} = 0 \approx \frac{\theta_1 - \theta_2}{L/2} \text{ (backward difference)} \tag{8.26}$$

requiring $\theta_1 \equiv \theta_2$ and reducing (8.25) to

$$\frac{4\rho}{GL^2}\frac{d^2\theta_2}{dt^2} + \theta_2 = 0 \tag{8.27}$$

which gives the single natural frequency

$$\omega_n = \frac{2}{L}\sqrt{\frac{G}{\rho}} \tag{8.28}$$

We see that in this particular example the simplest finite difference model is somewhat less accurate than the simplest model produced by physical lumping; however, this should not be considered a general rule.

8.3   FINITE ELEMENT METHODS

Rather than representing the medium under study by an array of grid points, the finite element technique utilizes small interconnected subregions or elements (see Fig. 8.6). The elements can be of various shapes and sizes, the triangular ones of Fig. 8.6 being quite common. Even with no mathematical evaluation whatsoever, the capability of more nearly matching the boundaries of the real object with a discretization of the finite element type is apparent. Also, *within each element*, the problem unknowns (temperature, stress, flow velocity, etc.) are allowed to vary in a simple known way approximating their actual variation. This more closely represents the real situation than do the stepwise changes assumed in physical lumping and finite difference methods. Furthermore, the method formulates, as an intermediate step, solutions for the individual elements, which are then assembled to generate equations for the complete problem, allowing the efficiency of a building block approach.

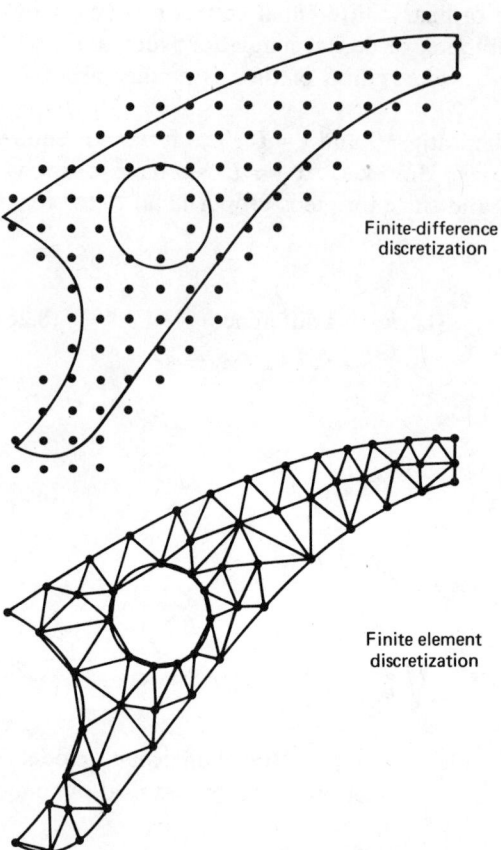

Finite-difference
discretization

Finite element
discretization

Figure 8.6   Comparison of finite-difference and finite element methods.

While there is no question that the recent development of the finite element method is the most important trend in numerical solution methods for many types of engineering problems, this does not mean that it is the preferred method for all applications. Development of a clear understanding of the basic principles of the various finite element techniques requires considerable effort, beyond our scope here, so we merely mention a few references.[2] We should also point out that while the capability of *creating* finite element computer programs requires the in-depth knowledge of a specialist, *use* of canned programs for certain broad classes of problems by nonspecialist engineers is quite common in industry. This is analogous to our use of CSMP for system dynamic simulation;

[2]Segerlind, L. J., *Applied Finite Element Analysis*, John Wiley and Sons, N.Y., 1976; Huebner, K. H., *The Finite Element Method for Engineers*, John Wiley and Sons, N.Y., 1975.

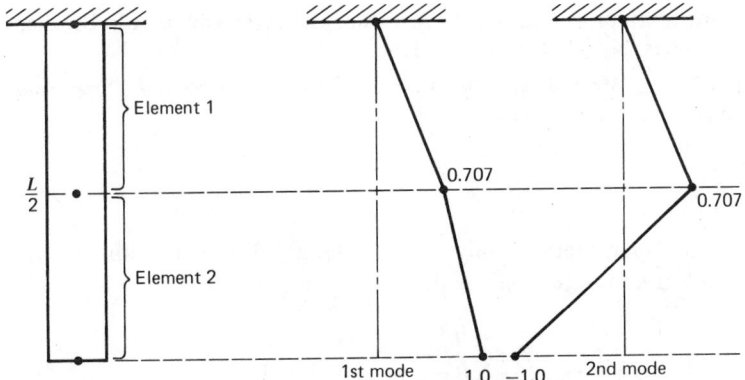

Figure 8.7   Two-element finite element model of rod vibration.

one does not have to be able to create a CSMP language in order to usefully apply it.

For the torsional vibration example used in our discussion of physical lumping and finite difference methods, a two-element model[3] gives natural frequencies of

$$\omega_1 = \frac{1.6}{L}\sqrt{\frac{G}{\rho}} \ , \qquad \omega_2 = \frac{5.6}{L}\sqrt{\frac{G}{\rho}} \tag{8.29}$$

and the mode shapes of Fig. 8.7 for comparison with our earlier results.

BIBLIOGRAPHY

1   Forsythe, G. E. and W. R. Wasow, *Finite-Difference Methods for Partial Differential Equations*, John Wiley and Sons, N.Y., 1960.

2   Firsch, H. P., A Spring-Mass Representation of a Free-Free Nonuniform Bar in Response to Longitudinal Forces, NASA TN D-2416, 1964.

3   Lin, Chi-Wen, How to Lump the Masses—A Guide to the Piping Seismic Analysis, ASME Paper 74-NE-7, 1974.

4   Rosenbrock, H. H. and C. Storey, *Computational Techniques for Chemical Engineers*, Oxford: Pergamon Press, 1966.

5   Segerlind, L. J., *Applied Finite Element Analysis*, John Wiley and Sons, N.Y., 1976.

6   Soifer, M. T. and A. W. Bell, "Reducing the Number of Mass Points in a Lumped Parameter System," *Shock Vib. Bull.*, No. 38, Part 2 (1968).

7   Tolani, S. K., and R. D. Rocke, A Strain Energy Comparison of Discrete Modeling for Vibrating Continuous Systems, ASME Paper 71-Vibr-5, 1971.

[3]Martin, H. C. and G. F. Carey, *Introduction to Finite Element Analysis*, McGraw-Hill, N.Y., 1973, p. 289.

8 Vatz, I. P., A General Mathematical Model for Beam and Plate Vibration in Bending Using Lumped Parameters, NASA TN D-3387, 1966.

9 von Rosenberg, D. V., *Methods for the Numerical Solution of Partial Differential Equations*, American Elsevier, N.Y., 1960.

PROBLEMS

8.1 Exact (distributed-parameter) analysis of the longitudinal vibrations of the rod of Fig. P8.1 gives the natural frequencies as

$$\omega = \frac{(2n+1)\pi}{2L}\sqrt{\frac{E}{\rho}} \ , \quad n = 0, 1, 2, 3, \ldots$$

Following a lumping rule analogous to that used in Section 8.1, formulate models, obtain natural frequencies, and compare with exact values for:

(a) one-lump   (b) two-lump   (c) three-lump
(d) four-lump   (e) five-lump
models.

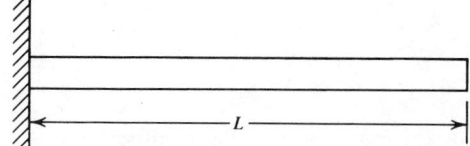

Figure P8.1

8.2 Repeat Problem 8.1 for the rod of Fig. P8.2, where

$$\omega = \frac{n\pi}{L}\sqrt{\frac{E}{\rho}} \ , \quad n = 1, 2, 3, \ldots$$

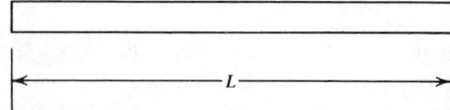

Figure P8.2

8.3 Repeat Problem 8.1 for the rod of Fig. P8.3, where

$$\omega = \frac{n\pi}{L}\sqrt{\frac{E}{\rho}} \ , \quad n = 1, 2, 3, \ldots$$

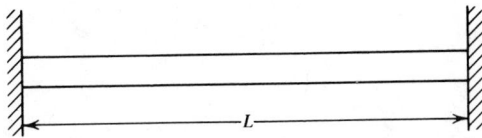

Figure P8.3

8.4   Find natural frequencies and mode shapes for the three-lump model of Fig. 8.1 and compare with the exact results.

8.5   Compare the frequency response $(\theta_i / T_i)(i\omega)$ for the three-lump model of Fig. 8.1 with $(\theta / T)(L, i\omega)$ in Eq. (4.28) for the distributed model.

PART III

Specific Applications of Modeling Techniques

CHAPTER 9
Hydraulic Conduits and Machinery

We now begin the final section of the text, which is devoted to discussion of specific practical applications of both theoretical and experimental modeling techniques. The emphasis is on theoretical methods and their experimental validation, however some purely experimental studies are included for cases where this is accepted practice or no adequate theory exists. Our level of sophistication and complexity is that of a "second course," following the introductory material typical of a physically oriented basic text such as the author's earlier work.[1] Of the many topics one might include in a selection such as that of our Part III, we choose some that:

1 Are *themselves* of considerable practical utility.
2 Are representative of *classes* of hardware of even wider interest.
3 Illustrate and illuminate a certain theoretical approach particularly well.
4 Are well accepted by practical engineers and have usually been experimentally verified.

9.1 HYDRAULIC CONDUIT DYNAMICS: DISTRIBUTED MODEL

Many natural and man-made processes involve the flow of fluids in pipes or conduits of some sort. Examples include blood flow in the human body, water flow in huge inlet pipes of hydroelectric power plants, oil flow in industrial and mobile hydraulic machinery systems, fuel and oxidizer flow in propellent feed systems of rocket engines, flow of various hydrocarbons and chemicals in refineries and chemical plants, cooling water flow in nuclear power plants, and so on. While the knowledge of steady pipe flow conveyed by the typical undergraduate fluid mechanics course is adequate for many applications, when dynamic problems arise we require models that comprehend the unsteady aspects of the behavior. Furthermore, such dynamic problems are usually *system* problems rather than being confined to the conduit itself; thus our models must be designed so as to allow easy coupling with those of other system components.

---

[1]Doebelin, E. O., *System Dynamics*, C. E. Merrill, Columbus, Ohio, 1972.

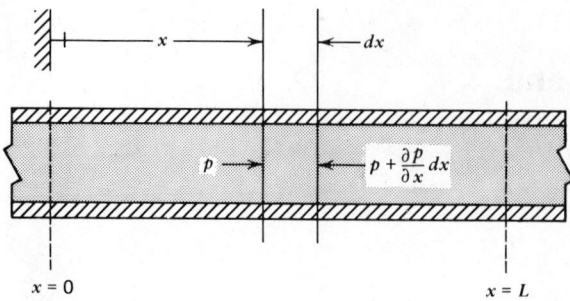

$x = 0$ $x = L$

Figure 9.1   Distributed-parameter model of hydraulic conduit.

A wide variety of such conduit dynamics models has been developed to deal with various flow regimes and application conditions. We now develop and explain the simplest practical distributed model.[2]

Consider a length of fluid conduit as in Fig. 9.1 under the assumptions:

1   one-dimensional flow;
2   no friction, work, or heat transfer;
3   no action-at-a-distance forces;
4   constant area, except for radial elastic deflection;
5   no longitudinal motion of pipe wall.

The variables of interest will be pressure $p(x,t)$ and velocity $v(x,t)$ and our goal is to develop a two-port model relating pressure and velocity at $x=0$ to pressure and velocity at $x=L$. Under our assumptions, the only force acting on the free body element dx in the longitudinal direction is that due to the pressure difference across dx:

$$f_p = \left[p - \left(p + \frac{\partial p}{\partial x} dx \right) \right] A_i = - \frac{\partial p}{\partial x} A_i dx \qquad (9.1)$$

where $A_i \triangleq$ instantaneous cross-sectional area. To apply Newton's law to this element we need the acceleration, and since velocity is a function of both x and t:

$$dv = \frac{\partial v}{\partial x} dx + \frac{\partial v}{\partial t} dt$$

$$a = \frac{dv}{dt} = \frac{\partial v}{\partial x} \frac{dx}{dt} + \frac{\partial v}{\partial t} \qquad (9.2)$$

[2]Regetz, J. D., Jr., An Experimental Determination of the Dynamic Response of a Long Hydraulic Line, NASA TN D-576, 1960.

Since dx is both the region of space we are observing and also the distance traveled by the fluid element in time dt, dx/dt is v and thus

$$a = v\frac{\partial v}{\partial x} + \frac{\partial v}{\partial t}$$ (9.3)

and Newton's law gives

$$-A_i\frac{\partial p}{\partial x}\,dx = \rho A_i\,dx\left(v\frac{\partial v}{\partial x} + \frac{\partial v}{\partial t}\right)$$ (9.4)

$$\frac{\partial p}{\partial x} + \rho\frac{\partial v}{\partial t} + \rho v\frac{\partial v}{\partial x} = 0$$ (9.5)

where $\rho \triangleq$ fluid mass density.

Since (9.5) involves two unknowns, we need another equation, which we obtain from conservation of mass for the space dx.

mass influx rate−mass efflux rate = mass storage rate

$$v\rho A_i - \left(v\rho A_i + \frac{\partial}{\partial x}(v\rho A_i)\,dx\right) = \frac{\partial}{\partial t}(\rho A_i\,dx)$$ (9.6)

$$A_i\frac{\partial \rho}{\partial t} + \rho\frac{\partial A_i}{\partial t} + \rho v\frac{\partial A_i}{\partial x} + \rho A_i\frac{\partial v}{\partial x} + vA_i\frac{\partial \rho}{\partial x} = 0$$ (9.7)

In (9.7), ρ and A_i are not really additional unknowns but can be related to p as we now show. For liquids the bulk modulus M_B is given by

$$M_B \triangleq -\frac{\Delta P}{\Delta V/V}$$ (9.8)

and since $\rho \triangleq$ mass/volume $= M/V$,

$$d\rho = \frac{V\,dM - M\,dV}{V^2} = -\frac{M\,dV}{V^2} = -\frac{M}{V}\frac{dV}{V} = \rho\frac{dp}{M_B}$$ (9.9)

In (9.9) we took $dM = 0$ since the density change $d\rho$ is accomplished by squeezing a fixed mass into a smaller volume by applying a pressure change dp. It is also appropriate at this point to define steady-state operating-point values (subscript zero) and small perturbation values (subscript p) for system variables as an aid in linearization. We assume an initial steady state with constant velocity v_0, pressure p_0, area A_0, and density ρ_0. We can then write instantaneous density ρ as

$$\rho = \rho_0(1 + p_p/M_B)$$ (9.10)

where p_p is a small dynamic perturbation away from p_0. This relates ρ to known constants ρ_0 and M_B, and unknown pressure p_p.

To similarly relate A_i and p_p we next consider the radial deflection of the pipe wall. Since this involves motion of a solid body, inertia, friction, and spring effects may be significant. If no intentional damping is added, only the internal hysteresis of the stressed metal dissipates energy, and this effect is known to be very small. The radial elastic and inertia effects will define radial modes of vibration, each behaving as a second-order undamped system. When operating frequencies are well below the natural frequency, we know that such a model is in its "flat" frequency range and is behaving essentially as a massless spring. Fortunately, formulas for the radial natural frequencies of circular rings (pipe of any length is the same, since both inertia and elasticity increase the same way with length) are available. The lowest natural frequency of a thin-wall ring is given by

$$f_n = \frac{1}{2\pi} \sqrt{\frac{E}{\rho_t r_m^2}} \quad \text{Hz} \tag{9.11}$$

where

$$E \triangleq \text{pipe material modulus of elasticity}$$

$$\rho_t \triangleq \text{pipe material mass density}$$

$$r_m \triangleq \text{mean radius of ring}$$

For a one-inch diameter stainless steel tube used in experimental tests to be discussed later, $f_n \approx 60{,}000$ Hz. Since experience with fluid piping systems shows that pressure and velocity fluctuations rarely exceed 1000 Hz, it is reasonable to model the pipe wall as a pure spring. A straightforward "hoop stress" analysis for a thin ring gives

$$A_i = \pi r_0^2 \left[1 + \frac{2r_0}{Et_w} p_p + \left(\frac{r_0}{Et_w} \right)^2 p_p^2 \right] \tag{9.12}$$

where

$$r_0 \triangleq \text{inside radius for } p = p_0$$

$$t_w \triangleq \text{wall thickness}$$

To maintain linearity we neglect the (small) term in p_p^2 to get

$$A_i = A_0 \left(1 + \frac{2r_0}{Et_w} p_p \right) \tag{9.13}$$

which can now be used to replace A_i in favor of p_p.

We now return to Eq. (9.5), which we wish to linearize and write in terms of perturbations:

$$p \triangleq p_0 + p_p, \qquad \frac{\partial p}{\partial x} = \frac{\partial p_0}{\partial x} + \frac{\partial p_p}{\partial x} = \frac{\partial p_p}{\partial x} \tag{9.14}$$

The term $\partial p_0 / \partial x$ is zero since there is no pressure gradient in a steady, frictionless, constant-area flow. The term $\rho(\partial v / \partial t)$ is linearized as the product of two variables:

$$\rho \frac{\partial v}{\partial t} \approx \rho_0 \frac{\partial v}{\partial t}\bigg|_0 + \frac{\partial v}{\partial t}\bigg|_0 p_p + \rho_0 \frac{\partial v}{\partial t}\bigg|_p = \rho_0 \frac{\partial v_p}{\partial t} \tag{9.15}$$

We have here taken $v \triangleq v_0 + v_p$ and noted that for the initial steady flow, $\partial v / \partial t$ is zero. The last term in (9.5) is similarly linearized as the product of three variables, using $\partial v / \partial x = 0$ for the initial steady flow:

$$\rho v \frac{\partial v}{\partial x} \approx \rho_0 v_0 \frac{\partial v}{\partial x}\bigg|_0 + \rho_0 v_0 \frac{\partial v_p}{\partial x} + \rho_0 \frac{\partial v}{\partial x}\bigg|_0 v_p + v_0 \frac{\partial v}{\partial x}\bigg|_0 p_p = \rho_0 v_0 \frac{\partial v_p}{\partial x} \tag{9.16}$$

Equation (9.5) now becomes

$$\frac{1}{\rho_0} \frac{\partial p_p}{\partial x} + \frac{\partial v_p}{\partial t} + v_0 \frac{\partial v_p}{\partial x} = 0 \tag{9.17}$$

Turning to Equations (9.7), (9.10) and (9.13):

$$A_0 \left(1 + \frac{2r_0}{Et_w} p_p\right) \frac{\rho_0}{M_B} \frac{\partial p_p}{\partial t} + \left(\rho_0 + \frac{\rho_0}{M_B} p_p\right) \frac{2A_0 r_0}{Et_w} \frac{\partial p_p}{\partial t}$$

$$+ \left(\rho_0 + \frac{\rho_0}{M_B} p_p\right)(v_0 + v_p)\left(\frac{2A_0 r_0}{Et_w} \frac{\partial p_p}{\partial x}\right)$$

$$+ \left(\rho_0 + \frac{\rho_0}{M_B} p_p\right)\left(A_0 + \frac{2A_0 r_0}{Et_w}\right) \frac{\partial v_p}{\partial x}$$

$$+ (v_0 + v_p)\left(A_0 + \frac{2A_0 r_0}{Et_w} p_p\right)\left(\frac{\rho_0}{M_B} \frac{\partial p_p}{\partial t}\right) = 0 \tag{9.18}$$

This could be linearized using Taylor series as was (9.5); the same result is obtained by simply neglecting products of perturbation quantities:

$$\frac{1}{K_c} \frac{\partial v_p}{\partial x} + \frac{\partial p_p}{\partial t} + v_0 \frac{\partial p_p}{\partial t} = 0 \tag{9.19}$$

where total compliance $\triangleq$ fluid compliance + pipe wall compliance,

$$K_c \triangleq \frac{1}{M_B} + \frac{2r_0}{Et_w} \qquad (9.20)$$

Equation pair (9.17), (9.19) are a complete description of the basic problem and, being linear, are amenable to a closed-form solution.[3] This solution shows that disturbances are propagated *relative* to the fluid with a velocity equal to the speed of sound, which generally exceeds 1000 m/sec. Since the average flow velocity v_0 in liquid systems is usually designed to be less than 10 m/sec to minimize friction losses, the effect of v_0 (called "convective" effect) is negligible and we can simplify our equations considerably by taking $v_0 \equiv 0$. We thus get an "exact" solution for a pipe filled with stationary liquid, and a very good approximation for a flowing liquid, by solving the equations

$$\frac{1}{\rho_0} \frac{\partial p_p}{\partial x} + \frac{\partial v_p}{\partial t} = 0 \qquad (9.21)$$

$$\frac{1}{K_c} \frac{\partial v_p}{\partial x} + \frac{\partial p_p}{\partial t} = 0 \qquad (9.22)$$

These are called the classical "waterhammer" equations.

Using the methods of Section 4.1, we Laplace transform t to s since we want sinusoidal transfer functions:

$$\frac{1}{\rho_0} \frac{\partial P_p(x,s)}{\partial x} + sV_p(x,s) - v_p(x,0) = \frac{1}{\rho_0} \frac{dP_p(x,s)}{dx} + sV_p(x,s) = 0 \quad (9.23)$$

$$\frac{1}{K_c} \frac{\partial V_p(x,s)}{\partial x} + sP_p(x,s) - p_p(x,0) = \frac{1}{K_c} \frac{dV_p(x,s)}{dx} + sP_p(x,s) = 0 \quad (9.24)$$

The initial conditions $v_p(x,0), p_p(x,0)$ are taken as zero as is usual when we derive transfer functions. Equations (9.23), (9.24) are a pair of simultaneous ordinary differential equations that could be solved by an additional Laplace transformation using a new transform variable; we prefer instead to use the classical method. Eliminating P_p we get

$$\frac{d^2V_p}{dx^2} - s^2 K_c \rho_0 V_p = 0 \qquad (9.25)$$

---

[3] Astleford, W. J., J. L. Holster and C. R. Gerlach, Analysis of Propellant Feedline Dynamics, Project 02-2889, Southwest Res. Inst., San Antonio, Tex., 1972, p. 7.

which has the solution

$$V_p = C_1 e^{s\sqrt{K_c\rho_0}\,x} + C_2 e^{-s\sqrt{K_c\rho_0}\,x} \tag{9.26}$$

Applying boundary conditions $V_p(0,s)$ and $V_p(L,s)$, respectively, at $x=0$ and L:

$$V_p(0,s) = C_1 + C_2 \tag{9.27}$$

$$V_p(L,s) = C_1 e^{s\sqrt{K_c\rho_0}\,L} + C_2 e^{-s\sqrt{K_c\rho_0}\,L} \tag{9.28}$$

$$C_1 = \frac{-V_p(0,s)e^{-s\sqrt{K_c\rho_0}\,L} + V_p(L,s)}{2\sinh\left(s\sqrt{K_c\rho_0}\,L\right)} \tag{9.29}$$

$$C_2 = \frac{V_p(0,s)e^{s\sqrt{K_c\rho_0}\,L} - V_p(L,s)}{2\sinh\left(s\sqrt{K_c\rho_0}\,L\right)} \tag{9.30}$$

Algebraic manipulation leads to

$$V_p(x,s) = \frac{1}{2\sinh(T_p s)} \left\{ V_p(L,s)\left[2\sinh\left(s\sqrt{K_c\rho_0}\,x\right)\right] \right.$$
$$\left. + V_p(0,s)\left[2\sinh\left(s\sqrt{K_c\rho_0}\,(L-x)\right)\right]\right\} \tag{9.31}$$

A similar process for P_p leads to

$$P_p(x,s) = \frac{1}{2\sinh(T_p s)} \left\{ P_p(L,s)\left[2\sinh\left(s\sqrt{K_c\rho_0}\,x\right)\right] \right.$$
$$\left. + P_p(0,s)\left[2\sinh\left(s\sqrt{K_c\rho_0}\,(L-x)\right)\right]\right\} \tag{9.32}$$

The parameter combination

$$T_p \triangleq L\sqrt{K_c\rho_0} \tag{9.33}$$

appears "naturally" in the above manipulations but also can be shown to have physical significance. The propagation velocity or speed of sound c is given by

$$c = \sqrt{1/K_c\rho_0} \tag{9.34}$$

thus the time for a pressure or velocity disturbance to travel the distance L is

$$T_p = \frac{L}{\sqrt{1/K_c\rho_0}} = L\sqrt{K_c\rho_0} \triangleq \text{propagation time}$$

If we are interested only in relating variables at $x=0$ and $x=L$ (and not in between), (9.31) and (9.32) must be further manipulated. From (9.24) and (9.31):

$$-sK_c P_p(x,s) = \frac{d}{dx} V_p(x,s)$$

$$= \frac{1}{2\sinh(T_p s)} \left\{ 2s\sqrt{K_c\rho_0}\ V_p(L,s)\cosh\left(s\sqrt{K_c\rho_0}\ x\right) \right.$$

$$\left. - 2s\sqrt{K_c\rho_0}\ V_p(0,s)\cosh\left(s\sqrt{K_c\rho_0}\ (L-x)\right) \right\} \qquad (9.35)$$

Now set $x=0$ to get

$$\left[\cosh(T_p s)\right] V_p(0,s) + \left(-\frac{1}{Z_0}\sinh(T_p s)\right) P_p(0,s) = V_p(L,s) \qquad (9.36)$$

where the parameter combination $\sqrt{\rho_0/K_c}$ is called Z_0, the *characteristic impedance*. Similar manipulations for (9.23) and (9.32) give

$$\left[-Z_0\sinh(T_p s)\right] V_p(0,s) + \left[\cosh(T_p s)\right] P_p(0,s) = P_p(L,s) \qquad (9.37)$$

Using transfer matrix notation:

$$\begin{bmatrix} V_p(L,s) \\ P_p(L,s) \end{bmatrix} = \begin{bmatrix} \cosh(T_p s) & -\frac{1}{Z_0}\sinh(T_p s) \\ -Z_0\sinh(T_p s) & \cosh(T_p s) \end{bmatrix} \begin{bmatrix} V_p(0,s) \\ P_p(0,s) \end{bmatrix} \qquad (9.38)$$

9.1.1 DETERMINATION OF TRANSFER FUNCTIONS FOR SPECIFIED INLET AND OUTLET CONDITIONS

Equations (9.38) should be thought of as extremely useful basic equations that can be combined with others (describing "devices" that might be attached at $x=0$ or $x=L$) to form models of practical fluid systems. Since (9.38) involves four variables, if we can invoke an additional relation between V_p and P_p at either $x=0$ or $x=L$, we can obtain *single* equations relating any pair of the four variables and thus define useful transfer functions. Let us illustrate this procedure with the simplest example, a sharp-edged orifice located at $x=L$ as in Fig. 9.2. The amount of fluid involved in the orifice flow process itself is so small that inertia and compressibility effects are insignificant except at extremely high frequencies, making the orifice nearly a pure fluid resistance.[4] Linearizing the familiar square root relation around the operating point V_0, P_0, we can write

$$v_p(L,t) = K_{or} p_p(L,t), \qquad V_p(L,s) = K_{or} P_p(L,s) \qquad (9.39)$$

---

[4]Funk, J. E., D. J. Wood, and S. P. Chao, The Transient Response of Orifices and Very Short Lines, ASME Paper 71-WA/FE-14, 1971.

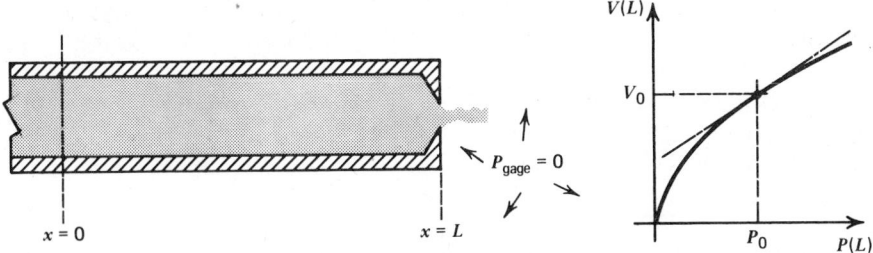

Figure 9.2 Conduit terminated in sharp-edged orifice.

where K_{or} is the local slope of the v–p curve, $v_p(L)$ refers to pipe area (not orifice area), and we take p_p as a gage pressure. In substituting (9.39) into (9.38) we could choose to retain either $V_p(L,s)$ or $P_p(L,s)$. Suppose our interest is in $V_p(L,s)$:

$$\left[\cosh(T_ps)\right]V_p(0,s)+\left(-\frac{1}{Z_0}\sinh(T_ps)\right)P_p(0,s)=V_p(L,s) \qquad (9.40)$$

$$\left[-Z_0\sinh(T_ps)\right]V_p(0,s)+\left[\cosh(T_ps)\right]P_p(0,s)=\frac{1}{K_{or}}V_p(L,s) \qquad (9.41)$$

Since (9.40) and (9.41) involve three variables, by combining them to eliminate one variable, we can get relations (transfer functions) involving various pairs of variables. For example:

$$\frac{V_p(L,s)}{V_p(0,s)}=\frac{1}{\cosh(T_ps)+\dfrac{1}{K_{or}Z_0}\sinh(T_ps)} \qquad (9.42)$$

$$\frac{V_p(L,s)}{P_p(0,s)}=\frac{K_{or}}{\cosh(T_ps)+K_{or}Z_0\sinh(T_ps)} \qquad (9.43)$$

$$\frac{P_p(L,s)}{P_p(0,s)}=\frac{1}{\cosh(T_ps)+K_{or}Z_0\sinh(T_ps)} \qquad (9.44)$$

If we desire the line "input impedance":

$$\text{impedance} \triangleq \frac{\text{effort}}{\text{flow}}=\frac{\text{pressure}}{\text{volume flow rate}}=\frac{P}{A_0V}$$

$$\frac{P_p(0,s)}{A_0V_p(0,s)}=\frac{\left[\cosh(T_ps)+K_{or}Z_0\sinh(T_ps)\right]}{A_0K_{or}\left(\cosh(T_{ps})+\dfrac{1}{K_{or}Z_0}\sinh(T_ps)\right)} \qquad (9.45)$$

There are in fact 12 transfer functions (6 are just reciprocals) that can be defined, since each of the four variables can be paired with any one of the other three.

Consider next the branched system of Fig. 9.3a. We can apply Equations (9.38) "three times" to, respectively, sections 1–2, 3–4, and 5–6 to generate six equations. At location 6 we have an orifice pressure-flow relation, while the dead end at location 4 gives us $V_4 \equiv 0$. Finally, at the junction of the three pipes we assume a common pressure and require conservation of volume. The complete equation set is thus:

$$\left[\cosh(T_{p12}s)\right]V_{p1} + \left(-\frac{1}{Z_{012}}\sinh(T_{p12}s)\right)P_{p1} = V_{p2} \tag{9.46}$$

$$\left[-Z_{012}\sinh(T_{p12}s)\right]V_{p1} + \left[\cosh(T_{p12}s)\right]P_{p1} = P_{p2} \tag{9.47}$$

$$\left[\cosh(T_{p34}s)\right]V_{p3} + \left(-\frac{1}{Z_{034}}\sinh(T_{p34}s)\right)P_{p2} = V_{p4} = 0 \tag{9.48}$$

$$\left[-Z_{034}\sinh(T_{p34}s)\right]V_{p3} + \left[\cosh(T_{p34}s)\right]P_{p2} = P_{p4} \tag{9.49}$$

$$\left[\cosh(T_{p56}s)\right]V_{p5} + \left(-\frac{1}{Z_{056}}\sinh(T_{p56}s)\right)P_{p2} = V_{p6} \tag{9.50}$$

$$\left[-Z_{056}\sinh(T_{p56}s)\right]V_{p5} + \left[\cosh(T_{p56}s)\right]P_{p2} = P_{p6} \tag{9.51}$$

$$V_{p6} = K_{or}P_{p6} \tag{9.52}$$

$$A_{012}V_{p2} = A_{034}V_{p3} + A_{056}V_{p5} \tag{9.53}$$

This set of eight equations in nine variables can be manipulated to generate transfer functions relating any selected pair of variables. If only numerical frequency response results are wanted (the usual case), this manipulation can all be avoided by applying a matrix frequency response computer program *directly* to the equation set (see Section 5.12).

Under our assumptions, the system of Fig. 9.3b would follow *exactly* the same set of equations as that of 9.3a given above. At first glance this does not seem correct (and, of course, it is not *exactly* true for the *real* systems); however, the two systems actually do behave very nearly the same under conditions often satisfied in practice. These conditions are:

1 Pipe motion effects are negligible. It is clear that the fluid forces on the two piping systems are *not* the same, and thus pipe vibrations (which influence fluid pressures and velocities) would be different. We assume no longitudinal pipe vibrations (often, but not always, a good assumption); thus this effect is suppressed.

2 Fluid oscillation frequencies are "sufficiently low." In Section 9.2 we find that at, say, 30 Hz in a typical system, the spacewise wavelength of pressure and velocity

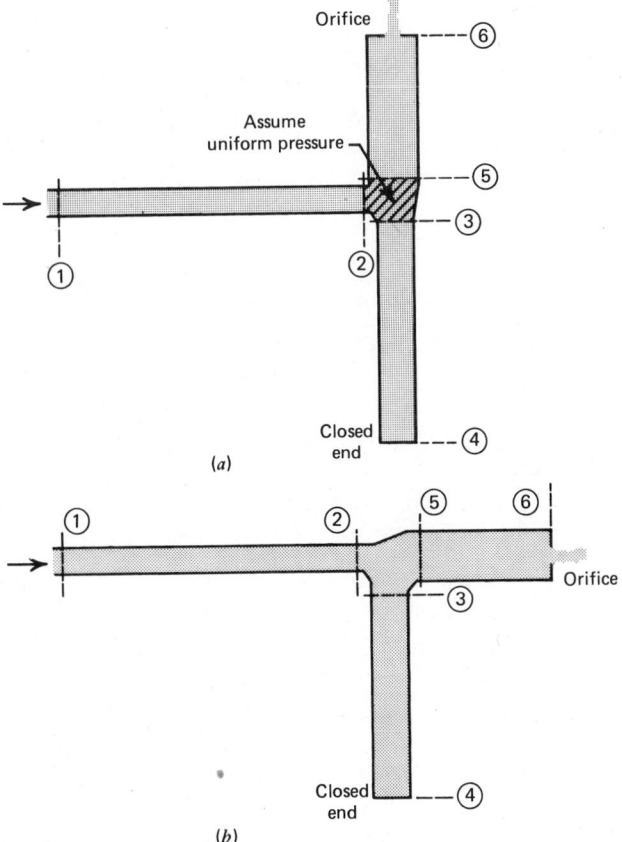

Figure 9.3   Branched piping systems.

gradients is about 40 meters. This means that "local" geometrical details, such as the exact "shape" of the three-pipe junction, cannot be "detected" by the system since the length scale of pressure and velocity changes is so much greater. At higher frequencies the wavelength gets proportionately shorter and at some high frequency the dynamic response of the two systems *will* be significantly different. Fortunately, the *predominant* dynamic effects in many systems occur at frequencies below this range.

A final example shows the combination of lumped with distributed fluid elements and coupling of fluid and mechanical components (see Fig. 9.4). We again use (9.38) for the two line sections 1–2 and 3–4 and assume conservation of volume at the junction of the lines with the accumulator, which is modeled as a lumped fluid compliance. Conservation of volume is applied at the hydraulic cylinder to relate pipe flow velocity to piston velocity (oil in cylinder assumed

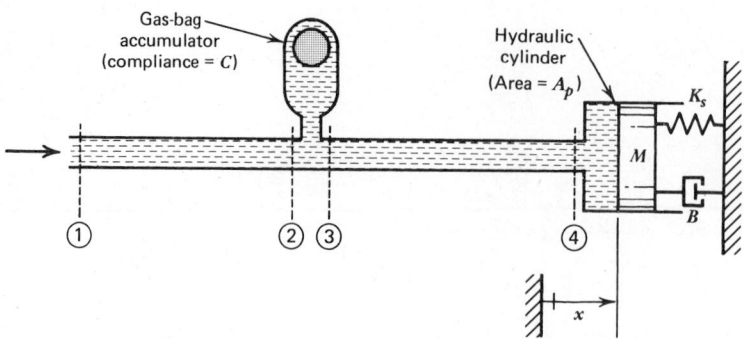

Figure 9.4 Combination of lumped and distributed models.

incompressible), while Newton's law relates pressure force to piston motion:

$$\left[\cosh(T_{p12}s)\right]V_{p1}+\left(-\frac{1}{Z_{012}}\sinh(T_{p12}s)\right)P_{p1}=V_{p2} \tag{9.54}$$

$$\left[-Z_{012}\sinh(T_{p12}s)\right]V_{p1}+\left[\cosh(T_{p12}s)\right]P_{p1}=P_{p2} \tag{9.55}$$

$$\left[\cosh(T_{p34}s)\right]V_{p3}+\left(-\frac{1}{Z_{034}}\sinh(T_{p34}s)\right)P_{p2}=V_{p4} \tag{9.56}$$

$$\left[-Z_{034}\sinh(T_{p34}s)\right]V_{p3}+\left[\cosh(T_{p34}s)\right]P_{p2}=P_{p4} \tag{9.57}$$

$$CsP_{p2}=A_0(V_{p2}-V_{p3}) \tag{9.58}$$

$$A_0V_{p4}=A_psX \tag{9.59}$$

$$P_{p4}A_p-K_sX-BsX=Ms^2X \tag{9.60}$$

9.1.2 FREQUENCY AND STEP RESPONSE CHARACTERISTICS OF AN ORIFICE-TERMINATED LINE

While we have found it rather easy to write equations for the above systems, this does not give one a physical feeling for the general nature of hydraulic line dynamics. To get this we must work out the details for specific cases, the orifice-terminated line of Fig. 9.2 being a good choice for this purpose. As we have mentioned before, it is usually not possible to inverse transform the transcendental transfer functions obtained from distributed models, so we will work mainly with frequency response, which never creates any difficulties. Let us examine

$$\frac{P_p(L,i\omega)}{P_p(0,i\omega)} = \frac{1}{\cosh(i\omega T_p) + K_{or}Z_0\sinh(i\omega T_p)} \tag{9.61}$$

$$= \frac{1}{\cos(\omega T_p) + iK_{or}Z_0\sin(\omega T_p)} \tag{9.62}$$

$$= \frac{1}{\sqrt{[\cos(\omega T_p)]^2 + [K_{or}Z_0\sin(\omega T_p)]^2}} \Big/ -\tan^{-1}[K_{or}Z_0\tan(\omega T_p)] \tag{9.63}$$

An interesting special case arises if the line is terminated with an impedance numerically equal to its characteristic impedance Z_0, that is, let $K_{or}=1/Z_0$. Then

$$\frac{P_p(L,i\omega)}{P_p(0,i\omega)} = e^{-i\omega T_p}, \text{ a pure dead time of } T_p \text{ seconds} \tag{9.64}$$

and a pressure pulse put in at $x=0$ arrives undistorted at $x=L$, but T_p seconds later.

Turning to the general case it appears that peaks and valleys will occur in the amplitude ratio. To find these let $\alpha \triangleq \omega T_p$ and $\beta \triangleq (K_{or}Z_0)^2$. Then peaks or valleys will occur for

$$\frac{d}{d\alpha}(\cos^2\alpha + \beta\sin^2\alpha) = -2\cos\alpha\sin\alpha + 2\beta\sin\alpha\cos\alpha = 0 \tag{9.65}$$

For $\beta=1$ we get the above special case. If $\beta\neq 1$, $\sin\alpha\cos\alpha=0$, giving $\alpha = 0, 90°, 180°, 270°, \dots$ and thus peaks or valleys occur at $\omega=0, \pi/2T_p$, $\pi/T_p, 3\pi/2T_p, 2\pi/T_p$, and so on. By plotting a few numerical values the shape of the curves is established as in Fig. 9.5. Note that true resonance (peaks higher than static response) occurs only for $K_{or}Z_0<1.0$, that the peak height is $1/K_{or}Z_0$, and that peaks of this height reoccur every π/T_p rad/sec on out to infinite frequency. The finite height of the resonant peaks (in a model that assumes frictionless fluid) is explained by the presence of the orifice, a fluid resistance element. If K_{or} is set to zero (closed-end pipe), then the peaks *do* go up

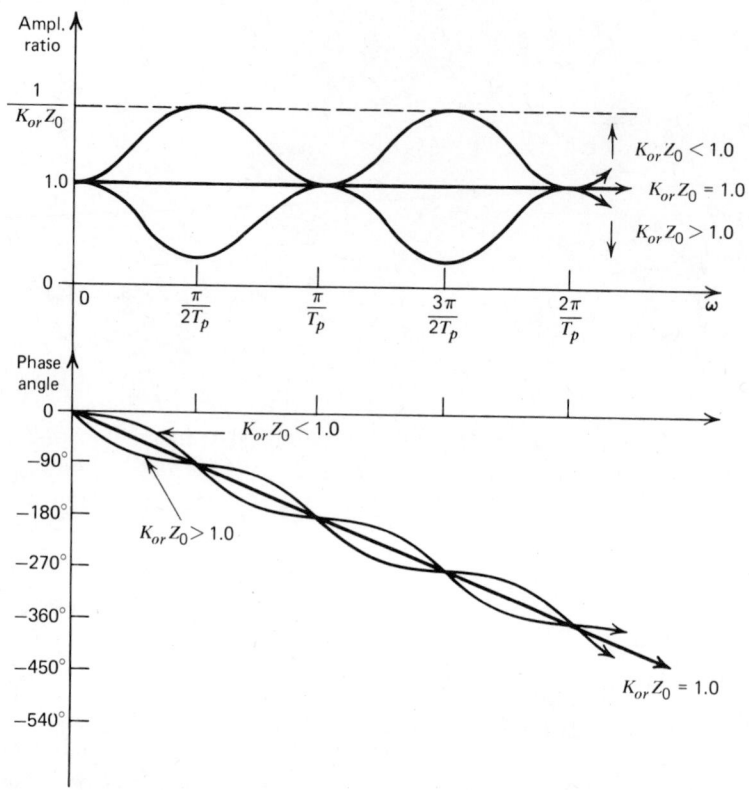

Figure 9.5   Frequency response of hydraulic line terminated with an orifice.

to infinity as one expects in an undamped system. Finally, note that for all values of $K_{or}Z_0$ the phase angle increases negatively without bound.

While the inverse transformation needed to compute step response for distributed parameter models is often not possible, in some cases a suitable trick[5] *does* work. Consider

$$\frac{P_p(0,s)}{V_p(0,s)} = \frac{\cosh(T_p s) + K_{or}Z_0 \sinh(T_p s)}{K_{or}\left[\cosh(T_p s) + \dfrac{1}{K_{or}Z_0}\sinh(T_p s)\right]} \tag{9.66}$$

[5]Oldenburger, R. and R. E. Goodson, Simplification of Hydraulic Line Dynamics by Use of Infinite Products, ASME Paper 62-WA-55, 1962.

Let

$$v_p(0,t) = v_{ps}u(t), \qquad V_p(0,s) = v_{ps}/s \tag{9.67}$$

$$P_p(0,s) = \frac{e^{T_p s} + e^{-T_p s} + K_{or}Z_0(e^{T_p s} - e^{-T_p s})}{e^{T_p s} + e^{-T_p s} + \frac{1}{K_{or}Z_0}(e^{T_p s} - e^{-T_p s})} \frac{v_{ps}}{K_{or}s} \tag{9.68}$$

$$= \frac{v_{ps}}{K_{or}s} \frac{(1 + K_{or}Z_0)e^{T_p s} + (1 - K_{or}Z_0)e^{-T_p s}}{\frac{1}{K_{or}Z_0}\left[(1 + K_{or}Z_0)e^{T_p s} - (1 - K_{or}Z_0)e^{-T_p s}\right]} \tag{9.69}$$

Divide by $e^{T_p s}$

$$P_p(0,s) = \frac{v_{ps}Z_0}{s} \frac{1 - \dfrac{K_{or}Z_0 - 1}{K_{or}Z_0 + 1}e^{-2T_p s}}{1 + \dfrac{K_{or}Z_0 - 1}{K_{or}Z_0 + 1}e^{-2T_p s}} \triangleq \frac{v_{ps}Z_0}{s} \frac{1 - be^{-2T_p s}}{1 + be^{-2T_p s}} \tag{9.70}$$

Dividing out the right-hand fraction gives the infinite sequence:

$$1 - 2be^{-2T_p s} + 2b^2 e^{-4T_p s} - 2b^3 e^{-6T_p s} + 2b^4 e^{-8T_p s} - \cdots \tag{9.71}$$

Using the delay theorem, $P_p(0,s)$ can now be inverse transformed one term at a time and it is not necessary to sum an infinite number of terms to calculate $p_p(0,t)$ out to any finite time since the later terms in the sequence are *zero* until their delayed step function occurs:

$$p_p(0,t) = v_{ps}Z_0 u(t) - 2v_{ps}Z_0 bu(t - 2T_p) + 2v_{ps}Z_0 b^2 u(t - 4T_p)$$

$$- 2v_{ps}Z_0 b^3 u(t - 6T_p) + \cdots \tag{9.72}$$

As an example, take $Z_0 = 1.0$ and $K_{or} = 2.0$ to get Fig. 9.6. If the change v_{ps} were applied slowly (steady state), we know from Eq. (9.39) that the pressure would rise to v_{ps}/K_{or}, so the fact that the oscillatory pressure in Fig. 9.6 decays to this value should not be surprising. The initial large pressure jump (twice the steady-state value) is potentially damaging and should make clear why engineers are interested in line dynamics. Finally, the meaning of T_p as a propagation time is again evident in that a disturbance $v_{ps}u(t)$ takes $2T_p$ seconds to travel from $x = 0$ to $x = L$ and back again before the pressure undergoes its next step change due to the reflected wave.

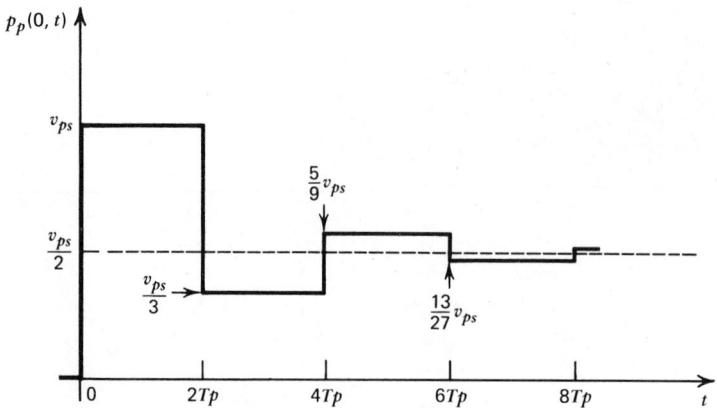

Figure 9.6   Step response of hydraulic line terminated with an orifice.

9.2   LUMPED MODELS AND EXPERIMENTAL RESULTS

Consider a section of line as in Fig. 9.7. Since we are again neglecting fluid friction we need to define only inertance and compliance for a typical lump. As in the distributed model we assume an initial steady flow and then analyze small dynamic perturbations. The mass of a lump of length l is clearly $\rho_0 A_0 l$. From Eq. (9.20) the total compliance of a lump of length l is $l A_0 K_c$ since K_c is $m^3/(m^3\text{-}pa)$. Over a time interval dt, conservation of mass for the nth lump gives

$$(\rho_0 A_0 v_{n-1} - \rho_0 A_0 v_n)\,dt = (\rho_0 A_0 l/K_c)\,dp_n \tag{9.73}$$

Laplace transforming with zero initial conditions:

$$V_{n-1} - V_n = l K_c s P_n \tag{9.74}$$

Newton's law for the nth lump gives

$$A_0 P_{n-1} - A_0 P_n = \rho_0 A_0 l \frac{dv_n}{dt} \tag{9.75}$$

$$P_{n-1} - P_n = l \rho_0 s V_n \tag{9.76}$$

Relationships among pressures and velocities of the $(n-2)$th, $(n-1)$th, and nth lumps are now obtained by algebraic manipulation of (9.74) and (9.76):

$$V_n = \frac{P_{n-1} - P_n}{l \rho_0 s}, \qquad V_{n-1} - \frac{P_{n-1} - P_n}{l \rho_0 s} = l K_c s P_n$$

$$P_{n-1} - l \rho_0 s V_{n-1} - P_n\left(1 - l^2 \rho_0 K_c s^2\right) = 0 \tag{9.77}$$

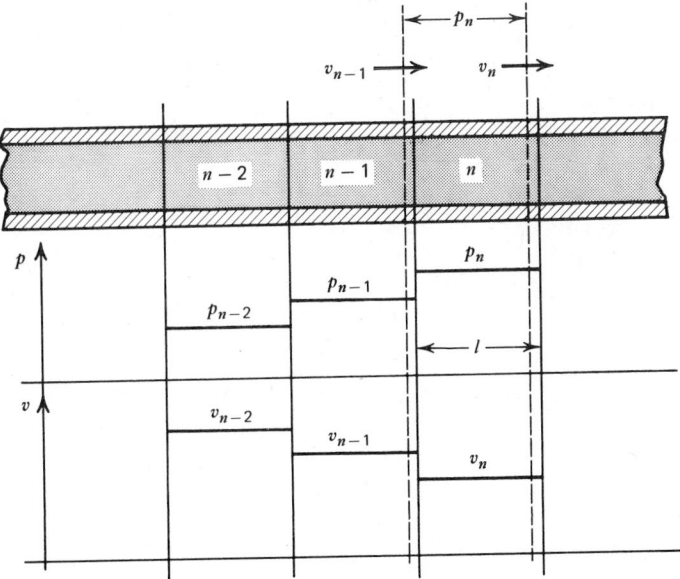

Figure 9.7 Lumped-parameter model of hydraulic conduit.

In Eq. (9.76) the subscripts can all be incremented by -1 without violating the physical law:

$$V_{n-1} = \frac{P_{n-2} - P_{n-1}}{l\rho_0 s}$$

We can now eliminate V_{n-1} from (9.77):

$$2P_{n-1} - P_{n-2} - P_n\left(1 - l^2\rho_0 K_c s^2\right) = 0 \tag{9.78}$$

Similarly,

$$P_n = \frac{V_{n-1} - V_n}{lK_c s}$$

$$P_{n-1} - \frac{V_{n-1} - V_n}{lK_c s} = l\rho_0 s V_n, \qquad P_{n-1} = \frac{V_{n-2} - V_{n-1}}{lK_c s}$$

$$2V_{n-1} - V_{n-2} - V_n\left(1 - l^2\rho_0 K_c s^2\right) = 0 \tag{9.79}$$

Equations (9.78) and (9.79) are considered the basic working relations and by letting the subscripts take on appropriate numerical values, one can generate a set of equations for a line model of any number of lumps. Separate equations giving pressure/velocity relations for any "devices" connected to the line must be combined with these to get a complete set of system equations.

We now do a numerical example to compare the distributed and lumped models, and since we choose conditions to agree with those of a published experimental test,[6] we will also check both theories against reality. The pipe was a 68-foot length of 1-inch OD ($\frac{1}{16}$ inch wall) stainless steel tube terminated in an orifice made of a plate with 34 drilled holes of 0.040 inch diameter. Kerosene (JP-4 fuel) with density 7.26×10^{-5} $lb_f\text{-}sec^2/in.^4$ and bulk modulus 173,600 psi at 77°F was the fluid. To choose an appropriate number of lumps we need an estimate of the highest frequency for which the model is to be accurate; take this arbitrarily at 30 Hz. Using the "10 lumps per wavelength" rule of Chapter 8 we compute:

$$\text{propagation velocity} = \frac{1}{\sqrt{K_c \rho_0}} = \frac{1}{\sqrt{(6.28 \times 10^{-6})(7.26 \times 10^{-5})}}$$

$$= 46800 \text{ in./sec} = 3900 \text{ ft/sec}$$

$$\text{wavelength at 30 Hz} = \frac{3900 \text{ ft/sec}}{30 \text{ cycles/sec}} = 130 \text{ ft/cycle}$$

This gives a lump length of 13 ft; so for the 68-ft line, five lumps is about right. The experimental test was run at a mean flow velocity of 62.7 in./sec and the orifice constant, obtained from the slope of the experimental steady pressure-/flow curve at this velocity, was 0.495 (in./sec)/psi.

In Fig. 9.8 the available measured data was for $(P_p/V_p)(0, i\omega)$ so we must get this frequency response from our lumped model. Letting n take on the value 5 in Equations (9.78) and (9.79):

$$2P_4 - P_3 - P_5(1 - l^2 \rho_0 K_c s^2) = 0 \tag{9.80}$$

$$2V_4 - V_3 - V_5(1 - l^2 \rho_0 K_c s^2) = 0 \tag{9.81}$$

and for $n = 4, 3,$ and 2, respectively,

$$2P_3 - P_2 - P_4(1 - l^2 \rho_0 K_c s^2) = 0 \tag{9.82}$$

$$2V_3 - V_2 - V_4(1 - l^2 \rho_0 K_c s^2) = 0 \tag{9.83}$$

$$2P_2 - P_1 - P_3(1 - l^2 \rho_0 K_c s^2) = 0 \tag{9.84}$$

$$2V_2 - V_1 - V_3(1 - l^2 \rho_0 K_c s^2) = 0 \tag{9.85}$$

$$2P_1 - P_p - P_2(1 - l^2 \rho_0 K_c s^2) = 0 \tag{9.86}$$

$$2V_1 - V_p - V_2(1 - l^2 \rho_0 K_c s^2) = 0 \tag{9.87}$$

[6]Regetz, *op. cit.*

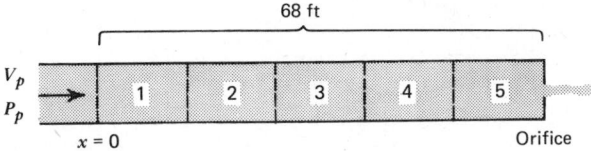

Figure 9.8 Five-lump model of pipeline.

The above eight equations were each generated by considering lumps 1, 2, 3, and 4, respectively as "sandwiched" between their adjacent lumps. Lump 5 cannot be treated this way since it is the "last" one, so we revert to Equations (9.74) and (9.75) which deal with only two adjacent lumps:

$$V_4 - V_5 = lK_c s P_5 \tag{9.88}$$

$$P_4 - P_5 = l\rho_0 s V_5 \tag{9.89}$$

Our final equation is the orifice relation:

$$V_5 = K_{or} P_5 \tag{9.90}$$

Since we are after $(P_p / V_p)(i\omega)$, V_p is presumed a given input, and we thus have 11 equations in the 11 unknowns $P_1 \to P_5$, $V_1 \to V_5$, and P_p. If only numerical frequency response were wanted, a matrix frequency response program could be applied directly to our present equation set. In fact this could have been done with a simpler set of equations obtained entirely from (9.74) and (9.76) rather than (9.78) and (9.79).

We prefer here, however, to work out a letter form of $(P_p / V_p)(s)$ so that we can see its general nature, rather than just a numerical special case. Using our present set of equations, we combine (9.90) and (9.89) to get

$$P_4 = P_5(1 + l\rho_0 K_{or} s) \tag{9.91}$$

and then substitute successively into (9.80), (9.82), (9.84), and (9.86) to get $(P_p / P_5)(s)$, and then $(P_p / V_5)(s)$ using (9.90). A similar procedure for the velocities gets us $(V_5 / V_p)(s)$, the final result being:

$$\frac{P_p}{V_p}(s) = \frac{P_p}{V_5}(s)\frac{V_5}{V_p}(s) = \frac{\left(l^5\rho_0^3 K_c^2 K_{or}\right)s^5 + \left(5l^4\rho_0^2 K_c^2\right)s^4}{\left(l^5\rho_0^2 K_c^3\right)s^5 + \left(5l^4\rho_0^2 K_c^2 K_{or}\right)s^4}$$

$$\frac{+\left(10l^3\rho_0^2 K_c K_{or}\right)s^3 + \left(10l^2\rho_0 K_c\right)s^2 + \left(5l\rho K_{or}\right)s + 1}{+\left(10l^3\rho_0 K_c^2\right)s^3 + \left(10l^2\rho_0 K_c K_{or}\right)s^2 + \left(5lK_c\right)s + K_{or}}$$

$$\tag{9.92}$$

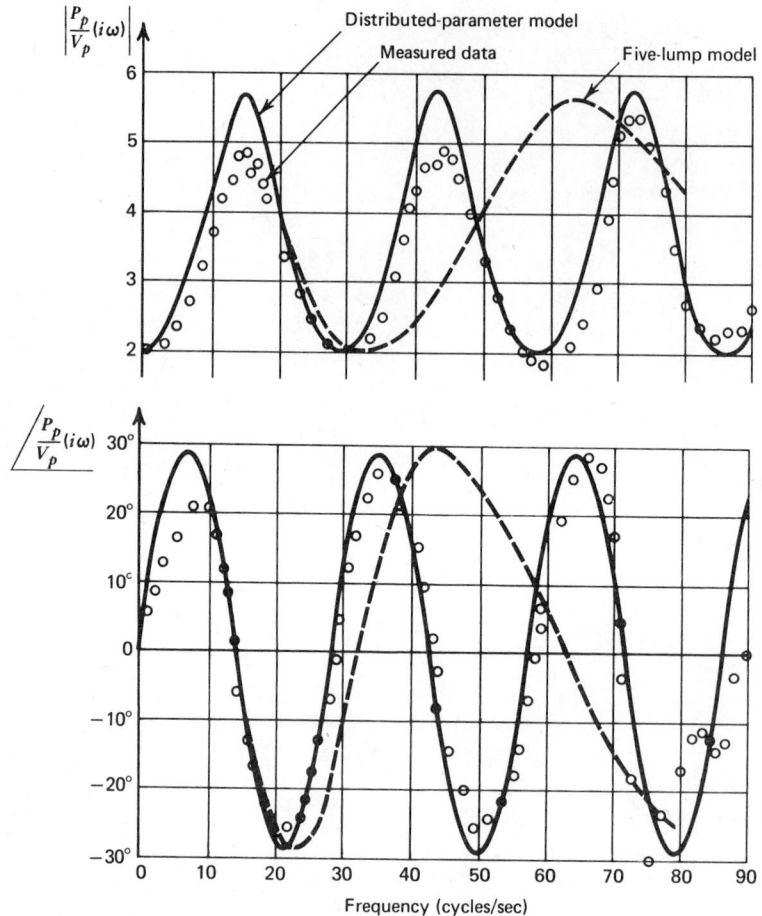

Figure 9.9 Comparison of distributed and lumped models with experimental results.

This ratio of polynomials in s is of course easily inverse transformable by routine methods but cannot exhibit the infinite number of resonant peaks present in the distributed model. Recall, however, that we only required an accurate model out to 30 Hz. Substituting numerical values:

$$\frac{P_p}{V_p}(s) = \frac{5.74(s^5 + 854s^4 + 8.28 \times 10^5 s^3 + 1.415 \times 10^8 s^2 + 3.43 \times 10^{10}s + 1.174 \times 10^{12})}{s^5 + 2421s^4 + 8.28 \times 10^5 s^3 + 4.02 \times 10^8 s^2 + 3.43 \times 10^{10}s + 3.33 \times 10^{12}}$$

$$(9.93)$$

Figure 9.9 shows superimposed graphs of the distributed model, lumped model, and experimental data[7] for this system. Sinusoidal testing using an

---

[7]Regetz, *op. cit.*

analyzer of the Fourier filter type (Section 6.1.2) was employed. We see that the lumped and distributed models agree very well up to the 30-Hz frequency limit for which the lumped model was designed. Above this the agreement gets progressively worse since the lumped model denominator, going up only to s^5, can provide only two resonant peaks. The agreement with experiment is also quite acceptable; the slightly lower resonant peaks being due mainly to fluid friction, neglected in both the distributed and lumped theories. Friction becomes more important as fluid viscosity increases and pipe diameter decreases. For our one-inch line flowing with kerosene, friction effects are not large.

9.3 DISTRIBUTED MODELS INCLUDING FRICTION

Frictional effects can be included in both distributed and lumped models in various ways. We briefly review one model[8] whose predictions agree well with measurements. By making assumptions that allow neglecting various terms, the Navier–Stokes equations are reduced to:

$$\rho \frac{\partial u}{\partial t} = -\frac{\partial p}{\partial x} + \mu \left(\frac{\partial^2 u}{\partial r^2} + \frac{1}{r} \frac{\partial u}{\partial r} \right) \tag{9.94}$$

$$\frac{1}{M_B} \frac{\partial p}{\partial t} + \frac{\partial v}{\partial r} + \frac{v}{r} + \frac{\partial u}{\partial x} = 0 \tag{9.95}$$

where

$$u \triangleq \text{axial velocity } (x \text{ direction})$$

$$v \triangleq \text{radial velocity } (r \text{ direction})$$

$$\mu \triangleq \text{fluid viscosity}$$

This model assumes laminar flow and neglects pipe wall radial elasticity compared to fluid compressibility (good assumption for small diameter pipe). Pressure is assumed uniform over any cross section (no r-variation), but velocity is allowed to vary with r. A space-averaged value of u called $\bar{u}$ is defined by integrating over the cross section. It is thus possible to speak of $p(x,t)$ and $\bar{u}(x,t)$ just as we used $p(x,t)$ and $v(x,t)$ in Section 9.1.

After some rather complex manipulations, Laplace transform techniques produce equation solutions in the s domain analogous to (9.38):

$$\begin{bmatrix} \bar{U}(L,s) \\ P(L,s) \end{bmatrix} = \begin{bmatrix} \cos \dfrac{s\beta L}{c} & -\dfrac{1}{\beta \rho c} \sin \dfrac{s\beta L}{c} \\ \beta \rho c \sin \dfrac{s\beta L}{c} & \cos \dfrac{s\beta L}{c} \end{bmatrix} \begin{bmatrix} \bar{U}(0,s) \\ P(0,s) \end{bmatrix} \tag{9.96}$$

[8]D'Souza, A. F. and R. Oldenburger, "Dynamic Response of Fluid Lines," *ASME J. Basic Eng.*, Sept. (1964), p. 589.

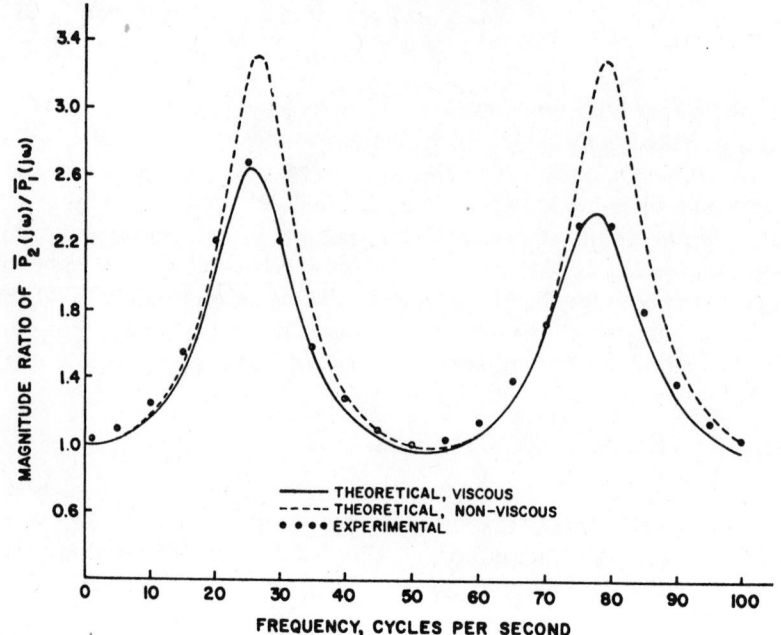

Magnitude ratio of pressure deviation at section 2 to pressure deviation at section 1 versus frequency

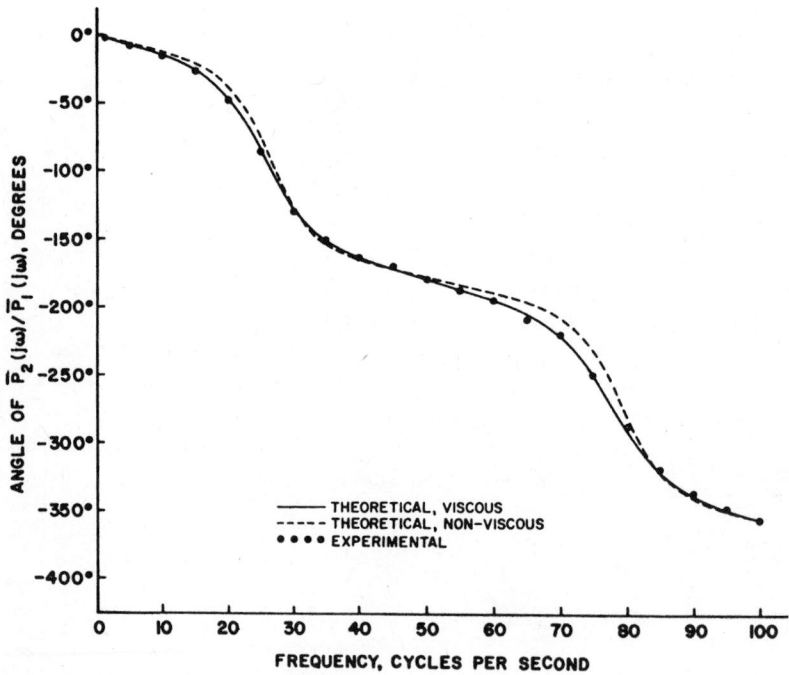

Figure 9.10 Comparison of distributed model with friction to frictionless model and experimental results.

For frequency response calculations the quantity β is a complicated, complex-valued function of frequency, pipe radius, viscosity, and density that requires use of Bessel functions in its calculation. Assuming that this computation is computerized, Equations (9.96) can be employed to analyze different types of fluid systems just as we used (9.38) in Section 9.1.1.

Frequency response tests were run with a 0.495-in. ID (0.065 in. wall) orifice-terminated tube and Mil-0-5606 hydraulic oil to check the predictions of this model, with the excellent results of Fig. 9.10. Note that the small diameter and high viscosity make our model of Section 9.1 (called theoretical, nonviscous) overestimate the peak height considerably. Figure 9.11 shows the test setup and

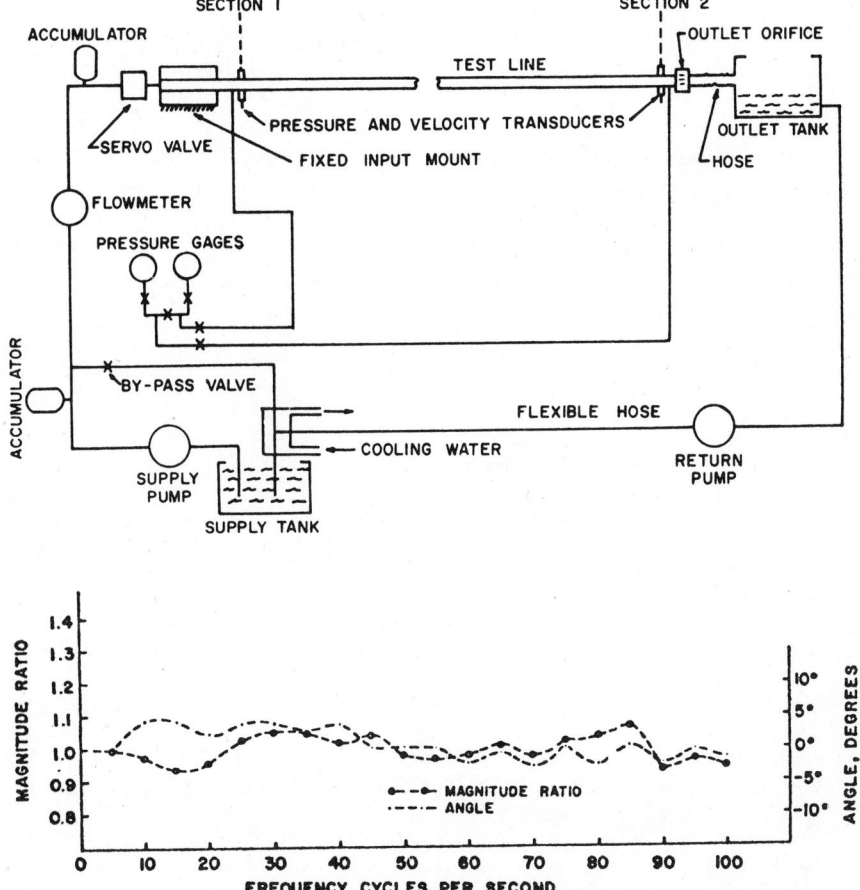

Normalized magnitude ratio of velocity to pressure deviations and their relative phase angle at section 2 versus frequency

Figure 9.11 Experimental test setup and orifice dynamics results.

some additional results that substantiate the instantaneous dynamic model usually assumed for an orifice.

9.3.1 LUMPED APPROXIMATION USING INFINITE PRODUCTS

When transfer functions in the form of a ratio of polynomials in s (rather than the unwieldy transcendental forms) are desired, one need not proceed by physical lumping (as we did in Section 9.2) if the transcendental form is available. Rather, we can look for *mathematical* ways to approximate these expressions with polynomials. For hydraulic line dynamics such methods have been worked out[9] and put in convenient form.[10] Using the results of footnote 10 the viscous line model is of the form

$$\begin{bmatrix} V_p(L,s) \\ P_p(L,s) \end{bmatrix} = \begin{bmatrix} \cosh\Gamma & -(1/Z_0)\sinh\Gamma \\ -Z_0\sinh\Gamma & \cosh\Gamma \end{bmatrix} \begin{bmatrix} V_p(0,s) \\ P_p(0,s) \end{bmatrix} \tag{9.97}$$

and the hyperbolic functions are approximated by the product forms

$$\cosh\Gamma = \prod_{n=0}^{\infty} \left(\frac{s^2}{\omega_{cn}^2} + \frac{2\zeta_{cn}}{\omega_{cn}}s + 1 \right) \tag{9.98}$$

$$\sinh\Gamma = \frac{sL}{c_0} \prod_{n=1}^{\infty} \left(\frac{s^2}{\omega_{sn}^2} + \frac{2\zeta_{sn}}{\omega_{sn}}s + 1 \right) \tag{9.99}$$

where

$$c_0 \triangleq \sqrt{M_B/\rho_0}$$

The ζ_{cn}, ζ_{sn}, ω_{cn}, and ω_{sn} are found from Fig. 9.12,[11] where

$$D_n \triangleq \text{damping number}$$

$$v \triangleq \text{kinematic viscosity}$$

By choosing $n = 0, 1, 2$, and so on, "lumped" models of various degrees of accuracy and complexity may be formed. For example, the system of Fig. 9.11

[9]Oldenburger, R. and R. E. Goodson, "Simplification of Hydraulic Line Dynamics by use of Infinite Products," *ASME J. Basic Eng.*, **86** (1964), p. 1.
[10]Astleford *et al.*, *op. cit.*, p. A-1.
[11]Astleford *et al.*, *op. cit.*, p. A-1.

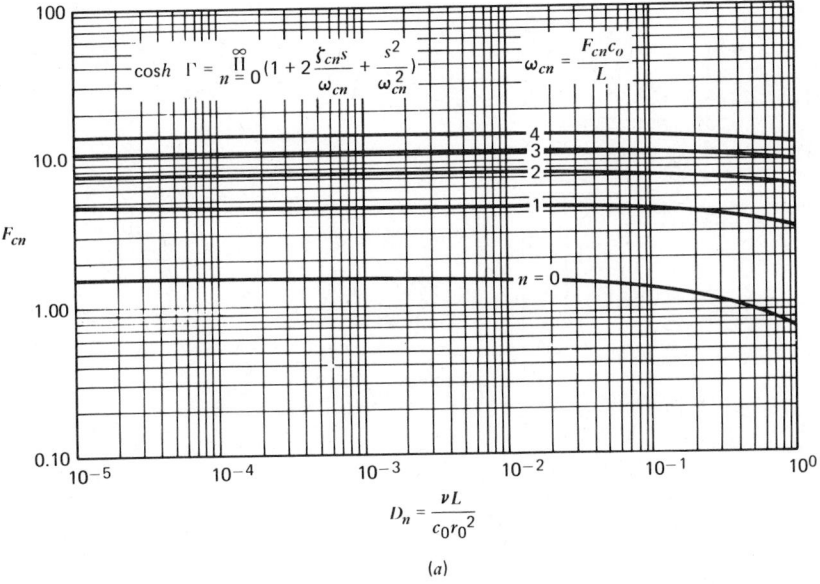

(a)

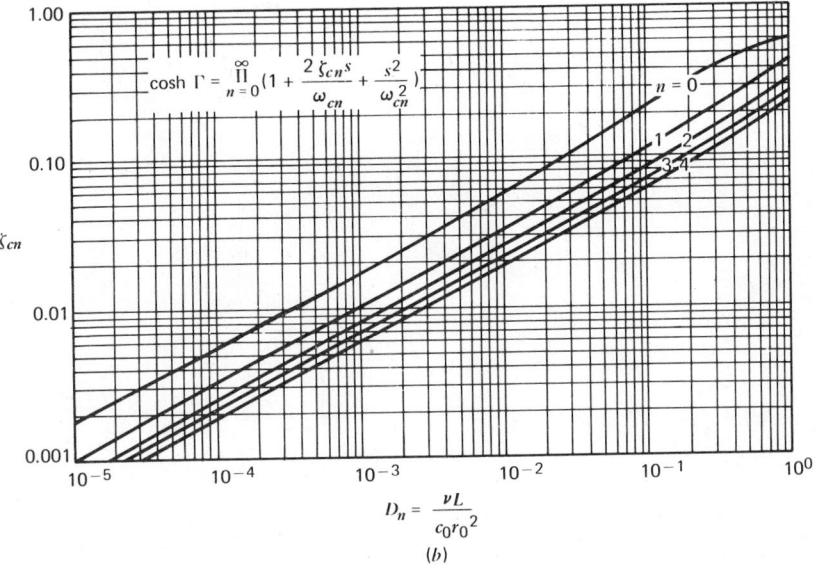

(b)

Figure 9.12 Working charts for line dynamics models; numerical example.

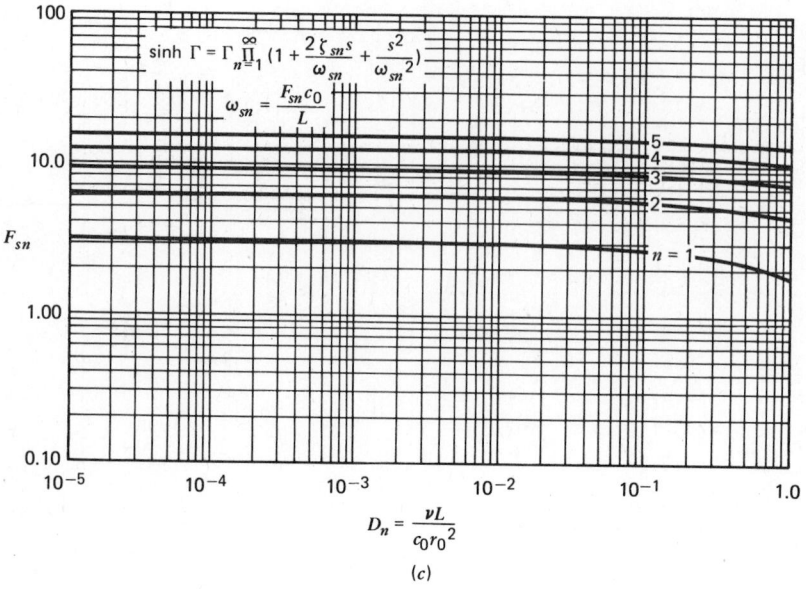

$$\sinh \Gamma = \Gamma \prod_{n=1}^{\infty} (1 + \frac{2 \zeta_{sn} s}{\omega_{sn}} + \frac{s^2}{\omega_{sn}^2})$$

$$\omega_{sn} = \frac{F_{sn} c_0}{L}$$

$$D_n = \frac{\nu L}{c_0 r_0^2}$$

(c)

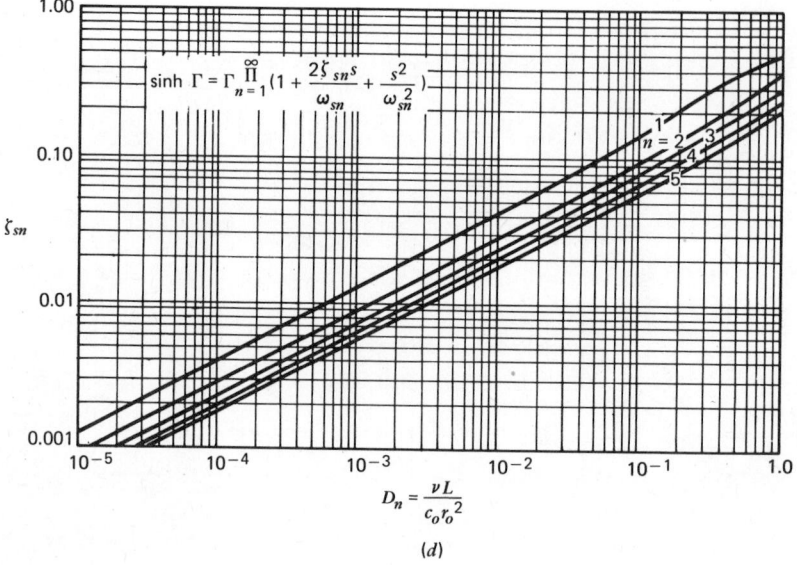

$$\sinh \Gamma = \Gamma \prod_{n=1}^{\infty} (1 + \frac{2 \zeta_{sn} s}{\omega_{sn}} + \frac{s^2}{\omega_{sn}^2})$$

$$D_n = \frac{\nu L}{c_0 r_0^2}$$

(d)

Figure 9.12   *continued*

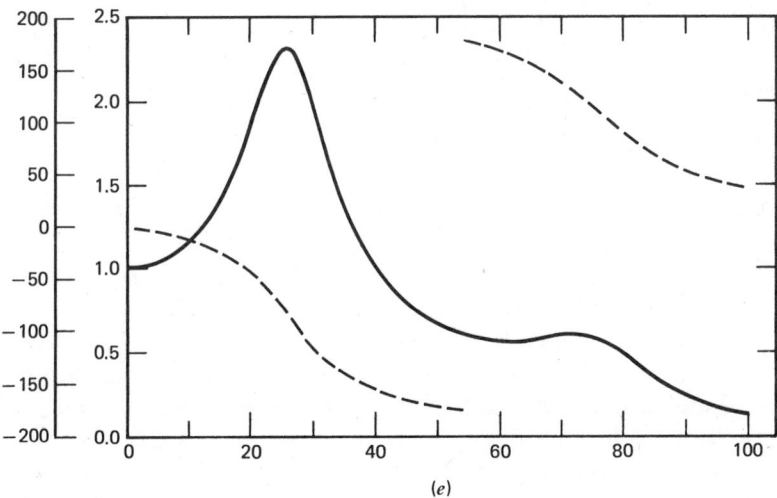

(e)

Figure 9.12   *continued*

has

$$\nu = 1.97 \times 10^{-4} \text{ ft}^2/\text{sec} \qquad L = 40.25 \text{ ft}$$
$$c_0 = 4250 \text{ ft/sec} \qquad r_0 = 0.02063 \text{ ft}$$
$$K_{or} = 4.27 \times 10^{-5} \text{ ft}^3/(\text{lb}_f\text{-sec}) \qquad Z_0 = 8.50 \times 10^3 (\text{lb}_f\text{-sec})/\text{ft}^3$$
$$D_n = 0.00438 \qquad M_B = 4.32 \times 10^7 \text{ lb}_f/\text{ft}^2$$

Suppose we try a model with $n = 1$, giving

$$\cosh \Gamma \approx \left(\frac{s^2}{169^2} + \frac{2 \times 0.04}{169} s + 1 \right) \left(\frac{s^2}{496^2} + \frac{2 \times 0.023}{496} s + 1 \right) \qquad (9.100)$$

$$\sinh \Gamma \approx 0.00948 s \left(\frac{s^2}{338^2} + \frac{2 \times 0.028}{338} + 1 \right) \qquad (9.101)$$

Due to the similarity of (9.38) and (9.97) we can use (9.100) and (9.101) in (9.44) to get:

$$\frac{P_p(L,s)}{P_p(0,s)} \approx \frac{1}{(3.5 \times 10^{-5} s^2 + 4.73 \times 10^{-4} s + 1)(4.06 \times 10^{-6} s^2 + 9.27 \times 10^{-5} s + 1)}$$

$$\frac{}{+ 2.86 \times 10^{-3} s (8.75 \times 10^{-6} s^2 + 1.65 \times 10^{-4} s + 1)}$$

$$\frac{P_p(L,s)}{P_p(0,s)} \approx \frac{1}{1.42 \times 10^{-10} s^4 + 3.01 \times 10^{-8} s^3 + 3.94 \times 10^{-5} s^2 + 3.42 \times 10^{-3} s + 1}$$

$$(9.102)$$

This is graphed in Fig. 9.12e and shows good agreement with Fig. 9.10 up to about 40 Hz. Note that the scale of Figs. 9.12a–9.12d makes accurate reading difficult in some regions.

9.4 EFFECT OF PIPE LONGITUDINAL VIBRATION (FLUID/SOLID INTERACTION)

The experimental results of Fig. 9.10 were obtained with the line supported every six feet by Styrofoam blocks with sandbags piled on top to suppress pipe motion. To see any possible effects of pipe vibration, the support system was changed so that, except for the left end which remained fixed, the pipe hung freely from thin steel wires, giving the results of Fig. 9.13. Note that there is little effect except for the range 70–90 Hz, where radical changes appear. To modify the system model to comprehend these pipe vibration effects, the reference proceeds as follows. Longitudinal vibrations of a slender rod are governed by a wave equation of form[12]

$$\frac{\partial^2 y}{\partial t^2} = \frac{E}{\rho_t}\frac{\partial^2 y}{\partial x^2} \triangleq b^2 \frac{\partial^2 y}{\partial x^2} \tag{9.103}$$

where E and ρ_t are as in Eq. (9.11) and y is the longitudinal pipe wall displacement. Laplace transforming with zero initial conditions:

$$\frac{d^2 Y}{dx^2}(x,s) - \frac{s^2}{b^2}Y(x,s)=0 \tag{9.104}$$

This ordinary differential equation has the solution

$$Y(x,s)=B_1(s)\cosh\frac{sx}{b} + B_2(s)\sinh\frac{sx}{b} \tag{9.105}$$

where constants of integration B_1 and B_2 can be found from the boundary conditions. At $x=0$ the pipe is rigidly held ($y(0,t)\equiv0$), so B_1 must be zero.

At $x=L$ the orifice is held in a "rigid" block of mass m. This mass feels the following forces:

1 Pressure p_2 exerts on the block a longitudinal force of p_2A_n, where A_n is the net area of the orifice plate, that is, pipe flow area minus orifice hole area.
2 A rubber hose connecting orifice block to tank exerts a "viscous" damping force of $c_d(\partial y/\partial t)(L,t)$.
3 The internal elastic force due to strain $\partial y/\partial x$ in the pipe is $A_w E(\partial y/\partial x)(L,t)$, where A_w is the pipe wall area.

[12]Doebelin, E. O., *System Dynamics*, C. E. Merrill, Columbus, Ohio, 1972, p. 451.

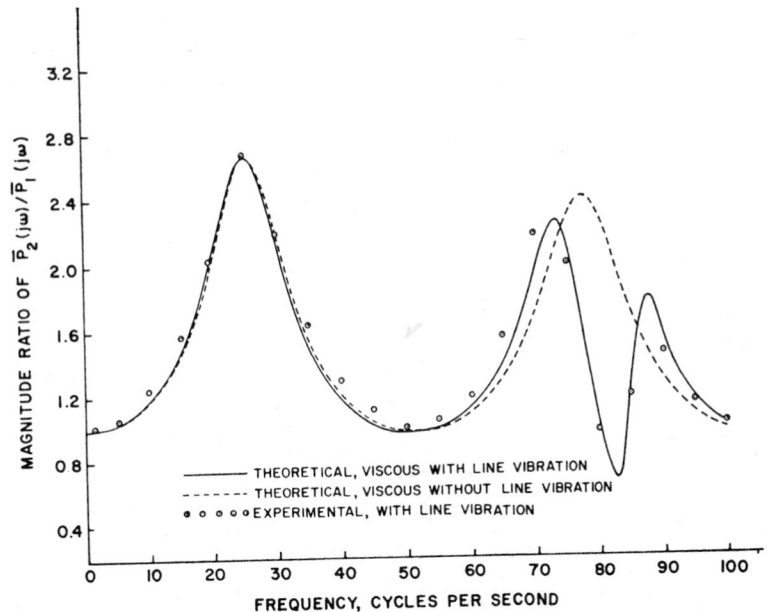

Magnitude ratio of pressure deviation at section 2 to pressure deviation at section 1 versus frequency, with line vibrations

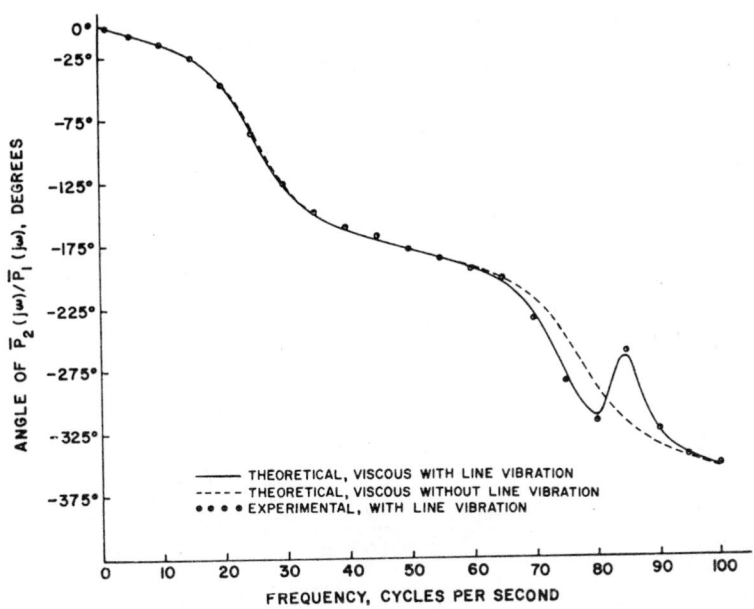

Phase angle of pressure deviation at section 2 relative to pressure deviation at section 1 versus frequency, with line vibrations

Figure 9.13 Theoretical and experimental results for conduit dynamics with pipe vibration included.

Applying Newton's law to the block:

$$P_2 A_n - c_d \frac{\partial y}{\partial t}(L,t) - A_w E \frac{\partial y}{\partial x}(L,t) = m \frac{\partial^2 y}{\partial t^2}(L,t) \qquad (9.106)$$

which Laplace transforms to

$$P_2 A_n - c_d s Y(L,s) - A_w E \frac{dY}{dx}(L,s) = ms^2 Y(L,s) \qquad (9.107)$$

From (9.105), $Y(x,s) = B_2(s)\sinh(sx/b)$, so (9.107) becomes:

$$P_2 A_n - c_d s B_2(s)\sinh\frac{sL}{b} - A_w E \frac{s}{b} B_2(s)\cosh\frac{sL}{b} = ms^2 B_2(s)\sinh\frac{sL}{b}$$

allowing solution for $B_2(s)$ as

$$B_2(s) = \frac{P_2 A_n}{A_w E \dfrac{s}{b} \cosh\dfrac{sL}{b} + ms^2 \sinh\dfrac{sL}{b} + c_d s \sinh\dfrac{sL}{b}} \qquad (9.108)$$

and finally giving $Y(L,s)$ as

$$Y(L,s) = \frac{P_2 A_n}{s\left(ms + c_d + \dfrac{A_w E}{b}\coth\dfrac{sL}{b}\right)} \qquad (9.109)$$

We must now modify the orifice pressure/velocity relation since the orifice plate now has a velocity $sY(L,s)$ and the linearized orifice pressure drop is proportional to the *relative* velocity of fluid and orifice plate:

$$V_2 - sY(L,s) = K_{or} P_2 \qquad (9.110)$$

$$V_2 = \left[K_{or} + \frac{A_n}{ms + c_d + \dfrac{A_w E}{b}\coth\dfrac{sL}{b}} \right] P_2 \qquad (9.111)$$

Equations (9.109) and (9.111) could now be included in the set of simultaneous equations for any of the distributed or lumped fluid models to study the effect of pipe vibration since they simply add one unknown ($Y(L,s)$) and one equation. The reference used the complicated fluid model of Eq. (9.96) to produce the excellent results of Fig. 9.13.

When a piping system is not rigidly clamped at any point (the above example *was* clamped), then the pipe is more likely to vibrate as a rigid body against its flexible supports, rather than by elastic wave propagation within the pipe. This situation is fairly common since flexible supports are often used to

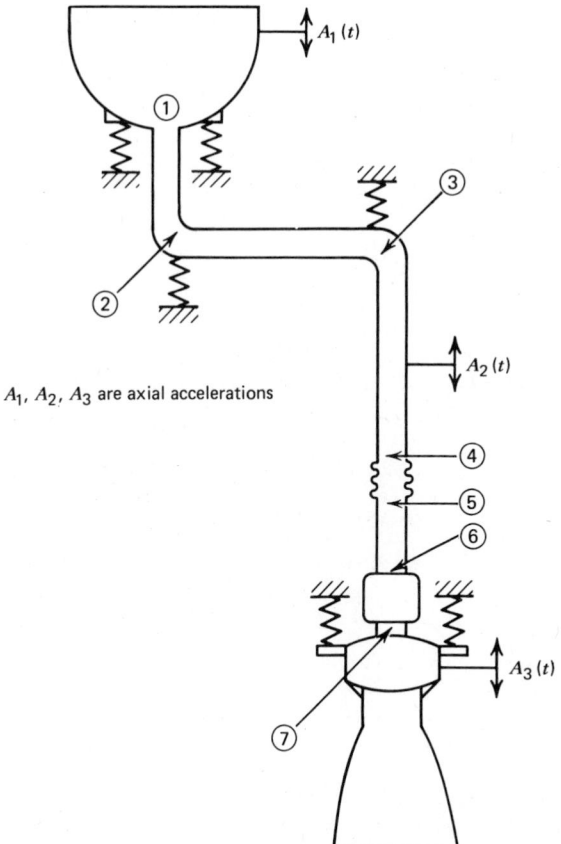

A_1, A_2, A_3 are axial accelerations

Figure 9.14   Rocket engine feed system.

accommodate thermal expansion. An early NASA study[13] employed the friction-free fluid line model or our Eq. (9.38) and combined this with a spring/mass/damper model of the moving pipe section to get an overall system model that showed excellent agreement with sinusoidal test results. Extensive effort since this time has resulted in a number of general purpose models (and associated computer programs) for handling a wide variety of problems of this general type. One[14] of these will be briefly discussed.

Figure 9.14 shows a model for a rocket engine feed system that includes mechanical and fluid features typical of many practical systems. The computer program allows routine treatment of such features as straight lengths of fluid

[13]Blade, R. J., W. Lewis and J. Goodykoontz, Study of a Sinusoidally Perturbed Flow in a Line Including a 90° Elbow with Flexible Supports, NASA TND1216, 1962.
[14]Astleford, *et al.*, *op. cit.*

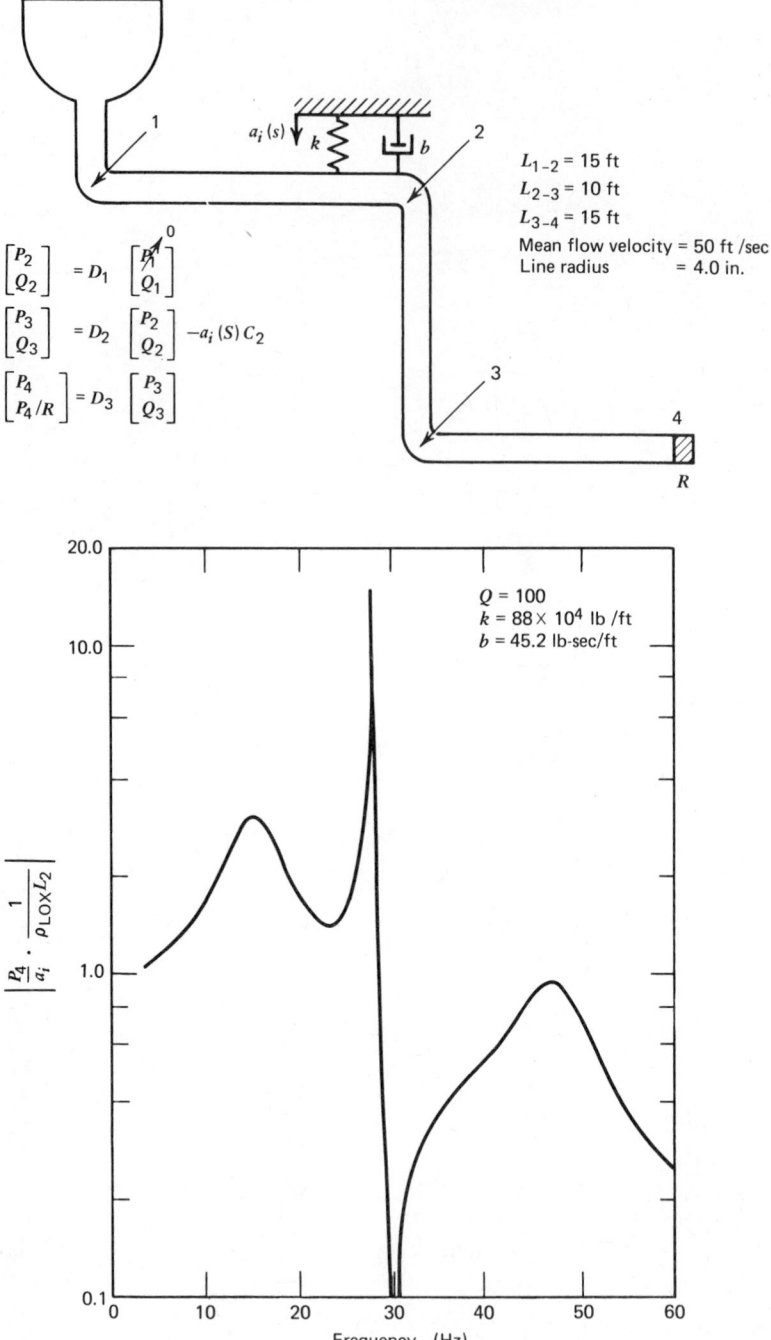

$$L_{1-2} = 15 \text{ ft}$$
$$L_{2-3} = 10 \text{ ft}$$
$$L_{3-4} = 15 \text{ ft}$$
Mean flow velocity = 50 ft /sec
Line radius   = 4.0 in.

$$\begin{bmatrix} P_2 \\ Q_2 \end{bmatrix} = D_1 \begin{bmatrix} P_1 \\ Q_1 \end{bmatrix}^0$$

$$\begin{bmatrix} P_3 \\ Q_3 \end{bmatrix} = D_2 \begin{bmatrix} P_2 \\ Q_2 \end{bmatrix} - a_i(S) C_2$$

$$\begin{bmatrix} P_4 \\ P_4/R \end{bmatrix} = D_3 \begin{bmatrix} P_3 \\ Q_3 \end{bmatrix}$$

$Q = 100$
$k = 88 \times 10^4$ lb /ft
$b = 45.2$ lb-sec/ft

Figure 9.15 Fluid/mechanical system and its frequency response. The line length from Station 1 to the tank exit is assumed to be negligible. Tank exit conditions exist at Station 1.

line with damping, dynamics of local gas bubbles in the liquid, effects of flexible bellows at pipe joints, side branches, rigid body motions of pipe sections, external force and motion inputs, and so on. Several specific numerical problems (Fig. 9.15 shows one) are worked out to explain the use of the program. Many of these programs were developed at government expense and are thus available to the general public for modest fees.

9.5   VARIABLE-DISPLACEMENT PUMP SYSTEMS

Of the many different types of hydraulic equipment used in open-loop and feedback applications, we now examine two that employ variable-displacement pumps.

9.5.1   PUMP-CONTROLLED HYDRAULIC MOTOR (HYDROSTATIC TRANSMISSION)

Variable-displacement hydraulic pumps, driven at constant speed, have a stroke control that allows their displacement (cm^3/(shaft rotation)) to be varied from full flow in one direction through zero flow, to full flow in the reverse direction. This control of displacement is accomplished in axial-piston pumps by changing the angle of a swash plate or similar mechanism (Volvo Corp., Fig. 9.16a), and in radial piston or vane pumps by changing the eccentricity of a cam ring with respect to the rotating piston (or vane) carrier (Racine Co., Fig. 9.16b). By connecting such a pump to a fixed displacement hydraulic motor or cylinder one can smoothly control the motion of a mechanical load in either open-loop or feedback applications. Figure 9.16c shows the feedback configuration (used when maximum accuracy is required), while the portion labeled "hydrostatic transmission" is alone sufficient for open-loop systems. Figure 9.17 shows a web-offset printing press that uses seven such pumps and six motors[15] to control and integrate various motions, torques, and air flow rates in this complex machine.

We now develop a simplified linear model of the hydrostatic transmission, useful for preliminary design work on both open-loop and feedback applications. While the simple configuration of Fig. 9.18a appears adequate in principle, fluid leakage and possible cavitation require the addition in practice of the replenishing and supercharging system as in Fig. 9.18b. The small replenishing pump runs continually and maintains a system base pressure (a few hundred psi) determined by the relief valve setting. If the pressure in either line drops below this setting, the check valves allow makeup fluid to enter the lines. Since the check valves do not allow reverse flow, the variable displacement pump can develop whatever pressure is needed to drive the motor and load. Cross-port

---

[15]The Oilgear Co., Milwaukee, Wisc.

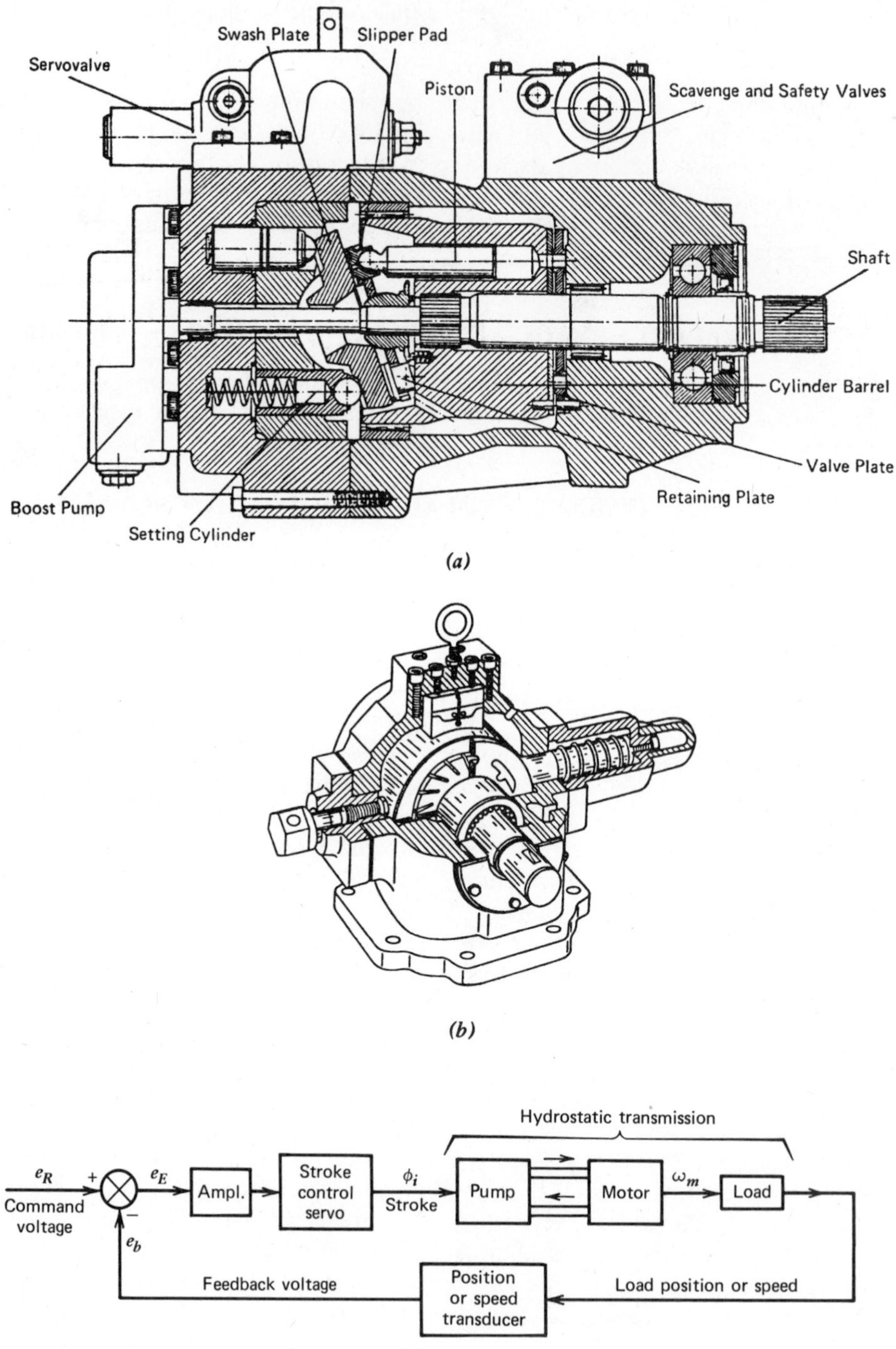

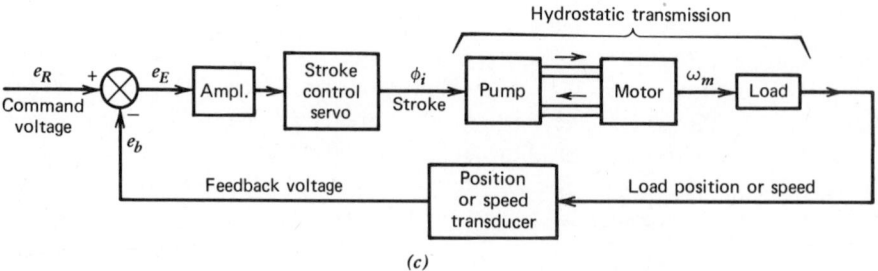

Figure 9.16 Variable-displacement pumps and servo-system application.

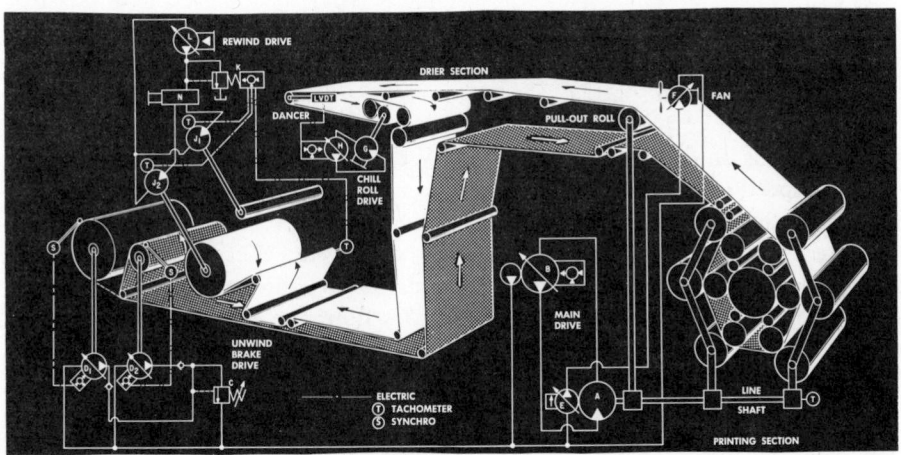

Figure 9.17 Hydraulically controlled printing press.

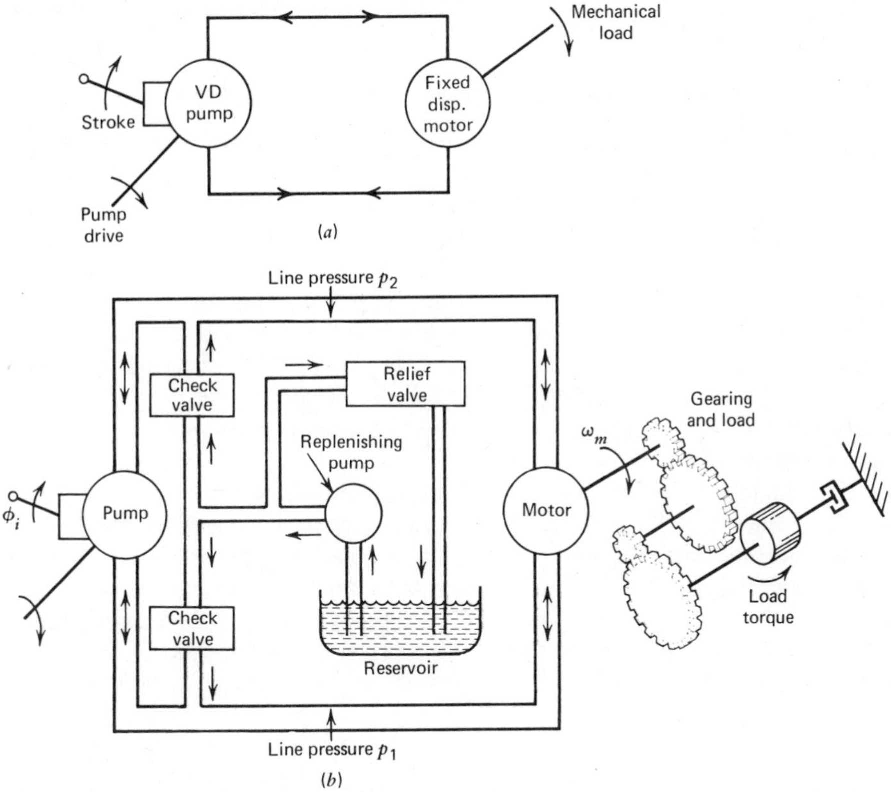

Figure 9.18 Pump/motor drive system.

relief valves (not shown) protect the system from excessive pressures such as might be caused if we set the pump at zero displacement when a large inertial load had attained high speed.

The number of parameters needed to characterize a variable-displacement pump depends on the assumptions made. We first assume the pump is driven at constant speed ω_p, making its rotary inertia irrelevant and avoiding any consideration of the dynamics of the prime mover that drives the pump. When at maximum stroke, the pump displacement is d_p cm^3/rad. Let ϕ_i be a generalized stroke such that $-1.0 \leqslant \phi_i \leqslant 1.0$ corresponds to the full range of positive and negative flows, and pump gross instantaneous flow rate is $\omega_p d_p \phi_i$ cm^3/sec. We take the *motion* ϕ_i as a given input; thus inertia, spring, damping, and pressure forces associated with this motion are irrelevant for the pump analysis but *would* be considered in the analysis of the *previous* device (such as the stroke-control servo of Fig. 9.16). Internal leakage occurs from the high-pressure to low-pressure side of the pump while leakage ("case drain") from either side to the reservoir is called external leakage. Since these leakage passages are "long and narrow," flow is nearly laminar and linear pressure/flow relations are good approximations:

$$\text{pump internal leakage} = K_{ilp}(p_1 - p_r) \text{ cm}^3/\text{sec} \tag{9.112}$$

$$\text{pump external leakage} = K_{elp}(p_1 - 0) \text{ cm}^3/\text{sec} \tag{9.113}$$

We are here considering a time instant when p_1 is the "driven" line and are assuming p_2 is kept *fixed* at replenishing pressure p_r. Also, the lines are assumed short enough to neglect line dynamic effects other than compliance, which will be included. Pressure p_2 will not explicitly appear in our analysis. However, when p_1 mathematically goes negative, this really means that p_1 and p_2 have "exchanged labels" and p_2 has become the dynamically varying (positive) pressure while p_1 has become fixed at p_r. Accurate values of the leakage coefficients require pressure/flow experiments (pump rotor clamped) on the machine in question.

For the motor, its fixed displacement is d_m and leakage coefficients are K_{ilm} and K_{elm}. Since motor speed varies, mechanical dynamic properties are needed. We assume no elasticity or backlash in the gearing and shafts so that the motor/load dynamics can be represented by a single equivalent inertia J_e and equivalent viscous damping B_e referred to the motor shaft (equivalent dynamic system for a gear train[16]). We also include an external load torque $T_l(t)$, referred to the motor shaft. A simple energy analysis for loss-free energy conversion in

---

[16]Doebelin, E. O., *Dynamic Analysis and Feedback Control*, McGraw-Hill, N.Y., 1962, p. 237.

any positive-displacement pump or motor gives the well-known result:

pump or motor "fluid torque" = (machine displacement)(pressure drop)

$$(9.114)$$

Experiments show that motors also exhibit a "dry friction" torque proportional to $(p_1 + p_r)$, but since this is nonlinear (and usually fairly small) we neglect it.

Now, by expressing the conservation of volume for the pressurized side (we choose p_1) of the system we can write:

pump gross flow − motor/pump leakage flow

− compressibility "flow" = motor displacement flow (9.115)

Taking all variables as dynamic perturbations away from an operating point (p_r then disappears since it does not change):

$$\omega_p d_p \phi_i - (K_{ilp} + K_{elp})p_1 - (K_{ilm} + K_{elm})p_1 - \frac{V}{M_B}\frac{dp_1}{dt} = d_m\omega_m \quad (9.116)$$

$M_B \triangleq$ fluid effective bulk modulus (working value

$\approx 100,000 \; psi \; (68,900 \; N/cm^2)$)

$V \triangleq$ volume exposed to pressure p
(one side of pump, one line, one side of motor)

Newton's law for the motor shaft gives:

$$T_l(t) + d_m p_1 - B_e\omega_m = J_e \frac{d\omega_m}{dt} \quad (9.117)$$

Eliminating p_1 in favor of ω_m and defining $K_{tl} \triangleq K_{ilp} + K_{elp} + K_{ilm} + K_{elm}$, we get:

$$\omega_m(s) = \frac{K_\phi \phi_i + K_T(\tau s + 1)T_l}{s^2/\omega_n^2 + 2\zeta s/\omega_n + 1} \quad (9.118)$$

where

$$K_\phi \triangleq \frac{\omega_p d_p d_m}{d_m^2 + K_{tl}B_e} \frac{rad/sec}{\% \; stroke}, \qquad K_T \triangleq \frac{K_{tl}}{d_m^2 + K_{tl}B_e} \frac{rad/sec}{N\text{-}cm} \quad (9.119)$$

$$\omega_n \triangleq \sqrt{\frac{M_B(d_m^2 + K_{tl}B_e)}{V}} \; \frac{rad}{sec}, \qquad \tau \triangleq \frac{V}{M_B K_{tl}} \; sec \quad (9.120)$$

$$\zeta \triangleq \frac{K_{tl}J_e M_B + VB_e}{2}\sqrt{\frac{1}{VJ_e M_B(d_m^2 + K_{tl}B_e)}} \quad (9.121)$$

Often $K_{tl}B_e$ can be neglected relative to d_m^2, allowing further simplification of the above relations for preliminary design purposes. Note that damping ζ comes from mechnical damping B_e plus leakage K_{tl}. If damping is too small, intentional adjustable leakage in the form of a cross-port needle valve may be used; however, this method reduces gain K_ϕ, so it cannot be carried to extreme.

9.5.2 ENERGY-SAVING HYDRAULIC POWER SUPPLY

When a constant-pressure supply is desired to power a hydraulic system, a fixed-displacement pump and relief valve provide the simplest system, which also has a low first cost. However, such supplies are very inefficient in most applications, and with energy costs continually rising, the high operating costs may outweigh the low initial investment and make more complex (but energy-saving) alternative power supplies attractive. Most such supplies are based on variable-displacement pumps and we now analyze a simple system of this type. In Fig. 9.19, when the mechanical load requires no power, the manual control valve is shut (as shown). Since the pump is driven at constant speed, pressure p rapidly rises. However, this causes, through the stroke-regulating piston shown, a reduction in pump displacement. The mechanism is adjusted so that when pressure reaches the desired value (say 1000 psi (689 N/cm^2)) the pump flow just matches the small leakage in the system. Thus the desired supply pressure is always ready when the mechanical load requests power; however, little energy is wasted in the standby condition since pump flow is very small. When the manual control valve is opened to drive the load, pressure starts to drop, but this

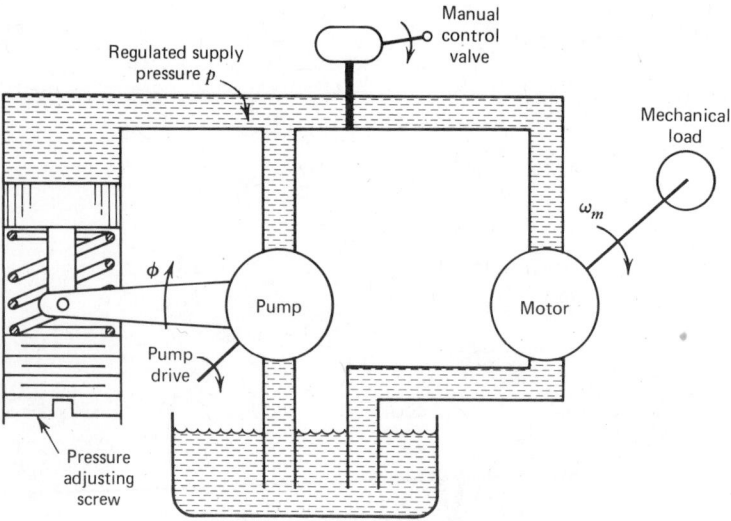

Figure 9.19   Pressure-compensated variable-displacement pump system.

causes an increase in pump displacement, returning pressure toward its set value. The system is thus a self-regulating (feedback) type that adjusts pump flow to keep pressure near the set value for any load demands that might occur.

Let us analyze this system for a situation where the manual valve is initially shut and system pressure is steady at the set value, when at $t=0$ the valve is opened wide and left open, driving a mechanical load of inertia and viscous friction toward an equilibrium speed. We model the pump stroke-adjusting mechanism with inertia, viscous friction, and linear spring effects driven by a torque ALp_{net}:

$$ALp_{net} - B_i\dot{\phi} - K_{si}\phi = J_i\ddot{\phi}_i \tag{9.122}$$

$$p_{net} \triangleq p_{set} + p_{za} - p \tag{9.123}$$

$$AL \triangleq \text{product of piston area and stroke arm length} \tag{9.124}$$

The quantity p_{net} is a fictitious pressure related to the desired pressure p_{set} and the actual pressure p by Eq. (9.123). The desired pressure is set into the system with the pressure adjusting screw. However, with the manual valve closed in this proportional control system, when $p \equiv p_{set}$, p_{net} and ϕ would be zero in steady state (giving no flow to balance pump leakage) unless we introduce p_{za}, the "zero adjustment" as in (9.123). In effect, if we desire a pressure p_{set} of, say, 1000 psi, the adjusting screw must be set for a *higher* value $(p_{set}+p_{za})$ so that at equilibrium we have a torque ALp_{za} producing a pump stroke ϕ just sufficient to balance leakage at 1000 psi.

We again assume that the lines in the system are short enough that fluid inertia and friction effects are negligible and only compliance need be modeled. By conservation of volume:

$$\omega_p d_p \phi - K_{pl}p - C_f\dot{p} - K_{ml}p = d_m\omega_m \tag{9.125}$$

$$\text{fluid compliance} \triangleq C_f \triangleq V/M_B \tag{9.126}$$

Newton's law for the motor shaft gives:

$$T_l(t) + d_mp - B_e\omega_m = J_e\dot{\omega}_m \tag{9.127}$$

The equation set obtained is sufficient to solve for the unknowns ϕ, p, and ω_m and can be represented in block diagram form as in Fig. 9.20. Combination of the equations into single equations in single unknowns leads to a rather complicated fourth order characteristic equation which could be stability analyzed using Routh criterion. Digital simulation might be a preferred method for both performance and stability studies on proposed designs. A CSMP

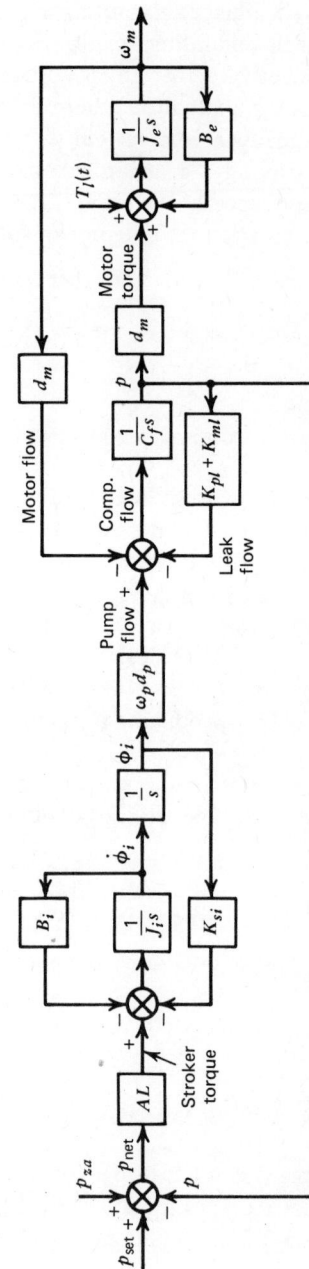

Figure 9.20 Block diagram of pressure-compensated pump and load.

program for some typical values might go as follows:

```
PARAM    JI = .1, BI = 50., KSI = 100.
PARAM    JM = .043, BM = 20.
PARAM    CF = .0001, DM = .486, DP = 1.0, AL = .5
PARAM    KPL = .0006, OMEGP = 180., TLOAD = 0.0
PARAM    PSET = 1000., PZA = 1.11
         KML = .0004*STEP(.003)
         PNET = PSET + PZA − P
         PHIDT2 = (AL*PNET − BI*PHIDT1 − KSI*PHI)/JI
         PHIDT1 = INTGRL(0.0, PHIDT2)
         PHI = INTGRL(.003335, PHIDT1)
         PDOTA = (OMEGP*DP*PHI − OMEGM*DM − (KML +
         KPL)*P)/CF
         PDOT = PDOTA*STEP(.003)
         P = INTGRL(1000., PDOT)
         OMEGM1 = (P*DM − BM*OMEGM + TLOAD)/JM
         OMEG1A = OMEGM1*STEP(.003)
         OMEGM = INTGRL(0.0, OMEG1A)
TIMER    FINTIM = .15, DELT = .0001, OUTDEL = .003
```

In this program the manual valve is assumed closed for $t < 0.003$, so the initial values of PHI and P and the value of PZA have been hand computed so that the system is in equilibrium at $P = 1000$, with only pump leakage active. At $t = 0.003$ (a value selected so that we get one plotted point for the initial steady state at $t = 0.0$) the valve is opened wide, the KML = ... statement now making motor leakage active. To "hold the motor still" for $0 < t < 0.003$, OMEG1A uses STEP(.003) to keep velocity at zero until the valve is opened. Pressure P is similarly maintained at 1000 using STEP(.003) in PDOT. Figure 9.21 shows a response that stabilizes in about 0.1 sec at a pressure of 987, so the pressure regulation for this load application is accurate to 1.3%. The fluid compliance of 0.0001 corresponds to about 10 in.[3] of oil volume for $M_B = 100,000$ psi. A second run with longer lines (giving CF = .001) and with reduced pump-stroker damping (BI = 20.) gave the absolutely unstable response of Fig. 9.22, emphasizing the importance of careful dynamic analysis in such systems.

When conditions warrant, the pump system models we have here developed can be augmented with additional linear and/or nonlinear fluid and mechanical details. Long lines may require use of distributed or lumped line models, pump prime mover dynamics may need to be included, dry friction effects in pumps and motors may require consideration, and so on. Digital simulation becomes increasingly attractive as model complexity rises.

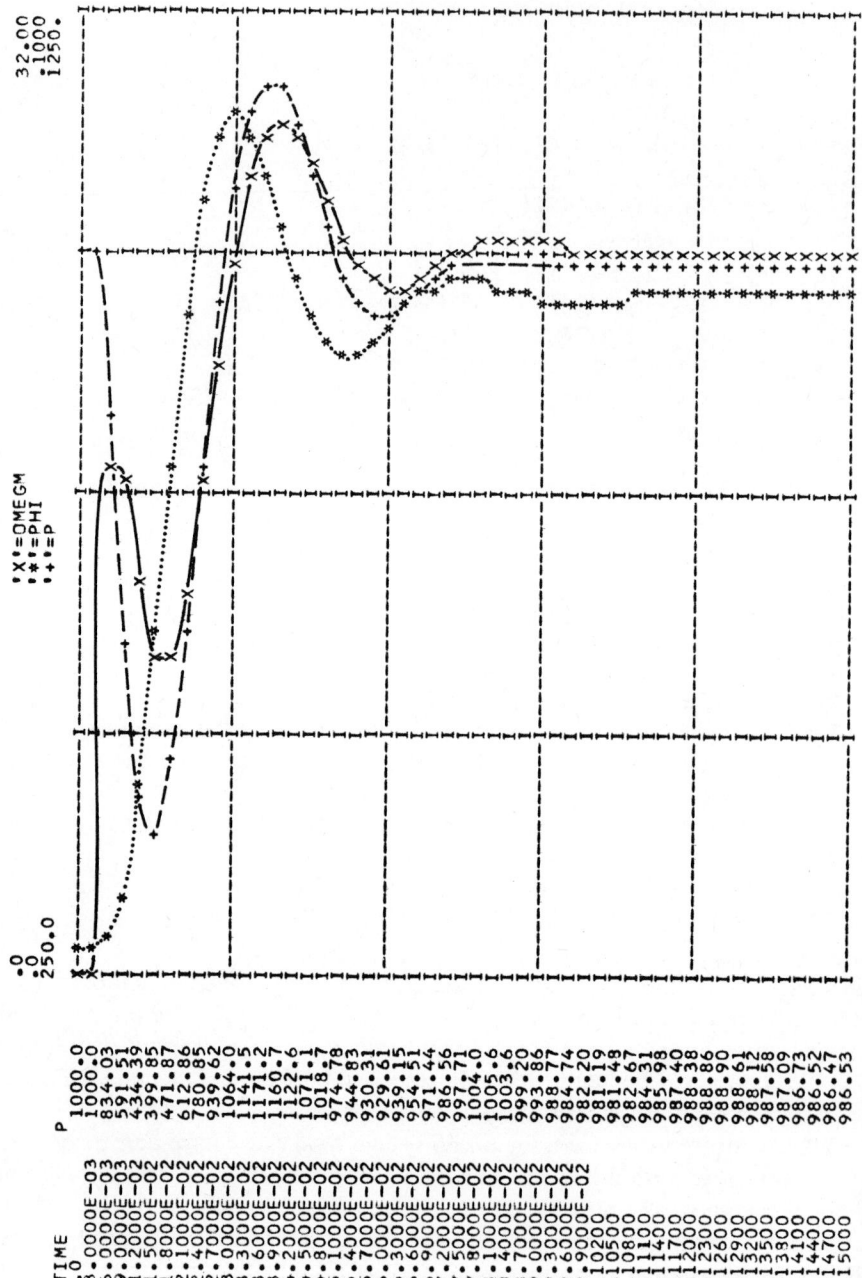

Figure 9.21 Response of stable system.

404

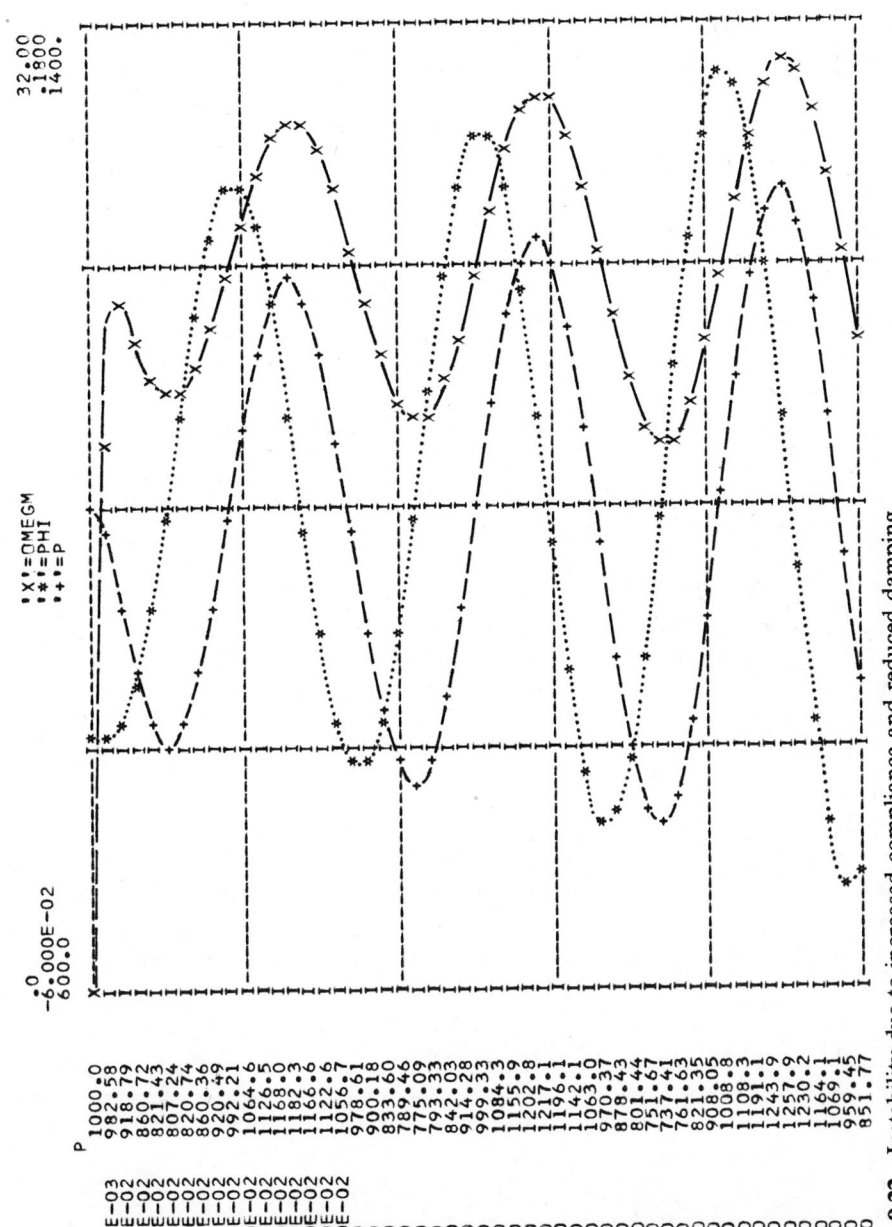

Figure 9.22 Instability due to increased compliance and reduced damping.

405

BIBLIOGRAPHY

1 Andersen, B. W., *The Analysis and Design of Pneumatic Systems*, John Wiley and Sons, N.Y., 1967.

2 Anderson, D. A., Blade, R. J., and W. Stevans, Response of a Radial-Bladed Centrifugal Pump to Sinusoidal Disturbances for Noncavitating Flow, NASA TN D-6556, 1971.

3 Batterton, P. G. and J. R. Zeller, Dynamic Performance Analysis of a Fuel-Control Valve for use in Airbreathing Engine Research, NASA TN D-5331, 1969.

4 Besant, R. W. and V. Srinivas, Transient-Response Study of Fluids in Pipeline Systems, ASME Paper 67-WA/FE-25, 1967.

5 Blackburn, J. F., Reethof, G., and J. L. Shearer, *Fluid Power Control*, MIT Press, Cambridge, 1960.

6 Boothe, W. A., "A Lumped-Parameter Technique for Predicting Analog Fluid Amplifier Dynamics," *ISA Trans.*, **4**, No. 1 (1965), pp. 84–92.

7 Brennen, C., The Dynamic Behavior and Compliance of a Stream of Cavitating Bubbles, ASME Paper 73-FE-34, 1973.

8 Brennen, C. and A. J. Acosta, The Dynamic Transfer Function for a Cavitating Inducer, ASME Paper 75-WA/FE-16, 1975.

9 Dustin, M. O., Analog Computer Study of Design Parameter Effects on the Stability of a Direct-Acting Gas Pressure Regulator, NASA TN D-6267, 1971.

10 Funk, J. E. and D. J. Wood, Frequency Response of Fluid Lines with Turbulent Flow, ASME Paper 74-FE-21, 1974.

11 Gerlach, C. R., Study of Fluid Transients in Closed Conduits, Fluid Power Lab, Okla. State U., Rept. 65-4, 1966.

12 Goodson, R. E. and R. G. Leonard, A Survey of Modeling Techniques for Fluid Line Transients, ASME Paper 71-WA/FE-9, 1971.

13 Healey, A. J. and J. A. Nicholson, Dynamic Characteristics of Annular Fluid Transmission Lines, ASME Paper 74-FE-16, 1974.

14 Iberall, A. S., "Attenuation of Oscillatory Pressures in Instrument Lines," *Nat'l. Bur. Stds. J. Res.*, **45** (1950).

15 Johnson, B. L. and D. E. Wandling, Transfer Functions and Input Impedances of Pressurized Piping Systems, ASME Paper 66-WA/Aut-13, 1966.

16 Karam, J. T. and R. G. Leonard, A Simple Yet Theoretically Based Time Domain Model for Fluid Transmission Line Systems, ASME Paper 73-FE-27, 1973.

17 Lewis, W., Simplified Analytical Model for Use in Design of Pump-Inlet Accumulators for Prevention of Liquid-Rocket Longitudinal Oscillation, NASA TN D-5394, 1969.

18 Lewis, W., Blade, R. J., and R. G. Dorsch, Study of the Effect of a Closed-End Side Branch on Sinusoidally Perturbed Flow of Liquid in a Line, NASA TN D-1876, 1963.

19 Lorenzo, C. F., Centaur Feedline Dynamics Study Using Power Spectral Methods, NASA TM X-3111, 1974.

20 Luoma, R. A. and C. L. McDearmont, Simulation of the Dynamics of a Large

Crawler Platform Leveling System, General Electric Apollo Support Dept. Rept. 66 SIMAR 97/8, GE Tech. Publ. DB66J061, 1966.

21 Mercier, O. L. and D. Wright, A Dynamic Modeling Method of Unsteady Flows in Long Fluid Lines with Turbulent Bulk Velocities, ASME Paper 73-FE-18, 1973.

22 Merritt, H. E., *Hydraulic Control Systems*, John Wiley and Sons, N.Y., 1967.

23 Nahavandi, A. N. and A. Batenburg, "Steam Generator Water-Level Control," *ASME J. Basic Eng.*, June (1966).

24 Ng, Sheung-Lip, Dynamic Response of Cavitating Turbomachines, Cal. Inst. Of Tech. Rept. E 183.1, 1976.

25 Ohashi, H., Analytical and Experimental Study of Dynamic Characteristics of Turbopumps, NASA TN D-4298, 1968.

26 Phillips, J. W., Reflection and Transmission of Fluid Transients at Elbows, Univ. of Ill. T. & A. M. Rept. No. 425, 1978.

27 Sankar, S. and M. O. M. Osman, On the Dynamic Accuracy of Machine Tool Hydraulic Copying Systems, ASME Paper 73-DET-95, 1973.

28 Sewall, J. L., Wineman, D. A. and R. W. Herr, An Investigation of Hydraulic-Line Resonance and its Attenuation, NASA TM X-2787, 1973.

29 Stevans, W. and R. J. Blade, Experimental Evaluation of a Pump Test Facility with Controlled Perturbations of Inlet Flow, NASA TN D-6543, 1971.

30 Streeter, V. L. and E. B. Wylie, Hydraulic Transients Caused by Reciprocating Pumps, ASME Paper 66-WA/FE-29, 1966.

31 Wallace, F. J., "Pressure Pulsations in Reciprocating Compressor Delivery Systems," *J. Mech. Eng. Sci.*, **8**, No. 2 (1966).

32 Wovd, D. J., Dorsch, R. G., and C. Lightner, Digital Distributed Parameter Model for Analysis of Unsteady Flow in Liquid-Filled Lines, NASA TN D-2812, 1965.

33 Wylie, E. B. and V. L. Streeter, *Fluid Transients*, McGraw-Hill, N.Y., 1978.

34 Yao, H., Goodson, R. E., and R. G. Leonard, State Space Representations of Distributed Fluid Line Dynamics, ASME Paper 74-FE-19, 1974.

PROBLEMS

9.1 Using Equations (9.38) for the conduit dynamics, write the equations for the system of Fig. P9.1. The chamber at $x = L$ has springiness but no

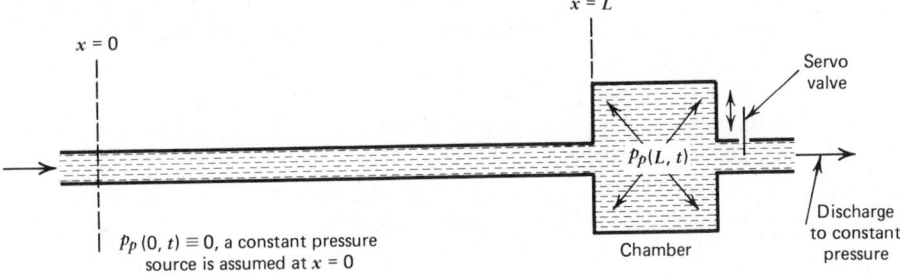

Figure P9.1

inertia and all points within the chamber have pressure $p_p(L,t)$. The chamber volume is given by

$$V_c = V_{co} + K_{vp}p_p(L,t)$$

where V_{co} and K_{vp} are constants. The servo-valve discharge area is given by

$$A_v = A_{vo} + A_{vp}(t)$$

where A_{vo} is a constant and $A_{vp}(t)$ is any function of time.
(a) Find the transfer functions $P_p(L,s)/A_{vp}(s)$, $V_p(L,s)/A_{vp}(s)$, and $V_p(0,s)/A_{vp}(s)$.
(b) Sketch the shape of $P_p(L,i\omega)/A_{vp}(i\omega)$.
(c) Find the response of p_p to a step change in A_{vp}.

9.2 Repeat Problem 9.1 using a lumped parameter conduit model with:
(a) One lump.   (b) Two lumps.   (c) Three lumps.

9.3 For the orifice-terminated line of Section 9.1.1:
(a) Find $V_p(L,s)/P_p(L,s)$ and sketch its frequency response.
(b) Find $P_p(L,s)/V_p(0,s)$ and sketch its frequency response.

9.4 Modify the equations for the system of Fig. 9.3a to account for an orifice at location 4.

9.5 Model the system of Fig. 9.3a with two-lump models for each line.

9.6 Modify the equations for the system of Fig. 9.4 to account for an orifice at location 2.

9.7 In NASA report TN-D-576, frequency-response tests were run on a 68-foot length of 1 inch OD stainless steel pipe (wall thickness $\frac{1}{16}$ inch) terminated with an orifice made of a thin plate with 34 drilled holes of 0.040 inch diameter. Fluid was JP-4 fuel with density 7.26×10^{-5} $lb_f\text{-sec}^2/in.^4$ and bulk modulus 173,600 psi.
(a) Using standard fluid mechanics formulas (include a discharge coefficient) get a "theoretical" expression for K_{or}. Compute a numerical value if the orifice pressure drop is 49.0 psi and compare with a measured K_{or} obtained from Fig. P9.2.
(b) Compute Z_o and T_p. Find the operating point (pressure, velocity) at which $Z_o = 1/K_{or}$, using the experimental curve given. For this operating point, sketch $P_p(0,i\omega)/V_p(0,i\omega)$ for $0 < \omega < 200$ rad/sec.
(c) Repeat the sketching of part b for an orifice pressure drop of 218 psi. Show *numerical* values for peaks and valleys.

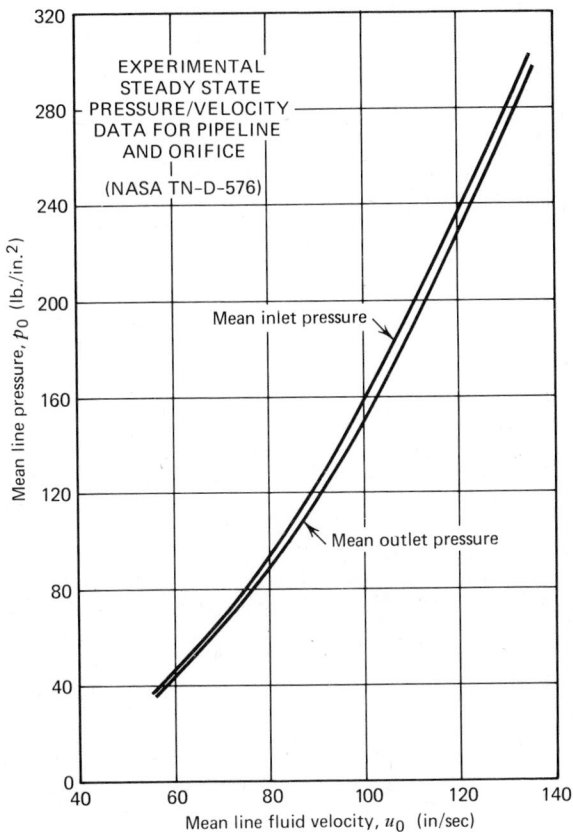

Figure P9.2

(d) Repeat part c for an orifice pressure drop of 49 psi except plot *accurate* graphs using SPEAKEASY.

(e) For conditions as in part d, suppose an air bubble of volume 0.54 in.^{3} is trapped at the orifice. Replace Eq. (9.39) by one that models this situation, explaining your reasoning. Use this new equation to find $P_p(0,s)/V_p(0,s)$ for such a system, then graph frequency response as in part d and comment on results.

9.8 In Fig. P9.3 we wish to model longitudinal pipe vibrations in our fluid pipeline model by treating the pipe itself as a one-lump spring/mass system. Combine the distributed-parameter fluid model of Equations 9.38 with the lumped pipe model to get an overall system model for the transfer function $V_p(L,s)/V_p(0,s)$. Be careful to note that the orifice pressure drop

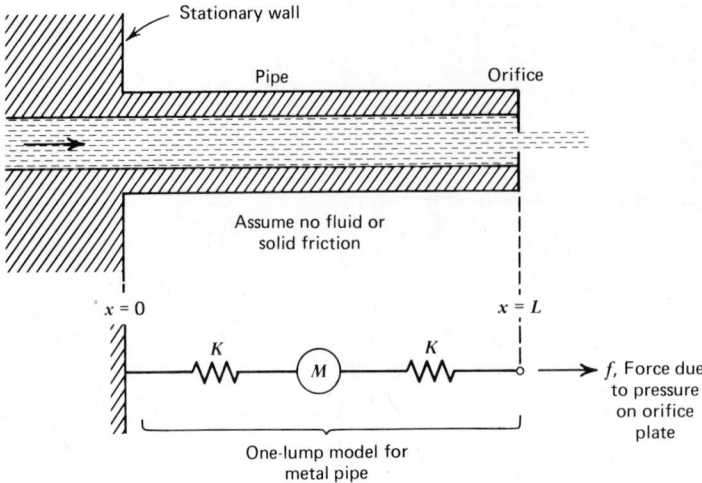

Figure P9.3

is proportional to the *relative* velocity of fluid and orifice plate. For $K_{or}=0$ (closed-end pipe) get $V_p(L,i\omega)/V_p(0,i\omega)$ and develop a graphical method for finding the frequencies at which infinite peaks occur.

9.9 For the system of Fig. 9.20, obtain the system characteristic equation needed for analytical stability studies.

9.10 In Eq. (9.125) the portion of pump outflow used by the stroke-adjusting piston has been neglected. Modify the model to include this effect for those systems where it might not be negligible and then redraw Fig. 9.20.

9.11 In the hydraulic drive system of Fig. P9.4, inertia J_1 represents a rotating machine part whose angular velocity should faithfully follow the com-

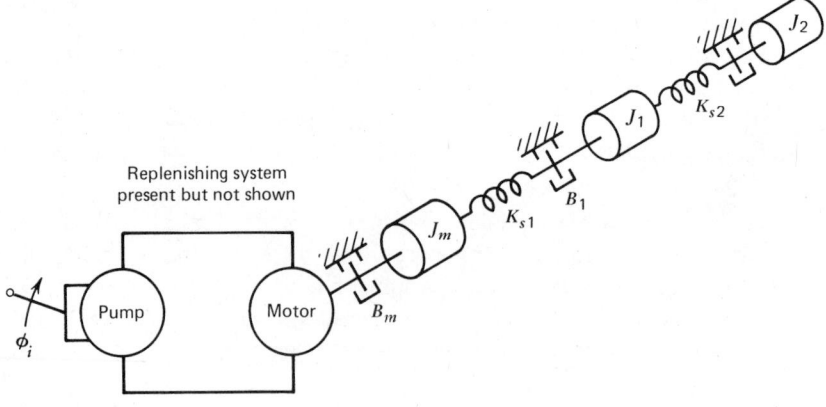

Figure P9.4

mands of pump stroke input ϕ_i. We wish to find the frequency range for which such accurate response is possible by examining the sinusoidal transfer function relating the angular velocity of J_1 to ϕ_i.

(a) Derive the complete set of differential equations that describe this system.

(b) Obtain the transfer function $(\omega_1/\phi_i)(i\omega)$.

(c) If only numerical results are wanted, the tedious algebra of part b can be avoided by use of a matrix frequency-response program (see Section 5.12). Using the numerical values given below, use such a program to compute and graph the desired frequency response over the frequency range 2 to 3450 rad/sec.

$$V = 5.0 \text{ in.}^3, \quad M_B = 100,000 \text{ psi}$$

$$J_l = J_2 = 1.0 \text{ in.-lb}_f\text{-sec}^2, \quad B_1 = B_2 = 0.0$$

$$B_M = 0.02 \text{ in.-lb}_f/(\text{rad/sec})$$

$$K_{tl} = 0.0003 \; (\text{in.}^3/\text{sec})/\text{psi}$$

$$\omega_p D_p = 10.0 \; (\text{in}^3/\text{sec})/\text{rad}, \quad d_m = 0.4 \text{ in}^3/\text{rad}$$

$$J_m = 0.0011 \text{ in.-lb}_f\text{-sec}^2$$

$$K_{s1} = K_{s2} = 3940 \; (\text{in.-lb}_f)/\text{rad}$$

(d) The highly resonant behavior observed in part c is due to light damping. Damping in this system arises from B_1, B_2, B_M, and leakage K_{tl}. A common practice in hydraulic systems, when more damping is needed, is to pipe in a needle valve across the motor parts to serve as an intentional (and easily adjustable) "leak." Too much leakage, of course, wastes power and reduces accuracy. Rerun the frequency response of part c with $K_{tl} = 0.003$ to determine the effect of this design change.

9.12 In the system of Fig. P9.5, the pump speed cannot be assumed constant and the motor is also variable displacement with stroke ϕ_{im}. Assume that the pump drive torque T_D, pump stroke ϕ_{ip}, and motor stroke ϕ_{im} have been held constant to establish a steady-state operating point. Now let T_D remain constant but let ϕ_{ip} and ϕ_{im} undergo small dynamic perturbations. Make necessary approximations to obtain a set of linear, constant-coefficient equations for this system.

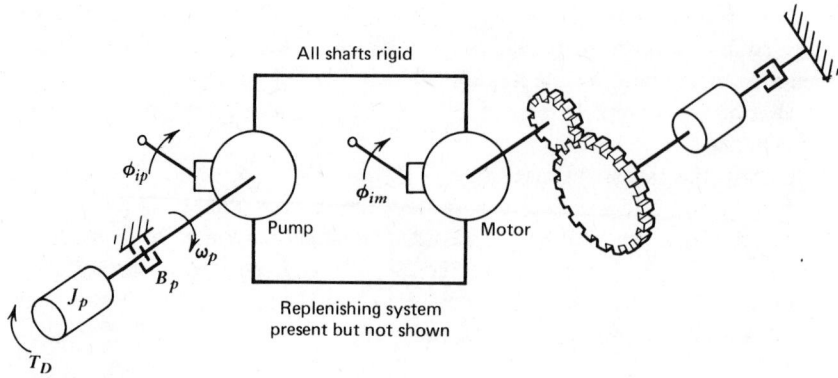

Figure P9.5

9.13 Consider the dual-motor drives of Fig. P9.6.

(a) In some applications we desire the two motors to run at nominally equal speeds. Which configuration (A or B) would you suggest and why? In other applications we wish the two motors to produce equal torques. Which configuration would you suggest now, and why?

(b) Derive linear differential equation models for System A.

(c) Derive linear differential equation models for System B.

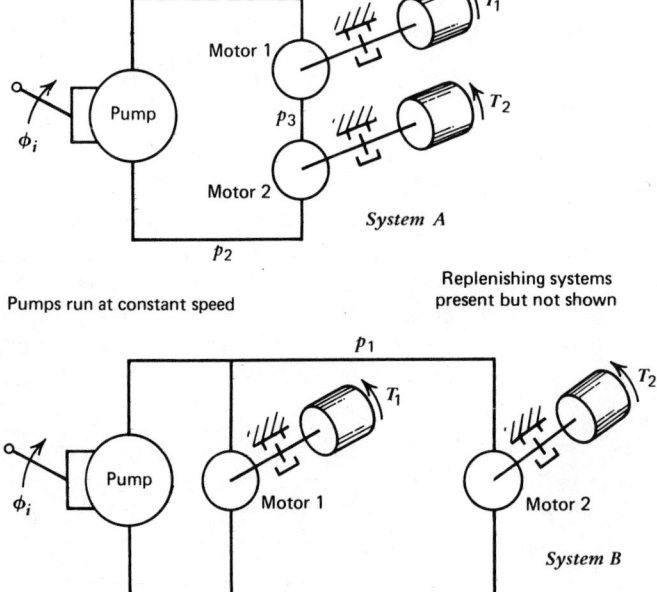

Figure P9.6

CHAPTER 10

Electromechanical Systems

In a single short chapter it is of course not possible to develop the engineering science background needed for a thorough and general treatment of the interactions between electric/magnetic fields and moving solids and fluids. Such a background is necessary to theoretically relate material properties and machine dimensions to coefficients that appear in system equations, and may be obtained from specialist texts[1] in the field. However, system engineers often need not deal at this level of detail, simply accepting the existence of a physical effect relating mechanical and electrical quantities and perhaps experimentally determining numerical values of coefficients for use in dynamic models. We accept this viewpoint and now proceed to analyze several simple but useful devices.

10.1 THE ELECTRODYNAMIC VIBRATION SHAKER

When an electrical command signal is to be transduced into mechanical force and/or motion, the "moving-coil" type of device is probably the most common. While we will concentrate on its application to vibration shakers, our model will be applicable to many other closely related devices such as rotary DC motors of various kinds, loudspeakers, high-speed galvanometers for oscillographic recorders and optical mirror scanners, torqueing coils for gyroscopic instruments, linear motors for positioning heads on computer disk memories, and so on. In all these cases a current-carrying coil is located in a steady magnetic field provided by permanent magnets in small devices and electrically excited wound coils in large ones. Two electromechanical effects ("motor" and "generator") are observed in such configurations. Motion of the coil through the field causes a voltage proportional to velocity to be induced into the coil, while passage of current through the coil causes it to experience a magnetic force proportional to the current. These two effects should be familiar from basic physics courses, and when the coil is of circular configuration, of n turns of radius r meters and in a

[1]Woodson, H. H. and J. R. Melcher, *Electromechanical Dynamics*, John Wiley and Sons, N.Y., 1968.

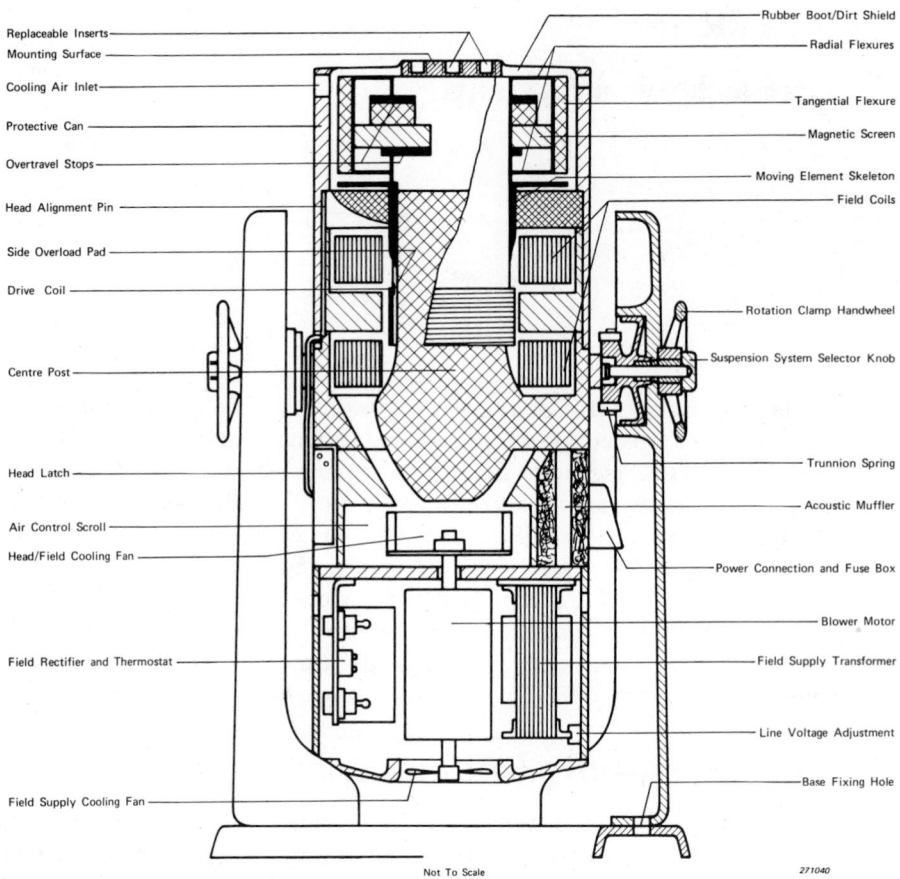

Figure 10.1 Construction details of electrodynamic vibration shaker.

radial magnetic field of magnetic flux density B webers/m^2 = (volt-sec)/m^2, one can estimate the coefficients from:

$$K_{f/i} \triangleq \text{N/amp} = K_{e/\dot{x}} \triangleq \text{V}/(\text{m/sec}) = 2\pi nrB \qquad (10.1)$$

(For permanent magnets, maximum values of B are the order of 1.0.) Note that in the metric system the force/current and voltage/velocity coefficients are numerically equal; in the British unit system they would be proportional.

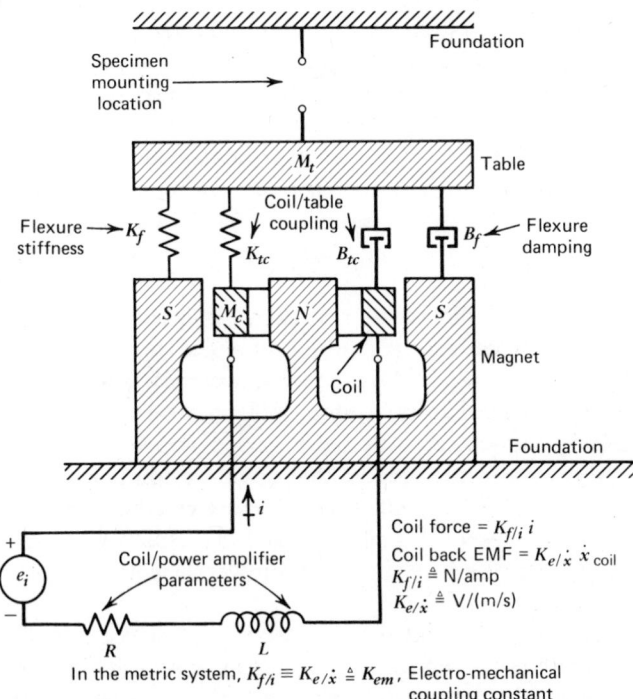

Figure 10.2   Electromechanical model of vibration shaker.

· Turning now to vibration shakers specifically, Fig. 10.1 shows some construction details of a practical machine[2] while Fig. 10.2 is the idealized model. Full-scale force capabilities for different models range from about 20 to 150,000 N, with permanent magnet fields being used below about 250 N. The flexure K_f is an intentional soft spring (stiff, however, in the *radial* direction) that serves to guide the axial motion of the coil and table without the need for conventional sliding or rolling bearings whose friction, wear, and lost motion are unacceptable in this type of application. Flexure damping B_f is usually intentional, fairly strong, and obtained not by mechanical dashpots but by laminated construction of the flexure spring, using layers of metal, elastomer, plastic, and so on. The coupling of the coil to the shaker "table" would ideally be rigid so that magnetic force is transmitted undistorted to the mechanical load. Thus K_{tc} and B_{tc}

---

[2]B. and K. Instruments, Cleveland, Ohio.

represent "parasitic" effects rather than intentional spring and damper elements. Spring stiffness K_{tc} is generally large while B_{tc}, due to internal losses in stressed metal, coil potting compounds, and so on is quite small. When the shaker is used as a force-producer to excite a specimen or mechanical structure, the table is really unnecessary and then represents simply the unavoidable mass (kept as small as possible) of this portion of the shaker. When the shaker is used as a motion-producer (such as when testing an automobile radio for a certain vibratory motion environment), then an actual table or fixture to support the tested package may be necessary.

When a specimen such as the radio mentioned above is mounted on the table to move freely with it, or when exciting force is applied to a structure mounted between the table and the foundation, these mechanical subsystems become part of the overall electromechanical system and must be included in the analysis. By characterizing the shaker with an unloaded transfer function and an output impedance, a general shaker model that could be combined with any specimen model (characterized by its input impedance) can be obtained using impedance methods of Chapter 7. This approach is explored in the problems at the end of this chapter. We here take the narrower view and analyze the system for a *specific* specimen.

If a package to be shaken is rigidly bolted to the table and behaves mainly as a mass element, we can analyze the entire system just as it appears in Fig. 10.2 since the package mass just becomes part of M_t. Similarly, if a specimen is essentially a single spring and/or damper and is fastened between the table and foundation, it experiences the same relative motions as K_f and B_f and may thus be included with them in the analysis. We could also in this case lump the specimen's mass at the table attachment point and include it in M_t. Our analysis will thus be useful for the bare shaker as shown in Fig. 10.2 or for the two kinds of specimens just described. Application of Newton's law to the table and coil masses gives:

$$- K_f x_t - B_f \dot{x}_t - K_{tc}(x_t - x_c) - B_{tc}(\dot{x}_t - \dot{x}_c) = M_t \ddot{x}_t \tag{10.2}$$

$$K_{tc}(x_t - x_c) + B_{tc}(\dot{x}_t - \dot{x}_c) + K_{f/i} i = M_c \ddot{x}_c \tag{10.3}$$

In the electric circuit, e_i is the output voltage of the electronic power amplifier that drives the shaker, while R and L are the total circuit resistance and inductance, including contributions from both the shaker coil and the amplifier output circuit. Kirchoff's voltage loop law gives:

$$- e_i + iR + L\frac{di}{dt} + K_{e/\dot{x}}\dot{x}_c = 0 \tag{10.4}$$

Of the three unknowns (outputs) x_t, x_c, and i, the one of most obvious interest is x_t, so we solve for it, finally getting:

$$\frac{X_t}{E_i}(s) = \frac{\dfrac{K_{f/i}}{K_f R}\left(\dfrac{B_{tc}}{K_{tc}}s+1\right)}{\dfrac{M_t M_c L}{K_f K_{tc} R}s^5 + \left[\dfrac{L}{R}\dfrac{B_{tc}}{K_{tc}}\dfrac{(M_t+M_c)}{K_f} + \dfrac{M_c M_t}{K_f K_{tc}} + \dfrac{L}{R}\dfrac{B_f}{K_f}\dfrac{M_c}{K_{tc}}\right]s^4 + \cdots}$$

$$\left[\dfrac{L}{R}\left(\dfrac{M_c}{K_{tc}} + \dfrac{M_c+M_t}{K_f}\right) + \dfrac{M_t K_{f/i}^2}{K_f K_{tc} R} + B_{tc}\left(\dfrac{M_t+M_c}{K_f K_{tc}}\right) + B_f\left(\dfrac{LB_{tc}}{K_f K_{tc} R} + \dfrac{M_c}{K_f K_{tc}}\right)\right]s^3 + \cdots$$

$$\left[\dfrac{M_c}{K_{tc}} + \dfrac{M_c+M_t}{K_f} + B_f\left(\dfrac{L}{RK_f} + \dfrac{B_{tc}}{K_f K_{tc}} + \dfrac{K_{f/i}^2}{RK_f K_{tc}}\right) + B_{tc}\left(\dfrac{L}{RK_{tc}} + \dfrac{K_{f/i}^2}{RK_f K_{tc}}\right)\right]s^2 + \cdots$$

$$\left[\dfrac{L}{R} + \dfrac{K_{f/i}^2}{R}\left(\dfrac{1}{K_{tc}} + \dfrac{1}{K_f}\right) + \dfrac{B_f}{K_f} + \dfrac{B_{tc}}{K_{tc}}\right]s+1$$

$$(10.5)$$

If only numerical frequency-response values are of interest, the tedious and error-prone algebraic reduction needed to obtain (10.5) can be avoided by applying a matrix frequency-response program directly to Equations (10.2)–(10.4). If transient response is of interest (as it is when shakers are used for shock testing), digital simulation applied directly to the original three equations again avoids the algebraic manipulation. Using available data from the literature[3] and estimating those coefficients not directly quoted, typical numerical values for a 5300 N peak force shaker of ±1.2 cm stroke with no specimen attached are:

$$L = 0.0012 \text{ h}, \qquad R = 3.0\ \Omega, \qquad K_{tc} = 8.16\times10^8 \text{ N/m}$$

$$K_{f/i} = 190 \text{ N/amp} = K_{e/\dot{x}} = 190 \text{ V/(m/sec)}, \qquad B_{tc} = 3850 \text{ N/(m/sec)}$$

$$M_c = 1.815 \text{ Kg}, \qquad M_t = 6.12 \text{ Kg}, \qquad K_f = 6.3\times10^5 \text{ N/m}, \qquad B_f = 1120 \text{ N/(m/sec)}$$

Using these values, the frequency response $\ddot{x}_t/e_i$ is as in Fig. 10.3. We computed this, rather than x_t/e_i, for comparison with Fig. 10.4,[3] which used the same

---

[3]Chapman, C. P., Derivation of the Mathematical Transfer Function of a Electrodynamic Vibration Exciter, Jet Propulsion Lab., Pasadena, Calif., Rept. 32-934, 1966.

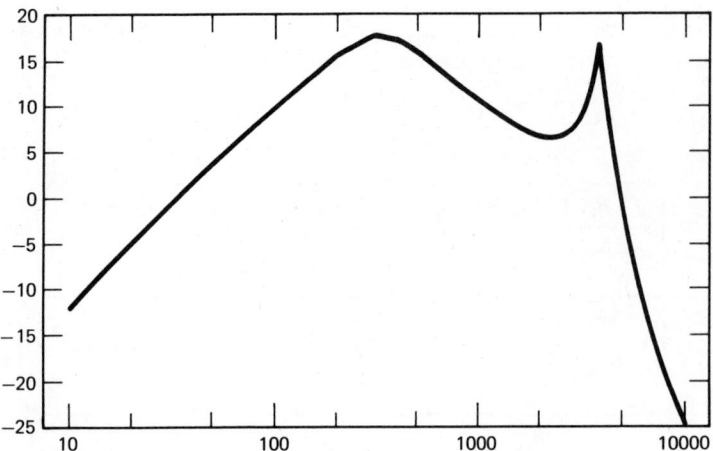

Figure 10.3   Theoretical frequency response of vibration shaker.

model as Eq. (10.5) but assumed B_f and B_{tc} to be zero, while we attempted to estimate B_f from a flexure damping ratio of 0.25 given elsewhere in the literature and B_{tc} from a "guessed" value of $\zeta = 0.05$ for the coil damping ratio. It appears that B_f and/or B_{tc} are too high. One might try $B_{tc} \approx 800$ ($\zeta_{coil} \approx 0.01$) in attempting to match the height of the high-frequency peak. The model also overestimates the frequency of this peak; one might try lowering K_{tc} and/or raising M_c to better fit this feature. Except for these numerical details, the model clearly predicts the shape of the measured response.

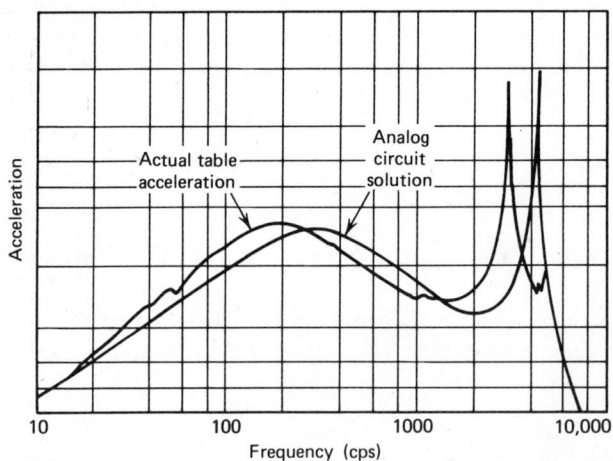

Figure 10.4   Comparison of theoretical and measured shaker frequency response.

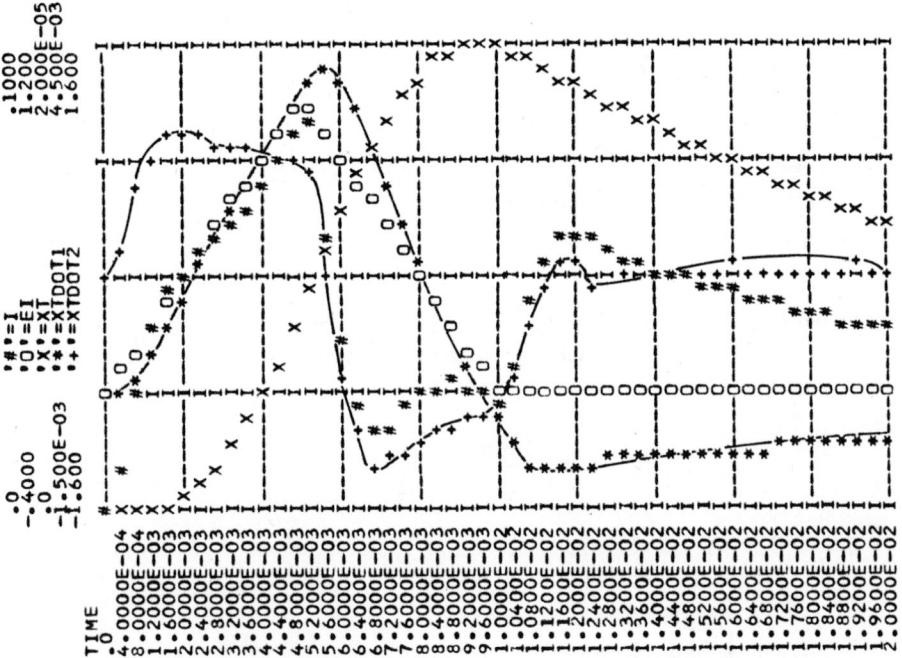

Figure 10.5   Pulse response of shaker.

When using a shaker to generate transient shock pulses, it is not usually obvious what waveform e_i should have if we desire a certain waveform in x_t, $\dot{x}_t$, or $\ddot{x}_t$. Once one has a valid shaker model, digital simulation allows easy investigation of this question and tailoring of e_i to obtain the desired motion. As an example, Equations (10.2)–(10.4) (using numerical values as originally given) were run on CSMP for the triangular pulse $e_i(t)$ shown in Fig. 10.5. We see that table velocity most nearly follows the triangular shape, and one can also estimate how the shape of $e_i(t)$ might be changed to cause $\dot{x}_t(t)$ to follow the desired triangular form even better. These ideas can be quickly tried out on CSMP and adjustments made until good correspondence is achieved.

10.1.1   DYNAMICS OF SHAKER PUSH ROD

When a shaker is used to excite a structure or machine for dynamic test purposes, the shaker and structure are often connected by a relatively slender "quill shaft" or push rod. Since such axial force members also occur frequently as elements of machines and structures, a general dynamic model may be of considerable utility. We will now develop a two-port model for such an element using the concepts of Chapter 7. This "mechanical transmission line" will share many behavioral features with the fluid transmission lines of Chapter 9.

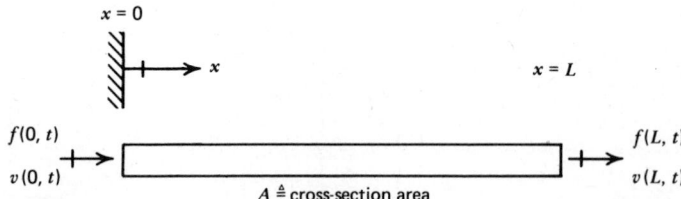

Figure 10.6 Longitudinal rod vibration.

Using methods analogous to those used on the torsional system of Section 4.1, one can obtain[4] the partial differential equation for longitudinal vibration of the slender rod in Fig. 10.6 as

$$\frac{\partial^2 u}{\partial x^2} = \frac{\rho}{E}\frac{\partial^2 u}{\partial t^2} \tag{10.6}$$

$u \triangleq u(x,t) \triangleq$ axial displacement of any plane in the rod, away from its neutral position

$\rho \triangleq$ material mass density

$E \triangleq$ material modulus of elasticity

$f(0,t), f(L,t) \triangleq$ forces of external "devices" on the rod

$v(0,t), v(L,t) \triangleq$ velocities of rod cross sections

$\frac{\partial u}{\partial x} \triangleq$ strain at any cross section

We can relate f and v to u by:

$$v(0,t) = \frac{\partial u}{\partial t}(0,t), \qquad f(0,t) = -EA\frac{\partial u}{\partial x}(0,t)$$

$$v(L,t) = \frac{\partial u}{\partial t}(L,t), \qquad f(L,t) = EA\frac{\partial u}{\partial x}(L,t) \tag{10.7}$$

Laplace transforming eq. (10.6) with zero initial conditions:

$$\frac{d^2 U}{dx^2} - \frac{\rho s^2}{E}U = 0 \tag{10.8}$$

$$U(x,s) = C_1 e^{-s\sqrt{\rho/E}\,x} + C_2 e^{s\sqrt{\rho/E}\,x} \tag{10.9}$$

[4]Doebelin, E. O., *System Dynamics*, C. E. Merrill, Columbus, Ohio, 1972, p. 451.

Manipulation of (10.7) and (10.9) ultimately leads to:

$$F(L,s) = Z_0 \sinh(T_p s) V(0,s) - \cosh(T_p s) F(0,s)$$

$$V(L,s) = \cosh(T_p s) V(0,s) - \frac{1}{Z_0} \sinh(T_p s) F(0,s)$$

(10.10)

where

$$\sqrt{E/\rho} = \text{propagation velocity}$$

$$T_p \triangleq L/\sqrt{E/\rho} = \text{propagation time}$$

$$Z_0 \triangleq A\sqrt{\rho E} = \text{characteristic impedance}$$

Equations (10.10) could be made identical in form to the fluid line equations of Chapter 9 simply by redefining $F(L,s)$ to be the force *of* the rod *on* the external device, thereby changing its sign.

Just as the fluid line equations yielded specific useful results when combined with those of attached devices, so also will our present relations. Suppose we attach at $x = L$ a rigid lumped mass M. This gives us the equation

$$-F(L,s) = MsV(L,s)$$

(10.11)

Note that a positive (to the right) acceleration of the mass corresponds to a negative (to the left) value of force *by* the mass *on* the rod. Combining (10.11) with (10.10) allows calculation of several useful transfer functions such as

$$\frac{V(L,s)}{V(0,s)} = \frac{1}{\cosh(T_p s) + \dfrac{Ms}{Z_0} \sinh(T_p s)}$$

(10.12)

Going to frequency response:

$$\frac{V(L,i\omega)}{V(0,i\omega)} = \frac{1}{\cos(\omega T_p) - \dfrac{M\omega}{Z_0} \sin(\omega T_p)}$$

(10.13)

This response will exhibit infinite peaks (natural frequencies) when

$$\cos(\omega T_p) = \frac{M\omega}{Z_0} \sin(\omega T_p)$$

(10.14)

or

$$\tan(\omega T_p) = \frac{Z_0}{M\omega}$$

(10.15)

This transcendental equation has no analytical solution. However, for specific numerical values a graphical solution as in Fig. 10.7 is possible. Note that for high frequencies the natural frequencies come closer and closer to $n\pi/T_p$, where $n = 1,2,3,\ldots$. This graphical solution technique is implemented numerically in the SPEAKEASY routine called ROOTS (F:X). We merely define any function F whose zeros we desire and stipulate a range of independent variable X within which we wish to search for zeros, giving also the increment in which X is to change. The routine then evaluates the function at all X values, watching for sign changes. When a sign change is found, it interpolates between the two neighboring X values to get a more accurate location of the zero. In our current problem, using some specific numerical values, the complete program could be as follows.

$$W = GRID(0,4E5,200)$$
$$F = TAN(5.0892E-5*W) - 15268/W$$
$$ROOTS(F:W)$$

Here frequency W goes from 0 to 400,000 rad/sec in steps of 200. The statement ROOTS(F:W) returns all the zeros found in the specified range. If higher accuracy is required, one can rerun the program with finer increments, covering only the neighborhoods of the first set of roots found.

Of course one can just calculate the amplitude ratio and phase angle of Eq. (10.13) without first trying to find the natural frequencies separately, giving the

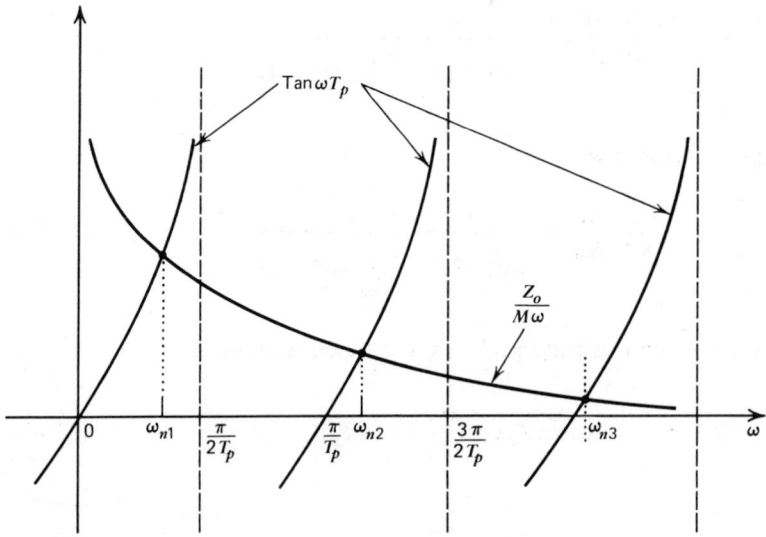

Figure 10.7 Graphical solution for natural frequencies.

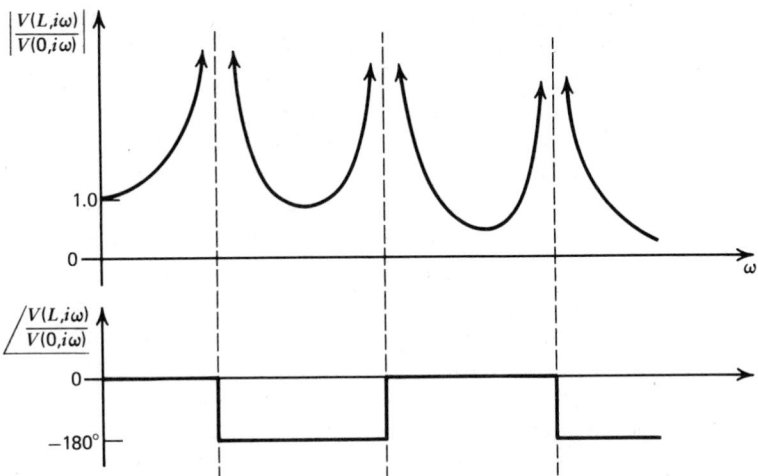

Figure 10.8   Frequency response of longitudinal vibrations.

curves of Fig. 10.8. The physical meaning of $V(L,i\omega)/V(0,i\omega)$ is of course that location $x=0$ is driven with a fixed sinusoidal *motion* (velocity) and we ask for the motion (velocity) caused by this excitation at location $x=L$. Natural frequencies revealed by this analysis will be identical with those of free vibration for a rod *fixed* at $x=0$ and free at $x=L$. If we decide to find $V(L,i\omega)/F(0,i\omega)$, this corresponds to applying a prescribed *force* at $x=0$ (motion *not* prescribed) and asking for the resulting motion at $x=L$. This gives a *different* set of natural frequencies since it corresponds to *both* ends being free to vibrate.

It should be clear that Equations (10.10) can be combined with force/velocity (impedance) relations for *any* mechanical subsystem connected at the rod ends, not just simple elements like the mass studied above. Our rod model is thus applicable to a very wide variety of practical situations, including those involving coupling of fluid and mechanical effects as described in Chapter 9.

10.2   MAGNETIC LEVITATION SYSTEM FOR AN EXPERIMENTAL RAIL VEHICLE

Considerable research and development has been carried out, and currently continues, on rail transport systems using magnetic levitation for support and a linear induction motor for propulsion. A recent paper[5] gives design details on a test vehicle constructed to obtain operating experience with a low-speed system

[5]Fruechte, R. D., Nelson, R. H., and T. A. Radomski, Power Conditioning Systems for a Magnetically Levitated Test Vehicle, General Motors Res. Labs. Report GMR-2431, 1977.

intended for personal rapid transit (people mover) applications. Levitation and lateral guidance are provided by eight U-shaped, DC-controlled electromagnets that attract toward a U-shaped steel rail. Each of the four pairs of magnets (one pair at each vehicle corner) is individually controlled to maintain a vertical air gap of 7.5 mm and provide lateral support and damping. Vertical support of vehicle weight by an attractive (rather than repulsive) constant-current magnet is inherently unstable since attractive force decreases with air gap, and thus the slightest disturbance from equilibrium causes the vehicle to either fall or clamp to the rail. A feedback system is thus employed to stabilize this motion. The two magnets of each pair are laterally offset equal amounts (see Fig. 10.9); thus for equal currents the lateral forces cancel each other. For lateral motion control one current is increased and the other decreased by the same amount, giving the desired lateral force but *not* changing the vertical force.

Formulas for magnetic forces are often obtained by use of virtual work methods. For vertical force and motion (Fig. 10.10) assume current is constant

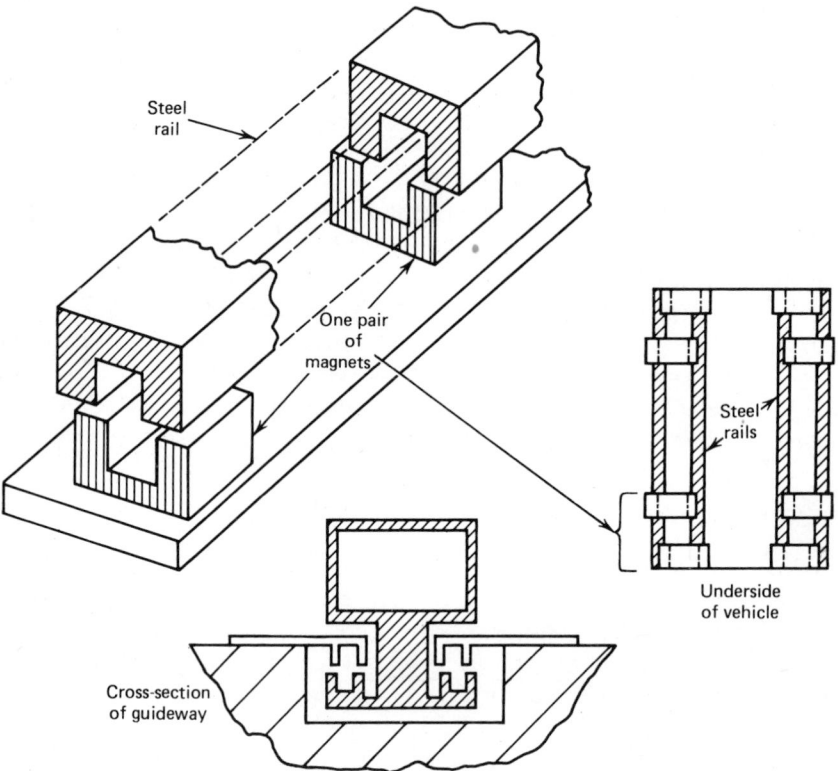

Figure 10.9   Magnetic levitation system.

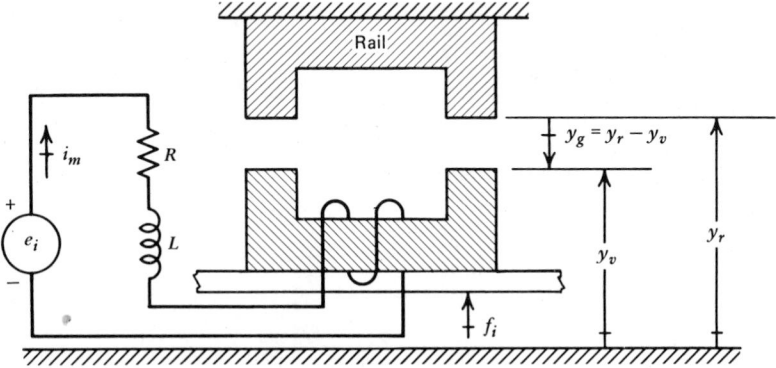

Figure 10.10   Model for vertical motion.

at i_{m0}, air gap is y_{g0}, and coil inductance is L_0. Note that when the air gap changes, inductance will also change, so L is a function of y_g. This function can be estimated from magnetic circuit theory or can be found numerically by measurements of L on an existing device. From basic physics we know that the energy stored in the inductor under the above conditions is:

$$E = \frac{L(y_{g0}) i_{m0}^2}{2} \tag{10.16}$$

Now let an infinitesimal displacement dy_g occur while the current is held constant. If the vertical magnetic force is f_{my}, mechanical work in amount $f_{my}\, dy_g$ is done. This work must equal the *change* in energy stored in the inductor:

$$\text{change in stored energy} = \frac{L(y_{g0}) i_{m0}^2}{2} - \frac{L(y_{g0} + dy_g) i_{m0}^2}{2}$$

$$= -\frac{(dL) i_{m0}^2}{2} \tag{10.17}$$

$$f_{my}\, dy_g = -\frac{(dL) i_{m0}^2}{2} \tag{10.18}$$

$$f_{my} = -\frac{i_{m0}^2}{2} \frac{\partial L}{\partial y_g}\bigg|_{y_g = y_{g0}} \tag{10.19}$$

We use a partial derivative for the rate of change of L with respect to y_g since L also depends on horizontal motions, which we study later. Using subscript p for small perturbations about an operating point:

$$f_{myp} = \frac{\partial f_{my}}{\partial i_m} i_{mp} + \frac{\partial f_{my}}{\partial y_g} y_{gp} = -i_{m0} \frac{\partial L}{\partial y_g} i_{mp} - \frac{i_{m0}^2}{2} \frac{\partial^2 L}{\partial y_g^2} y_{gp} \qquad (10.20)$$

Measured values for an air gap of 7.5 mm were:

$$i_{m0} = 1.85 \text{ amp}, \qquad \frac{\partial L}{\partial y_g} = -1300 \text{ h/m}, \qquad \frac{\partial^2 L}{\partial y_g^2} = 322{,}000 \text{ h/m}^2$$

Newton's law thus gives:

$$f_{vip} + 2410 i_{mp} - 550{,}000 y_{gp} = M_v \ddot{y}_{vp} = M_v(\ddot{y}_{rp} - \ddot{y}_{gp}) \qquad (10.21)$$

Here, f_{vi} is an external disturbing force (such as the weight of an entering passenger) and M_v is the share of the total vehicle mass allotted to a single magnet. Since total mass is 1800 kg and there are eight magnets, $M_v = 225$ kg.

Turning now to the electrical circuit, the inductance must here be treated with more care than most nonelectrical engineers are accustomed to, since it is not a simple constant. A fundamental relation for induced voltage e in a coil of area A exposed to uniform magnetic flux density B is:

$$e = \frac{d\lambda}{dt} = \frac{d}{dt}(BA) \qquad (10.22)$$

The quantity λ is called the flux linkage. When there is a linear relation beteen λ and current i (approximately true for our small-signal model), then

$$\lambda = Li \qquad (10.23)$$

where L is defined as the inductance. We can now write the current/voltage relation for inductance as

$$e_L = \frac{d\lambda}{dt} = \frac{d}{dt}(Li) = L\frac{di}{dt} + i\frac{dL}{dt} \qquad (10.24)$$

where the rightmost term is usually zero, but not in our present application. In Fig. 10.10 e_i represents the output voltage of an electronic power amplifier that drives the magnet coil, while R and L represent the total resistive and inductive effects of the amplifier output circuit and coil. Kirchoff's voltage loop law gives

$$e_{ip} - i_{mp}R - L\frac{di_{mp}}{dt} - i_{mo}\frac{\partial L}{\partial y_g}\frac{dy_{gp}}{dt} = 0 \qquad (10.25)$$

Measured values at the operating point are $L = 18$ h and $R = 120$ Ω, giving

$$e_{ip} - 120 i_{mp} - 18 \frac{di_{mp}}{dt} + 2410 \frac{dy_{gp}}{dt} = 0 \tag{10.26}$$

Treating y_{gp} as the desired output and eliminating i_{mp}, combination of the electrical and mechanical equations leads to:

$$Y_{gp} = \frac{-1.82 \times 10^{-6}(0.15s + 1)F_{vip} - 3.66 \times 10^{-5}E_{ip} + 4.08 \times 10^{-4}(.15s + 1)s^2 Y_{rp}}{6.14 \times 10^{-5}s^3 + 4.10 \times 10^{-4}s^2 - 6.20 \times 10^{-2}s - 1} \tag{10.27}$$

The sign changes in the characteristic equation mathematically substantiate the instability earlier anticipated intuitively. Of the various methods possible to stabilize this system, the one chosen employed a feedback scheme with an inductive noncontact gap sensor to measure y_g and an accelerometer to measure $\ddot{y}_v$. These signals were processed and combined such that the power amplifier output was given by:

$$E_{ip} = K_p y_{gp} + \frac{K_v s}{\tau s + 1} y_{gp} - \frac{K_v \tau}{\tau s + 1} \ddot{y}_{vp} - K_a \ddot{y}_{vp} \tag{10.28}$$

Desired system response was obtained with

$$K_p = 38300 \text{ V/m}, \qquad K_a = 22 \text{ V/(m/sec}^2)$$
$$K_v = 4100 \text{ V/(m/sec)}, \qquad \tau = 0.068 \text{ sec}$$

Working from Equations (10.27) and (10.28), or from the block diagram of Fig. 10.11, one obtains the equation of the stabilized system as:

$$Y_{gp} = \frac{-1.82 \times 10^{-6}(0.068s + 1)(0.15s + 1)F_{vip}}{(0.068s + 1)(0.208s + 1)(7.23 \times 10^{-4}s^2 + 0.0109s + 1)}$$

$$\frac{+1.13 \times 10^{-2}(3.67 \times 10^{-4}s^2 + 1.27 \times 10^{-2}s + 1)s^2 Y_{rg}}{(0.068s + 1)(0.208s + 1)(7.23 \times 10^{-4}s^2 + 0.0109s + 1)} \tag{10.29}$$

(A root-finder has been used to factor the characteristic polynomial in the denominator.) The system is now clearly stable with two negative real roots and an oscillatory term with frequency about 6 Hz and 0.2 damping ratio. Experimental tests on the actual vehicle verified these analytical results (see Fig. 10.12a).

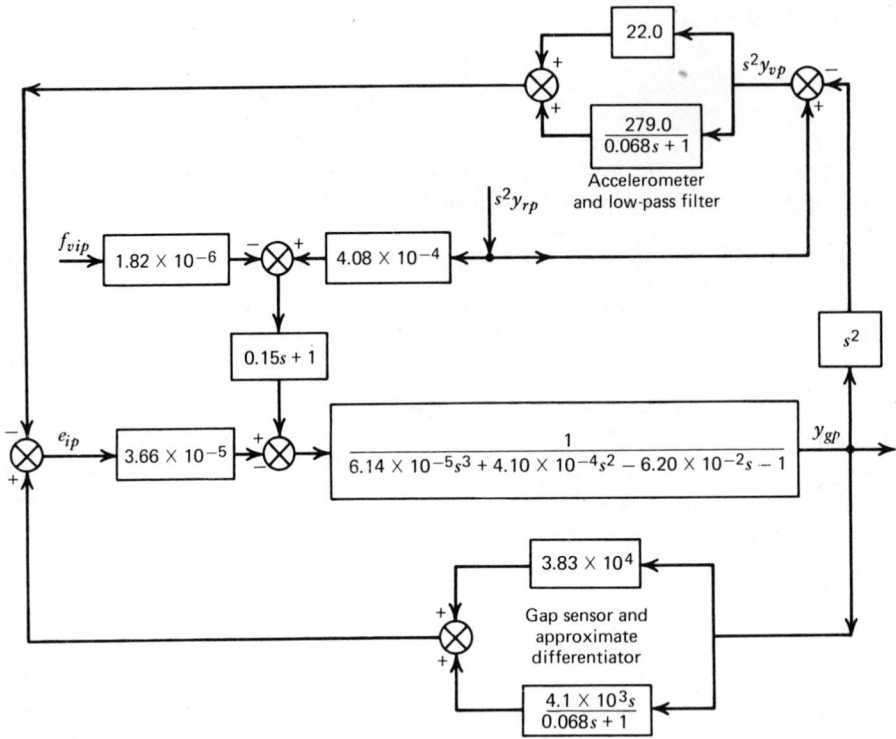

Figure 10.11   Feedback scheme to stabilize levitation system.

With the levitation system functioning properly we now turn our attention to the lateral guidance system. Figure 10.13 shows the results of experiments on the effect of lateral displacement on magnet vertical and lateral forces. We see that vertical force drops off as lateral displacement increases and that the ratio of lateral to vertical force varies linearly with displacement. For lateral control the magnets are considered in pairs, and Newton's law gives

$$f_{mx1} + f_{mx2} + f_{hi} = M_p \ddot{x} \qquad (10.30)$$

where

$$f_{hi} \triangleq \text{horizontal disturbance force}$$

$$f_{mx1} = -\frac{i_{m1}^2}{2} \frac{\partial L_1}{\partial x}, \qquad f_{mx2} = -\frac{i_{m2}^2}{2} \frac{\partial L_2}{\partial x} \qquad (10.31)$$

$$M_p \triangleq \tfrac{1}{4} \text{ of total mass}$$

The horizontal forces are related to the vertical forces through the results of

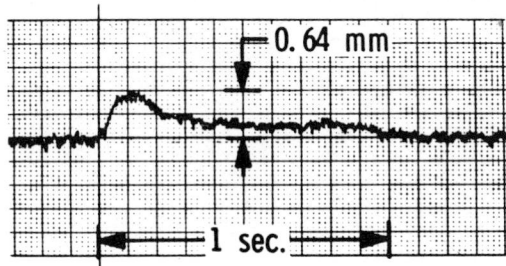

Magnet Gap Response to 90kg Passenger
Boarding Test Vehicle.

(a)

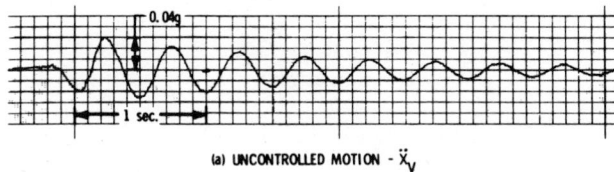

(a) UNCONTROLLED MOTION - $\ddot{X}_V$

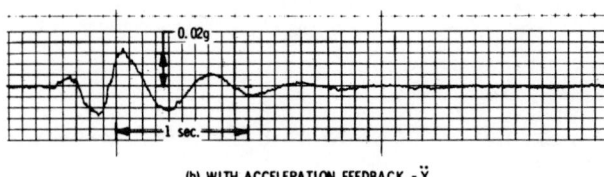

(b) WITH ACCELERATION FEEDBACK - $\ddot{X}_V$

Lateral Response to Initial Lateral
Displacement.

(b)

Figure 10.12 Measured response of levitation and lateral guidance systems.

Fig. 10.13:

$$f_{mx1} + f_{mx2} = 22(x_0 - x_p)f_{my1} + 22(-x_0 - x_p)f_{my2} \tag{10.32}$$

where x_0 is the "built-in" initial offset of each magnet. The vertical forces are given by:

$$f_{my1} = f_{my10} + \frac{\partial f_{my1}}{\partial y_g} y_{gp} + \frac{\partial f_{my1}}{\partial x} x_p + \frac{\partial f_{my1}}{\partial i_{m1}} i_{m1p} \tag{10.33}$$

$$f_{my2} = f_{my20} + \frac{\partial f_{my2}}{\partial y_g} y_{gp} + \frac{\partial f_{my2}}{\partial x} x_p + \frac{\partial f_{my2}}{\partial i_{m2}} i_{m2p} \tag{10.34}$$

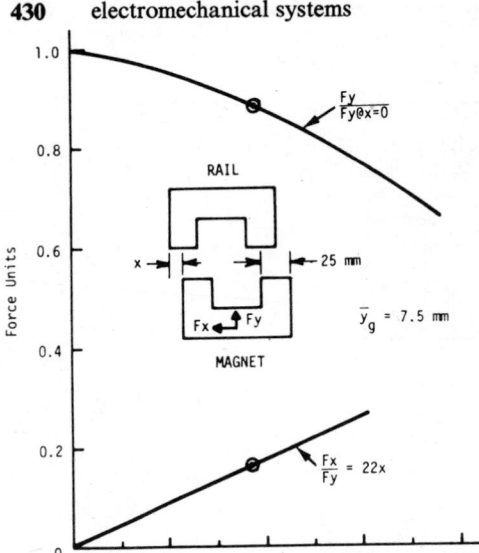

Figure 10.13   Measured magnetic force characteristics.

Due to symmetry:

$$\frac{\partial f_{my1}}{\partial i_{m1}} = \frac{\partial f_{my2}}{\partial i_{m2}}, \qquad \frac{\partial f_{my1}}{\partial y_g} = \frac{\partial f_{my2}}{\partial y_g}, \qquad \frac{\partial f_{my1}}{\partial x} = -\frac{\partial f_{my2}}{\partial x}$$

$$f_{my10} = f_{my20} \qquad (10.35)$$

Using these relations and neglecting all products of perturbation quantities gives:

$$44\left(-f_{my10} + x_0\frac{\partial f_{my1}}{\partial x}\right)x_p + 22x_0\frac{\partial f_{my1}}{\partial i_{m1}}(i_{m1p} - i_{m2p}) + f_{hi} = M_p\ddot{x}_p \qquad (10.36)$$

Numerical values are:

$$M_p = 450 \text{ kg}, \qquad x_0 = 0.0075 \text{ m}, \qquad \frac{\partial f_{my1}}{\partial x} = 59400 \text{ N/m}$$

$$f_{my10} = 2200 \text{ N}, \qquad \frac{\partial f_{my1}}{\partial i_{m1}} = 2400 \text{ N/amp}$$

giving

$$-7.73 \times 10^4 x_p + 396(i_{m1p} - i_{m2p}) + f_{hi} = 450\ddot{x}_p \qquad (10.37)$$

For the electric circuits:

$$e_{i1} - i_{m1}R_1 - L_1\frac{di_{m1}}{dt} - i_{m1}\frac{\partial L_1}{\partial x}\dot{x} - i_{m1}\frac{\partial L_1}{\partial y_g}\dot{y}_g = 0 \qquad (10.38)$$

$$e_{i2} - i_{m2}R_2 - L_2\frac{di_{m2}}{dt} - i_{m2}\frac{\partial L_2}{\partial x}\dot{x} - i_{m2}\frac{\partial L_2}{\partial y_g}\dot{y}_g = 0 \qquad (10.39)$$

The term in $\ddot{y}_g$ will be neglected since the levitation control is much faster and thus $y_g \approx$ const in the horizontal model. Going to perturbations and again making use of the symmetry:

$$e_{i1p} - R_1 i_{m1p} - L_1 \frac{di_{m1p}}{dt} - i_{m10} \frac{\partial L_1}{\partial x} \dot{x}_p = 0 \tag{10.40}$$

$$e_{i2p} - R_1 i_{m2p} - L_1 \frac{di_{m2p}}{dt} + i_{m20} \frac{\partial L_1}{\partial x} \dot{x}_p = 0 \tag{10.41}$$

Experimental measurements gave

$$\frac{\partial L}{\partial x} = 212 \text{ h/m}$$

which makes the circuit equations

$$e_{i1p} - 120 i_{m1p} - 18 \frac{di_{m1p}}{dt} - 392 \dot{x}_p = 0 \tag{10.42}$$

$$e_{i2p} - 120 i_{m2p} - 18 \frac{di_{m1p}}{dt} + 392 \dot{x}_p = 0 \tag{10.43}$$

Combining the mechanical and electrical equations leads to:

$$X_p = \frac{4.26 \times 10^{-5}(E_{i1p} - E_{i2p}) + 1.29 \times 10^{-5}(0.15s + 1)F_{hi}}{(0.179s + 1)(0.00488s^2 + 0.00522s + 1)} \tag{10.44}$$

This system's response was considered satisfactory except for the low damping ($\zeta \approx 0.04$) of the quadratic (see Fig. 10.12b). Feedback from a lateral accelerometer was used to augment stability by making

$$e_{i1p} - e_{i2p} = K_{ax}\ddot{x}_p = -208\ddot{x}_p \tag{10.45}$$

Combining this with Eq. (10.44) gives for the new closed-loop system

$$X_p = \frac{1.29 \times 10^{-5}(0.15s + 1)}{(0.12s + 1)(0.00725s^2 + 0.0637s + 1)} F_{hi} \tag{10.46}$$

which has improved damping of about 0.4, as seen in the time record of Fig. 10.12b. While the cited reference includes additional useful material on control of the propulsion motor, use of a programmable controller to manage overall system operation and safety features, and so on, we conclude our treatment here. We have seen how a relatively simple approach combining basic theory with experimental measurements has led to a valid model, useful both in understanding basic system response and also for designing augmentation systems where needed.

BIBLIOGRAPHY

1 Azmoon, M., Bohorquez, J. H., and R. A. Warp, "Desktop Plotter/Printer Does Both Vector Graphic Plotting and Fast Text Printing," *Hewlett–Packard Jour*, **29**, No. 13 (1978).

2 Bailey, E. M., *et al.*, "Optimizing the Performance of an Electromechanical Print Mechanism," *Hewlett–Packard Jour.*, Nov. (1978).

3 Blackburn, J. F., *et al.*, *Fluid Power Control, Electromagnetic Actuators*, John Wiley and Sons, N.Y., 1960, p. 322.

4 Blecke, W. F., "Designing PM Linear Actuators," *Machine Design*, April 21 (1977).

5 Brunetti, L., *et al.*, "A New Family of Intelligent Multi-Color X-Y Plotters," *Hewlett–Packard Jour.*, 1977.

6 Buehner, W. L., *et al.*, "Ink Jet Printer Technology," *IBM J. Res. Dev.*, **21**, No. 1 (1977).

7 Chi, C. C., LIM Guidance System Dynamics, Airesearch Corp. Rept. 73-9065, 1973.

8 Coffey, H. T., *et al.*, An Evaluation of the Dynamics of a Magnetically Levitated Vehicle, Fed. Rail. Adm. Rept. FRA-ORD & D-74-41, 1974.

9 Coffey, H. T., Colton, J. D. and K. D. Mahrer, Study of a Magnetically Levitated Vehicle, Fed. Rail. Adm. Rept. FRA-RT-73-24, 1973.

10 *DC Motors, Speed Controls, Servo Systems*, ElectroCraft Corp., Hopkins, Minn., 1972.

11 Drysdale and Jolley, *Electrical Measurement Instrumentation, Eddy-Current Damping*, John Wiley and Sons, N.Y., 1952, p. 136.

12 Fenoglio, J. A., *et al.*, "A High-Quality Digital X-Y Plotter Designed for Reliability, Flexibility and Low Cost," *Hewlett–Packard Jour.*, Feb. (1979).

13 Gehmlich, D. K. and S. B. Hammond, *Electromechanical Systems*, McGraw-Hill, N.Y., 1967.

14 Gourishankar, V., *Electromechanical Energy Conversion*, Int. Textbook Co., Scranton, PA, 1966.

15 Holzbock, W. F., "Simplified Equations Speed Moving Coil Design," *Contr. Eng.*, May (1961).

16 Jayawant, B. V., *et al.*, "Development of 1-Ton Magnetically Suspended Vehicle Using Controlled DC Electromagnets," *Proc. IEE*, **123**, No. 9, Sept. (1976).

17 Johnson, H. L., "Permanent Magnet Design Guidelines for Linear Actuators," *Electromechanical Design*, Dec. (1970).

18 Koenig, H. E. and W. A. Blackwell, *Electromechanical Systems Theory*, McGraw-Hill, N.Y., 1961.

19 Kuo, B. C., ed., Sym. on Incremental Motion Control Systems and Devices, Univ. of Ill. EE Dept., 1972.

20 Majmudar, H., *Electromechanical Energy Converters*, Allyn and Bacon, Boston, 1965.

21 Meisenholder, S. G. and T. C. Wang, Dynamic Analysis of an Electromagnetic

Suspension System for a Suspended Vehicle System, Fed. Rail. Adm. Rept. FRA-RT-73-1, 1972.

22 Merritt, H. E., *Hydraulic Control Systems*, Permanent Magnet Torque Motors, John Wiley and Sons, N.Y., 1967, p. 177.

23 Nelson, R. H., *et. al.*, Electric Vehicle Simulation Program, Gen'l. Mtrs. Res. Lab. Rept. GMR-2758, 1978.

24 Neubert, H. K., Bilateral Electromechanical Transducers; A Unified Theory, Roy. Aircraft Est. Tech. Rept. 68248, 1968.

25 Raumann, N. A., Application of Eddy Current Clutches to Tracking Antenna Drive Systems, NASA TM X-55084, 1964.

26 Reich, S., A Review of Electromechanical Beam Choppers, Electro-Optical Systems Design, Nov. 1976.

27 Reitz, J. R., *et al.*, Preliminary Design Studies of Magnetic Suspensions for High-Speed Ground Transportation, Fed. Rail. Adm. Rept. FRA-RT-73-27, 1973.

28 Robinson, D. J., Dynamic Analysis of Permanent Magnet Stepping Motors, NASA TN D-5094, 1969.

29 Schmidt, H. A., Stepover Needlebar Positioner, Tech. Bull. 129, Moog Inc., E. Aurora, N.Y., 1978.

30 Schmitz, N. L. and D. W. Novotny, *Introductory Electromechanics*, Ronald Press, N.Y., 1965.

31 Simpson, H., "Printers '77," *Digital Design*, Oct. (1977).

32 Stickel, H. P., "New 50-Megabyte Disc Drive," *Hewlett–Packard Jour*. (1977).

33 Wieselman, I. L., "Trends in Computer Printer Technology," *Computer Design*, Jan. (1979).

34 Woodson, H. H. and J. R. Melcher, *Electromechanical Dynamics*, John Wiley and Sons, N.Y., 1968.

35 Wulfing, G., "Speeding a Reel-to-Reel Tape Drive," *Digital Design*, Oct. (1976).

36 Zschau, E. V. W., "The Quiet Revolution in Ink Jet Printing," *Digital Design*, Sept. (1978).

PROBLEMS

10.1 Using methods of Section 7.3, treat the vibration shaker of Fig. 10.2 as a two-port device with ports at e_i, i, and f_t, v_t (where f_t, v_t are respectively, the external force and velocity at mass M_t) and find the four-pole parameters. Check your results using Eq. (7.3). Now obtain the four-pole parameters for the system of Fig. P10.1 and use these together with the previous result to obtain $(f_b/e_i)(s)$ when point "a" is connected to M_t and point b to the foundation.

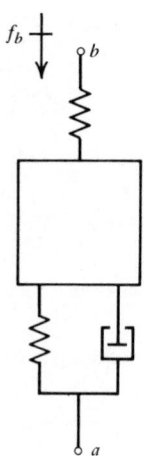

Figure P10.1

10.2 Repeat Problem 10.1 but now use the impedance methods of Section 7.4 to deal with the coupling of the shaker to the load of Fig. P10.1.

10.3 Simplified models of the vibration shaker of Fig. 10.2 may give physical insight into its behavior, which is obscured by the complexity of Eq. (10.5). Consider the following.
  (a) Assume an "infinitely stiff" specimen is mounted between table and foundation, preventing table motion. Also assume K_{tc} infinitely stiff so that coil cannot move. Application of $e_i(t)$ now produces a specimen force $f_s(t)$, but no motion. Derive $(f_s/e_i)(i\omega)$ and find the frequency range for which amplitude ratio is "flat" $(+0.0, -3$ db), using numerical values from Section 10.1. Sometimes power amplifiers for shakers are designed to operate as current (i_i) sources rather than as voltage (e_i) sources. For such a case, find $(f_s/i_i)(s)$ and comment on the significance of L and R.
  (b) Now consider table motion zero, coil mass negligible, but K_{tc} finite, allowing coil to move. Find $(x_c/e_i)(s)$ and $(f_s/e_i)(s)$, sketch frequency-response curves and compare $(f_s/e_i)(i\omega)$ (numerical values) with results of part a.
  (c) This part is the same as part b except now M_c is not negligible. In reducing equations to final form, be on the lookout for a cubic polynomial in s being formed by a product of form $(as+b)(cs^2+ds+e)$ plus an additional term of form gs that adds to the s term in the cubic. Show that if $B_{tc}R + K_{tc}L \gg K_{f/i}^2$, then the term of form gs is negligible and the cubic can be carried forward already factored, in the form $(\tau s+1)(s^2/\omega_n^2+2\zeta s/\omega_n+1)$. Show the two requested transfer functions each in two forms; one with the cubic factored (good approximation when the above inequality holds) and one with the

cubic expanded out ("*gs*" term included). To check on the accuracy of the approximation for the numerical values given in the text, use available computer programs to calculate and graph $|(f_s/e_i)(i\omega)|$ for both exact and approximate formulas.

10.4   For the rod of Fig. 10.6, if the right end is held fixed, find $V(0,s)/F(0,s)$, $F(L,s)/F(0,s)$, and $F(L,s)/V(0,s)$ and sketch the form of the frequency-response curves.

10.5   For the rod of Fig. 10.6, if the right end is free, find $V(0,s)/F(0,s)$, $V(L,s)/F(0,s)$, and $V(L,s)/V(0,s)$ and sketch the form of the frequency-response curves.

10.6   Repeat Problem 10.5 if a lumped ideal spring is attached between the rod's right end and the foundation.

10.7   Repeat Problem 10.6 but replace spring with an ideal damper.

10.8   (a)   Define the four-pole parameters for the rod of Fig. 10.6.
         (b)   Using four-pole coupling methods, find $V(0,s)/F(0,s)$ if the system of Fig. P10.1 is coupled to the rod with point "*a*" at $x=L$ and point *b* attached to the foundation.

10.9   For the simplified shaker model of Problem 10.2(a), it has been suggested that adding a capacitor to the circuit might extend the range of flat frequency response for $(f_s/e_i)(i\omega)$. Assuming L is fixed at 0.0012, $R=3.0$, $C=$ any value, analyze the circuits of Figs. P10.2a and P10.2b to see if any improvement is possible.

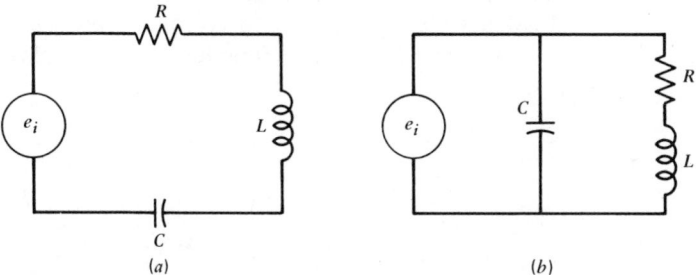

(a)   (b)

Figure P10.2

10.10   Vibration shakers such as that of Fig. 10.2 are sometimes spring mounted to isolate them from the foundation. Assume the magnet has mass M_m and is mounted with a spring K_m and damper B_m between it and the foundation. Write new system equations (analogous to (10.2)–(10.4)) for this configuration.

10.11   Using virtual work methods and the basic laws $E=Ce^2/2$, $C=\epsilon A/x$ for a parallel-plate capacitor, show that such a capacitor develops an attractive force between the plates given by $f=\epsilon Ae^2/2x^2$. Use this result to find

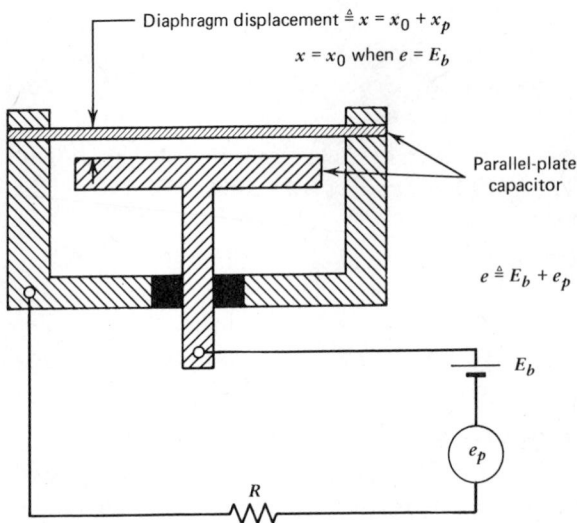

Diaphragm displacement $\triangleq x = x_0 + x_p$

$x = x_0$ when $e = E_b$

Parallel-plate capacitor

$e \triangleq E_b + e_p$

E_b

e_p

R

Figure P10.3

the transfer function $(x_p/e_p)(s)$ for the electrostatic "loadspeaker" of Fig. P10.3. Use linearized perturbation analysis and model the diaphragm as a second-order spring/mass/damper mechanical system.

10.12 In the electric motor drive of Fig. P10.4 we wish to find $(T_{o1}, e_{i1})(s)|_{\text{loaded}}$ using the impedance coupling methods of Section 7.4. Define the proper Z_{go}, Z_{gi}, and unloaded transfer function to solve this problem. Then actually compute these three transfer functions and display $(T_{o1}/e_{i1})(s)|_{\text{loaded}}$ in terms of them.

Motor torque = $K_{MT} i_{i1}$

Motor back EMF = $K_\omega \omega_{o1}$

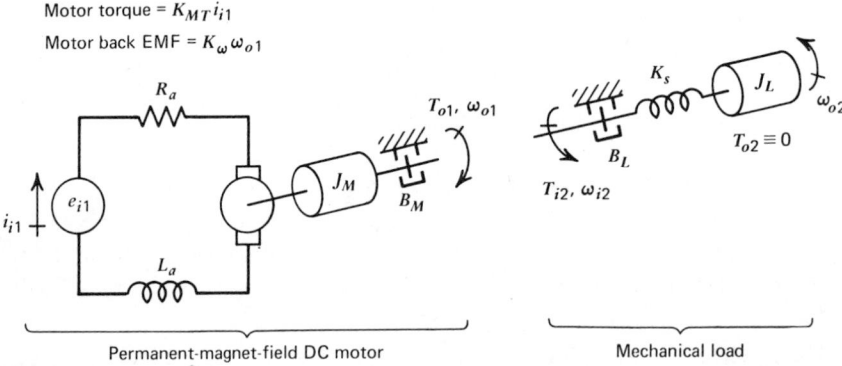

Permanent-magnet-field DC motor   Mechanical load

Figure P10.4

CHAPTER 11

Heat Exchanger Dynamics

Heat exchangers of various types are among the most common components encountered in the design and analysis of thermofluid systems. They are vital to the operation of all kinds of power plants (fossil and nuclear steam plants, internal combustion engines and gas turbines, etc.), heating and air conditioning systems, refineries and chemical plants, and myriad specialized industrial processes and machines. When interest centers on behavior pertinent to startup, shutdown, emergency situations, or control, the steady-state analyses familiar from introductory heat transfer courses are inadequate and dynamic models must be developed. This has been done for most of the common classes of heat exchanger and many results are available in the literature.[1,2] We will here present two examples that illustrate the methods of analysis employed and provide some exposure to the nature of heat exchanger dynamic behavior.

11.1 DISTRIBUTED-PARAMETER DYNAMICS OF A SINGLE-FLUID HEAT EXCHANGER

As part of a larger study[3] on control optimization, Watts and Schoenhals present a dynamic analysis of a simple tubular heat exchanger we now develop in somewhat more detail. In Fig. 11.1, fluid enters the tube at constant velocity V and arbitrarily varying temperature $T_i(t)$. At the outer surface of the tube is applied a timewise-controllable heat flux $q_i(t)$ watt/m^2, which is assumed at any instant uniform spacewise over the surface of the tube. (An example would be

[1]Williams, T. J. and H. J. Morris, "A Survey of the Literature on Heat Exchanger Dynamics and Control," *Chem. E. Prog. Symp. Series*, **57**, No. 36 (1961), pp. 20–33.

[2]Hsu, J. P. and N. Gilbert, "Transfer Functions of Heat Exchangers," *AIChE J.*, **8**, No. 5, Nov. (1962), pp. 593–598.

[3]Watts, R. G. and R. J. Schoenhals, Feedback Control Optimization of a Single Fluid Heat Exchanger, ASME Paper 66-WA/HT-56, 1966.

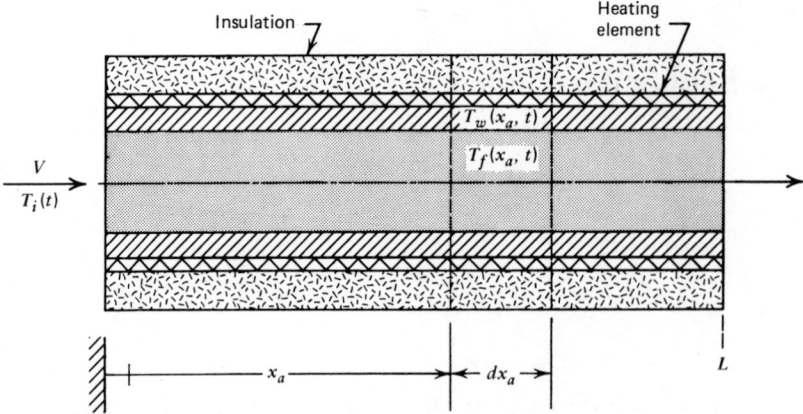

Figure 11.1 Distributed-parameter heat exchanger model.

spirally wound electric resistance heating tape.) Insulation, assumed perfect, forces all of this heat flux into the metal tube wall whose temperature is $T_w(x_a, t)$. Heat flows by convection from the tube wall into the flowing fluid at temperature $T_f(x_a, t)$. Our main interest is to develop transfer functions relating inputs T_i and q_i to output $T_f(x_a, t)$.

Neglecting the radial temperature variation in the tube wall is reasonable for a thin wall made of a good thermal conductor, while the assumption of rectangular fluid velocity and temperature profiles should be quite accurate for turbulent flow conditions. We also assume an incompressible fluid and neglect axial heat conduction in both the wall and the fluid. Axial conduction in the tube wall should be small since the wall is thin (giving a high thermal resistance) and also since there is no intentional heat sink provided to drain off axial heat flow. In the fluid, axial conduction is relatively small because of the poor conductivity of most fluids and the existence of the "low resistance" heat flow path offered by the fluid transport. Note also that such a heat exchanger is *intentionally designed* to be efficient in transferring heat from the heating element to the fluid (a *radial* heat flow) and thus it should not be surprising that axial heat flows are small.

Observing the above assumptions and applying conservation of energy to the fluid control volume of length dx_a over a time interval dt gives:

$$hP_i \, dx_a (T_w - T_f) \, dt - \left(V A_f \rho_f c_f \frac{\partial T_f}{\partial x_a} dx_a \right) dt = A_f \rho_f c_f \, dx_a \left(\frac{\partial T_f}{\partial t} dt \right) \quad (11.1)$$

where

$h \triangleq$ convection heat transfer coefficient

watts$/(m^2\text{-}C^\circ)$ (assumed constant in time and uniform over space)

$P_i \triangleq$ perimeter of inside tube cross section

$A_f \triangleq$ fluid cross-section area

$\rho_f, c_f \triangleq$ fluid mass density and specific heat, assumed constant

Dividing through by $dx_a\, dt$ gives:

$$\rho_f c_f A_f \frac{\partial T_f}{\partial t} + \rho_f c_f A_f V \frac{\partial T_f}{\partial x_a} = h P_i (T_w - T_f) \tag{11.2}$$

Similarly, for the metal tube control volume:

$$\rho_w c_w A_w \frac{\partial T_w}{\partial t} = q_i(t) P_0 - h P_i (T_w - T_f) \tag{11.3}$$

where

$\rho_w, c_w \triangleq$ tube mass density and specific heat, assumed constant

$A_w \triangleq$ tube annular cross-section area

$P_0 \triangleq$ perimeter of outside tube cross section

These two equations, plus the boundary conditions, are a complete description of our model. Note that input q_i appears explicitly but T_i does not; it will show up later when we apply the boundary conditions.

Before proceeding with the equation solution, two common preliminary operations, nondimensionalization and introduction of steady-state and perturbation variables, will be carried out. The purpose of nondimensionalization is generally to reduce to a minimum the number of essential parameters required to define the system. At this point, Equations (11.2) and (11.3) include ten parameters; we will be able to reduce this to two, an obvious advantage when exploring the effect of parameters on system behavior. Introduction of steady-state and perturbation variables facilitates definition of transfer functions and defines operating points for any linearizations that might be employed. Equations (11.2) and (11.3) are general in that they comprehend both steady-state and dynamic behavior. To define an initial steady state, assume T_i and q_i have been constant for a long time at T_{fio} and q_{io}, making all time rates of change equal to zero and collapsing (11.2) and (11.3) into

$$\rho_f c_f A_f V \frac{dT_f}{dx_a} = hP_i\left[T_{wo}(x_a) - T_{fo}(x_a)\right] \tag{11.4}$$

$$q_{io}P_o = hP_i\left[T_{wo}(x_a) - T_{fo}(x_a)\right] \tag{11.5}$$

These are easily solved to get:

$$T_{fo}(x_a) = \frac{q_{io}P_o}{\rho_f c_f A_f V} x_a + T_{fio} \tag{11.6}$$

$$T_{wo}(x_a) = \frac{q_{io}P_o}{\rho_f c_f A_f V} x_a + T_{fio} + \frac{q_{io}P_o}{hP_i} \tag{11.7}$$

giving the steady-state temperature distributions of Fig. 11.2. When we later get dynamic perturbations of the temperatures as functions of x_a, each will vary around these steady-state values.

Turning now to the nondimensionalization, recall that there are always several different ways of defining nondimensional parameters and variables and that the ultimate choice is somewhat subjective. Thus, in the scheme used by the reference and given below, some choices are more-or-less obvious while for others we simply assume that several possibilities were tried and that the one chosen gave the most convenient results. The treatment of the independent

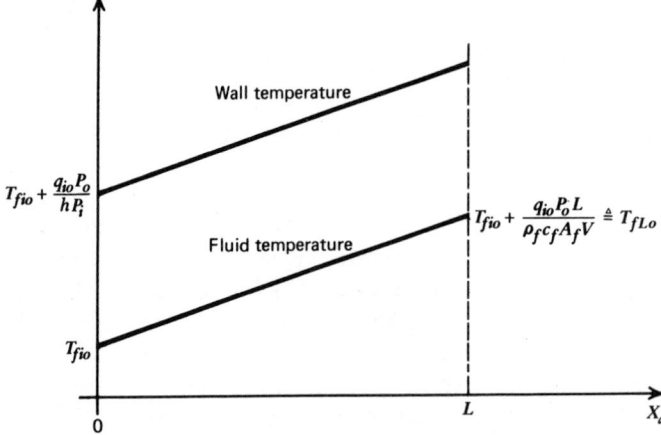

Figure 11.2 Steady-state operating point conditions.

length and time variables is conventional:

$$\text{nondimensional length} \triangleq x \triangleq x_a / L \qquad (11.8)$$

$$\text{fluid residence time} \triangleq t_r \triangleq L / V \qquad (11.9)$$

$$\text{nondimensional time} \triangleq \tau \triangleq t / t_r \qquad (11.10)$$

Recalling the definition of time constant for convection heating of a solid body immersed in fluid ($\tau = Mc/hA$), we can here define for a unit length of pipe and fluid two such time constants in terms of groups of parameters that naturally occur in our equations:

$$\tau_f \triangleq \frac{\rho_f c_f A_f}{h P_i}, \qquad \tau_w \triangleq \frac{\rho_w c_w A_w}{h P_i} \qquad (11.11)$$

Two "nondimensional time constants" (those ultimately used to describe the heat exchanger) are then:

$$\sigma \triangleq t_r / \tau_f, \qquad \lambda \triangleq t_r / \tau_w \qquad (11.12)$$

A nondimensional heat flux input quantity is defined by

$$H_t \triangleq \frac{q_i(t) P_0 L}{\rho_f c_f A_f V (T_{fLo} - T_{fio})} \qquad (11.13)$$

while nondimensional fluid and wall temperatures are taken as:

$$\theta_{ft}(x_a, t) \triangleq \frac{T_f(x_a, t) - T_{fio}}{T_{fLo} - T_{fio}} \qquad (11.14)$$

$$\theta_{wt}(x_a, t) \triangleq \frac{T_w(x_a, t) - T_{wo}(x_a)}{T_{fLo} - T_{fio}} \qquad (11.15)$$

Using the above definitions and recalling that

$$\frac{\partial f}{\partial t} = \frac{\partial f}{\partial \tau} \frac{\partial \tau}{\partial t} = \frac{\partial f}{\partial \tau} \frac{1}{t_r}, \qquad \frac{\partial f}{\partial x_a} = \frac{\partial f}{\partial x} \frac{\partial x}{\partial x_a} = \frac{\partial f}{\partial x} \frac{1}{L} \qquad (11.16)$$

we can reduce (11.2) and (11.3) to

$$\frac{\partial \theta_{ft}}{\partial \tau} + \frac{\partial \theta_{ft}}{\partial x} = \sigma(\theta_{wt} - \theta_{ft}) + \sigma \frac{T_{wo} - T_{fio}}{T_{fLo} - T_{fio}} \qquad (11.17)$$

$$\frac{\partial \theta_{wt}}{\partial \tau} = \frac{\lambda}{\sigma} H_t - \lambda(\theta_{wt} - \theta_{ft}) - \lambda \frac{T_{wo} - T_{fio}}{T_{fLo} - T_{fio}} \qquad (11.18)$$

Equations (11.17) and (11.18) are now reduced to their simplest final form by defining the total quantities θ_{ft}, θ_{wt}, and H_t each as the sum of an initial steady-state value and a dynamic perturbation:

$$\theta_{ft} \triangleq \theta_{fo}(x) + \theta_{fp}(x, \tau) \tag{11.19}$$

$$\theta_{wt} \triangleq \theta_{wo}(x) + \theta_{wp}(x, \tau) \tag{11.20}$$

$$H_t \triangleq H_o + H_p(\tau) \tag{11.21}$$

Substituting in (11.17) gives:

$$\frac{\partial \theta_{fp}}{\partial \tau} + \frac{\partial \theta_{fo}}{\partial x} + \frac{\partial \theta_{fp}}{\partial x} = \sigma(\theta_{wo} - \theta_{fo}) + \sigma(\theta_{wp} - \theta_{fp}) + \sigma \frac{T_{wo} - T_{fio}}{T_{fLo} - T_{fio}} \tag{11.22}$$

In steady state, (11.22) becomes

$$\frac{\partial \theta_{fo}}{\partial x} = \sigma(\theta_{wo} - \theta_{fo}) + \sigma \frac{T_{wo} - T_{fio}}{T_{fLo} - T_{fio}} \tag{11.23}$$

which we subtract from (11.22) to get our final result:

$$\frac{\partial \theta_{fp}}{\partial \tau} + \frac{\partial \theta_{fp}}{\partial x} = \sigma(\theta_{wp} - \theta_{fp}) \tag{11.24}$$

A similar procedure for (11.18) yields:

$$\frac{\partial \theta_{wp}}{\partial \tau} = \frac{\lambda}{\sigma} H_p - \lambda(\theta_{wp} - \theta_{fp}) \tag{11.25}$$

Note that our final equations require only two numerical parameters, σ and λ, to completely characterize a particular heat exchanger.

To get transfer functions we now Laplace transform τ to s:

$$s\theta_{fp}(s) - \theta_{fp}(0, x) + \frac{d\theta_{fp}(s)}{dx} = \sigma(\theta_{wp}(s) - \theta_{fp}(s)) \tag{11.26}$$

$$s\theta_{wp}(s) - \theta_{wp}(0, x) = \frac{\lambda}{\sigma} H_p(s) - \lambda(\theta_{wp}(s) - \theta_{fp}(s)) \tag{11.27}$$

As usual for transfer functions, the initial conditions $\theta_{fp}(0, x)$ and $\theta_{wp}(0, x)$ are zero. Our main interest is in fluid temperature so we eliminate θ_{wp} to get

$$\frac{d\theta_{fp}}{dx} + \left(s + \sigma - \frac{\sigma \lambda}{\lambda + s}\right)\theta_{fp} = \frac{\lambda}{\lambda + s} H_p \tag{11.28}$$

Solving this ordinary differential equation by classical methods (note that right-hand side does not involve x, so is treated as a *constant* in getting the particular solution):

$$\theta_{fp}(x,s) = C_1 e^{-(s+\sigma-\sigma\lambda/(s+\lambda))x} + \frac{\lambda/(\lambda+s)}{s+\sigma-\sigma\lambda/(\lambda+s)} H_p(s) \qquad (11.29)$$

Invoking the boundary condition at $x=0$:

$$\theta_{fp}(s,0) = \theta_{fip}(s) = C_1 + \frac{1}{(s/\lambda)(s+\lambda+\sigma)} H_p(s) \qquad (11.30)$$

$$C_1 = \theta_{fip}(s) - \frac{1}{(s/\lambda)(s+\lambda+\sigma)} H_p(s)$$

$$\theta_{fp}(s,x) = \left(\theta_{fip}(s) - \frac{H_p(s)}{(s/\lambda)(s+\lambda+\sigma)} \right) e^{-(s+\sigma-\sigma\lambda/(s+\lambda))x} + \frac{H_p(s)}{(s/\lambda)(s+\lambda+\sigma)}$$

$$\qquad (11.31)$$

This equation contains the superimposed effects of both the heat flux input H_p and the inlet temperature θ_{fi}, so we let each in turn be zero to separate the individual transfer functions:

$$\frac{\theta_{fp}(s,x)}{\theta_{fip}(s)} = e^{-(s(s+\sigma+\lambda)/(s+\lambda))x} \qquad (11.32)$$

$$\frac{\theta_{fp}(s,x)}{H_p(s)} = \frac{1 - e^{-(s(s+\sigma+\lambda)/(s+\lambda))x}}{(s/\lambda)(s+\sigma+\lambda)} \qquad (11.33)$$

Even for simple types of $\theta_{fip}(t)$ or $H_p(t)$ such as step functions, inverse transforms to get $\theta_{fp}(t)$ are not available, so we again rely on frequency response. Since our interest is usually in exit temperature, we take $x=1.0$ and for $s=i\omega$ get:

$$\frac{\theta_{fp}(i\omega, 1.0)}{\theta_{fip}(i\omega)} = e^{-\sigma\omega^2/(\omega^2+\lambda^2)} \bigg/ \underline{-\frac{\omega^3+\lambda(\sigma+\lambda)\omega}{\omega^2+\lambda^2}} \qquad (11.34)$$

$$\left| \frac{\theta_{fp}(i\omega, 1.0)}{H_p(i\omega)} \right| = \frac{\sqrt{\left|1 + e^{-2\sigma\omega^2/(\omega^2+\lambda^2)} - 2e^{-2\sigma\omega^2/(\omega^2+\lambda^2)}\cos\dfrac{\omega^3+\lambda(\sigma+\lambda)\omega}{\omega^2+\lambda^2}\right|}}{(\omega/\lambda)\sqrt{\omega^2+(\sigma+\lambda)^2}}$$

$$\phi = \tan^{-1}\frac{\sin\dfrac{\omega^2+\lambda(\sigma+\lambda)\omega}{\omega^2+\lambda^2}}{e^{\sigma\omega^2/(\omega^2+\lambda^2)} - \cos\dfrac{\omega^2+\lambda(\sigma+\lambda)\omega}{\omega^2+\lambda^2}} - \tan^{-1}\frac{\omega}{\sigma+\lambda} - 90° \qquad (11.35)$$

Equations (11.34) and (11.35) are useful for establishing the general shapes of the curves (see Fig. 11.3). However, for numerical computations the complex number capability of a language like SPEAKEASY allows much easier calculation directly from (11.32) and (11.33).

To indicate at least one useful application of these results we now discuss briefly the control system study of the reference. In Fig. 11.4, fluid enters at average temperature zero, but with random fluctuations of flat spectral density from frequency zero to Ω^*, but zero above Ω^*. Such "band-limited white noise" is convenient mathematically but actually is a fairly good representation of random fluctuations observed in practice. (The method to be developed can easily handle *any* spectral density. However, the optimization to be performed requires that some definite spectrum be available, so one might as well choose a

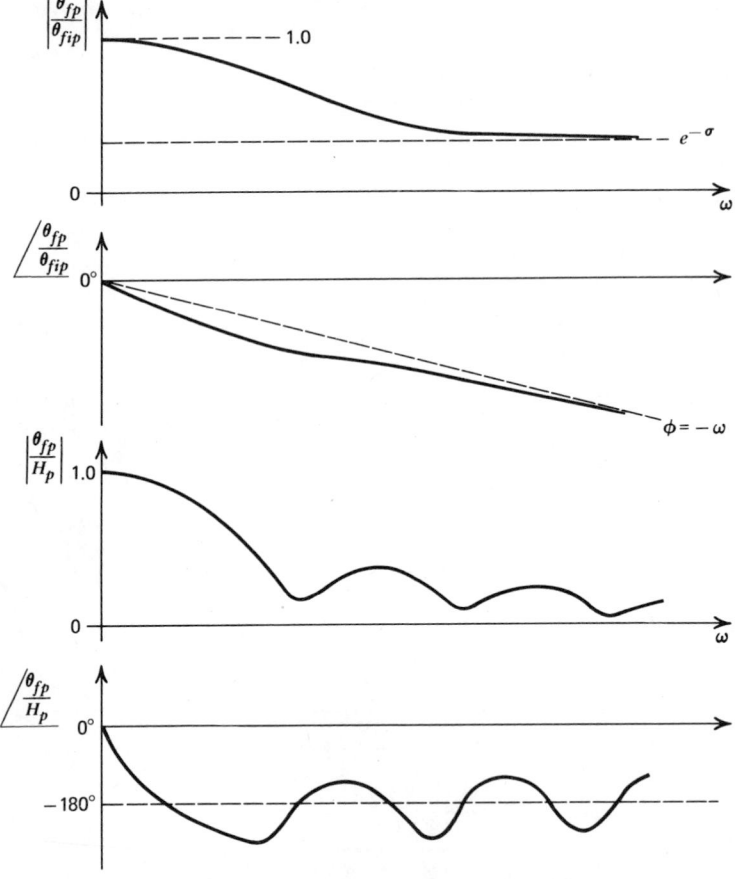

Figure 11.3 Heat exchanger frequency response.

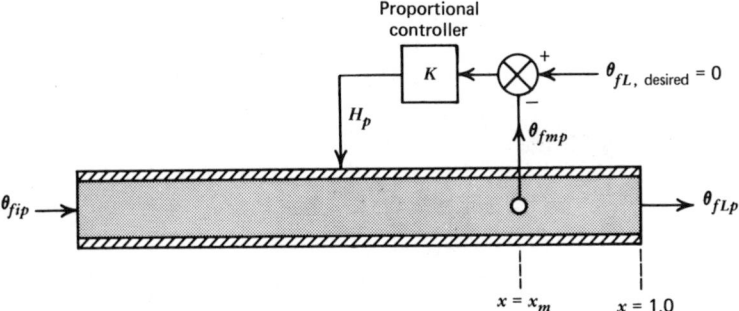

Figure 11.4 Feedback controlled heat exchanger.

simple one to demonstrate the technique.) The purpose of the control system is to apply heating $H_p(t)$ in such a way as to smooth out the random fluctuations by the time the fluid reaches the exit, the ideal performance corresponding to the mean-square value of θ_{fLp} being zero. Simple proportional control with gain K is to be employed, thus the measured value of temperature is compared with its desired value (0.0) and the difference multiplied by K to generate the corrective heat input H_p. The obvious location for the temperature sensor is $x = 1.0$ since that is where the controlled variable θ_{fLp} exists. Experience has shown, however, that better performance is obtained when the sensor is located somewhat upstream, at $x = x_m$. This seems plausible since a fluid particle of incorrect temperature at x_m can still be influenced by a change in H_p *before* it leaves the heater. Once one accepts this concept, it is then natural to wonder whether there is an *optimum* location for the sensor. The referenced study establishes where this location is and further finds the optimum value of K to minimize the mean-square value of θ_{fLp}.

To accomplish the above results we need to obtain a formula for $\overline{\theta_{fLp}^2}$ as a function of x_m and K and then minimize this function. From Fig. 11.4:

$$H_p(s) = -K\theta_{fmp}(s) \tag{11.36}$$

Writing (11.32) and (11.33) as

$$\frac{\theta_{fp}(s,x)}{\theta_{fip}(s)} = e^{-h(s)x} \tag{11.37}$$

$$\frac{\theta_{fp}(s,x)}{H_p(s)} = \frac{1 - e^{-h(s)x}}{g(s)} \tag{11.38}$$

and noting that θ_{fmp} is the result of a superposition of the effects of θ_{fip} and H_p, we get:

$$\theta_{fmp}(s) = \theta_{fip}(s)e^{-h(s)x_m} + H_p(s)\frac{1 - e^{-h(s)x_m}}{g(s)} \tag{11.39}$$

From (11.36) the heat flux is then

$$H_p(s) = \frac{-Kg(s)e^{-h(s)x_m}}{g(s) + K\left[1 - e^{-h(s)x_m}\right]}\,\theta_{fip}(s) \tag{11.40}$$

Controlled variable θ_{fLp} is similarly obtained by a superposition of the effects of θ_{fip} and H_p, leading to the feedback system "closed-loop transfer function":

$$\frac{\theta_{fLp}}{\theta_{fip}}(s) = e^{-h(s)} - \frac{Ke^{-h(s)x_m}(1 - e^{-h(s)})}{g(s) + K(1 - e^{-h(s)x_m})} \tag{11.41}$$

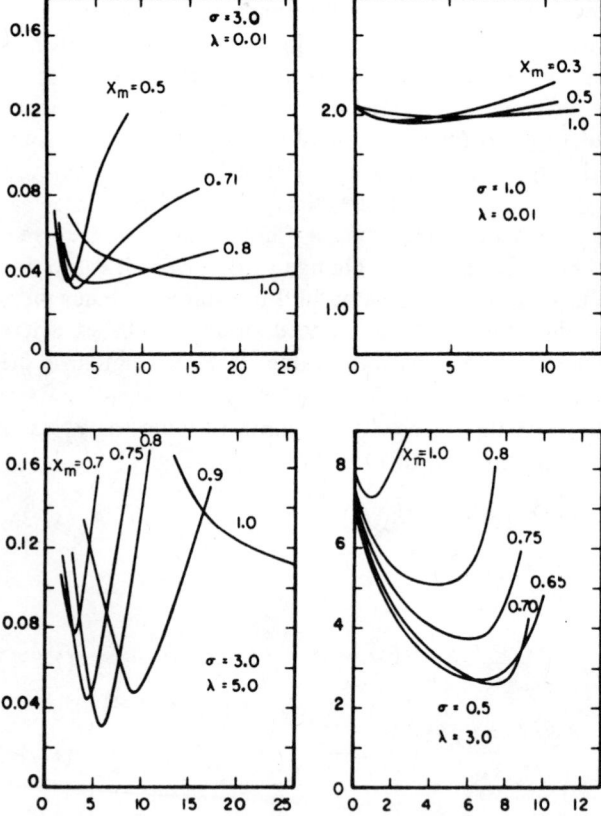

Figure 11.5 Optimum gain and sensor location for various heat exchangers.

Using standard random-signal results from Chapter 3, and taking the power spectral density of θ_{fip} as 1.0 for convenience since for a flat spectrum the existence of a relative minimum value for $\overline{\theta_{fLp}^2}$ is not influenced by this numerical value, we get:

$$\overline{\theta_{fLp}^2} = \int_0^{\Omega^*} \left| e^{-h(i\omega)} - \frac{Ke^{-h(i\omega)x_m}\left(1 - e^{-h(i\omega)}\right)}{g(i\omega) + K\left(1 - e^{-h(i\omega)x_m}\right)} \right|^2 d\omega \qquad (11.42)$$

The search for optimum values of K and x_m to minimize $\overline{\theta_{fLp}^2}$ was accomplished by strictly numerical means, as follows. First, choose a numerical λ and σ, thus fixing the heat exchanger. Then let $x_m = 1.0$, and evaluate (11.42) for a range of K's starting at $K = 0$, watching for a minimum. Then let x_m equal, say, 0.9 and repeat the above calculations. Repeat for enough different x_m's to locate the minimum, if any. These results can be displayed graphically as in Fig. 11.5, where K is the abscissa and $\overline{\theta_{fLp}^2}$ the ordinate. With $\lambda = 3$ and $\sigma = 0.5$, for example, the optimum x_m is about 0.7, and if one uses the optimum K of about 8, the value of $\overline{\theta_{fLp}^2}$ is reduced to about one-third of the value that can be achieved for $x_m = 1.0$. For certain combinations of λ and σ, the reference quotes even more impressive improvements of 30 or 40 to 1.

11.2   THERMOFLUID DYNAMICS OF A CONDENSER FOR A SPACE VEHICLE NUCLEAR POWER PLANT

In the 1960s, a number of projects intended to develop the technology of nuclear-based auxiliary power systems for space vehicles were initiated and carried to various stages of completion. One of these, called SNAP-8,[4] used the Rankine cycle scheme of Fig. 11.6 to generate 35 Kw of usable electrical power for about a one-year period after a remote automatic startup in space. This system was designed, built, and thoroughly ground tested, establishing the technology base for multikilowatt Rankine-cycle space power systems. However, due to changes in space mission goals, which for the time being de-emphasized high electric power requirements, the project was terminated in 1970 and has thus not actually been operated in space. Much theoretical and experimental work carried out in developing this system has been reported in detail and is available for application to related systems. Our interest here centers on the dynamic behavior of the condenser, and we draw mainly on two published reports[5] in presenting the material below.

---

[4]Hodgson, J. N., Geimer, R. G. and A. H. Kreeger, "Energy Systems in Space Vehicles," *Mech. Eng.*, Nov. (1967), p. 42.

[5]Schoenberg, A. A., Mathematical Model with Experimental Verification for the Dynamic Behavior of a Single-Tube Condenser, NASA TN D-3453, 1966; Fisher, R. C., Experimental Study of Dynamics of Mercury-Vapor Condensing, NASA TM X-1375, 1967.

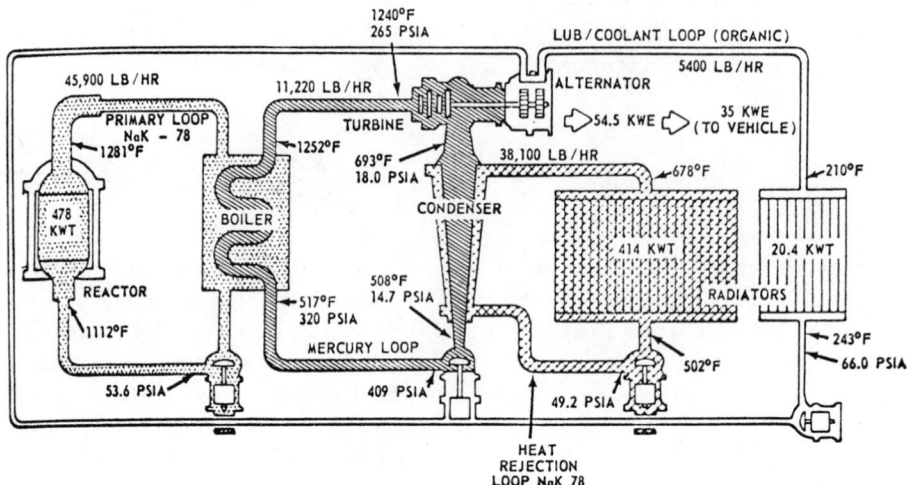

Figure 11.6 SNAP-8 power generating system.

While there is of course a wealth of operating experience available with earth-bound Rankine cycle systems (steam power plants), the SNAP-8 system presents many unusual problems due mainly to the zero-g space environment and the use of exotic fluid media such as mercury, sodium/potassium liquid metal alloy, and organic lube/coolant liquid (polyphenyl ether). Thus detailed analysis, experimentation, and development testing of all system components was necessary to achieve successful operation. For our condenser study, interest centers on the condenser pressure since if inlet pressure (turbine back pressure) gets too high the turbine will not produce rated power, while if condenser discharge pressure is too low, cavitation may occur at the inlet of the mercury-loop pump. A model of condenser dynamics is needed to predict whether system disturbances will cause unacceptable pressure excursions and, should this be the case, to allow a rational design of a feedback control system for condenser pressure.

While the actual condenser is a counterflow tube-in-shell type with 73 tapered tubes, initial analysis and testing was done on a simplified single-tube device, not tapered, and cooled by nitrogen gas rather than liquid metal. Such simplifications are not unusual in the exploratory stages of new-device development when illumination of basic characteristics in reasonable time and cost frames requires sacrifice of detail in modeling. The theoretical model will thus correspond to the experimental test rig of Fig. 11.7, the portion actually analyzed being shown schematically in Fig. 11.8. The input variables of main interest are the inlet mixture mass flow rate w_H, the receiver pressure p_r, and the

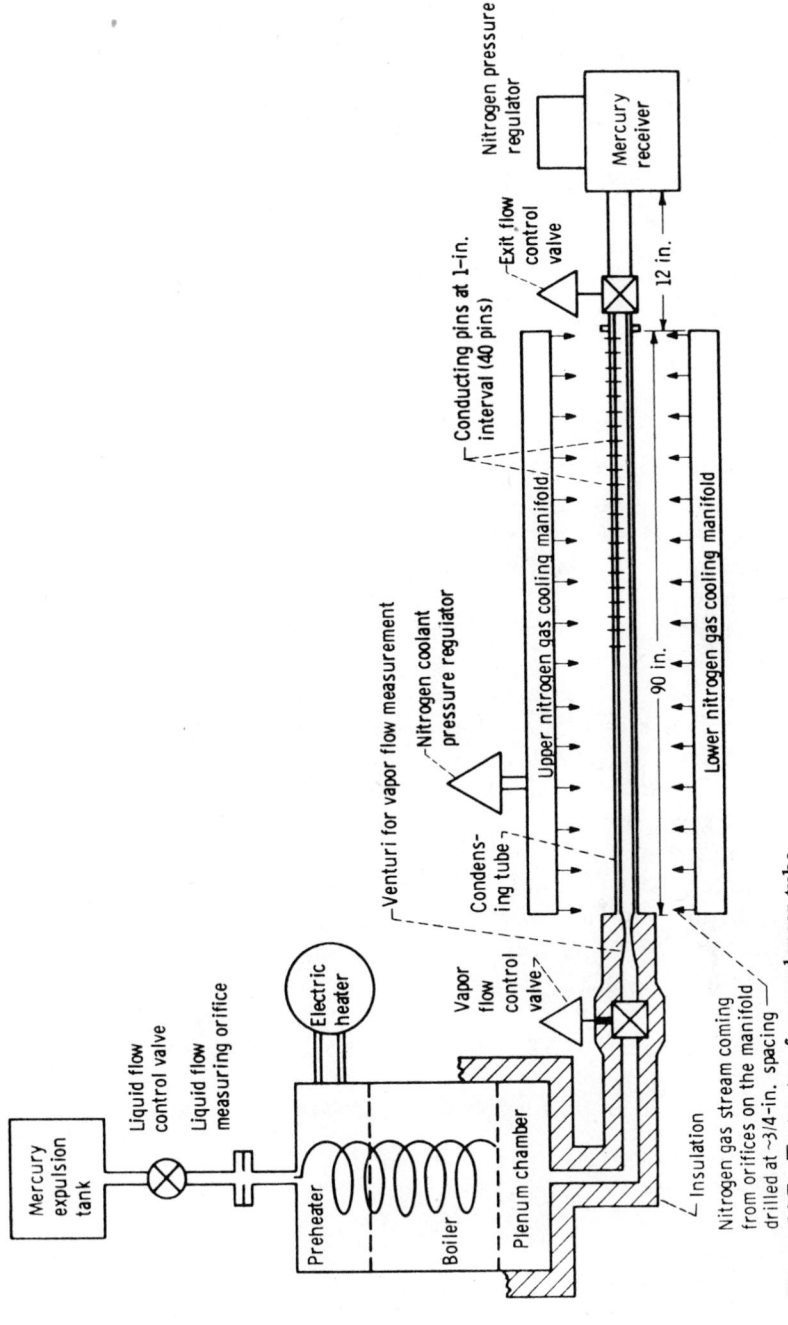

Figure 11.7 Test setup for condenser tube.

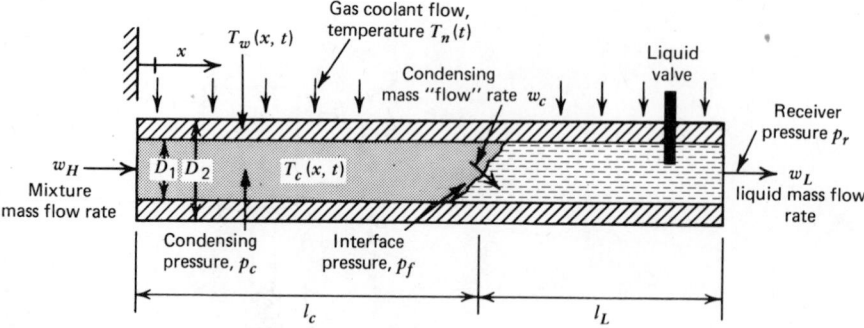

Figure 11.8   Model of single-tube condenser.

gas coolant temperature T_n, while output responses were desired for condensing pressure p_c and the location l_c of the vapor/liquid interface.

It was concluded that small-signal linearized models would be obtained for various operating points, thus we begin by describing how the apparatus is manipulated to establish an initial equilibrium state. Referring to Fig. 11.7, the upstream pressure of the vapor flow control valve (which always operates with choked flow and is thus insensitive to downstream conditions) is regulated constant. This allows proportional manipulation of mass flow rate by simply changing valve flow area with an electro-pneumatic actuator. The flow rate of nitrogen gas coolant is kept constant by regulating manifold pressure constant. With the vapor flow control valve set at a desired fixed position, if we shut the exit liquid valve, vapor condenses and liquid begins to fill up the tube. By opening the exit valve and adjusting the receiver pressure with its nitrogen pressure regulator, the liquid outflow can be just balanced with the condensation rate at any desired interface position, giving an equilibrium operating point. Perturbations can now be induced in any of the input variables to dynamically exercise the apparatus.

In the theoretical model developed below, metric units will be used when defining general terms. However, when quoting specific numerical values we will revert to British to allow easy correlation with the original references, which use only British units. The analysis consists of the following basic relations:

1. Mass conservation for the liquid phase.
2. Mass conservation for the gas phase.
3. Energy conservation for vapor condensation.
4. Energy conservation for tube wall.
5. Newton's law for motion of liquid slug in tube.

Conservation of mass for the liquid phase gives:

$$[(1-X_q)w_H(t)+w_c(t)-w_L(t)]dt=\rho_L\,dV_L=-\rho_L\,dV_v=-\rho_L A_c\,dl_c$$

$$(11.43)$$

where

$$X_q \triangleq \text{vapor quality at condenser inlet, kg vapor/kg mixture}$$

$$w_H(t) \triangleq \text{mixture mass flow rate, kg/sec}$$

$$w_c(t) \triangleq \text{rate of condensation, kg/sec}$$

$$w_L(t) \triangleq \text{liquid mass flow rate to receiver, kg/sec}$$

$$\rho_L \triangleq \text{density of liquid condensate, kg/m}^3$$

$$V_v \triangleq \text{volume of gas phase, m}^3$$

$$V_L \triangleq \text{volume of liquid phase, m}^3$$

$$A_c \triangleq \text{tube cross-section flow area, m}^2$$

$$l_c(t) \triangleq \text{length of condensing region, m}$$

A number of assumptions are implied in the above equation. Since for mercury at 16 psia (a typical operating point) the density ratio of liquid to gas is 3000 to 1, we assume liquid droplets in l_c take up insignificant volume and thus we take mixture and gas volume in l_c as equal. We further assume the existence of a distinct liquid/vapor interface which moves in a pistonlike fashion when the condenser liquid inventory changes. Finally, to avoid the great complexity of developing a valid model of droplet formation and motion, we assume that liquid condensate collects at the interface without significant time delay. (Later experimentation revealed the need for a dead time in the model and also allowed calculation of its numerical value, but no theory was available to predict this.)

Considering now the gas phase, conservation of mass gives:

$$X_q w_H(t)-w_c(t)=\frac{d}{dt}(\rho_v V_v)$$

$$(11.44)$$

where

$$\rho_v \triangleq \text{gas phase density, kg/m}^3$$

The perfect gas law can be used to relate density, pressure and temperature of the gas phase:

$$\rho_v = p_c / RT_c \tag{11.45}$$

where

$$p_c \triangleq \text{condensing pressure, Pa absolute}$$

$$R \triangleq \text{gas constant for the vapor, } \frac{\text{N-m}}{\text{kg-}^\circ\text{K}}$$

$$T_c \triangleq \text{condensing temperature, } ^\circ\text{K}$$

In these equations, ρ_v, p_c, and T_c should be average quantities over the volume V_v. However, for simplicity they are taken as point values at $x=0$. Validation of these (and other) assumptions of course rests on experimental verification of the overall model. A further relation between condensing pressure and temperature for saturation conditions is the Clausius–Clapeyron equation:

$$\left. \frac{dp_c}{dT_c} \right|_{\text{saturation}} = \frac{H_L}{T_c v_{fg}} = \frac{H_L p_c}{RT_c^2} \tag{11..46}$$

where

$$H_L \triangleq \text{enthalpy of vaporization, joule/kg}$$
$$v_{fg} \triangleq \text{specific volume difference between gas and liquid, m}^3/\text{kg}$$

$$v_{fg} \approx \frac{1}{\rho_v} = \frac{RT_c}{p_c}$$

If an experimental vapor-pressure curve such as Fig. 11.9 is available it may be used in place of Eq. (11.46) to get a numerical value of slope dp_c/dT_c and thereby relate perturbations in T_c and p_c. We will later use (11.45) and (11.46) to eliminate ρ_v and T_c from our equations in favor of p_c.

Assuming that all the heat given up by the condensing vapor flows into the metal wall of the tube:

$$H_L w_c(t) = \int_0^{l_c(t)} h_{cw} \pi D_1 \left[T_c(x,t) - T_w(x,t) \right] dx \tag{11.47}$$

where

$$h_{cw} \triangleq \text{heat transfer coefficient at inside wall, watts/(m}^2\text{-3}^\circ\text{C)}$$

This equation assumes that energy changes other than that associated with the condensation are negligible (insignificant superheat energy, for example), that

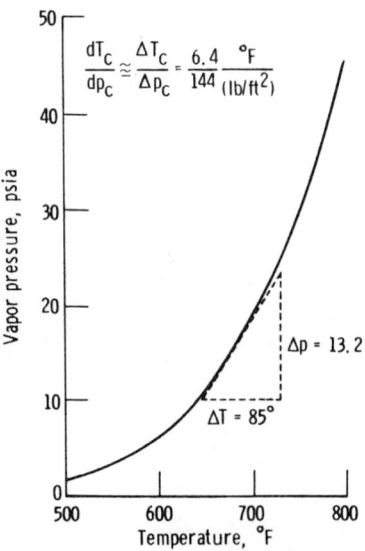

Figure 11.9   Vapor-pressure curve for mercury.

axial heat flow can be ignored, and that h_{cw} can be taken as a constant and uniform average value over l_c. A heat balance on the tube wall now gives

$$\{(\pi D_1 dx)h_{cw}[T_c(x,t)-T_w(x,t)]-(\pi D_2 dx)h_{nw}[T_w(x,t)-T_n(t)]\}\, dt$$

$$= \frac{\rho_w c_w \pi}{4}(D_2^2-D_1^2)\, dx\, dT_w(x,t) \tag{11.48}$$

$$\tau_w \frac{\partial T_w(x,t)}{\partial t}+T_w(x,t)=K_{1w}T_c(x,t)+K_{2w}T_n(t) \tag{11.49}$$

where

$$\rho_w \triangleq \text{tube density, kg/m}^3$$

$$c_w \triangleq \text{tube specific heat, joule/(kg-°C)}$$

$$h_{nw} \triangleq \text{heat transfer coefficient at outside wall, watts/(m}^2\text{-°C)}$$

$$\tau_w \triangleq \text{wall time constant} \triangleq \frac{\rho_w c_w (D_2^2-D_1^2)}{4(h_{cw}D_1+h_{nw}D_2)},\ \sec \tag{11.50}$$

$$K_{1w} \triangleq \text{wall/condensate gain} \triangleq \frac{h_{cw}D_1}{h_{cw}D_1+h_{nw}D_2} \tag{11.51}$$

$$K_{2w} \triangleq \text{wall/coolant gain} \triangleq \frac{h_{nw}D_2}{h_{cw}D_1+h_{nw}D_2} \tag{11.52}$$

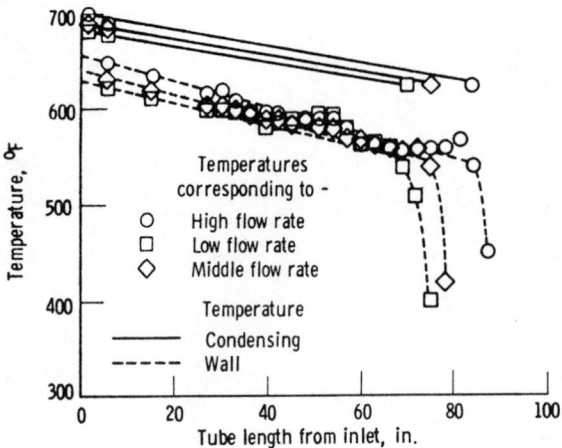

Figure 11.10   Steady-state temperature gradients.

To avoid partial differential equation solution without cutting l_c into lumps, we now assume a *known* spacewise variation of T_c. This approach is suggested by the results of preliminary steady-state experiments shown in Fig. 11.10. These showed essentially straight-line gradients whose slope was relatively insensitive to operating conditions. The assumption was made that the spacewise variation of *dynamic* perturbations in T_c would follow this linear pattern with the same numerical slope as for steady state. We thus take

$$T_c(x,t) = T_c(t)(1 - mx) \qquad (11.53)$$

where

$$T_c(t) \triangleq T_c(0,t)$$

and slope m is taken from the steady-state graphs. Note that these graphs also could be used to get the variation of T_w with x, but this should not be done since Eq. (11.49) will *enforce* a certain variation on T_w once T_c is assumed as in (11.53).

Before using Newton's law to model the motion of the liquid slug, all the equations already derived are expressed in terms of steady-state operating values and dynamic perturbations, and, where necessary, linearized. The notation, using inlet flow rate as an example, will follow the form:

$$\underbrace{w_H(t)}_{\text{total value}} = \underbrace{\overline{w_H}}_{\substack{\text{constant operating-} \\ \text{point value}}} + \underbrace{\Delta w_H(t)}_{\substack{\text{small dynamic} \\ \text{perturbation}}}$$

Since our goal is transfer functions, we will also Laplace transform with zero initial conditions as we go. From Eq. (11.43):

$$(1-X_q)(\bar{w}_H+\Delta w_H(t))+\bar{w}_c+\Delta w_c(t)-\bar{w}_L-\Delta w_L(t)=-\rho_L A_c\frac{d}{dt}\Big[\bar{l}_c+\Delta l_c(t)\Big]$$

(11.54)

Since in steady state

$$(1-X_q)\bar{w}_H+\bar{w}_c-\bar{w}_L=0 \tag{11.55}$$

we can subtract (11.55) from (11.54) to get

$$(1-X_q)\Delta w_H(s)+\Delta w_c(s)-\Delta w_L(s)=-\rho_L A_c s\,\Delta l_c(s) \tag{11.56}$$

From Eq. (11.44):

$$X_q(\bar{w}_H+\Delta w_H(t))-(\bar{w}_c+\Delta w_c(t))=\frac{d}{dt}\Big[(\bar{\rho}_v+\Delta\rho_v(t))(\bar{V}_v+\Delta V_v(t))\Big]\quad(11.57)$$

Subtracting the steady-state equation and neglecting the product of perturbations gives:

$$X_q\Delta w_H(s)-\Delta w_c(s)=\bar{\rho}_v s\,\Delta V_v(s)+\bar{V}_v s\,\Delta\rho_v(s) \tag{11.58}$$

Using the perfect gas law to eliminate ρ_v in favor of p_c gives:

$$\rho_v=\frac{p_c}{RT_c}\approx\frac{\bar{p}_c}{R\bar{T}_c}+\frac{1}{R\bar{T}_c}\Delta p_c(t)+\frac{\bar{p}_c}{R}\Big(-\frac{1}{\bar{T}_c^2}\Big)\Delta T_c(t) \tag{11.59}$$

Using Fig. 11.9 to get a numerical value for dp_c/dT_c at $\bar{p}_c$, we can eliminate T_c in favor of p_c also:

$$\Delta\rho_v(s)=\frac{1}{R\bar{T}_c}\Delta p_c(s)-\frac{\bar{p}_c}{R\bar{T}_c^2}\Delta T_c(s)=\frac{1}{R\bar{T}_c}\Delta p_c(s)-\frac{\bar{p}_c}{R\bar{T}_c^2\dfrac{dp_c}{dT_c}\Big|_{\bar{p}_c}}\Delta p_c(s)$$

(11.60)

Using this and

$$\Delta V_v(s)=A_c\,\Delta l_c(s)$$

in (11.58) leads to:

$$X_q\Delta w_H(s)-\Delta w_c(s)=\bar{\rho}_v A_c s\,\Delta l_c(s)+\left[\frac{\bar{V}_v}{R\bar{T}_c}\left[1-\frac{\bar{\rho}_v R}{\dfrac{dp_c}{dT_c}\Big|_{\bar{p}_c}}\right]\right]s\,\Delta p_c(s) \quad(11.61)$$

In dealing with Eq. (11.47) we will use (11.53) to carry out the integration, although further approximations will also be necessary in order to arrive at a linear differential equation with constant coefficients:

$$
\begin{aligned}
H_L(\bar{w}_c + \Delta w_c(t)) & \\
&= \int_0^{\bar{l}_c + \Delta l_c(t)} h_{cw} \pi D_1 \left[\left(\bar{T}_c(x,0) + \Delta T_c(x,t) \right) - \left(\bar{T}_w(x,0) + \Delta T_w(x,t) \right) \right] dx \quad (11.62) \\
&= \int_0^{\bar{l}_c} h_{cw} \pi D_1 \left[\bar{T}_c(x,0) - \bar{T}_w(x,0) \right] dx + \int_0^{\bar{l}_c} h_{cw} \pi D_1 \left[\Delta T_c(x,t) - \Delta T_w(x,t) \right] dx \\
&\quad + \int_{\bar{l}_c}^{\bar{l}_c + \Delta l_c(t)} h_{cw} \pi D_1 \left[\left(\bar{T}_c(x,0) + \Delta T_c(x,t) \right) - \left(\bar{T}_w(x,0) + \Delta T_w(x,t) \right) \right] dx
\end{aligned}
$$

$$(11.63)$$

In steady state

$$
H_L \bar{w}_c = \int_0^{\bar{l}_c} h_{cw} \pi D_1 \left[\bar{T}_c(x,0) - \bar{T}_w(x,0) \right] dx \quad (11.64)
$$

so we can subtract (11.64) from (11.63) to remove these two terms. Figure 11.11 may be helpful in understanding how the two remaining integrals are dealt with. The total change in heat transfer can be thought of as arising from two sources: change in heat transfer area due to $\Delta l_c(t)$, and change in the driving temperature difference $T_c - T_w$. By analogy to a Taylor series linearization, we consider the integral

$$
\int_0^{\bar{l}_c} h_{cw} \pi D_1 \left[\Delta T_c(x,t) - \Delta T_w(x,t) \right] dx
$$

as the contribution, for constant heat transfer area, of the dynamic perturbation

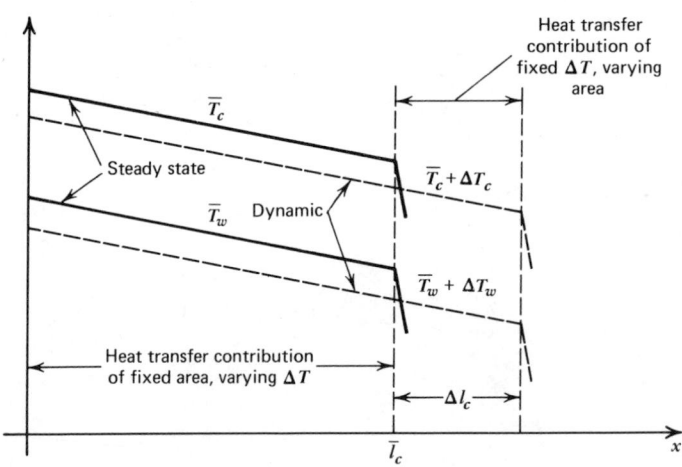

Figure 11.11 Approximate separation of heat transfer into two portions.

in temperature difference. The other integral

$$\int_{\bar{l}_c}^{\bar{l}_c+\Delta l_c(t)} h_{cw} \pi D_1 \left[\left(\bar{T}_c(x,0) + \Delta T_c(x,t) \right) - \left(\bar{T}_w(x,0) + \Delta T_w(x,t) \right) \right] dx$$

is mainly the contribution of area change, for a nearly constant temperature difference. In Fig. 11.11 we assume that the actual temperature difference can be approximated by its original steady-state value at the interface, $\bar{T}_c(\bar{l}_c,0) - \bar{T}_w(\bar{l}_c,0)$, since the perturbation $\Delta T_c(x,t) - \Delta T_w(x,t)$ would be "multiplied" by the area (length) perturbation, and we conventionally neglect products of perturbations. We thus write approximately

$$\int_{\bar{l}_c}^{\bar{l}_c+\Delta l_c(t)} h_{cw} \pi D_1 \left[\left(\bar{T}_c(x,0) + \Delta T_c(x,t) \right) - \left(\bar{T}_w(x,0) + \Delta T_w(x,t) \right) \right] dx$$

$$\approx \int_{\bar{l}_c}^{\bar{l}_c+\Delta l_c(t)} h_{cw} \pi D_1 \left[\bar{T}_c(\bar{l}_c,0) - \bar{T}_w(\bar{l}_c,0) \right] dx$$

$$= h_{cw} \pi D_1 \left[\bar{T}_c(\bar{l}_c,0) - \bar{T}_w(\bar{l}_c,0) \right] \Delta l_c(t) \quad (11.65)$$

Equation (11.63) thus finally becomes

$$H_L \Delta w_c(s) = \int_0^{\bar{l}_c} h_{cw} \pi D_1 \left[\Delta T_c(x,s) - \Delta T_w(x,s) \right] dx$$

$$+ h_{cw} \pi D_1 \left[\bar{T}_c(\bar{l}_c,0) - \bar{T}_w(\bar{l}_c,0) \right] \Delta l_c(s) \quad (11.66)$$

Since T_w is not of primary interest, we now use (11.49) in (11.66) to eliminate it in favor of T_c and T_n:

$$\Delta T_w(x,s) = \frac{K_{1w}(1-mx)}{\tau_w s+1} \Delta T_c(s) + \frac{K_{2w}}{\tau_w s+1} \Delta T_n(s) \quad (11.67)$$

$$H_L \Delta w_c(s) = \int_0^{\bar{l}_c} h_{cw} \pi D_1 \left(\Delta T_c(s)(1-mx) - \frac{K_{1w}}{\tau_w s+1} \Delta T_c(s)(1-mx) \right.$$

$$\left. - \frac{K_{2w}}{\tau_w s+1} \Delta T_n(s) \right) dx + h_{cw} \pi D_1 \left[\bar{T}_c(\bar{l}_c,0) - \bar{T}_w(\bar{l}_c,0) \right] \Delta l_c(s)$$

$$(11.68)$$

$$= h_{cw} \pi D_1 \bar{l}_c \left(1 - \frac{m\bar{l}_c}{2} \right)(1-K_{1w}) \left(\frac{\tau_{1w} s+1}{\tau_w s+1} \right) \Delta T_c(s)$$

$$- \frac{h_{cw} \pi D_1 K_{2w} \bar{l}_c}{\tau_w s+1} \Delta T_n(s) + h_{cw} \pi D_1 \left[\bar{T}_c(\bar{l}_c,0) - \bar{T}_w(\bar{l}_c,0) \right] \Delta l_c(s) \quad (11.69)$$

where

$$\tau_{1w} \triangleq \frac{\tau_w}{1 - K_{1w}} = \frac{\rho_w c_w (D_2^2 - D_1^2)}{4 h_{nw} D_2} \tag{11.70}$$

In preparation for nondimensionalization, we note the following steady-state relations. For heat flow into the tube wall:

$$\bar{w}_c H_L = h_{cw} \pi D_1 \bar{l}_c \left[\bar{T}_c(\bar{l}_c) - \bar{T}_w(\bar{l}_c) \right] \tag{11.71}$$

and for heat flow through the tube wall:

$$\bar{w}_c H_L = \frac{\text{average } \Delta T}{\sum \text{ heat transfer resistances}} = \frac{\bar{T}_c(\bar{l}_c)(1 - m\bar{l}_c/2) - \bar{T}_n}{\dfrac{1}{h_{cw} \pi D_1 \bar{l}_c} + \dfrac{1}{h_{nw} \pi D_2 \bar{l}_c}}$$

$$= \frac{h_{cw} h_{nw} D_1 D_2 \pi \bar{l}_c T_a}{h_{cw} D_1 + h_{nw} D_2} \tag{11.72}$$

where

$$Z_L \triangleq 1 - \frac{m\bar{l}_c}{2}, \qquad T_a \triangleq \bar{T}_c Z_L - \bar{T}_n$$

We now convert to nondimensional variables by (in most cases) dividing the perturbation variables by their steady-state values. Thus these new variables are also *percentage-type* quantities. Proceeding in this fashion we define

$$W_c \triangleq \frac{\Delta w_c(s)}{\bar{w}_c}, \qquad L_c \triangleq \frac{\Delta l_c(s)}{\bar{l}_c}, \qquad C_n \triangleq \frac{\Delta T_n(s)}{T_a}$$

and thus rewrite (11.69) as

$$W_c = Z_L \frac{\tau_{1w} s + 1}{\tau_w s + 1} \frac{\Delta T_c(s)}{T_a} - \frac{C_n}{\tau_w s + 1} + L_c \tag{11.73}$$

From Eq. (11.61), defining

$$W_H \triangleq \frac{\Delta w_H(s)}{\bar{w}_H}, \qquad P_c \triangleq \frac{\Delta p_c(s)}{\bar{p}_c}, \qquad \theta \triangleq \frac{\bar{\rho}_v A_c \bar{l}_c}{X_q \bar{w}_H}, \qquad \beta_x \triangleq \frac{\bar{p}_c}{\bar{\rho}_v H_L J} \tag{11.74}$$

we get

$$W_c = W_H - \theta s L_c - (1 - \beta_x) \theta s P_c \tag{11.75}$$

where $J \triangleq$ mechanical/thermal energy conversion factor $(\text{ft-lb}_f)/\text{BTU}$ and we have used the fact

$$\bar{w}_H = \bar{w}_c/X_q = \bar{w}_L, \text{ for steady state}$$

(Note that the conversion factor J would not be needed if a consistent set of metric units had been employed. The reference used R with ft-lb_f units and H_L with BTU, so required J as shown.) Combining (11.73) and (11.75):

$$\frac{Z_L(\tau_{1w}s+1)}{\tau_w s+1}\left[\frac{\bar{P}_c}{T_a \left.\frac{dp_c}{dT_c}\right|_{\bar{P}_c}}\right]P_c - \frac{C_n}{\tau_w s+1} + L_c = W_H - \theta s L_c - (1-\beta_x)\theta s P_c$$

$$(11.76)$$

Solving for P_c leads to

$$P_c = \frac{(1/K_t)(\tau_w s+1)}{a_s s^2 + b_s s+1}\left(W_H + \frac{C_n}{\tau_w s+1} - (\theta s+1)L_c\right) \qquad (11.77)$$

where

$$K_t \triangleq \frac{Z_L \bar{p}_c}{T_a}\left.\frac{dT_c}{dp_c}\right|_{\bar{p}_c}$$

$$a_s \triangleq \frac{\tau_w(1-\beta_x)\theta}{K_t}, \qquad b_s \triangleq \tau_{1w} + \frac{(1-\beta_x)\theta}{K_t}$$

We can now draw a block diagram representing Eq. (11.77) as in Fig. 11.12, where C_n (coolant temperature) and W_H (mixture inflow rate) are truly inputs, but L_c is actually an output quantity for which we must now develop modeling equations. We first add the linearized versions of Equations (11.43) and (11.44)

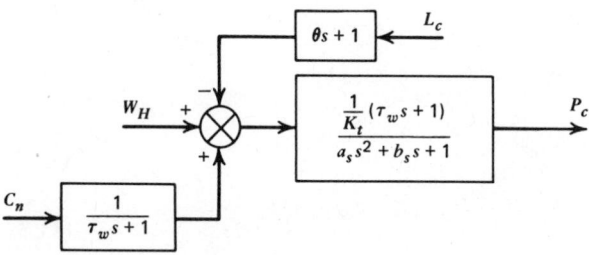

Figure 11.12   Partial system block diagram.

to get

$$L_c = \frac{W_L - W_H}{X_q \theta s(\rho_L/\rho_v - 1)} + \frac{(1 - \beta_x)}{\rho_L/\rho_v - 1} P_c \approx \frac{W_L - W_H}{X_q \theta s(\rho_L/\rho_v - 1)} \qquad (11.78)$$

where

$$W_L \triangleq \Delta w_L / \bar{w}_L$$

and the term in P_c, which represents the change in mass of the gas phase due to pressure changes, is assumed negligible relative to the contributions of W_L and W_H. Newton's law is now applied to the motion of the liquid slug of length l_L in Fig. 11.8:

$$(p_f - p_r)A_c - A_c q_L(w_L) = M_L \frac{dv_L}{dt} \qquad (11.79)$$

where

$$M_L \triangleq \rho_L A_c \bar{l}_L, \text{ assumed constant}$$

$$v_L \triangleq w_L / \rho_L A_c$$

$$q_L(w_L) \triangleq \text{frictional pressure drop due to pipe and valve,}$$
$$\text{assumed function of } w_L \text{ only}$$

The interface pressure p_f is related to the condensing pressure p_c at the inlet, and frictional pressure drop q_v in the condensing region by

$$p_f = p_c - q_v(w_H, l_c) \qquad (11.80)$$

Experimental data on the dependence of q_L on w_L and q_v on w_H and l_c are used to obtain numerical values of the partial derivatives in the linearized equation obtained by combining (11.79) and (11.80):

$$\Delta p_c - \frac{\partial q_v}{\partial w_H}\bigg|_{\bar{w}_H, \bar{l}_c} \Delta w_H - \frac{\partial q_v}{\partial l_c}\bigg|_{\bar{w}_H, \bar{l}_c} \Delta l_c - \Delta p_r - \frac{\partial q_L}{\partial w_L}\bigg|_{\bar{w}_L} \Delta w_L = \frac{\bar{l}_L}{A_c} \frac{d}{dt}(\Delta w_L)$$

$$(11.81)$$

$$W_L = \frac{K_p(P_c - P_r) - K_w W_H - K_v L_c}{\tau_p s + 1} \qquad (11.82)$$

where

$$K_p \triangleq \frac{\bar{p}_c}{\bar{w}_H \, \partial q_L / \partial w_L}, \qquad P_r \triangleq \frac{\Delta p_r}{\bar{p}_c}, \qquad K_w \triangleq \frac{\partial q_v / \partial w_H}{\partial q_L / \partial w_L}$$

$$K_v \triangleq \frac{\bar{l}_c \, \partial q_v / \partial l_c}{\bar{w}_H \, \partial q_L / \partial w_L}, \qquad \tau_p \triangleq \frac{\bar{l}_L}{A_c \, \partial q_L / \partial w_L}$$

The physical analysis of the condenser is now complete and we need only manipulate our equations into final form. Combining (11.78) and (11.82) to eliminate w_L and solving for L_c gives

$$L_c = \frac{K_p / K_v}{as^2 + bs + 1}(P_c - P_r) - \left(\frac{1 + K_v}{K_p}\right)(\tau_{1p}s + 1)\frac{K_p / K_v}{as^2 + bs + 1}(W_H) \quad (11.83)$$

where

$$\theta_r \triangleq X_q \theta\left(\frac{\rho_L}{\rho_v} - 1\right), \qquad a \triangleq \frac{\theta_r \tau_p}{K_v}, \qquad b \triangleq \frac{\theta_r}{K_v}, \qquad \tau_{1p} \triangleq \frac{\tau_p}{1 + K_w}$$

The block diagram of Fig. 11.12 can now be completed as in Fig. 11.13. Working directly from this diagram or from the equations on which it is based, one can obtain the differential equations relating outputs P_c and L_c to inputs

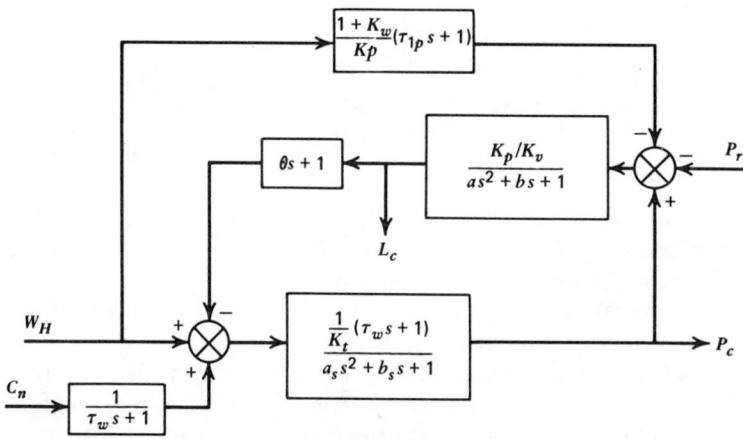

Figure 11.13 Complete system block diagram.

W_H, C_n, and P_r:

$$(Ds^4 + Cs^3 + Bs^2 + As + 1)P_c = K_{1p}[(\tau_w s + 1)(N_2 s^2 + N_1 s + 1)]W_H$$
$$+ K_{2p}(as^2 + bs + 1)C_n + K_{3p}(a_L s^2 + b_L s + 1)P_r$$
$$(11.84)$$

$$(Ds^4 + Cs^3 + Bs^2 + As + 1)L_c = K_{1L}(-M_{3L}s^3 - M_{2L}s^2 + M_{1L}s + 1)W_H$$
$$+ K_{2L}C_n - K_{3L}(a_s s^2 + b_s s + 1)P_r \qquad (11.85)$$

The new parameter combinations found in these equations will not here be defined; they can be found by carrying out the steps needed to get (11.84) or (11.85), or in the quoted reference.

Of the six transfer functions implied by Equations (11.84) and (11.85), experimental testing was carried out on two: $(P_c/W_H)(s)$ and $(L_c/W_H)(s)$. Sinusoidal testing with perturbations of $\pm 10\%$ around the steady-state operating point for input W_H was used (see Fig. 11.14 for some actual recorder traces). The steady-state gains K_{1p} and K_{1L} were measured by setting W_H constant at $+10\%$ and -10% and recording the resulting steady-state values of P_c and L_c. While the research facility had a frequency-response analyzer of the Fourier filter type, it was tied up on another project, and thus a two-channel recorder and manual data processing had to be used to get amplitude ratios and phase angles. Of the various operating points that were studied, we concentrate on two, one with a long condensing length (Case I) and one with a short condensing length (Case II). To get some feel for numerical values, the transfer functions predicted by the theoretical model for Case I were

$$\frac{P_c}{W_H}(s) = \frac{1.35(0.851s + 1)(0.075s^2 + 1.48s + 1)}{0.00203s^4 + 0.208s^3 + 3.40s^2 + 3.86s + 1} \qquad (11.86)$$

$$\frac{L_c}{W_H}(s) = \frac{0.732(-0.000035s^3 + 0.0081s^2 + 1.66s + 1)}{0.00203s^4 + 0.208s^3 + 3.40s^2 + 3.86s + 1} \qquad (11.87)$$

These predictions are compared with the measured behavior in Fig. 11.15; for the L_c response, measured data was available only for Case I. Note that the P_c response is well predicted by the model, including the changes caused by using a different steady-state operating point (condensing length). These changes are, of course, a manifestation of the nonlinearity of the system for large signals. The L_c response is well predicted in amplitude ratio but is clearly inaccurate for phase angle. By adding a two-second dead-time term e^{-2s} to the model, almost perfect correspondence (dashed line) is obtained with the measured phase. As discussed earlier with Eq. (11.43), this phase discrepancy is felt to be due mainly to the

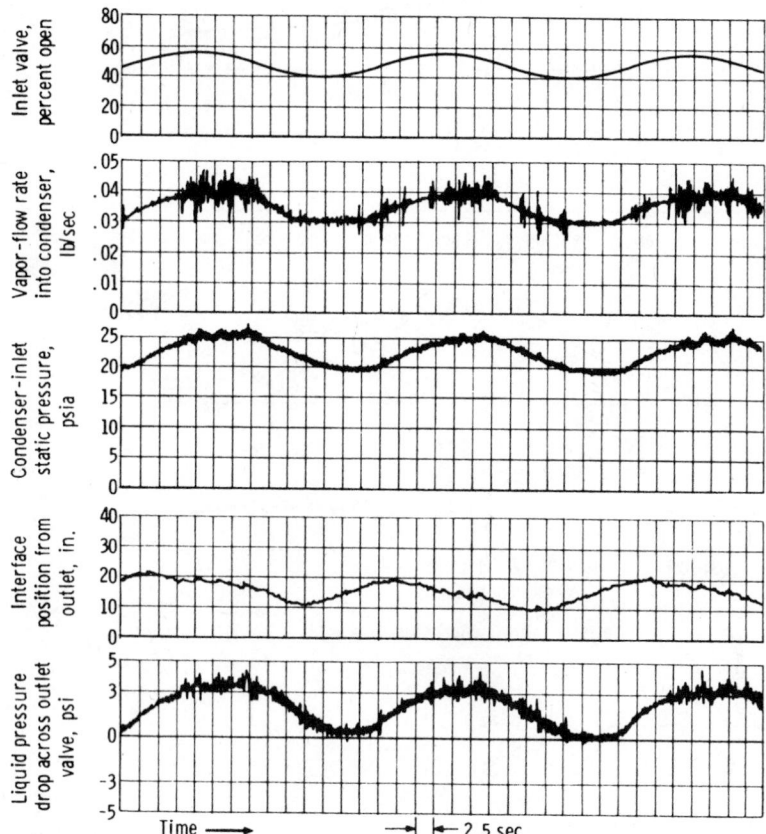

Figure 11.14   Measured sinusoidal test signals.

lack of a theoretical model for droplet formation and transport in the condensing space.

Since the theoretical model includes many parameters, some of which are difficult to measure, and since all these parameters are not of equal importance in determining the numerical values of the final transfer functions, it is useful to investigate the effect on the final results of changes in parameters. The results of such a *sensitivity study* are shown in Fig. 11.16. Such studies are quite useful in helping decide how much effort should be expended in trying to get accurate values for parameters, and also give guidance as to which parameters might be neglected in simplified models of the system. In the present example, our model has been found to quite accurately predict the actual behavior; but it is very complicated, making physical interpretation and design modification difficult.

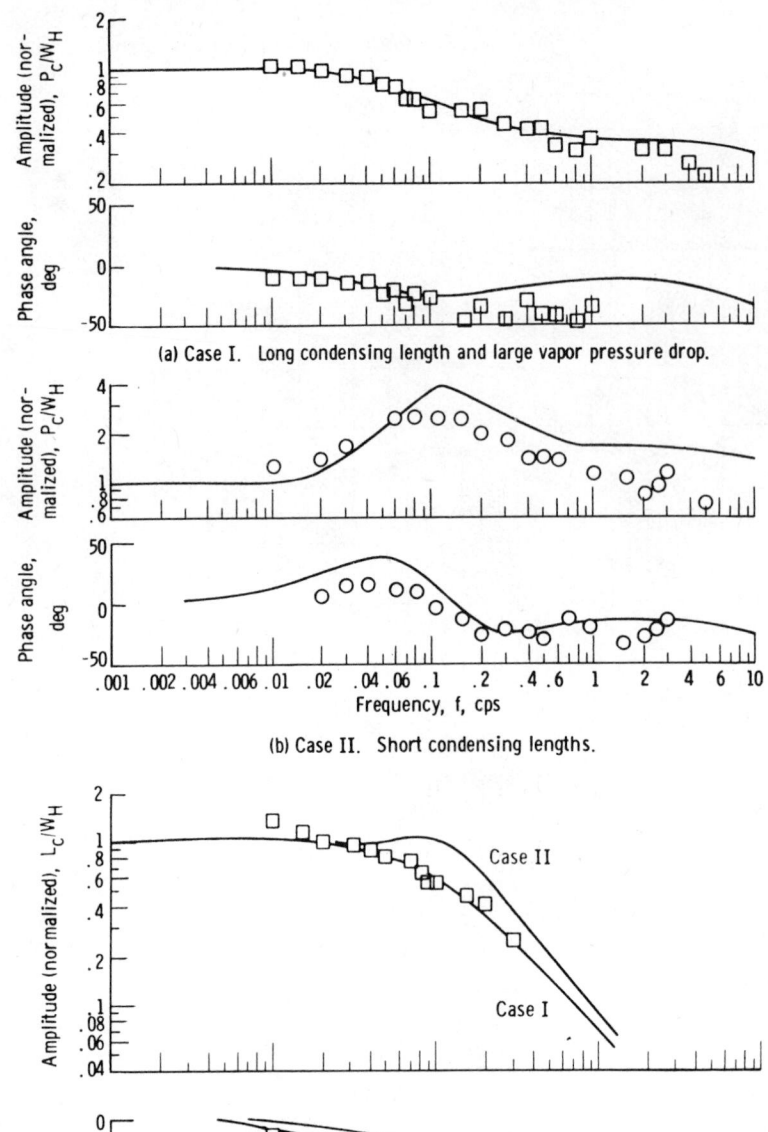

(a) Case I. Long condensing length and large vapor pressure drop.

(b) Case II. Short condensing lengths.

Figure 11.15 Comparison of theoretical and measured frequency response.

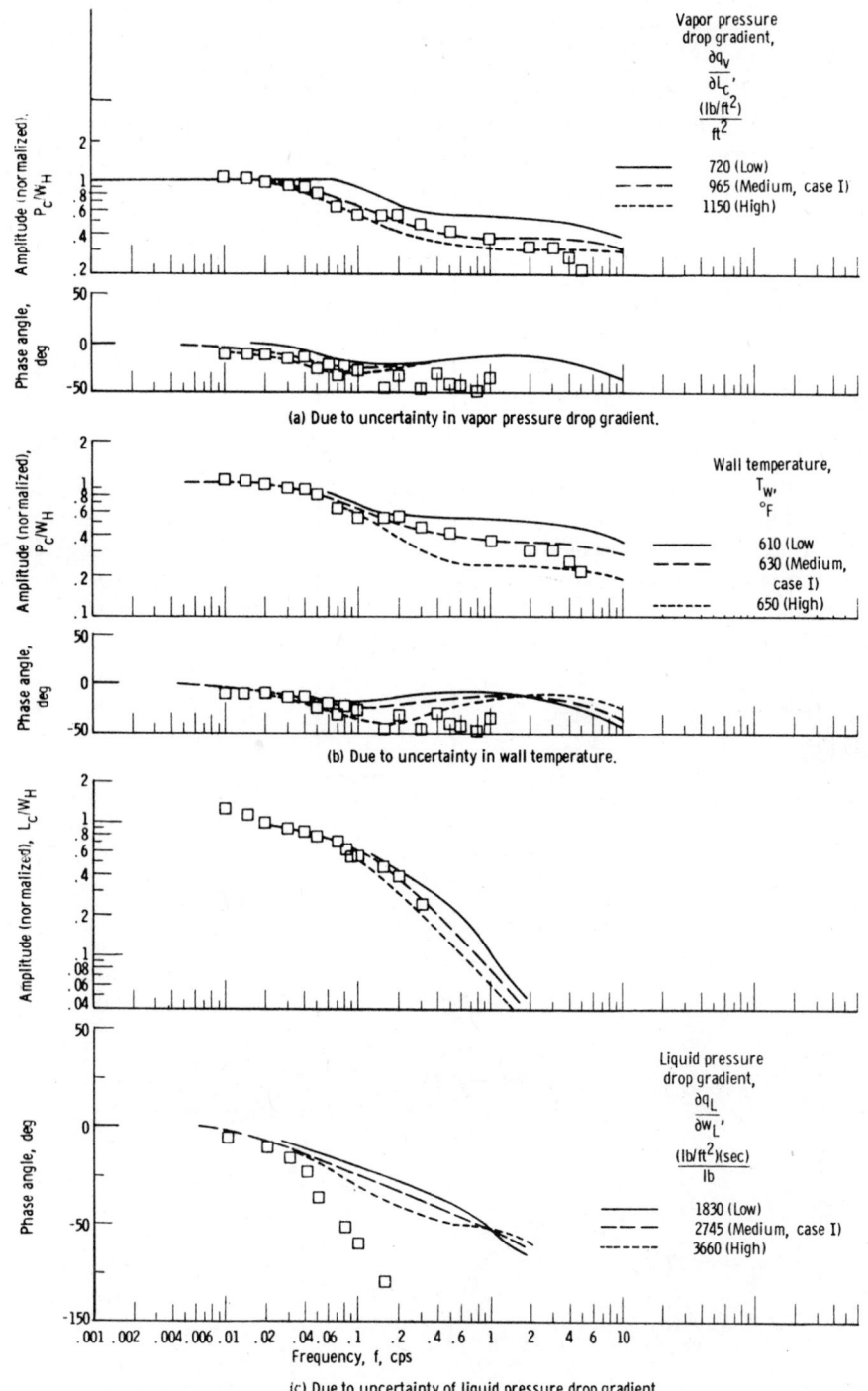

Figure 11.16 Sensitivity study for several critical parameters.

One should thus in general at least attempt to simplify such models. Guided by the above sensitivity study and also the relative size of coefficients in the final transfer functions, simplified forms of P_c/W_H and L_c/W_H were developed:

$$\frac{P_c}{W_H}(s) \approx \frac{K_{1p}(\tau_w s + 1)(N_1' s + 1)}{B' s^2 + A' s + 1} \tag{11.88}$$

$$\frac{L_c}{W_H}(s) \approx \frac{K_{1L}(M_{1L}' s + 1)}{B' s^2 + A' s + 1} \tag{11.89}$$

(We here again do not define the new parameters introduced; they may be found in the reference.) Figure 11.17 shows that these approximations do not seriously degrade the predictive qualities of the model for the P_c response, and similar good results are obtained for L_c, thus the condenser is now described by a model simple enough to comprehend without the aid of computers. Such

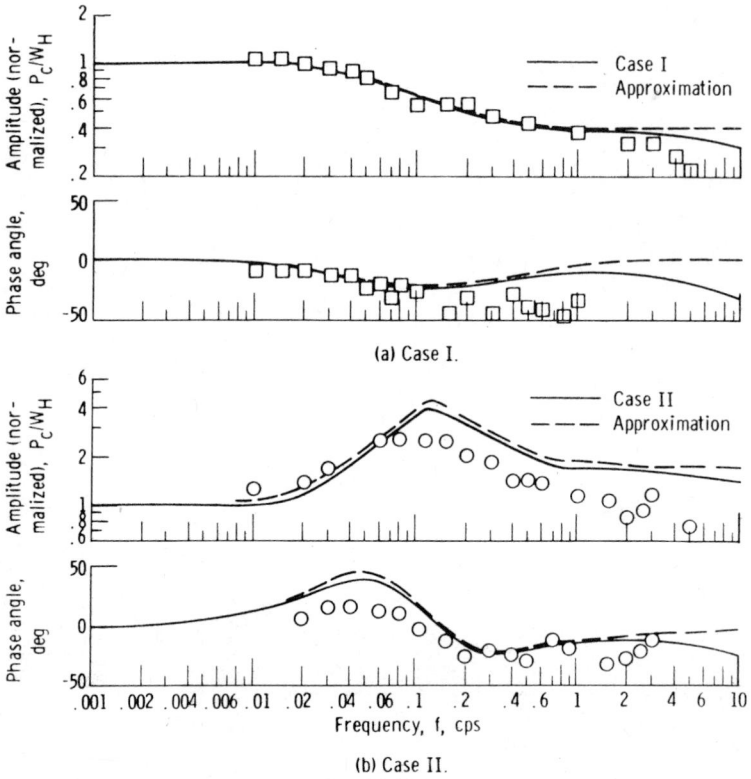

(a) Case I.

(b) Case II.

Figure 11.17 Quality of simplified models.

simplification becomes even more significant when we recall that the condenser is but one component in the entire power system and that all these component models must ultimately be coupled to study the entire system.

BIBLIOGRAPHY

1   Anderson, J. H., Superheater Dynamic Models, ASME Paper 66-WA/HT-57, 1966.

2   Baer, R. and R. Forman, The Prediction of Transient Performance of Pressurized Water Reactors, ASME Paper 67-WA/NE-7, 1967.

3   Ball, S. J., "Approximate Models for Distributed-Parameter Heat-Transfer Systems," *ISA Trans.*, **3**, No. 1 (1964).

4   Buckley, P. S., *Techniques of Process Control*, John Wiley and Sons, N. Y., 1964.

5   Chen, T. and K. R. Galle, "Transfer Functions for Steam-Generating Heat Exchangers," *ISA Trans.*, **3**, No. 2 (1964).

6   Chiu, P. C. and K. L. Poon, Frequency Response of Pool Boiling Plants, ASME Paper 71-HT-A, 1971.

7   Cohen, W. C. and E. F. Johnson, "Distributed Parameter Process Dynamics," Chem. Eng. Progress Symp. Series, *Process Dyn. Cont.*, **57**, No. 36, 1961.

8   Deyo, J. N. and W. T. Wintucky, Instrumentation of the SNAP-8 Simulator Facility, NASA TM X-1525, 1968.

9   Dolezal, R. and L. Varcop, *Process Dynamics*, American Elsevier, N. Y., 1970.

10   Duleba, G. S. and A. J. P. Lloyd, Modeling Heat Exchangers as Part of a Large System Simulation, ASME Paper 77-ENAS-9, 1977.

11   Fujii, S. and T. Nishida, "Dynamic Characteristics of Fluid Temperature in a Duct of Air Conditioning Systems," *Bull. JSME*, **15**, No. 79 (1972).

12   Grace, T. M. and E. A. Krejsa, Analytical and Experimental Study of Boiler Instabilities Due to Feed-System-Subcooled Region Coupling, NASA TN D-3961, 1967.

13   Guardabassi, G. and S. Rinaldi, Heat Exchanger Transfer Functions, ASME Paper 67-HT-5, 1967.

14   Hempel, A., "On the Dynamics of Steam Liquid Heat Exchangers," *ASME J. Basic Eng.*, June (1961).

15   Hess, H. L., Hooper, J. R. and S. L. Organ, Analytical and Experimental Study of the Dynamics of a Single-Tube Counterflow Boiler, NASA CR-1230, 1969.

16   Hsu, J. P. and N. Gilbert, "Transfer Functions of Heat Exchangers," *AIChE J.*, **8**, No. 5, Nov. (1962).

17   Iyer, J. S., Analog Simulation of Sodium-Heated Steam Generator, Babcock and Wilcox Res. Ctr. Rept. 5132, Alliance, Ohio, 1967.

18   Krejsa, E. A., Model for Frequency Response of a Forced-Flow, Hollow, Single Tube Boiler, NASA TM X-1528, 1968.

19   Krejsa, E. A., Goodykoontz, J. H., and G. H. Stevens, Frequency Response of a Forced-Flow Single-Tube Boiler, NASA TN-D-4039, 1967.

20 Leko, R. W. and K. S. Jefferies, Transfer Functions for Radiator Dynamics in the SNAP-8 System, NASA TN D-4946, 1968.

21 Masubuchi, M., "Dynamic Response and Control of Multipass Heat Exchangers," *ASME J. Basic Eng.*, March (1960).

22 Miles, J. H., Computer Method for Identification of Boiler Transfer Functions, NASA TM X-2436, 1971.

23 Miles, J. H., Identification of Boiler Inlet Transfer Functions and Estimation of System Parameters, NASA TM X-68075, 1972.

24 M'Pherson, P. K., *et al.*, "Dynamics Analysis of a Nuclear Boiler," *Proc. Inst. of Mech. Engrs.*, **180**, Pt. 1, No. 17 (1965–66).

25 Murray, W. D. and F. Landis, The Effect of Spacewise Lumping on the Solution Accuracy of the One-Dimensional Diffusion Equation, ASME Paper 62-WA-57, 1962.

26 Myers, G. E., Mitchell, J. W., and R. F. Norman, The Transient Response of Crossflow Heat Exchangers, Evaporators and Condensers, ASME Paper 66-WA/HT-34, 1966.

27 Stevens, G. H., Goodykoontz, J. H., and E. A. Krejsa, Exp. Evaluation of Four Transfer Functions for a Single Tube Boiler Which are Dynamically Independent of Exit Restrictions, NASA TM X-2247, 1971.

28 Stevens, G. H., Krejsa, E. A., and J. H. Goodykoontz, Frequency Response of a Forced-Flow Single-Tube Boiler with Inserts and Exit Restriction, NASA TN D-5023, 1969.

29 Thal-Larsen, H., "Dynamics of Heat Exchangers and Their Models," *ASME J. Basic Eng.*, June (1960).

30 Tripathi, S. S. and A. Ramachandran, Frequency Response of Counterflow Heat Exchangers with Flow Rate Inputs, ASME Paper 67-WA/AUT-5, 1967.

31 Williams, T. J. and H. J. Morris, "A Survey of the Literature on Heat Exchanger Dynamics and Control," Chem. Eng. Progress Symp. Series, *Process Dyn. Cont.*, **57**, No. 36, 1961.

PROBLEMS

11.1 Consider the heat exchanger system of Fig. 11.4 and use the symbol $\theta_{fL,D}$ for the desired value of θ_{fL_p}. Since settings of gain K based on minimization of mean-square error for random inputs sometimes give unacceptable step responses, we wish to find the response $\theta_{fL_p}(t)$ to a unit step of $\theta_{fL,d}$.

(a) Find the transfer function $(\theta_{fL_p}/\theta_{fL,d})(s)$ and give $\theta_{fL_p}(s)$ if $\theta_{fL,d}(t)$ is a unit step.

(b) Since this $\theta_{fL_p}(s)$ would probably not be analytically inverse transformable, explain clearly how frequency-response methods could be used to calculate $\theta_{fL_p}(t)$ numerically.

(c) Find $(\theta_{fLp}/\theta_{fL,d})(i\omega)$ and $\theta_{fL,d}(i\omega)$ for $\omega=0$ when $\theta_{fL,d}(t)$ is a unit step. What difficulty arises in carrying out the procedure of part b?

(d) Explain how $\theta_{fL,d}(t)$ can be modified to overcome the problem found in part c and still get a good approximation to the step response.

11.2 The heat exchanger shown in Fig. P11.1 is identical with that of Fig. 11.1 except that: (a) tube wall thickness is negligible; (b) no heating element is present; (c) the outside of the tube is surrounded by a fluid at fixed temperature T_0. It is desired to get the transfer function relating fluid temperature $T_f(x,t)$ at any location x to the inlet temperature $T_f(0,t)$.

(a) Set up the partial differential equation of this system, defining any new basic physical parameters needed, but otherwise using the nomenclature of Section 11.1.

(b) Rewrite the equation using the new variable $\theta \triangleq T_f(x,t) - T_0$. Define combinations of parameters to introduce a time constant τ and thus reduce the number of system parameters to a minimum.

(c) If inlet temperature $T_f(0,t)$ is constant at T_{f0}, find the steady-state solution for θ and sketch θ versus x.

(d) Represent θ as a sum of a steady-state part $\theta_0(x)$ and a perturbation $\theta_p(x,t)$, $\theta = \theta_0(x) + \theta_p(x,t)$. Let inlet temperature be $T_{f0} + T_{fp}(t)$. Find the transfer function $\theta_p(x,s)/T_{fp}(s)$ and put it into its most revealing and useful form.

(e) Find the response of $\theta_p(x,t)$ to a step input of $T_{fp}(t)$.

11.3 In the concentric pipe heat exchanger of Fig. P11.2, hot and cold fluids flow at constant velocities V_1 and V_2.

(a) Derive three partial differential equations involving temperatures of the two fluids and the inner pipe wall, using assumptions similar to those of Section 11.1 where appropriate. Define time constants and other appropriate parameters needed in the equations.

(b) Fig. P11.2 shows a counterflow exchanger (V_2 opposite to V_1). How

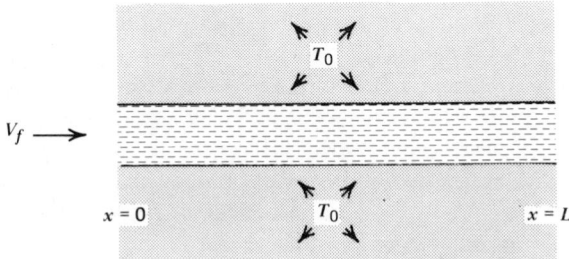

Figure P11.1

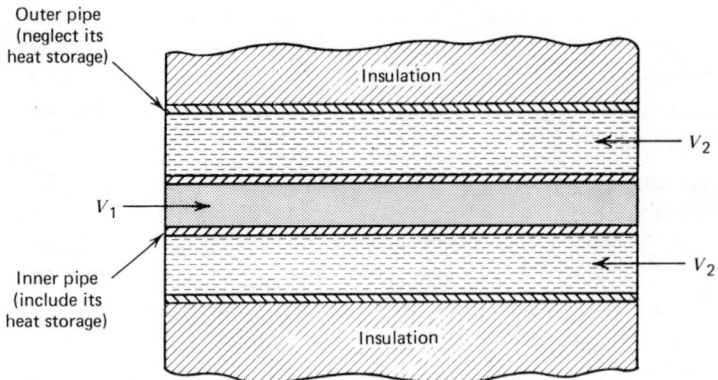

Figure P11.2

are the equations changed for a parallel flow configuration (V_2 same direction as V_1)?

(c) If the inner pipe wall is *very* thin, how do the equations of part a simplify?

(d) Simplify the equations of part a for steady-state operation (fluids enter at fixed temperatures and we wait for steady operation). Explain how these equations would be solved for the steady-state temperature distributions but do not actually solve them.

(e) For the dynamic model of part a, use Laplace transform to reduce equations to ordinary differential equations. Explain how these would be used to get a transfer function relating temperature of fluid number one at outlet to its inlet temperature.

11.4 We wish to study the step response of Eq. (11.32) for $x = 1.0$.

(a) Find $\theta_{fp}(s, 1)$ if $\theta_{fip}(t)$ is a unit step.

(b) Since the transcendental s function found in part a is probably not analytically inverse transformable, we must pursue other methods to find $\theta_{fp}(t, 1)$. By physical reasoning and application of other known data, certain *features* of $\theta_{fp}(t, 1)$ can be deduced *without* inverse transforming. Draw a clear sketch of $\theta_{fp}(t, 1)$ indicating two such useful *numerical* features and explaining how you obtained these numbers.

(c) It is suggested that if *numerical* values of σ and λ are known, we can compute a detailed numerical curve of $\theta_{fp}(t, 1)$ by use of Fourier transform methods. One problem with this approach is that the Fourier transform of a step function does not converge since a step function does not die out. Suggest a practical solution to this problem and explain in detail how you would carry it out.

(d)   Another suggestion notes the similarity between the general shapes of the frequency-response curves of Fig. 11.3 (upper pair) and for the form:

$$G_a(s) = \frac{\tau_1 s + 1}{\tau_2 s + 1} e^{-\tau_{DT} s} \quad (\tau_2 > \tau_1)$$

which *can* be easily inverse transformed. If we can "curve fit" the exact frequency response closely enough with $G_a(i\omega)$, then the *time* functions must also be close. Sketch the essential features of the $G_a(i\omega)$ curves on ordinary (not log) coordinates to check the validity of this suggestion and label significant features with their values. We now attempt to choose τ_{DT}, τ_1, and τ_2 so as to get the best match of frequency-response curves. What should the value of τ_{DT} be, and why? What should τ_1/τ_2 be and why? It now remains to determine τ_1 (or τ_2). Suppose we choose τ_1 such that $|G_a(i\omega)|$ matches the exact curve at the frequency ω_{MID} in Fig. P11.3. If $\sigma = 0.37$, $\lambda = 1.25$, and $t_r = 5$ sec, calculate τ_1, τ_2, and τ_{DT}.

(e)   Using the numerical values of part d, compute and compare the exact and approximate frequency-response curves. Comment on the accuracy expected from a step response calculated from $G_a(s)$.

11.5   Figure P11.4 shows a heat exchanger similar to that of Fig. 11.1 except that *microwave* heating is substituted for the heating element. Microwave

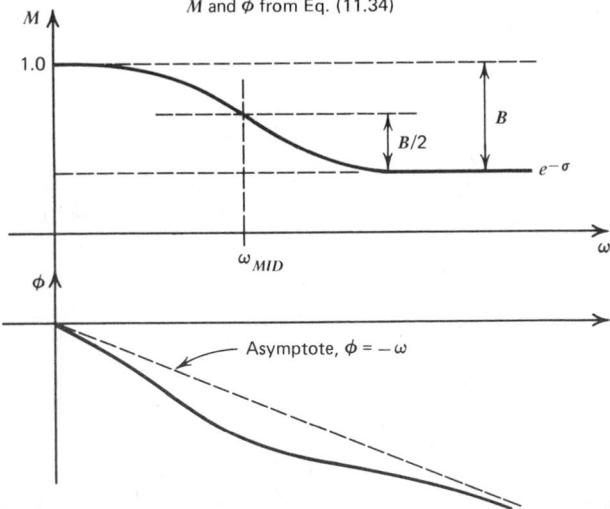

Figure P11.3

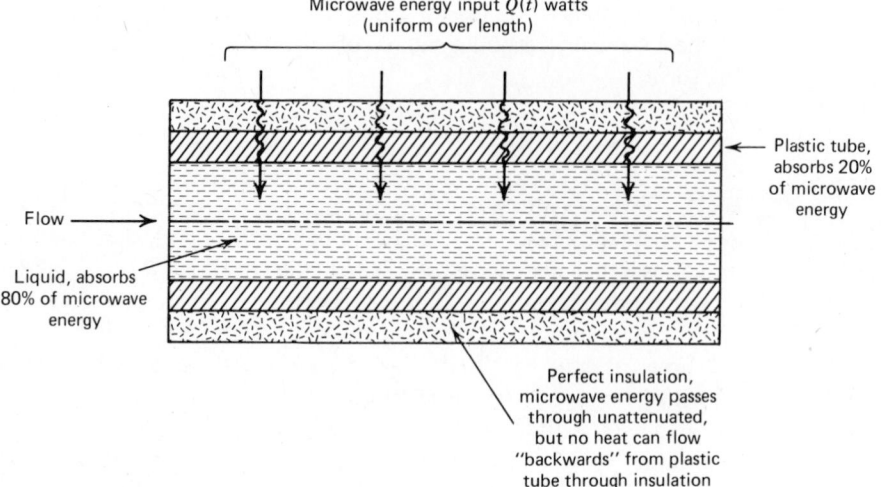

Figure P11.4

heating may be assumed to produce internal heat generation in proportion to the mass of material involved. Modify Equations (11.2) and (11.3) to account for these changes, explaining your methods and assumptions.

11.6 For the heat exchanger of Section 11.1, modify Equations (11.2) and (11.3) so that fluid velocity V can be a dynamic input variable $V(t)$. Then apply appropriate approximations to make the new equations linear with constant coefficients.

11.7 For the heat exchanger of Section 11.1, modify Equations (11.2) and (11.3) to account for imperfect insulation at the outer surface. Model the insulation as thermal resistance only (no energy storage) and let the temperature outside the insulation be constant.

11.8 The system shown in Fig. P11.5 is identical to the heat exchanger of Section 11.1 except for the addition of the porous tube that injects liquid (the same liquid as flowing in the heat exchanger) uniformly along its length. (Neglect any *thermal* effects of this inner tube.) The injection flow rate per unit length of tube is the same all along the length. The main flow Q_1 and injection flow Q_2 are both constant, but temperatures T_i and T_2 vary with time.

  (a) What is the volume flow rate Q_o at $x_a = L$? Get an expression for the volume flow rate $Q(x_a)$ and flow velocity $V(x_a)$ at any location x_a.

  (b) Modify Equations (11.2) and (11.3) to take into account the new conditions. Assume the injected liquid enters dx_a at T_2 and leaves at T_f.

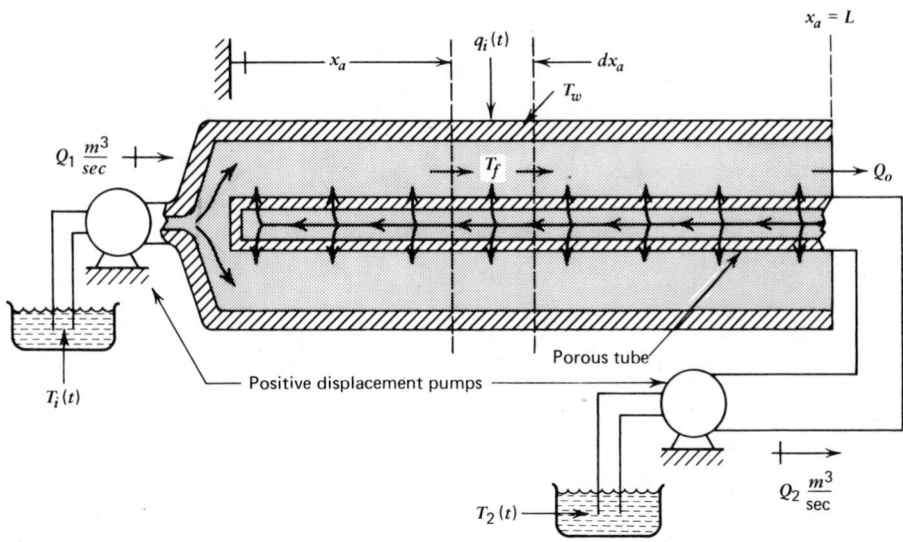

Figure P11.5

(c)   The partial differential equations of part b may not be solvable. Divide the heat exchanger into three equal-length lumps and derive the set of simultaneous ordinary differential equations for this lumped model.

11.9   For the heat exchanger of Fig. 11.1:
  (a)   What are the disadvantages of a lumped-parameter heat exchanger model in the control system application of Fig. 11.4?
  (b)   Formulate the simplest nontrivial lumped model of this heat exchanger and obtain transfer functions relating fluid outlet temperature to heat input rate and fluid inlet temperature.
  (c)   Repeat part b, adding one additional lump to the model. Compare the frequency response of the distributed parameter model with that of the lumped models of parts b and c.

1.10   For the system of Fig. 11.13:
  (a)   Augment Fig. 11.13 with the necessary blocks and signals to show how W_C and W_L are produced as outputs.
  (b)   Returning to the *basic* physical equations of the condenser, consider a case where the metal walls are *very* thin and modify all equations which would be affected by this fact.
  (c)   In Fig P11.6, assume the condenser is running in steady state with constant w_H, p_r, and T_n. Exit valve and pump are set to keep the receiver liquid level constant. With this steady-state established, suppose the exit valve fails and goes to its shut position. We wish to

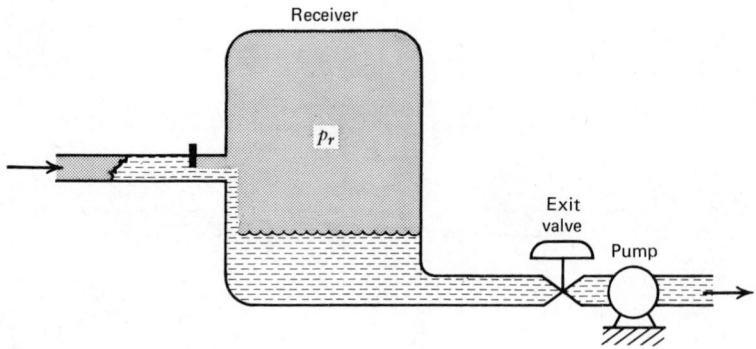

Figure P11.6

model the resulting transient behavior. Make and explain simplifying assumptions that must be added to the original set to deal with this situation. Explain in words what you intuitively expect will happen and then derive the needed new relations. Let equations be nonlinear if they need to be and show how they would be handled in a CSMP digital simulation.

11.11 Consider a smooth brass pipe, 1.0 inch ID, 0.08 inch wall, 5.0 feet long. Water flowing at 1.0 ft/sec enters at 50°F and leaves at 100°F in the initial steady state. Calculate and graph the frequency response curves of Fig. 11.3 for this numerical example, being careful to use the "real" frequency as abscissa, rather than the nondimensional ω of the text formulas.

CHAPTER 12

Vehicle Dynamics

If one begins to actually list the various "mobile machines" (vehicles) we use to transport people or goods (cars, trucks, trailers, bicycles, motorcycles, snowmobiles, conventional surface ships, hydrofoil craft, hovercraft, helicopters, aircraft, rockets, space vehicles, wheel/rail trains, monorails, proposed magnetic levitation or air-cushion trains), to perform some practical or entertainment function that requires mobility (road machinery, wheeled and crawler farm tractors, self-propelled farm equipment, mining machinery, satellites, submersibles for off-shore operations, Moon rovers, roller coasters), or to make war (tanks, submarines, missiles, aircraft carriers, bombers), the impact of vehicle technology on the development of civilization assumes major proportions. It would of course be misleading to suggest that the diverse devices listed above can be analyzed by some single unified vehicle dynamics modeling theory, and we will shortly hasten to retreat from the rather grandiose chapter title that suggests this to more modest and limited ground.

Examination of the textbook and periodical literature reveals that systematic and detailed attention to vehicle dynamics developed first in the aircraft industry and, in fact, most of the texts presently available are still addressed to this area. When the dynamics of land and sea vehicles later received attention, we find considerable borrowing of aeronautical methods and terminology wherever this was appropriate. Thus while a totally unified theory of vehicle dynamics does not exist, at least for certain broad classes of land, sea, air, and space vehicles, considerable commonality of approach may be seen. This is the basis for titling the chapter as it is, even though the detailed treatment concentrates on some specific aspects of automobile dynamics. That is, the automobile is used as a vehicle for introducing the reader to some elementary and basic concepts of general vehicle dynamics.

12.1 RIGID-BODY DYNAMICS

While sophisticated techniques for dealing with flexibility in vehicles are available, understanding of the rigid-body models is a prerequisite to their intelligent use. Also, many applications do not require flexible models to achieve acceptable accuracy, thus we concentrate on rigid-body models in this introductory treatment. Some readers may already be familiar with the dynamics of three-dimensional translation and rotation of a rigid body; others may not. Our approach will be to present and discuss the use of the basic equations without derivation. Thus those who have not been through the derivations are asked to accept the equations on faith and use them as a starting point in applications.

12.1.1 INERTIAL PROPERTIES OF A RIGID BODY
When we shortly develop a model for the handling dynamics of the automobile, we will find that actual solution of the equations requires numerical knowledge of many parameters: inertial, tire, and suspension. In fact, determination of such numerical values for real vehicles turns out to be a major undertaking requiring sophisticated apparatus, not just for automobiles but for all vehicles. When a new vehicle exists only as drawings, one must use whatever theory is available to estimate the needed parameter values; some may be quite inaccurate. Once a prototype has been built one can then make measurements to define parameters more accurately.

For a rigid-body model the following inertial parameters (or their equivalents) are necessary:

1 the total mass;
2 the location of the center of mass;
3 three moments of inertia, referred to specific axes;
4 three products of inertia, referred to specific axes.

Theoretical and experimental methods for finding each of these are available. Referring to Fig. 12.1:

$$\text{total mass} \triangleq M \triangleq \int_V \rho \, dV \tag{12.1}$$

Location of mass center:

$$\bar{x} \triangleq \frac{1}{M} \int_V x \rho \, dV \tag{12.2}$$

$$\bar{y} \triangleq \frac{1}{M} \int_V y \rho \, dV \tag{12.3}$$

$$\bar{z} \triangleq \frac{1}{M} \int z \rho \, dV \tag{12.4}$$

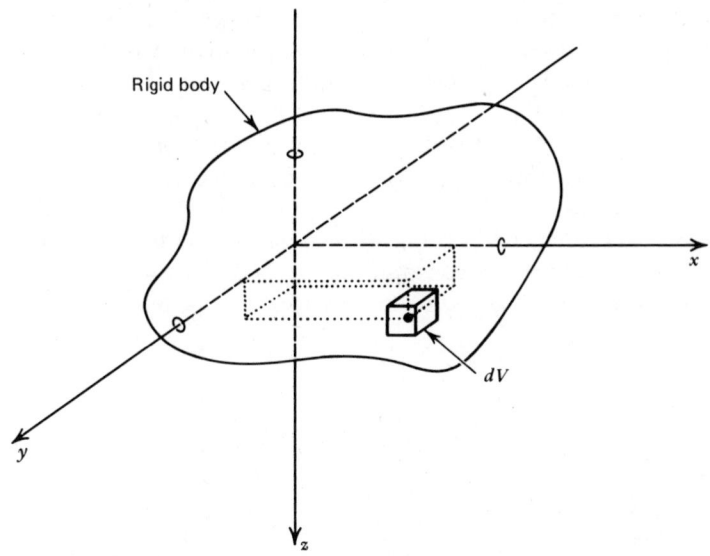

Figure 12.1   Rigid body.

Moments of inertia:

$$I_{xx} \triangleq \int_V (y^2 + z^2)\rho \, dV \tag{12.5}$$

$$I_{yy} \triangleq \int_V (z^2 + x^2)\rho \, dV \tag{12.6}$$

$$I_{zz} \triangleq \int_V (x^2 + y^2)\rho \, dV \tag{12.7}$$

Products of inertia:

$$I_{xy} = I_{yx} \triangleq \int_V xy\rho \, dV \tag{12.8}$$

$$I_{xz} = I_{zx} \triangleq \int_V xz\rho \, dV \tag{12.9}$$

$$I_{zy} = I_{yz} \triangleq \int_V zy\rho \, dV \tag{12.10}$$

For homogeneous bodies of simple geometrical shape these properties can be calculated fairly easily. Unfortunately, vehicles are assemblages of many components of peculiar shapes and diverse materials, making calculation tedious and possibly inaccurate, but nonetheless necessary, until hardware is built and measured values become accessible.

Details of the apparatus used to measure the various inertial properties may be found in the literature (see bibliography at end of chapter). Total mass and location of the mass center are found by supporting the body at several points on weighing platforms and/or by suspending it from various points. Moments of inertia are generally determined by constraining the body to rotation about the desired axis (using bearings, flexures, or knife edges), imposing a torsional

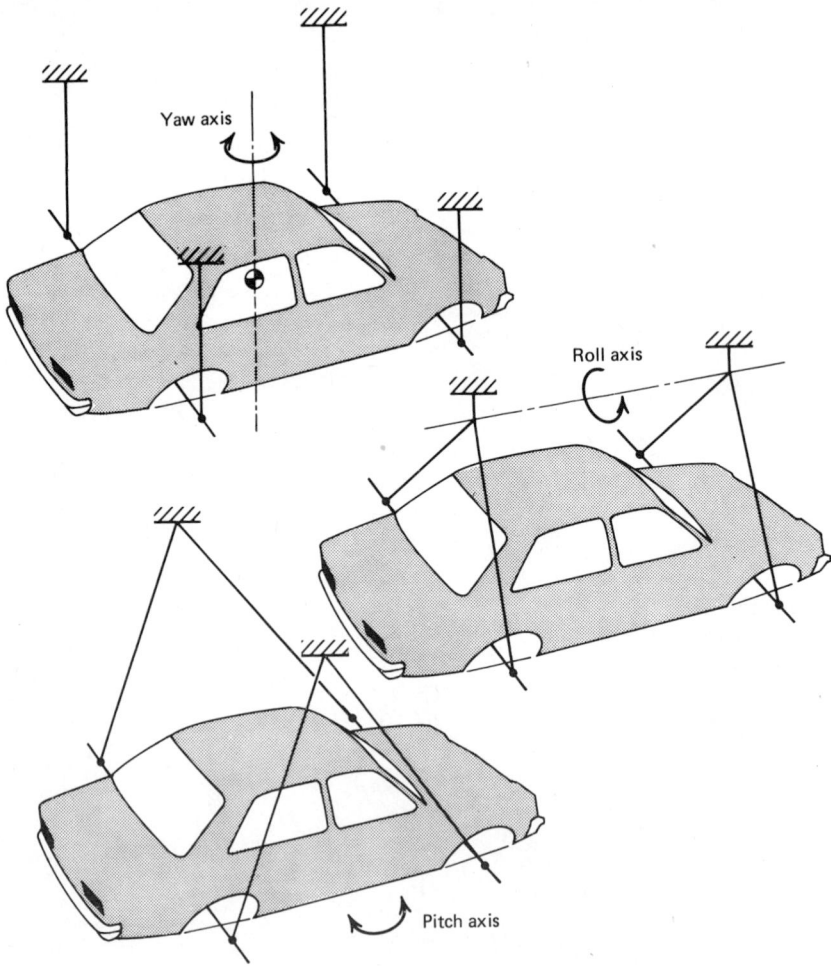

Figure 12.2   Experimental determination of vehicle moments of inertia.

elastic restraint of calibrated stiffness, and measuring the natural frequency of the torsional pendulum so produced. The multifilar pendulum approach using suspension cables and gravity effects (rather than elastic springs) to provide the restoring torques is also widely used (see Fig. 12.2). Products of inertia may be found with an oscillation apparatus allowing two rotational degrees of freedom or in a single-axis test by calculations from the change in frequency caused by tilting the body at known angles.

12.1.2   PRINCIPAL AXES OF A RIGID BODY

It is clear that the numerical values of the moments and products of inertia depend on both the origin location of our xyz axis set and its angular orientation. Thus one cannot speak of *the* moment of inertia of a body; it depends on the axes specified. If the origin of the xyz axes is fixed relative to the body, but the angular orientation of this set of mutually perpendicular axes is allowed to vary, it can be proven that one can always find a particular orientation for which all the products of inertia are exactly zero. This set of axes is called the *principal axis set* associated with the particular origin chosen. An especially useful set of principal axes has its origin at the mass center of the body.

Of the three moments of inertia (called principal moments of inertia) that one can calculate for a set of principal axes, it can be shown that the largest is also the largest possible for *any* orientation of axes with the same origin, and similarly for the smallest. If one knows by measurement or calculation the moments and products of inertia for any one origin location and axis orientation, he can calculate[1] these quantities for any other origin location and/or angular orientation by means of various axis-transfer and rotation theorems. To find[1] the orientation of the principal axes, one sets the expressions for the products of inertia equal to zero and solves the resulting set of equations for the unknown axis directions. In practical situations we often build up the complete object as a combination of simpler components whose inertial properties have been individually calculated or measured. Computer programs[2] for combining the component properties, finding the principal axis directions, and computing the principal moments of inertia of the composite body are available.

While the determination of principal axes is difficult for a body of arbitrary shape, when certain symmetry exists they may be located by inspection. If a uniform-density body has a plane of symmetry and we locate two of our axes (say x and z) in this plane, then I_{xy} and I_{zy} will be zero since for every mass element $dM = \rho\,dV$ (in Equations (12.8) and (12.10)) at a distance $+y$ from the plane there exists an "opposing" element at $-y$. The y axis is then a principal axis but the angular orientation of the x and z axes in the xz plane remains to be

[1]Housner, G. W. and D. E. Hudson, *Applied Mechanics: Dynamics*, Van Nostrand, N.Y., 1950, p. 151
[2]Cake, J. E., A Fortran Code for Computing the Principle Mass Moments of Inertia of Composite Bodies, NASA TM X-1754, 1969.

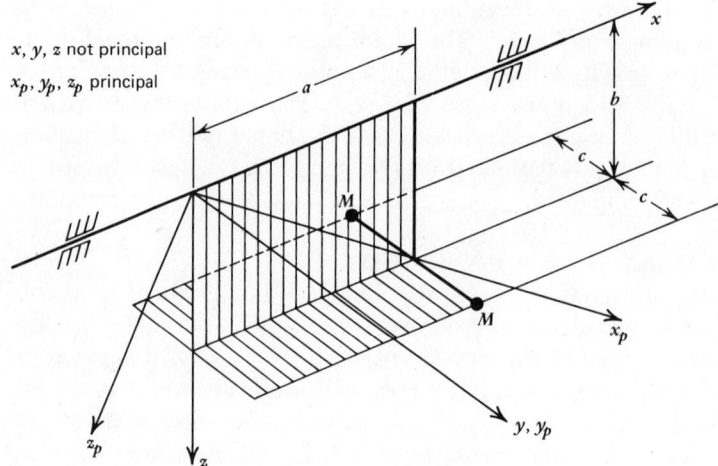

Figure 12.3   Point-mass illustration of principal axes.

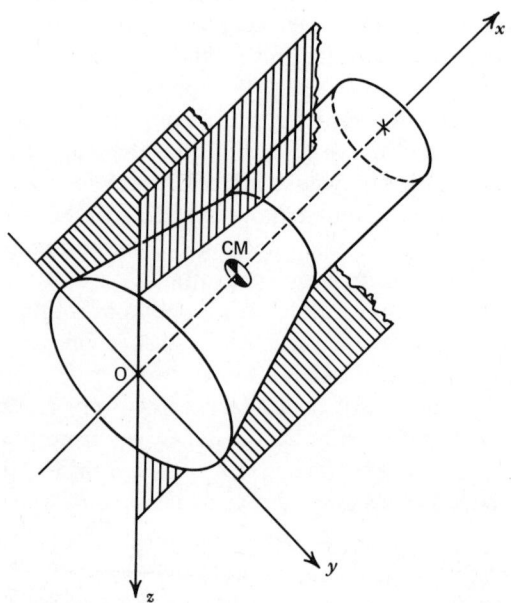

Figure 12.4   Principal axes for body of revolution.

found. Figure 12.3 shows a simple point-mass example (for which integrals may be replaced by summations $I_{xy} = \Sigma x_i y_i M_i$, etc.) that may be of help in visualizing these relations.

If the mass distribution (or, for a uniform-density body, the volume distribution) exhibits *two* perpendicular planes of symmetry, then two of the coordinate planes formed by the principle axes coincide with these planes of symmetry, thus determining all three principle axis directions. Bodies of revolution, such as that in Fig. 12.4 are perhaps the most common examples of this situation. There, *xyz* is the set of principal axes with origin at *O*. Principal axes with origin at the center of mass CM would have the same directions. The physical significance of the products of inertia is less familiar to most engineers than is that of the moments of inertia. When a body is caused to rotate about one of its principal axes (products of inertia are zero), there is *no* torque produced about either of the other two axes, whereas rotation about a non-principal axis causes torques (bearing reaction moments) about the other axes. For a body not constrained by bearings to rotate about a single axis, these torques cause a cross-coupling effect; torques about one axis can cause motions about a different axis.

12.1.3 EQUATIONS OF MOTION
In many, but not all, vehicle dynamics studies the earth may be taken as a flat and non-moving reference, such as the $X_0 Y_0 Z_0$ axes in Fig. 12.5. Motions of a rigid body relative to such "inertial" axes may be treated as absolute motions when applying Newton's law. We will specify such motion as a combination of

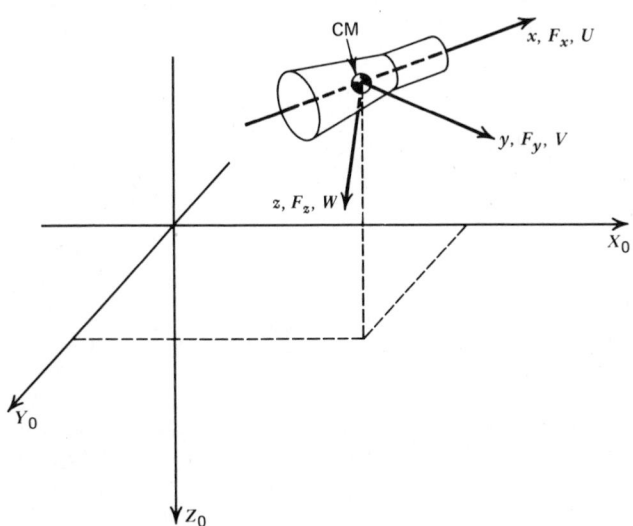

Figure 12.5 Axis systems for rigid-body motion.

translation (in three dimensions) of the body center of mass and rotation (in three dimensions) around the center of mass. The translatory motion of the center of mass is governed by the equations:

$$\sum F_{X_0} = M\ddot{X}_0, \qquad \sum F_{Y_0} = M\ddot{Y}_0, \qquad \sum F_{Z_0} = M\ddot{Z}_0 \qquad (12.11)$$

The forces $F_{X_0}, F_{Y_0}, F_{Z_0}$ are resolved along the space-fixed axes X_0, Y_0, Z_0, as are the resulting accelerations $\ddot{X}_0$, $\ddot{Y}_0$, and $\ddot{Z}_0$. These accelerations are thus *not* those that would be measured by vehicle accelerometers "bolted" to the vehicle and oriented along the body-fixed centroidal xyz axes.

The equations[3] describing the rotational motion are much more complicated:

$$\sum T_x = \dot{H}_x - \omega_z H_y + \omega_y H_z \qquad (12.12)$$

$$\sum T_y = \dot{H}_y - \omega_x H_z + \omega_z H_x \qquad (12.13)$$

$$\sum T_z = \dot{H}_z - \omega_y H_x + \omega_x H_y \qquad (12.14)$$

$H_x \triangleq x$ component of angular momentum

$$= I_{xx}\omega_x - I_{xy}\omega_y - I_{zx}\omega_z \qquad (12.15)$$

$H_y \triangleq y$ component of angular momentum

$$= -I_{xy}\omega_x + I_{yy}\omega_y - I_{yz}\omega_z \qquad (12.16)$$

$H_z \triangleq z$ component of angular momentum

$$= -I_{zx}\omega_x - I_{yz}\omega_y + I_{zz}\omega_z \qquad (12.17)$$

$$\dot{H}_x = I_{xx}\dot{\omega}_x - I_{xy}\dot{\omega}_y - I_{zx}\dot{\omega}_z \qquad (12.18)$$

$$\dot{H}_y = -I_{xy}\dot{\omega}_x + I_{yy}\dot{\omega}_y - I_{yz}\dot{\omega}_z \qquad (12.19)$$

$$\dot{H}_z = -I_{zx}\dot{\omega}_x - I_{yz}\dot{\omega}_y + I_{zz}\dot{\omega}_z \qquad (12.20)$$

The torques T_x, T_y, and T_z are about the body-fixed xyz axes and the angular velocity components ω_x, ω_y, and ω_z are similarly measured about these body-fixed axes, as by rate gyroscopes strapped to the body and oriented along x, y, and z. These rotational equations, derived for centroidal body axes, can also be shown to hold for noncentroidal body axes if:

1 the origin of xyz is fixed (translation suppressed but rotation still allowed);
2 the origin of xyz has a translational acceleration that always passes through the mass center.

[3] Housner and Hudson, *op. cit.*, p. 149.

Sometimes the translational equations are needed with forces resolved along the body axes *xyz* and with the motions referred to these same axes. These equations are:

$$\sum F_x = M\left(\dot{U} + W\omega_y - V\omega_z\right) \tag{12.21}$$

$$\sum F_y = M\left(\dot{V} + U\omega_z - W\omega_x\right) \tag{12.22}$$

$$\sum F_z = M\left(\dot{W} + V\omega_x - U\omega_y\right) \tag{12.23}$$

where U, V, and W are, respectively, the components of the absolute velocity of the mass center along x, y, and z. If one chooses the xyz axes to be not only centroidal but also principal axes, the translational equations are unchanged but the rotational equations are much simplified since all products of inertia are zero:

$$\sum T_x = I_{xx}\dot{\omega}_x + \left(I_{zz} - I_{yy}\right)\omega_y\omega_z \tag{12.24}$$

$$\sum T_y = I_{yy}\dot{\omega}_y + \left(I_{xx} - I_{zz}\right)\omega_x\omega_z \tag{12.25}$$

$$\sum T_z = I_{zz}\dot{\omega}_z + \left(I_{yy} - I_{xx}\right)\omega_x\omega_y \tag{12.26}$$

12.1.4   EULER ANGLES AND DIRECTION COSINES

The three "rotational unknowns" in the set of six motion equations are the body-axis angular velocity components ω_x, ω_y, and ω_z. In almost all practical applications we need to solve for the instantaneous angular *attitude* (orientation) of the body, not just its rotational velocity. Time integration of ω_x, ω_y, and ω_z certainly produces quantities whose dimensions are those of angles. However, these angles are *not* ones that give the body attitude since the *directions* of the x, y, and z axes are continually changing while the integration is being carried out. Several methods (Euler angles, direction cosines, quaternions, etc.) have been invented to deal with this problem and we now discuss some of them briefly.

Various sets of Euler angles have been defined, there being perhaps two main sets: those useful in analysis of gyroscopic devices and those often applied in vehicle studies. These latter Euler angles ϕ, θ, and ψ—are displayed in Fig. 12.6. Note that if one knew instantaneous values for ϕ, θ, and ψ, the angular attitude of the xyz axis set (and thus that of the attached rigid body) would be known since we could:

1   locate x axis from ψ and θ (ψ is measured from X_0' in the horizontal (X_0', Y_0') plane; then θ is measured in the vertical plane through line OR, as shown);
2   rotate around x by ϕ to locate y and z axes.

Or, another way:

1   locate xyz to coincide with X_0', Y_0', Z_0';
2   rotate xyz about Z_0' by ψ;

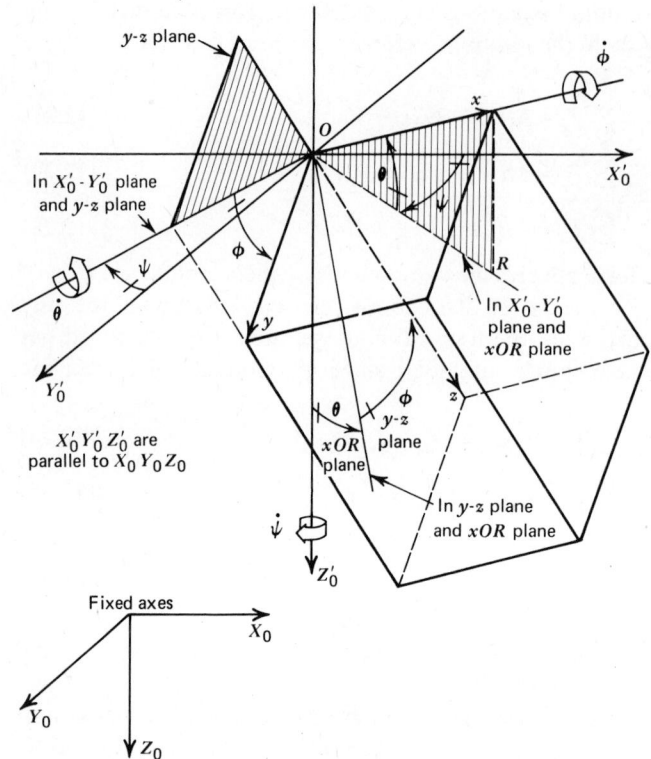

Figure 12.6   Definition of Euler angles.

3   rotate xyz about y by θ;
4   rotate xyz about x by ϕ.

From the geometry of Fig. 12.6 one can show:

$$\text{roll rate} \triangleq \omega_x \triangleq p = \dot{\phi} - \dot{\psi}\sin\theta \tag{12.27}$$

$$\text{pitch rate} \triangleq \omega_y \triangleq q = \dot{\theta}\cos\phi + \dot{\psi}\cos\theta\sin\phi \tag{12.28}$$

$$\text{yaw rate} \triangleq \omega_z \triangleq r = -\dot{\theta}\sin\phi + \dot{\psi}\cos\theta\cos\phi \tag{12.29}$$

$$\dot{\phi} = p + q\sin\phi\tan\theta + r\cos\phi\tan\theta \tag{12.30}$$

$$\dot{\theta} = q\cos\phi - r\sin\phi \tag{12.31}$$

$$\dot{\psi} = (q\sin\phi + r\cos\phi)/\cos\theta \tag{12.32}$$

Using the above relations we can convert the rotational equations of motion from the unknowns ω_x, ω_y, and ω_z to the new unknowns ϕ, θ, and ψ. By direct substitution and manipulation we finally arrive at:

$$\ddot{\phi} = \frac{1}{I_{xx}} \Big[\sum T_x + I_{xx}(\ddot{\psi}\sin\theta + \dot{\psi}\dot{\theta}\cos\theta)$$

$$+ (I_{yy} - I_{zz})(\dot{\theta}\cos\phi + \dot{\psi}\cos\theta\sin\phi)(\dot{\psi}\cos\theta\cos\phi - \dot{\theta}\sin\phi) \Big] \qquad (12.33)$$

$$\ddot{\theta} = \frac{1}{I_{yy}\cos\phi} \Big[\sum T_y - I_{yy}(\ddot{\psi}\cos\theta\sin\phi + \dot{\psi}\dot{\phi}\cos\theta\cos\phi$$

$$- \dot{\theta}\dot{\phi}\sin\phi - \dot{\psi}\dot{\theta}\sin\phi\sin\theta) + (I_{zz} - I_{xx})(\dot{\phi} - \dot{\psi}\sin\theta)(\dot{\psi}\cos\theta\cos\phi - \dot{\theta}\sin\phi) \Big]$$

$$(12.34)$$

$$\ddot{\psi} = \frac{1}{I_{zz}\cos\theta\cos\phi} \Big[\sum T_z + I_{zz}(\dot{\theta}\dot{\phi}\cos\phi + \ddot{\theta}\sin\phi$$

$$+ \dot{\psi}\dot{\phi}\cos\theta\sin\phi + \dot{\psi}\dot{\theta}\sin\theta\cos\phi) + (I_{xx} - I_{yy})(\dot{\phi} - \dot{\psi}\sin\theta)(\dot{\theta}\cos\phi + \dot{\psi}\cos\theta\sin\phi) \Big]$$

$$(12.35)$$

These equations are obviously thoroughly nonlinear and analytical solutions of the general case should not be expected. There are, in fact, even some difficulties with analog or digital simulation, since although $\ddot{\phi}$, $\ddot{\theta}$, and $\ddot{\psi}$ have been isolated on the left-hand sides (as is usual), $\ddot{\psi}$ appears on the *right-hand* side of (12.33) and (12.34) and $\ddot{\theta}$ on the *right-hand* side of (12.35),, creating a "computing loop" problem since $\ddot{\psi}$ must be known to compute $\ddot{\psi}$. This fortunately can be resolved by algebraically combining (12.34) and (12.35) (*before* simulation). That is, substitute $\ddot{\theta}$ into (12.35) from (12.34) and then gather the terms in $\ddot{\psi}$ on the left, giving a right-hand side that has *no* second derivatives in it. This then allows computation of $\ddot{\psi}$ in the usual way, and having $\ddot{\psi}$, $\ddot{\theta}$ can be computed from (12.34) and then $\ddot{\phi}$ from (12.33).

The other difficulty, while being a problem only for large angular motions (which are not often required in vehicle studies) is of a more fundamental nature. Note that in (12.34) and (12.35) if $\cos\theta$ and/or $\cos\phi$ go through zero, $\ddot{\theta}$ and $\ddot{\psi}$ can become infinite, causing *singularities* in the differential equations (the situation is called "gimbal lock" by engineers) and impeding both analytical and simulation solutions. This difficulty can be avoided in several ways by using methods other than Euler angles to obtain angular attitude. We briefly explain the direction cosine approach.[4] A matrix D of direction cosines l is defined as:

$$D \triangleq \begin{bmatrix} l_{xx} & l_{xy} & l_{xz} \\ l_{yx} & l_{yy} & l_{yz} \\ l_{zx} & l_{zy} & l_{zz} \end{bmatrix} \tag{12.36}$$

This matrix allows one to transform any vector (such as, say, the x axis, y axis, or z axis) from one coordinate system to another (such as the nonrotating X_0', Y_0', Z_0') and thereby locate xyz.

It can be shown that the direction cosines may be found by augmenting Equations (12.24–12.26) with a set of nine equations in the direction cosines:

$$\begin{aligned}
\dot{l}_{xx} &= l_{yx}\omega_z - l_{zx}\omega_y & \dot{l}_{xy} &= l_{yy}\omega_z - l_{zy}\omega_y \\
\dot{l}_{xz} &= l_{yz}\omega_z - l_{zz}\omega_y & \dot{l}_{yx} &= -l_{xx}\omega_z + l_{zx}\omega_x \\
\dot{l}_{yy} &= -l_{xy}\omega_z + l_{zy}\omega_x & \dot{l}_{yz} &= -l_{xz}\omega_z + l_{zz}\omega_x \\
\dot{l}_{zx} &= l_{xx}\omega_y - l_{yx}\omega_x & \dot{l}_{zy} &= l_{xy}\omega_y - l_{yy}\omega_x \\
\dot{l}_{zz} &= l_{xz}\omega_y - l_{yz}\omega_x
\end{aligned} \tag{12.37}$$

To integrate these $\dot{l}$ equations one needs initial conditions for each of the nine l's. We usually know initial values for the Euler angles; thus the following equations relating direction cosines to Euler angles are useful for computing numerical initial values for the l's.

$$\begin{aligned}
l_{xx} &= \cos\psi\cos\theta, & l_{xy} &= \sin\psi\cos\theta, & l_{xz} &= -\sin\theta \\
l_{yx} &= -\sin\psi\cos\phi + \cos\psi\sin\theta\sin\phi, & l_{yz} &= \cos\theta\sin\phi \\
l_{yy} &= \cos\psi\cos\phi + \sin\psi\sin\theta\sin\phi, & l_{zz} &= \cos\theta\cos\phi \\
l_{zx} &= \sin\psi\sin\phi + \cos\psi\sin\theta\cos\phi \\
l_{zy} &= -\cos\psi\sin\phi + \sin\psi\sin\theta\cos\phi
\end{aligned} \tag{12.38}$$

Once we have solved the differential equations for the direction cosines as functions of time, the Euler angles are available from:

$$\phi = \tan^{-1}\frac{l_{yz}}{l_{zz}}, \qquad \theta = -\sin^{-1}l_{xz}, \qquad \psi = \tan^{-1}\frac{l_{xy}}{l_{xx}} \tag{12.39}$$

[4]Gainer, T. G. and S. Hoffman, Summary of Transformation Equations and Equations of Motion, NASA SP-3070, 1972, p. 90; Rogers, A. E. and T. W. Connolly, *Analog Computation in Engineering Design*, McGraw-Hill, N.Y., 1960, p. 415.

Most practical problems can be adequately treated by employing Euler angles directly, the direction cosine method, or the quaternion[5] approach not here discussed.

12.1.5   A SPINNING SATELLITE: EXACT AND SIMPLIFIED MODELS

While analog, digital, or hybrid simulation can deal with the six-degree-of-freedom nonlinear equations discussed above, the cost, and complexity of interpretation dictate that this model be used only when absolutely necessary. The most common simplifications involve *linearization* for small angular perturbations and *decoupling* of certain degrees of freedom by consideration of restricted classes of motion. Both these techniques are explained in detail in the aircraft dynamics literature[6] and will be discussed in the context of the automobile example of Section 12.2. In addition we now discuss briefly another simple example: a spinning satellite. In Fig. 12.7 we assume the satellite is in earth orbit, with its center of mass following essentially a circular trajectory about the earth. For simplicity, assume our interest is only in the rotational motion of the satellite about its center of mass and that we can ignore other motions. In the top view of Fig. 12.7 the satellite has a constant spin angular velocity ω_{xo} about the x body axis and gas jets, fired in opposed pairs, can apply pure attitude-steering torques about the z body axis without causing any unbalanced force to disturb the trajectory. We are interested in the rotational motion that is caused by short bursts of jet torque.

To provide some benchmark results for comparison with the simplified model we will shortly develop, let us first get an "exact" solution using digital simulation on the nonlinear equations and utilizing the direction-cosine technique. Assuming some arbitrary numerical values, a CSMP program might go as follows.

```
PARAM   IXX=5.,IYY=3.,IZZ=3.
PARAM   WX0=1.,WY0=0.0,WZ0=0.0
PARAM   LXX0=1.0,LXY0=0.0,LXZ0=0.0
PARAM   LYX0=0.0,LYY0=1.0,LYZ0=0.0
PARAM   LZX0=0.0,LZY0=0.0,LZZ0=1.0
        TX=0
        TY=0
        TZ=4.*(STEP(0.0)−STEP(.1))
```

[5]Gainer and Hoffman, *op. cit.*; Fang, A. C. and B. G. Zimmerman, Digital Simulation of Rotational Kinematics, NASA TN D-5302, 1969; Mitchel, E. E. and A. E. Rogers, "Quaternion Parameters in the Simulation of a Spinning Rigid Body," *Simulation*, 4(6), June (1965), p. 390.

[6]McRuer, D., Ashkenas, I., and D. Graham, *Aircraft Dynamics and Automatic Control*, Princeton U. Press, Princeton, N.J., 1973; Etkin, B., *Dynamics of Atmospheric Flight*, John Wiley and Sons, N.Y., 1972; Blakelock, J. H., *Automatic Control of Aircraft and Missiles*, John Wiley and Sons, N.Y., 1965.

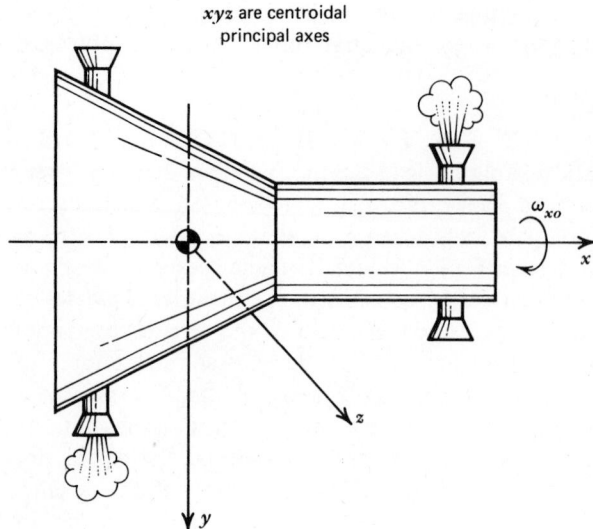

xyz are centroidal
principal axes

Figure 12.7 Spinning satellite.

$$WXDOT = (TX + (IYY - IZZ)*WY*WZ)/IXX$$
$$WYDOT = (TY + (IZZ - IXX)*WX*WZ)/IYY$$
$$WZDOT = (TZ + (IXX - IYY)*WX*WY)/IZZ$$
$$WX = INTGRL(WX0,WXDOT)$$
$$WY = INTGRL(WY0,WYDOT)$$
$$WZ = INTGRL(WZ0,WZDOT)$$
$$LXXDOT = WZ*LYX - WY*LZX$$
$$LXYDOT = WZ*LYY - WY*LZY$$
$$LXZDOT = WZ*LYZ - WY*LZZ$$
$$LYXDOT = -WZ*LXX + WX*LZX$$
$$LYYDOT = -WZ*LXY + WX*LZY$$
$$LYZDOT = -WZ*LXZ + WX*LZZ$$
$$LZXDOT = WY*LXX - WX*LYX$$
$$LZYDOT = WY*LXY - WX*LYY$$
$$LZZDOT = WY*LXZ - WX*LYZ$$
$$LXX = INTGRL(LXX0,LXXDOT)$$
$$LXY = INTGRL(LXY0,LXYDOT)$$
$$LXZ = INTGRL(LXZ0,LXZDOT)$$
$$LYX = INTGRL(LYX0,LYXDOT)$$
$$LYY = INTGRL(LYY0,LYYDOT)$$
$$LYZ = INTGRL(LYZ0,LYZDOT)$$
$$LZX = INTGRL(LZX0,LZXDOT)$$
$$LZY = INTGRL(LZY0,LZYDOT)$$
$$LZZ = INTGRL(LZZ0,LZZDOT)$$

```
            THETA = ARSIN(−LXZ)
            PSIA = ATAN2(LXY,LXX)
            PHIA = ATAN2(LYZ,LZZ)
            PHIAS = PHIA*.05
TIMER       FINTIM = 10.,DELT = .01,OUTDEL = .2
OUTPUT      PHIAS,PSIA,THETA
PAGE        GROUP,WIDTH = 50,NTAB = 0
END
```

In Fig. 12.8, PSIA and THETA are ψ and θ in radians, while PHIAS is 0.05ϕ in radians (scaled to fit on the graph for the best display). We see that ψ

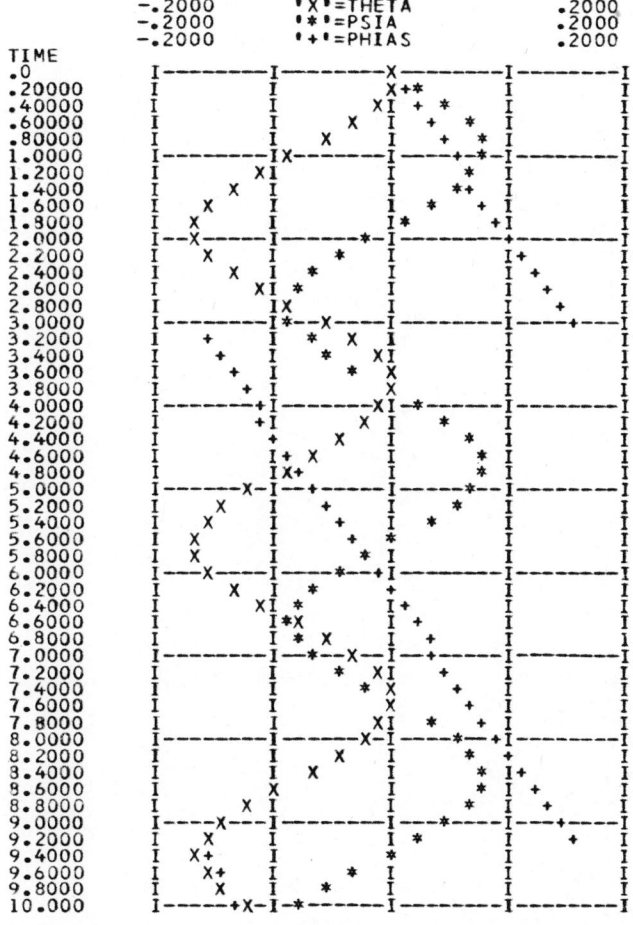

Figure 12.8 Results of direction-cosine analysis.

oscillates around a zero mean value, with amplitude of 0.08 rad and period 3.8 sec while θ has a similar oscillation but with mean value -0.08, thus never being positive. The apparent discontinuity in ϕ at 3 sec is caused by the ATAN2 (inverse tangent) function's behavior when its argument passes through π radians. That is, the "jump" from $+\pi$ to $-\pi$ is actually a perfectly smooth continuous rotation of ϕ since $+\pi$ and $-\pi$ are the same angle. With this interpretation the graph of PHIAS is seen to be a continuously rising straight line, indicative of a rotation at constant angular velocity. While the visualization of the total motion is probably not obvious to most, use of Figs. 12.6 and 12.8 to sketch a few instantaneous axis positions leads to the interpretation shown in Fig. 12.9. The x (spin) axis performs a conical motion (called nutation) in which the axis of the cone is in the $X_0'Z_0'$ plane but displaced downward (θ) by 0.08 rad. If the applied torque is thought of as an attempt to steer the spinning vehicle into a new angular attitude, this θ displacement (interpretable as a gyroscopic precession effect) shows that such steering is possible, but unfortunately accompanied by persistent oscillations. In an actual steering system, feedback control using body-mounted position gyros (for attitude sensing) and rate gyros (for stabilizing derivative-control signals) would be employed to manipulate the jet torques so as to suppress the oscillations while allowing precise angular positioning of the vehicle. Analysis and design of this control system obviously requires a vehicle dynamics model such as the one we have shown.

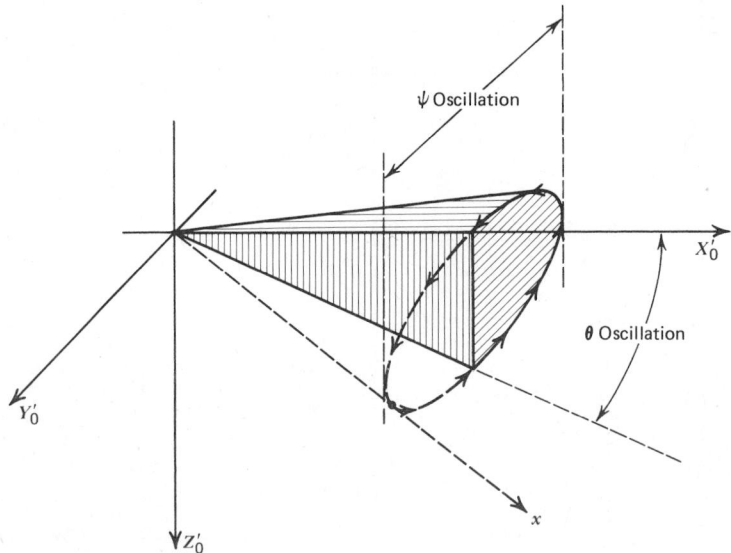

Figure 12.9 Satellite nutation motion.

If we try to apply the direct Euler angle method to the system of Fig. 12.7, Equations (12.34) and (12.35) have singularity problems since, due to vehicle spin, ϕ gets large enough to make $\cos\phi$ go through zero repeatedly. While this appears to rule out the method, a little reflection reveals a possible alternative viewpoint that could be used to at least study the nature of the motion. Since large ψ does *not* cause singularities in Equations (12.33)–(12.35), if we took z as the spin axis and applied torque about, say, the x axis, it would seem that the *character of the motion* produced would be the same as for x spin and z torque, even though the Euler-angle time histories might be different. Let us pursue this idea using a CSMP simulation.

```
INIT
PARAM IXX = 3.,IYY = 3.,IZZ = 5.
         A = (IXX − IYY)/IZZ
         B = (IZZ − IXX)/IYY
         C = (IYY − IZZ)/IXX
DYNAM
         TX = 4.*(STEP(0.0) − STEP(.1))
         TY = 0
         TZ = 0
         L = TZ/IZZ + (TY*TAN(PHI))/IYY +
         THETDT*PHIDT*COS(PHI)
         M = PSIDT*THETDT*SIN(THETA)*COS(PHI)
         N = A*(PHIDT − PSIDT*SIN(THETA))*(THETDT*COS(PHI) +
         PSIDT*COS(THETA)*SIN(PHI))
         O = THETDT*PHIDT*TAN(PHI)*SIN(PHI) +
         PSIDT*THETDT*SIN(THETA)*SIN(PHI)*TAN(PHI)
         P = B*SIN(PHI)*(PHIDT − PSIDT*SIN(THETA))*
         (PSIDT*COS(THETA) − THETDT*TAN(PHI))
         Q = COS(THETA)*COS(PHI) + COS(THETA)*SIN(PHI)*TAN(PHI)
         PSIDT2 = (L + M + N + O + P)/Q
         PSIDT = INTGRL(PSIDT0,PSIDT2)
INCON  PSIDT0 = 1.0
         PSI = INTGRL(PSI0,PSIDT)
INCON  PSI0 = 0.0
         R = TY/(IYY*COS(PHI)) − PSIDT2*COS(THETA)*
         TAN(PHI) − PSIDT*PHIDT*COS(THETA)
         S = THETDT*PHIDT*TAN(PHI) + PSIDT*
         THETDT*TAN(PHI)*SIN(THETA)
         T = B*(PHIDT − PSIDT*SIN(THETA))*
         (PSIDT*COS(THETA) − THETDT*TAN(PHI))
         THETD2 = R + S + T
         THETDT = INTGRL(THETD0,THETD2)
```

```
INCON   THETD0=0.0
        THETA=INTGRL(THETA0,THETDT)
INCON   THETA0=0.0
        U=TX/IXX+PSIDT2*SIN(THETA)+
        PSIDT*THETDT*COS(THETA)
        V=C*(THETDT*COS(PHI)+PSIDT*COS(THETA)*SIN(PHI))*
        (PSIDT*COS(THETA)*COS(PHI)-THETDT*SIN(PHI))
        PHIDT2=U+V
        PHIDT=INTGRL(PHIDT0,PHIDT2)
INCON   PHIDT0=0.0
        PHI=INTGRL(PHI0,PHIDT)
INCON   PHI0=0.0
TIMER   FINTIM=25.,DELT=.01,OUTDEL=0.5
OUTPUT  PHI,PSI,THETA
PAGE    WIDTH=50,NTAB=0
```

In Fig. 12.10 the behavior of PSI is clearly analogous to that of PHIAS in Fig. 12.8; however, the time histories of the other two angles show clearly different patterns in the two figures. It can be shown, however, that the z axis does have in fact a conical nutation motion analogous to that of the x axis in Fig. 12.9. This is not obvious from the results of Fig. 12.10 because θ and ϕ do not locate the z axis as directly as θ and ψ located the x axis when it was the spin axis. It appears thus that in this example, while the Euler-angle method can be applied, the need to re-orient the body to avoid singularities causes some difficulty in interpretation, making the direction cosine method the preferred approach.

Turning now to simplified models for our spinning satellite example, the most common case involves linearization under the assumption of small angles. It is possible to linearize either the direction cosine approach or the direct Euler-angle method. Even though we found that the exact Euler-angle equations required re-orienting the body so that z was the spin axis, let us linearize these equations for the actual configuration of Fig. 12.7 to see where it leads. We can work directly from the CSMP program equations if we take IXX=5., IYY=IZZ =3., $\dot{\phi}_0=1.$, and $\dot{\psi}_0=\dot{\theta}_0=\phi_0=\psi_0=\theta_0=0.0$. For increased generality we also let T_x, T_y, and T_z be small perturbations that are arbitrary functions of time. Using the usual Taylor series linearization methods (or alternatively using small-angle trigonometric approximations and neglecting products of perturbations), we can finally arrive at:

$$
\begin{array}{llll}
[I_{xx}s^2]\phi & +[0]\theta & +[0]\psi = T_x & \\
[0]\phi & +[I_{yy}s^2]\theta & +[\,(I_{yy}+I_{xx}-I_{zz})\dot{\phi}_0 s]\psi = T_y & (12.40) \\
[0]\phi & +[(I_{yy}-I_{xx}-I_{zz})\dot{\phi}_0\, s]\theta & +[I_{zz}s^2]\psi = T_z &
\end{array}
$$

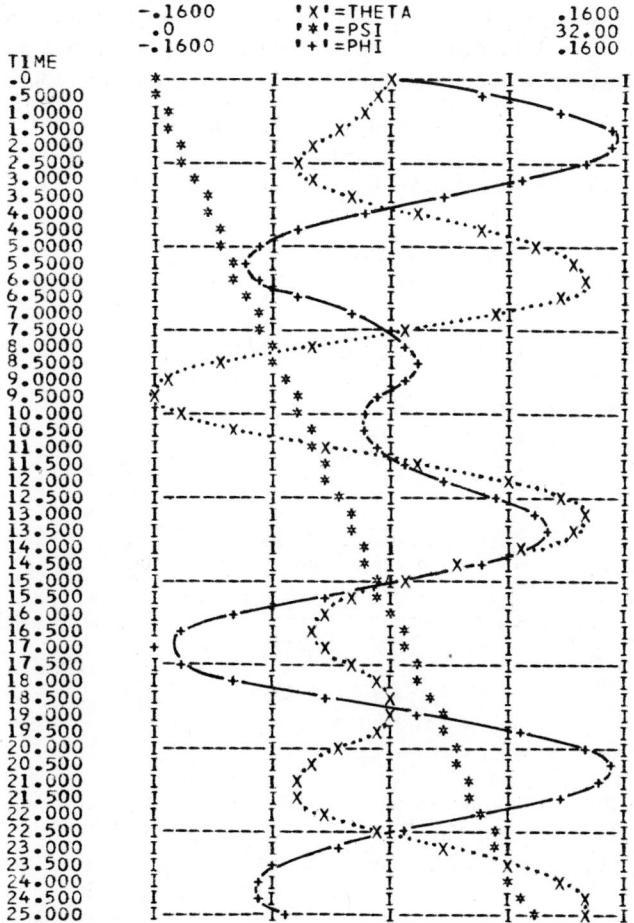

Figure 12.10 Results of Euler-angle analysis.

Note that in the above linearization ϕ is assumed small even though the x-axis spin is sure to cause large ϕ's. Since we have an exact result to compare with, let us retain this "invalid" assumption until we actually see its bad effect, if any.

In Equations (12.40) note first that the ϕ equation is completely "decoupled" from θ and ψ and can thus be solved all by itself as soon as $T_x(t)$ is given. This agrees almost perfectly with the numerical results for the exact model in Fig. 12.8, where ϕ is a ramp corresponding to $\dot\phi_0$ and is almost totally unaffected by T_z. (Further confirmation can be found in the CSMP program for the Euler-angle method, where, if one gets a printout of PSIDT2 (which is a long, complicated expression), one sees that PSIDT2≈0.0, making ψ a ramp function corresponding to $\dot{\psi}_0$.) Thus the first equation of our linearized model agrees well with the exact model when θ and ψ are small, even though ϕ gets large. The other two equations are coupled and must be solved simultaneously, leading to:

$$\theta(s) = \frac{K_{T_y}}{s^2/\omega_n^2 + 1} T_y(s) - \frac{K_{T_z}}{s\left[s^2/\omega_n^2 + 1\right]} T_z(s) \tag{12.41}$$

$$\psi(s) = \frac{C_{T_y}}{s\left[s^2/\omega_n^2 + 1\right]} T_y(s) + \frac{C_{T_z}}{s^2/\omega_n^2 + 1} T_z(s) \tag{12.42}$$

$$K_{T_y} \triangleq \frac{I_{zz}}{(I_{yy} + I_{xx} - I_{zz})(I_{xx} + I_{zz} - I_{yy})\dot{\phi}_0^2}$$

$$K_{T_z} \triangleq \frac{1}{(I_{xx} + I_{zz} - I_{yy})\dot{\phi}_0}$$

$$C_{T_z} \triangleq \frac{I_{yy}}{(I_{yy} + I_{xx} - I_{zz})(I_{xx} + I_{zz} - I_{yy})\dot{\phi}_0^2}$$

$$C_{T_y} \triangleq \frac{1}{(I_{xx} + I_{yy} - I_{zz})\dot{\phi}_0}$$

$$\omega_n^2 \triangleq \frac{(I_{yy} + I_{xx} - I_{zz})(I_{xx} + I_{zz} - I_{yy})\dot{\phi}_0^2}{I_{yy}I_{zz}}$$

Substituting numerical values, we can show a system block diagram as in Fig. 12.11. The torque pulse T_z is short enough, relative to the period of the oscillations at ω_n, to be treated as an impulse, allowing simple calculation of θ and ψ as:

$$\psi = 0.08\sin 1.67t, \qquad \theta = 0.08(\cos 1.67t - 1) \tag{12.43}$$

These results agree almost perfectly with the exact model results of Fig. 12.8, confirming that our simplified model is accurate under these conditions. Thus even though the exact Euler-angle model was *not* usable, its small-angle linearized version accurately predicts the motion and provides all the usual benefits of linear models.

We conclude this section by noting that if the reader has available a gyroscope mounted in two gimbals, as in Fig. 12.12, if the gimbal inertia is small compared to that of the spinning wheel, this system is quite a good model of the satellite we have been studying and can be used to get some physical feeling for the motions predicted by our equations. Pulses of x torque may be simulated by switching the wheel's electric drive motor on and off after full spin speed has

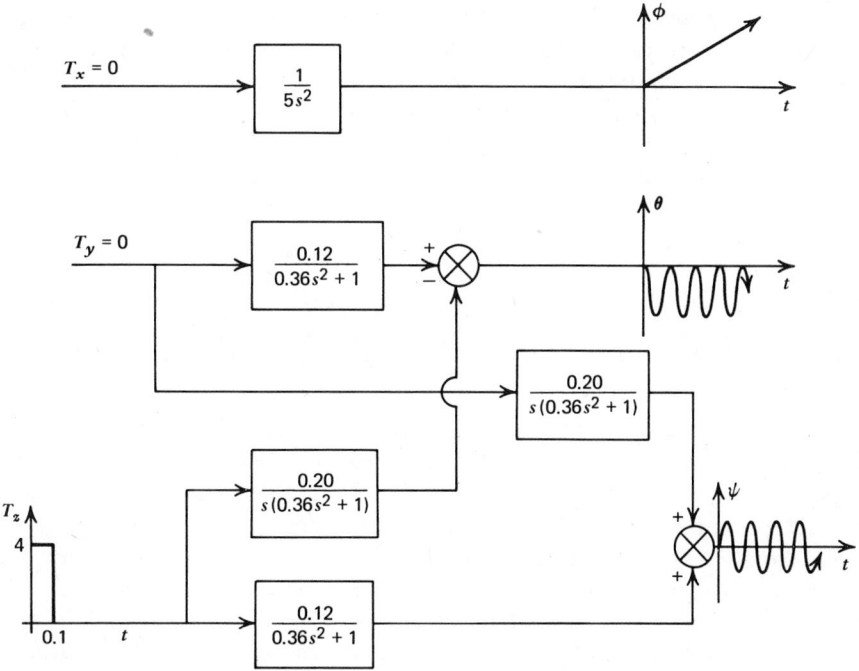

Figure 12.11 Block diagram of linearized satellite system.

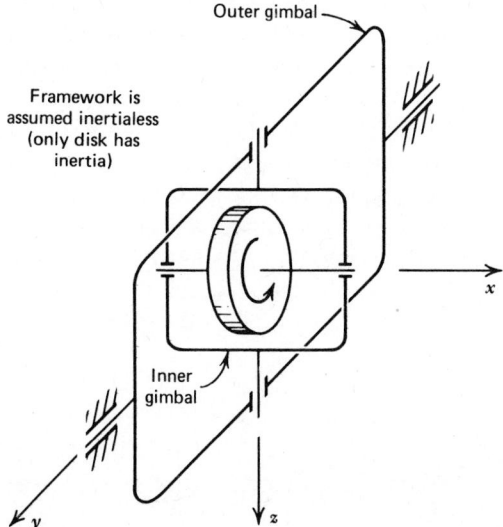

Figure 12.12 Demonstration model for satellite motions.

been reached. Pulses of y or z torque can be produced by momentarily "bumping" the appropriate gimbal with a finger. These simple experiments should produce the nutation and precession motions indicated in Fig. 12.11. Due to gimbal-bearing friction the oscillatory motions will of course gradually die out. However, in a high quality gyroscope this friction is small enough to allow quite a good demonstration.

12.2 AUTOMOBILE HANDLING DYNAMICS

It is generally accepted that the classic work in the area of automobile handling dynamics appears in a set of five papers[7] presented on the evening of October 9, 1956 at the Institution of Mechanical Engineers in London, England. The authors were staff members of the Cornell Aeronautical Laboratory Inc. of Buffalo, New York, which carried out the research program under the sponsorship of automotive and tire companies and federal agencies. Starting with a carefully developed historical perspective, these papers document in unusual detail the application of the "systems engineering approach" and "technology transfer" from the aeronautical industry to the problem of automobile handling dynamics. Few technical developments have been presented in such comprehensive form in the literature and these papers are highly recommended, not only to automotive specialists, but also to anyone with an interest in the history of technology. While extensions of this basic theory have of course appeared since 1956, the essential features remain unchanged today. We thus base our brief presentation on these papers and textbooks[8] which have since appeared.

The conventional operation of an automobile involves interaction between the vehicle, a human operator, and the road environment, as depicted in Fig. 12.13. (Chapter 13 will address the question of human operator dynamics, so we do not consider this in detail here.) Note in particular the feedback nature of the total system, in that vehicle responses caused by the driver are also sensed by the driver and influence further actions. "Minor" feedback loops involving braking, steering, and power (throttle and transmission) systems are also present. For example, with manual (rather than power) steering, one gets significant "road-feel" cues through the hands on the steering wheel. The diagram of Fig. 12.13 comprehends all the degrees of freedom of the automobile's possible motions

[7]Milliken, W. F. and D. W. Whitcomb, General Intro. to a Programme of Dynamic Research, Proc. I. Mech. E. (A.D.), 1956; Segel, L., Theoretical Prediction and Experimental Substantiation of the Response of the Automobile to Steering Control, Proc. I. Mech. E. (A.D.), 1956; Close, W. and C. L. Muzzey, A Device for Measuring Mechanical Characteristics of Tyres on the Road, Proc. I. Mech. E. (A.D.), 1956; Fonda, A. G., Tyre Tests and Interpretation of Experimental Data, Proc. I. Mech. E. (A.D.), 1956; Whitcomb, D. W. and W. F. Milliken, Design Implications of a General Theory of Automobile Stability and Control, Proc. Inst. Mech. E. (A.D.), 1956.

[8]Ellis, J. R., *Vehicle Dynamics*, Business Books Ltd, London, 1969; Steeds, W., *Mechanics of Road Vehicles*, Iliffe and Sons Ltd., London, 1960

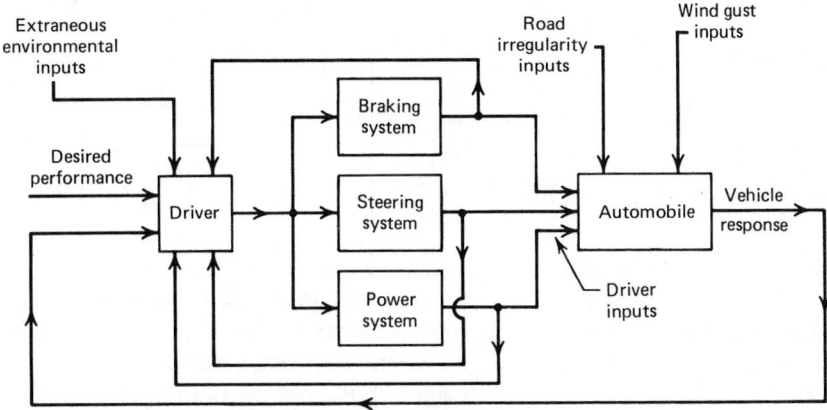

Figure 12.13 Automobile/Driver/Road system.

while we wish to focus interest only on the lateral-directional or "handling" degrees of freedom. We will ignore the so-called "ride" degrees of freedom, which consist of the verticle motion of the center of mass and the pitching rotation about the horizontal transverse axis. The "performance" degree of freedom (acceleration/deceleration) associated with the forward speed is also suppressed by stipulating that forward speed remain constant. We are thus left with sidewise motion of the mass center, yaw rotation about a vertical axis, and rolling rotation about a horizontal longitudinal axis, the three handling degrees of freedom. Figure 12.14 shows these various motions.

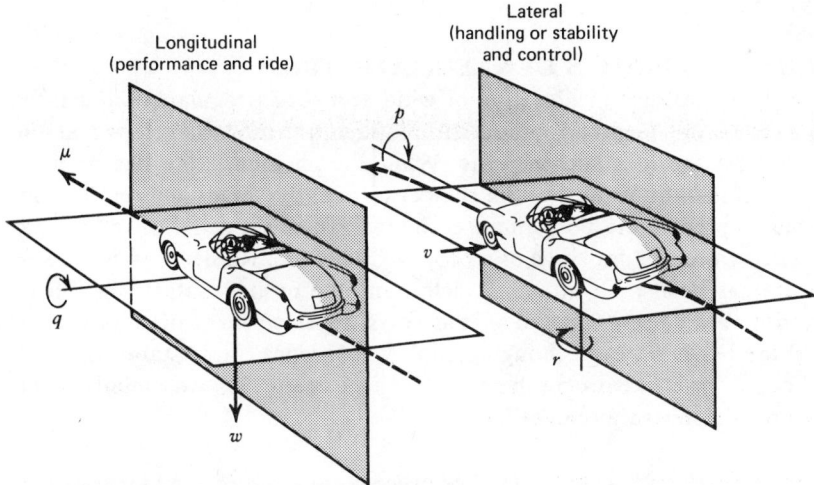

Figure 12.14 Automobile degrees of freedom.

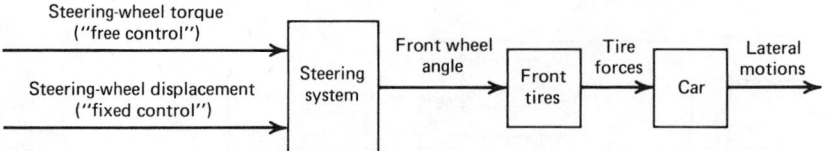

Figure 12.15 Simplified system model.

The assumption of constant forward speed is a reasonable one since most normal (rather than emergency) steering maneuvers do take place at nearly constant speed. Also, the longitudinal speed of response to throttle is relatively slow compared to steering motions, so lateral transients would tend to be completed before significant speed changes could occur. With regard to the pitching and bouncing degrees of freedom, experiments to be discussed later showed that normal steering maneuvers on smooth roads do not induce much response in these degrees of freedom. Finally, we will assume the handling motions are representable as a linear system, allowing use of superposition with regard to the three inputs. We choose to concentrate attention on the driver input and put off the wind and road inputs for later consideration, leading to the simplified diagram of Fig. 12.15. If the driver input at the steering wheel is considered to be a given *torque*, then the motion of the steering wheel is an unknown that must be solved for and the model is described as "free control." On the other hand, if the rotary *displacement* of the steering wheel is given, the model is called "fixed control." For the sake of simplicity, we go even further and take the front wheel angle to be a given input, thereby completely avoiding the need to model the steering system. (Later studies[9] included modeling of the steering system.)

12.2.1 CHARACTERISTICS OF PNEUMATIC TIRES
Since we have put aside consideration of wind and road irregularity inputs, the only forces available to cause lateral car motions are those developed at the interface between the tires and the road. While the pneumatic tire has been for many years (and continues to be) the subject of intensive investigation, it is not yet possible to predict the performance characteristics needed for vehicle dynamics models from the drawings of a proposed new tire. Rather, one must wait until the tire has been built and then determine the needed data from experimental testing. Such testing has, over the years, revealed most of the essential features of tire behavior, even though complete theoretical understanding is still lacking. In our brief treatment here we neglect many known details[10] and introduce only the most significant factors.[10, 11]

[9]Segel, L., "On the Lateral Stability and Control of the Automobile as Influenced by the Dynamics of the Steering System," *ASME J. Eng. Ind.*, Aug. (1966), p. 283.

[10]Ellis, *op. cit.*, Chapter 1.

[11]Steeds, *op. cit.*, Chapter 8.

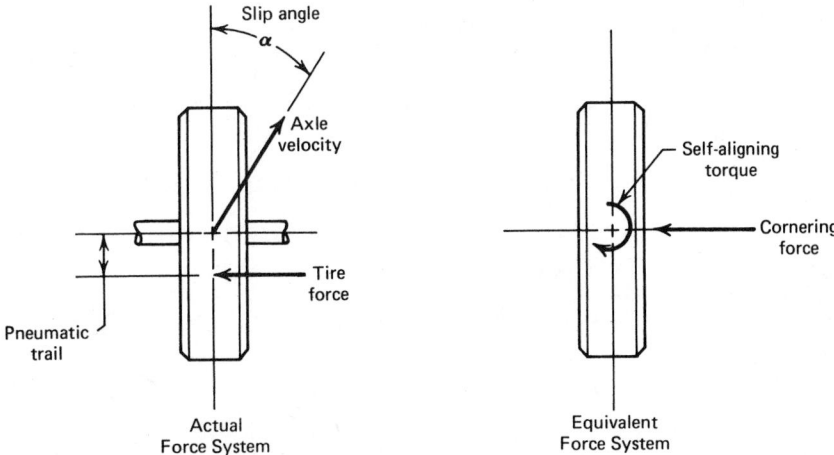

Figure 12.16 Tire forces and moments.

Figure 12.16, a top view of a vertically loaded rolling tire, shows that when the velocity vector of the tire axle is deflected from the tire center plane by a *slip angle* α, a sidewise tire force called the *cornering force* is developed. The line of action of this tire force, due to the rolling direction and elasticity of the tire, is somewhat behind the axle, this offset distance being called the *pneumatic trail*. It is conventional to represent the offset tire force by its equivalent, an axle-centered force plus a pure torque equal to the product of force and pneumatic trail. This torque is called the *self-aligning torque* since it tends to restore the slip angle to zero if the axle is not restrained. Figure 12.17 shows in simplified form the rotating-drum test apparatus often used to determine tire characteristics. (The curved drum surface gives somewhat different results than a flat road surface; however, correction factors allow reasonable correlation with road results so the much more convenient and inexpensive drum testing is usually employed.)

The tire under test is loaded against the drum with a normal load of magnitude similar to the portion of vehicle weight it would support, oriented at the slip angle desired, and force and torque transducers that support the tire axle measure the cornering force f_c and self-aligning torque T_{sa}. We find that cornering force is essentially linear with slip angle for the small slip angles typical of normal vehicle operation. Self-aligning torque is less linear, but since its contribution to total vehicle yaw torque is much less than that of the cornering forces, this is not a serious problem. Self-aligning torque *is* an important factor in steering-gear dynamics, but since we are taking front wheel angle as a given input, this is not pertinent to our study. Cornering force is seen to be sensitive to normal load; however the sensitivity is fortunately slight in the neighborhood of the operating load (about $\frac{1}{4}$ of vehicle weight).

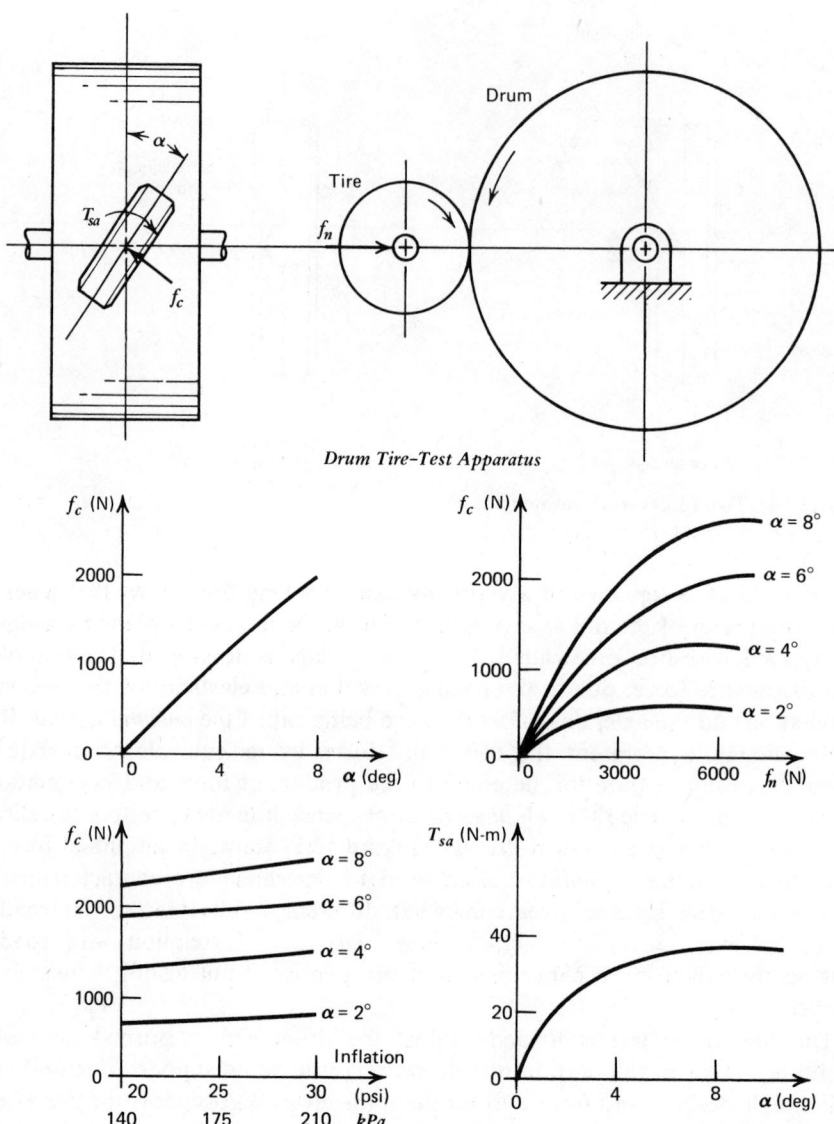

Drum Tire-Test Apparatus

Figure 12.17 Tire testing apparatus with typical results.

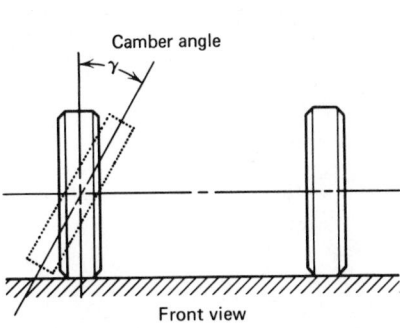

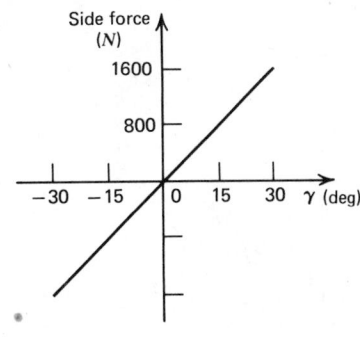

Figure 12.18 Camber thrust effect.

While slip angle is the main factor in determining cornering force, the camber angle of the tire also has an effect and this is usually also measured in a complete tire test program. Figure 12.18 defines the camber angle and shows the quite linear dependence of side force (called camber thrust) on camber angle. While much more could be said about tire behavior, we have now presented sufficient background for the vehicle dynamics models we wish to present.

12.2.2. NONROLLING, FIXED-AXIS MODEL

Our plan is to present two models; a relatively simple one that clearly reveals certain basic features of automobile behavior but suffers from some inaccuracy in predicting certain details, and a second, more accurate treatment that is, however, more difficult to interpret. In our first model the automobile is considered as a single rigid mass capable only of sidewise translation and yawing rotation. The rolling of the body ("sprung mass") relative to the unsprung mass is completely ignored. Also, we will set up the equations of motion using a space-fixed axis system rather than the body-fixed axis system explained earlier in this chapter and widely used in general dynamics studies.

Our treatment essentially follows the original work of Rocard[12] as presented by Ellis.[13] In Fig. 12.19, dimensions a and b locate the center of mass relative to the front and rear wheels, U is the forward velocity (assumed constant), ψ is the heading angle of the vehicle, and δ is the front wheel steer angle (assumed the same for both front wheels). At any instant, the car centerline and the velocity vector of the center of mass need not be aligned; the angle between them is called β, the sideslip angle. In normal highway steering maneuvers (lane changing, for example) all the angles remain quite small, allowing the usual approximations of sine and cosine functions.

[12]Rocard, Y., *Dynamic Instability*, Crosby Lockwood & Son Ltd., London.
[13]Ellis, *op. cit.*, Chapter 3.

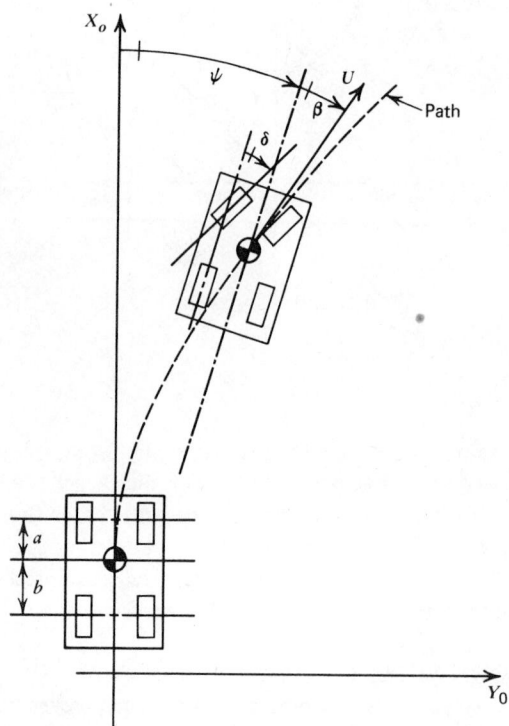

Figure 12.19   Diagram for nonrolling, fixed-axis model.

To implement the translational and rotational Newton's laws for our two degrees of freedom we need to determine the tire forces; thus the slip angle at each tire must be found first. At the rear wheels the sidewise velocity is given by a superposition of sidewise translation $\dot{Y}_0$ and the effect of rotation, $-b\dot{\psi}$. The term $\dot{Y}_0$ is along the Y_0 axis while the term $-b\dot{\psi}$ is perpendicular to the tire centerplane. However, for small angles, $-b\dot{\psi}$ is very nearly along Y_0 and we will simply sum these two terms to get the total sidewise velocity. The slip angle α_r for each rear tire is thus

$$\alpha_r \approx \tan \alpha_r = \frac{\text{lateral velocity}}{\text{forward velocity}} - \text{heading angle}$$

$$= \frac{\dot{Y}_0 - b\dot{\psi}}{U} - \psi \tag{12.44}$$

The cornering force produced by this slip angle is perpendicular to the tire centerplane, but again, because of the small angles, we treat this as a Y_0 force. At the front wheels, we need only superimpose the effect of steer angle δ on the rear-wheel analysis to get

$$\alpha_f \approx \tan \alpha_f = \frac{\dot{Y}_0 + a\dot{\psi}}{U} - \psi - \delta \tag{12.45}$$

Most steering gears are arranged so that the steer angles of left and right front wheels are slightly different. However, we neglect this small effect and take the slip angles of both front wheels to be the same.

Neglecting both tire aligning torque (of secondary importance) and camber thrust (our vehicle model does not include camber angle changes) we can write

$$-2C_r\left(\frac{\dot{Y}_0 - b\dot{\psi}}{U} - \psi\right) - 2C_f\left(\frac{\dot{Y}_0 + a\dot{\psi}}{U} - \psi - \delta\right) = M\ddot{Y}_0 \tag{12.46}$$

$$+2bC_r\left(\frac{\dot{Y}_0 - b\dot{\psi}}{U} - \psi\right) - 2aC_f\left(\frac{\dot{Y}_0 + a\dot{\psi}}{U} - \psi - \delta\right) = I_z\ddot{\psi} \tag{12.47}$$

where

$C_f \triangleq$ slope (at origin) of f_c/α curve (Fig. 12.17), front tires
$C_r \triangleq$ slope (at origin) of f_c/α curve (Fig. 12.17), rear tires

(Note: C_f and C_r are sometimes defined negative in the literature.) Manipulation of these equations leads to a system characteristic equation of second order. A stability analysis of this equation shows that unstable roots can occur only if $aC_f > bC_r$. Since front and rear tire coefficients are generally similar, it appears that if $a < b$, stability is assured, thus we should locate the center of mass forward of the midpoint of the wheelbase. If, however, $a > b$, then instability is a function of forward speed, unstable roots appearing when the speed exceeds a critical speed U_c given by

$$U_c = \sqrt{\frac{2C_fC_r(a+b)^2}{M(aC_f - bC_r)}} \tag{12.48}$$

The instability is associated with a positive real root rather than a complex pair with positive real part, so the motion is not oscillatory but rather an exponential divergence. That is, if the vehicle is proceeding in a straight path when the front wheels are momentarily deflected and then returned to zero, above the critical speed the car "steers itself" into a tighter and tighter turn, rather than proceeding along a new straight path as it would below the critical speed. Of course a human operator can stabilize the combined man/machine system by appropriate steering wheel motions. However, if the wheel is held *fixed*, the above-described unstable behavior does occur. Even though the driver can in most cases provide system stability, an unstable vehicle may keep the driver so busy with this task that attention to other important factors lapses and safety suffers.

The vehicle model of Equations (12.46) and (12.47) gives information on dynamic performance and stability in a relatively simple form since the characteristic equation is second order and one can define a natural frequency and damping ratio in letter form, giving useful design formulas. However, for certain vehicles and maneuvers, experiments reveal the presence of motions not predicted with accuracy sufficient for some purposes, so we will now develop a more comprehensive model.

12.2.3 TWO-MASS MODEL WITH BODY ROLL AND BODY AXES

Figure 12.20 shows a more correct representation of the automobile as a two-mass system with the body (sprung mass) free to roll with respect to the axles and wheels (unsprung mass). Our drawing shows the rolling motion constrained by an actual "axle" with "bearings" in the unsprung mass. A real vehicle is of course not constructed in this way; however, the *effect* is essentially the same. That is, the suspension system, whatever its detailed nature might be, defines an effective axis of roll that can be located by analysis of design drawings or experimental test of the actual vehicle. The suspension system of course also allows simultaneous pitching and bouncing motions (the "ride" degrees of freedom), but recall that we earlier decided to ignore these.

Our analysis essentially follows the original work of Segel,[14] which is presented also by Ellis[15] and Steeds.[16] We will need to utilize several axis systems as in Fig. 12.21. The automobile mass distribution is assumed symmetrical about a vertical fore-and-aft midplane and Fig. 12.21 lies in this plane. Axes $x''y''z''$ are assumed centroidal principle axes of the sprung mass, which is typically 85% of the total vehicle mass. The fact that the orientation $x''y''z''$ is principle may not be obvious, but measurements of actual vehicles confirm this to be a good approximation. Axes $x'y'z'$ at the vehicle total center of mass will be used to set up the original equations. However, these will then be transformed to the xyz axes whose origin is located by the intersection of the actual roll axis and a vertical line through M_s. In most vehicles the actual roll axis is slightly inclined with the horizontal. This inclination can be taken into account in the equations but since the effect is small it is common to take the x (horizontal) axis as the roll axis.

All the axis systems in Fig. 12.21 are body-fixed systems that move with the vehicle, thus we can apply the general equations of rigid-body motion discussed earlier in this chapter, the only complication being that our system has *two* rigid bodies involved. Our approach follows Segel's in that the dynamic equations are

[14]Segel, L., Theoretical Prediction and Experimental Substantiation of the Response of the Automobile to Steering Control, Proc. I. Mech. E. (A.D.), 1956.
[15]Ellis, *op. cit.*
[16]Steeds, *op. cit.*

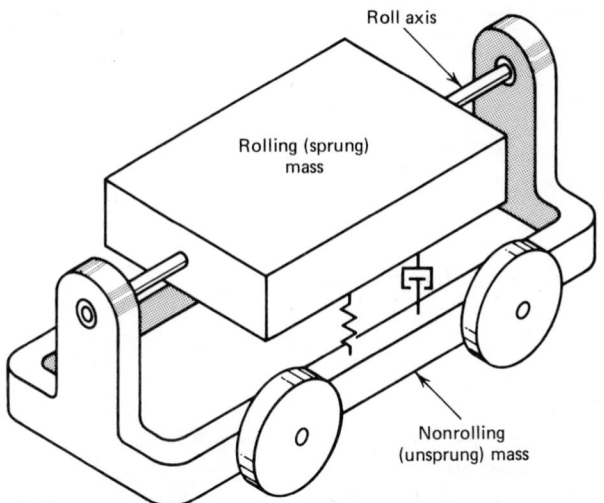

Figure 12.20 Two-mass model of automobile.

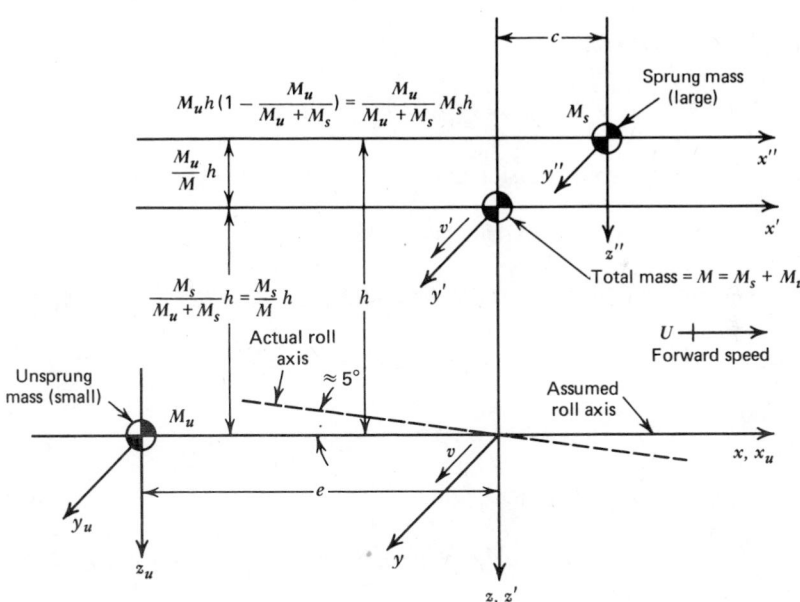

Figure 12.21 Automobile axis systems.

set up by first writing the inertial terms and then later introducing the external forces and moments. Considering first the lateral motion of the total mass center, Eq. (12.22)

$$\sum F_y = M(\dot{V} + U\omega_z - W\omega_x)$$

gives in this case

$$\sum Y' = M(\dot{v}' + Ur) \tag{12.49}$$

where r is the yaw rate of the entire vehicle, the vertical velocity W is zero, and $\sum Y'$ denotes the external forces in the y' direction.

In next considering the yawing rotation we observe that the unsprung mass feels the tire forces and also an inertial reaction due to the rolling mass. This inertial reaction is felt at the hypothetical "bearings" of the roll axis in Fig. 12.20. Thinking of the rolling mass as a separate freebody, it has angular velocity components ω_x and ω_z, but $\omega_y = 0$. To produce these angular velocities, the unsprung mass must exert certain torques on the rolling mass and the reverse of these torques is thus felt by the unsprung mass. Applying Eq. (12.14),

$$\sum T_z = (-I_{xz}\dot{\omega}_x - I_{yz}\dot{\omega}_y + I_{zz}\dot{\omega}_z) - \omega_y H_x + \omega_x(-I_{xy}\omega_x + I_{yy}\omega_y - I_{yz}\omega_z)$$

to the rolling mass only we get

$$\sum T_z = -I_{xz}\dot{\omega}_x + I_{zzs}\dot{\omega}_z - I_{xy}\omega_x^2 - I_{yz}\omega_x\omega_z \tag{12.50}$$

We now make use of the axis transfer theorem for products of inertia:

$I_{yz} = I_{y''z''} + M_s(y$ distance between origins of y and $y'')$

$$\times (z \text{ distance between origins of } z \text{ and } z'') \tag{12.51}$$

$I_{yz} = 0 + M_s(0)(-h) = 0$ $(I_{y''z''} = 0$ since $x''y''z''$ are principle axes) $\tag{12.52}$

Similarly:

$$I_{xy} = I_{x''y''} + M_s(c)(0) = 0 \tag{12.53}$$

$$I_{xz} = I_{x''z''} + M_s(c)(-h) = -M_s ch \tag{12.54}$$

Note that the "distances" in the axis transfer theorem are measured in the unprimed coordinate system and carry algebraic signs. Equation (12.50) thus simplifies to

$$\sum T_z = M_s ch\dot{\omega}_x + I_{zzs}\dot{\omega}_z \qquad (12.55)$$

Since the unsprung mass has only yaw rotation its Newton's law is simply

$$\sum \text{yaw torques} = \sum \text{tire torques} + \text{inertial reactions from sprung mass}$$

$$= I_{zzu}\dot{r} \qquad (12.56)$$

where

$$I_{zzu} \triangleq z' \text{ moment of inertia of unsprung mass}$$

$$\dot{r} \triangleq \text{yaw angular acceleration} = \dot{\omega}_z$$

Using $\sum N'$ for the tire torques and $\dot{p}$ for $\dot{\omega}_x$ we get

$$\sum N' - I_{zzs}\dot{r} - M_s ch\dot{p} = I_{zzu}\dot{r} \qquad (12.57)$$

$$\sum N' = (I_{zzs} + I_{zzu})\dot{r} + M_s ch\dot{p} = I_{z'}\dot{r} + M_s ch\dot{p} \qquad (12.58)$$

where

$$I_{z'} \triangleq \text{yaw moment of inertia of entire vehicle about } z'$$

Equation (12.58) demonstrates the yaw/roll coupling present in this model since roll motion $\dot{p}$ causes yaw torque $M_s ch\dot{p}$. A physical feeling for this effect can be obtained by consideration of the simple point-mass model of Fig. 12.22.

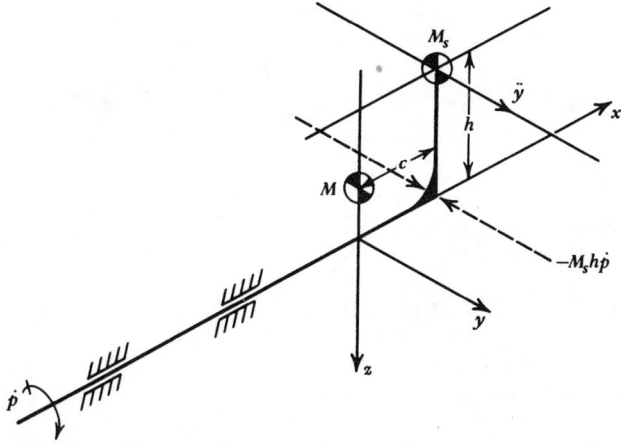

Figure 12.22 Point-mass interpretation of yaw/roll coupling.

The rolling mass M_s has a tangential acceleration $\ddot{y}$ given by $h\dot{p}$ and thus must feel a force $M_s h\dot{p}$, the reverse of this force being reacted back into the rotating frame. By "adding zero force" to this force system as shown by the dashed lines, we obtain a pure roll couple and the y force $-M_s h\dot{p}$. The moment of this force about the z axis is $-M_s hc\dot{p}$, a *yaw* torque produced by roll motion.

An analysis similar to that leading to Eq. (12.58) can be carried out for the roll torque equation, giving

$$\sum L' = I_{x'}\dot{p} + M_s ch\dot{r} \tag{12.59}$$

where

$$\sum L' \triangleq \text{roll torques other than inertial effects}$$

$$I_{x'} \triangleq \text{moment of inertia of rolling mass about } x'$$

$$M_s ch\dot{r} = \text{roll/yaw coupling torque due to product of inertia}$$

We now have expressed the inertial effects in the three equations of motion for our three chosen degrees of freedom, using axes centered at the center of mass of the entire vehicle $(x'y'z')$. These equations are next transformed so that the motion quantities are referred to the xyz axes, a more convenient location. Since the z and z' axes are identical, this equation requires no change except that the yaw torques are now called $\sum N$ rather than $\sum N'$ and I_z replaces $I_{z'}$:

$$\sum N = I_z \dot{r} + M_s ch\dot{p} \tag{12.60}$$

For the side force equation, $\sum Y' \equiv \sum Y$, however, the velocity v' is related to v by

$$v' = v + \frac{M_s}{M}hp, \qquad \dot{v}' = \dot{v} + \frac{M_s}{M}h\dot{p} \tag{12.61}$$

giving

$$\sum Y = M(\dot{v} + Ur) + M_s h\dot{p} \tag{12.62}$$

The roll torque equation has

$$\sum L = \sum L' + \frac{M_s}{M}h\sum Y' \tag{12.63}$$

$$\sum L = I_{x'}\dot{p} + M_s ch\dot{r} + M_s h\left(\dot{v} + Ur + \frac{M_s}{M}h\dot{p}\right) \tag{12.64}$$

$$\sum L = \left(I_{x'} + \frac{M_s^2 h^2}{M}\right)\dot{p} + M_s h(\dot{v} + Ur) + M_s ch\dot{r} \tag{12.65}$$

$$\sum L = I_x \dot{p} + M_s h(\dot{v} + Ur) + M_s ch\dot{r} \tag{12.66}$$

where

$$I_x \triangleq I_{x'} + \frac{M_s^2 h^2}{M} \tag{12.67}$$

Our next task is to develop expressions for the noninertial terms (tire, suspension and gravity forces, and torques) in our equations. Let us first find the slip angles of all the tires. In Fig. 12.23, using small-angle approximations we obtain the slip angle of each front wheel as

$$\alpha_1 = \frac{v + ar}{U} - \delta = \beta + \frac{a}{U} r - \delta \tag{12.68}$$

where $\beta \triangleq$ vehicle sideslip angle. At each rear wheel we now take into account a suspension property called "rear roll steer," not discussed earlier. While the rear wheels of a conventional car are not *intentionally* steered, a small steering motion, proportional to body roll angle ϕ, does occur in some vehicles. Using ϵ_2 as the proportionality factor (usually found by measurement) relating rear steer angle to roll angle, we can write the slip angle of each rear tire as

$$\alpha_2 = \frac{v - br}{U} - \epsilon_2 \phi = \beta - \frac{b}{U} r - \epsilon_2 \phi \tag{12.69}$$

The sidewise forces at front and rear tires can be written as

$$\sum Y = -2C_1 \alpha_1 - 2C_2 \alpha_2 + 2 \frac{\partial Y}{\partial \gamma} \frac{\partial \gamma}{\partial \phi} \phi \tag{12.70}$$

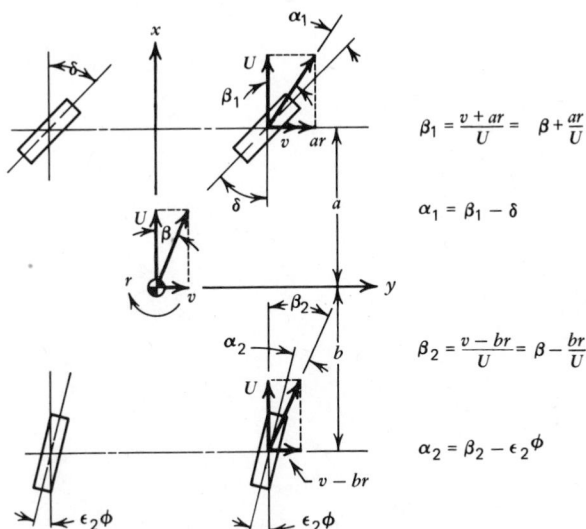

$$\beta_1 = \frac{v + ar}{U} = \beta + \frac{ar}{U}$$

$$\alpha_1 = \beta_1 - \delta$$

$$\beta_2 = \frac{v - br}{U} = \beta - \frac{br}{U}$$

$$\alpha_2 = \beta_2 - \epsilon_2 \phi$$

Figure 12.23 Tire slip-angle geometry.

The last term is the side force ("camber thrust") due to camber angle changes at the two front wheels. These camber angle changes are caused by and are proportional to body roll angle ϕ. Suspension experiments are used to obtain numerical values for the parameter $\partial\gamma/\partial\phi$, while tire experiments are used to find $\partial Y/\partial\gamma$.

Turning to the yaw torques ΣN, we need to account for the self-aligning torques due to slip angle at each tire, the moments about the vehicle center of mass of the side forces Y, and the moments due to changes in tire rolling resistance X caused by load transfer. This last effect has not been previously mentioned and requires discussion. Rolling resistance is the horizontal tire force that opposes forward rolling motion at the tire/road interface. Before any steering maneuvers are attempted, all four rolling resistance forces are exactly balanced by propulsive forces derived from the engine and thus no net force exists. When we initiate a turn, however, there will be some shift in weight between the left and right wheels ("load transfer") and since rolling resistance depends on normal load, the rolling resistance will increase on one side and decrease an equal amount on the other. The X (fore-and-aft) forces thus remain balanced but a component of yaw torque due to rolling resistance appears (see Fig. 12.24).

Since the yaw torque due to rolling resistance changes depends on the load transfer (change in vertical force Z on tire), we must get an expression for the ΔZ's. This is done by considering the forces and moments acting at the front and rear suspensions. Figure 12.25b shows the situation at the rear suspension.

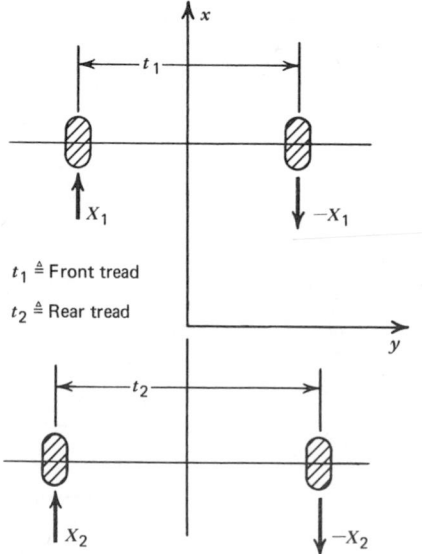

Figure 12.24   Tire rolling-resistance forces.

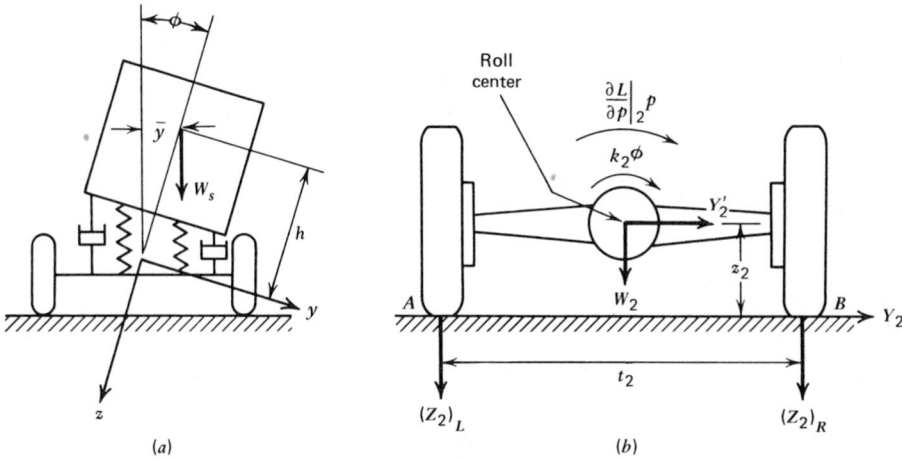

Figure 12.25 Roll torques and vehicle load transfer.

By neglecting the inertia of the unsprung mass (relative to that of the sprung mass), force Y_2' at the hypothetical pin joint between sprung and unsprung masses can be approximated as Y_2, the tire force of both rear tires. Taking moments about point B we can write

$$\Delta Z_2 = \frac{1}{t_2}\left(-Y_2 z_2 - k_2\phi - \frac{\partial L}{\partial p}\bigg|_2 p\right) \tag{12.71}$$

where

$$k_2 \triangleq \text{torsional (roll) spring constant at rear}$$

$$\frac{\partial L}{\partial p}\bigg|_2 \triangleq \text{torsional (roll) damping constant at rear}$$

$$z_2 \triangleq \text{height of rear roll center}$$

$$p \triangleq \dot{\phi} = \text{roll rate}$$

Similarly, at the front wheels:

$$\Delta Z_1 = \frac{1}{t_1}\left(-Y_1 z_1 - k_1\phi - \frac{\partial L}{\partial p}\bigg|_1 p\right) \tag{12.72}$$

We can now express the yaw torques ΣN as

$$\Sigma N = -2aC_1\alpha_1 + 2bC_2\alpha_2 - 2\frac{\partial T_{sa}}{\partial \alpha_1}\alpha_1 - 2\frac{\partial T_{sa}}{\partial \alpha_2}\alpha_2 - t_1\frac{\partial X_1}{\partial Z_1}\Delta Z_1 - t_2\frac{\partial X_2}{\partial Z_2}\Delta Z_2 \tag{12.73}$$

where

$$\frac{\partial T_{sa}}{\partial \alpha_1}, \frac{\partial T_{sa}}{\partial \alpha_2} \triangleq \text{slope of aligning torque versus slip-angle curves for front and rear tires}$$

$$\frac{\partial X_1}{\partial Z_1}, \frac{\partial X_2}{\partial Z_2} \triangleq \text{slope of rolling resistance versus normal load curves for front and rear tires (test data)}$$

Turning finally to the roll torques ΣL, these consist of suspension spring and damping effects plus a gravity term (see Fig. 12.25a):

$$\Sigma L = -k_1\phi - k_2\phi - \left.\frac{\partial L}{\partial p}\right|_1 p - \left.\frac{\partial L}{\partial p}\right|_2 p + W_s h\phi \qquad (12.74)$$

Equations have now been derived for all the external forces and moments included in this model, so these can now be combined with the inertial terms in Equations (12.60), (12.62), and (12.66) to obtain the final three equations. It is common to eliminate the side velocity (v) degree of freedom in favor of the vehicle sideslip angle β, using the relation $\beta \approx v/U$. Our equations then take the form:

$$MU(\dot{\beta} + r) + M_s h\dot{p} = Y_\beta \beta + Y_r r + Y_\delta \delta + Y_\phi \phi \qquad (12.75)$$

$$I_z \dot{r} + M_s ch\dot{p} = N_\beta \beta + N_r r + N_\delta \delta + N_\phi \phi + N_p p \qquad (12.76)$$

$$I_x \dot{p} + M_s hU(\dot{\beta} + r) + M_s ch\dot{r} = L_p p + L_\phi \phi \qquad (12.77)$$

Quantities such as Y_β, N_ϕ, L_p, and so on (called *stability derivatives*) are introduced to compactly express the coefficients of the variables since these coefficients are sums of as many as six individual terms. Expressions for the various stability derivatives are easily obtained by putting our earlier equations in the form of (12.75)–(12.77). The reader is cautioned that sign conventions differ from author to author, thus definitions of stability derivatives found in various books and papers may appear inconsistent; however, these can generally be resolved by comparison of sign conventions. Note also that there are only three basic unknowns (β, r, and ϕ); $p = \dot{\phi}$ and thus p is not a fourth unknown. Sometimes the sidewise displacement of the mass center relative to the road is of interest. This is given by

$$Y_0 = U \int_0^t (\sin \psi + \beta \cos \psi) \, dt \qquad (12.78)$$

or in approximate linear form for small heading angles as

$$Y_0 \approx U \int_0^t \int_0^t r \, dt \, dt + U \int_0^t \beta \, dt \qquad (12.79)$$

In Segel's[17] original work the model given by Equations (12.75)–(12.77) was substantiated by experimental testing of a 1953 Buick sedan. Since the model is linear with constant coefficients it is possible to manipulate (12.75)–(12.77) to obtain sinusoidal transfer functions for any output/input pair desired. Three transfer functions, r/δ, p/δ, and $\dot{v}/\delta$ were selected for study as being amenable to convenient measurement. Pulse testing was used to obtain transient records of δ, r, p, and $\dot{v}$, which were then transformed to the frequency domain to obtain the three desired transfer functions for comparison with theoretical predictions of the model. Figure 12.26 shows Segel's instrumentation and a typical pulse test record, while Fig. 12.27 compares measured sinusoidal transfer functions (open and solid circles) with theoretically predicted values (solid curves). Since the agreement of theory and experiment is seen to be very good, the model of Equations (12.75)–(12.77) is substantiated.

To study the stability question, the system characteristic equation must be obtained from the simultaneous equations (12.75)–(12.77). The characteristic equation turns out to be of fourth degree, with the coefficients being complicated combinations[18] of the inertial, suspension, and tire parameters. The table of Fig. 12.28 explores the variation in the four roots of the characteristic equation as influenced by vehicle "static margin" (SM) and forward speed. Static margin is a complex combination of parameters that plays a roll similar to that of the center-of-mass location in the simplified model of Section 12.2.2. There, instability at high speed was possible only if $a > b$, whereas in the present model instability generally is possible if the static margin is negative. This is shown by the specific numerical data for the 1953 Buick in Fig. 12.28. For a positive static margin of 0.0692, increase of forward speed causes a change from two real roots and one pair of complex roots to two pairs of complex roots whose damping decreases as speed increases; but unstable roots do not occur. For a negative static margin of -0.0543, increasing speed causes one of the negative real roots to gradually approach zero and then become positive (unstable), the speed of 146.7 ft/sec giving the root $+0.755$ in this example. For a stable vehicle, typical step responses (analog simulated from Equations (12.75)–(12.77)) are shown in Fig. 12.29.

Good modeling practice, as we have indicated in earlier examples, requires that complex models be examined for possible simplification. Segel pursues this with some success; we quote only a few basic results. Examination of the quartic characteristic equation reveals that it is roughly factorable into two quadratic terms, one of which is closely related to the second-order behavior of the nonrolling model of Section 12.2.2, while the other factor is associated mainly with rolling motions of the sprung mass. The total motion can thus be thought

---

[17]Segel, *op. cit.*
[18]Segel, *op. cit.*, p. 41.

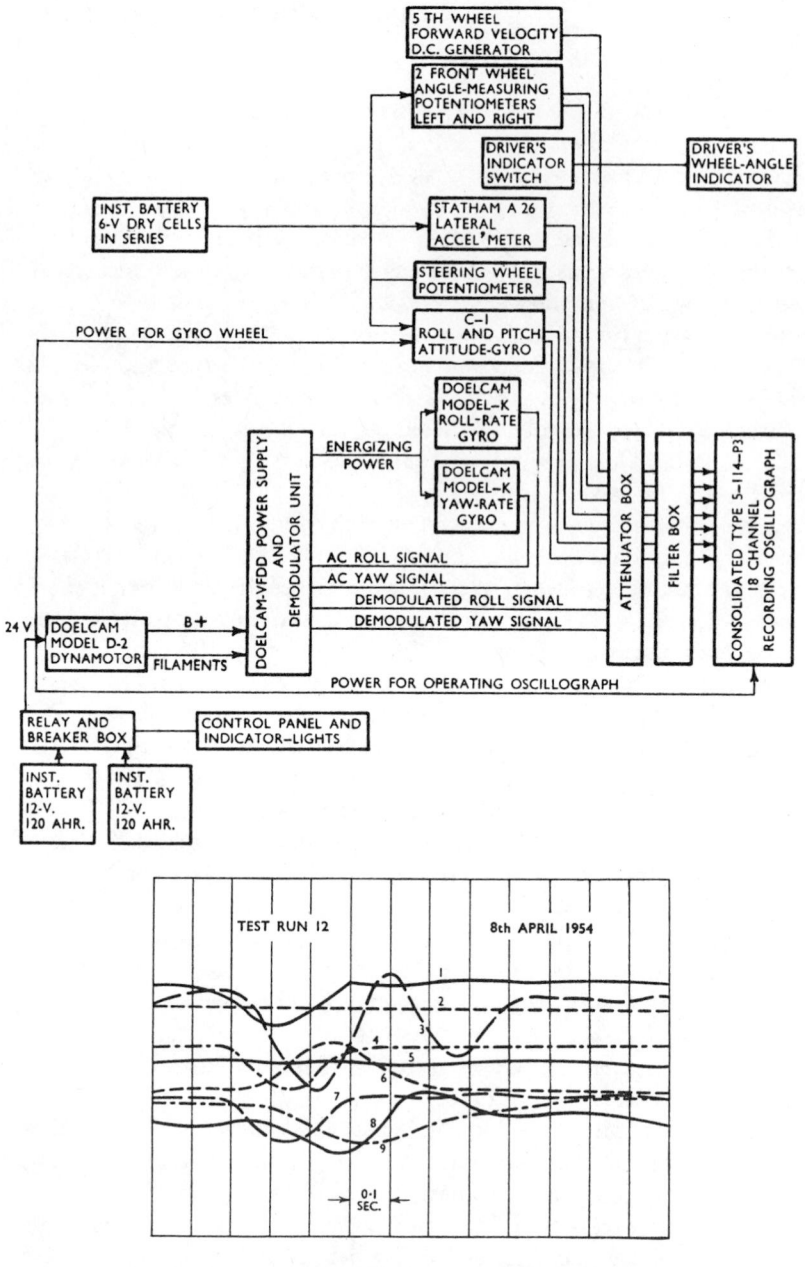

1. Steering wheel position, δ_{SW}.
2. Forward velocity, U.
3. Lateral acceleration, n_y.
4. Left front wheel, δ_L.
5. Pitch attitude, θ.
6. Yaw rate, r.
7. Right front wheel, δ_R.
8. Roll rate, p.
9. Roll angle, ϕ.

Figure 12.26 Instrumentation and typical data records for automobile pulse test.

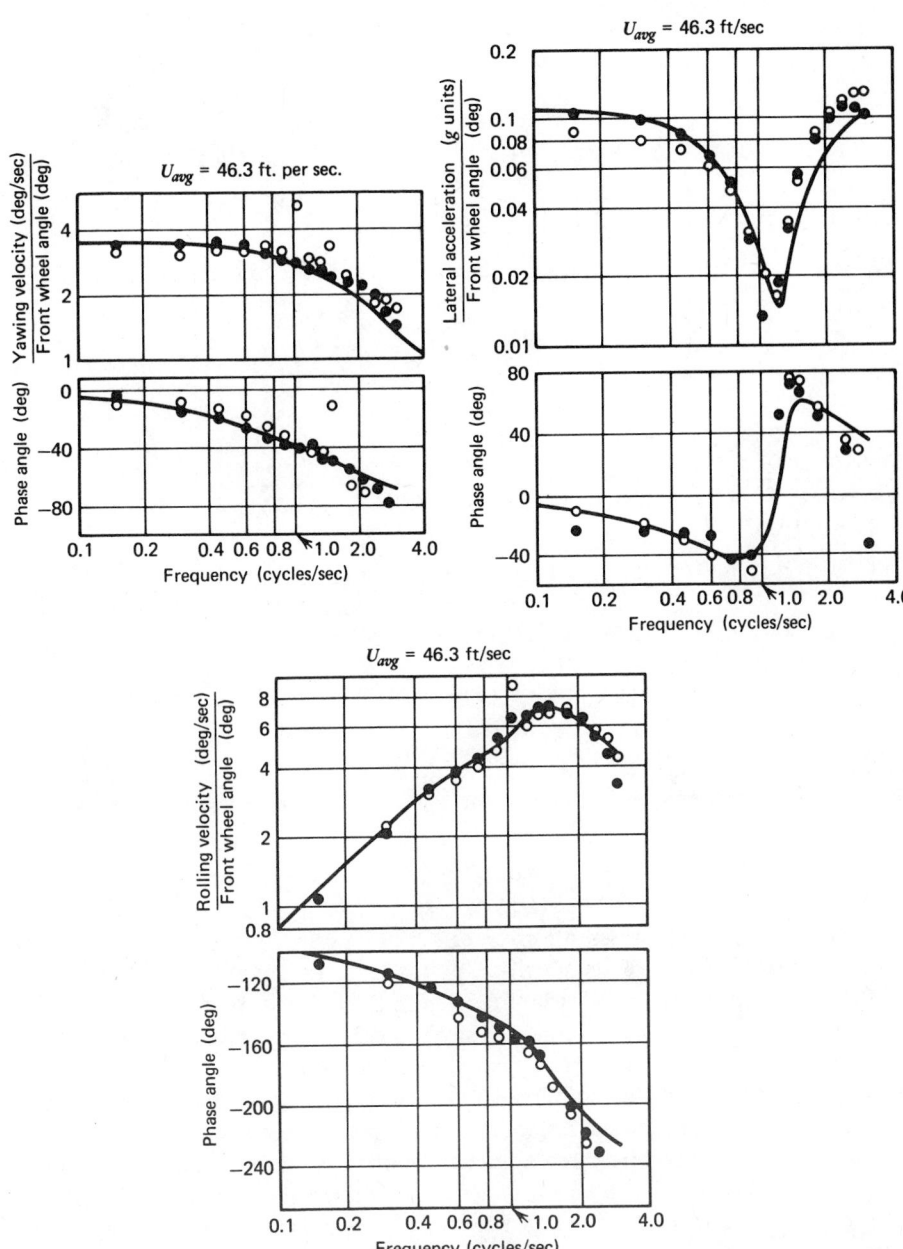

Figure 12.27 Pulse test data converted to frequency domain and compared with theoretical model.

CONFIGURATION	VELOCITY, (FT/SEC)	DIMENSIONAL ROOTS (1/SEC)	DAMPED NATURAL FREQUENCY (CYCLES/SEC)	PERCENT CRITICAL DAMP-ING
Understeer (SM = 0.0692)	25.4	$-1.673 \pm 6.09i$ -10.39 -16.89	0.97	0.265
	46.3	$-2.04 \pm 6.39i$ $-7.00 \pm 2.505i$	1.018 0.399	0.3045 0.941
	83.8	$-2.40 \pm 7.02i$ $-3.22 \pm 3.20i$	1.119 0.51	0.324 0.71
	146.7	$-1.796 \pm 7.53i$ $-1.671 \pm 2.78i$	1.199 0.443	0.232 0.516
Oversteer (SM = −0.0543)	44.5	$-3.165 \pm 8.10i$ -2.515 -9.09	1.29	0.364
	88	$-2.62 \pm 9.25i$ -0.1795 -5.05	1.473	0.273
	146.7	$-2.185 \pm 9.59i$ $+0.755$ -3.805	1.527	0.2165

Figure 12.28   Roots of vehicle characteristic equation for various conditions.

of as a combination of two basic modes; a yawing and side-slipping mode and a rolling mode.

We conclude our presentation of this automobile dynamics model with a brief discussion of one of its specific applications. In a study[19] of a proposed new automative option called adaptive steering, a vehicle dynamics model was required to be coupled with electronic, mechanical, and hydraulic system models for computer evaluation of overall system performance and feasibility. Adaptive steering borrowed vehicle stability-augmentation concepts from the aircraft industry to propose a new steering system of improved performance. A yaw rate gyroscope was to sense undesired vehicle motions caused by wind gusts and/or rough roads and signal a hydraulic steering actuator that would augment the operator's steering wheel inputs in such a way as to maintain a more steady vehicle course while minimizing the need for operator corrective action.

[19]Kasselmann, J. T. and T. W. Keranen, Adaptive Steering, Bendix Technical Journal, Autumn (1969), pp. 26–35.

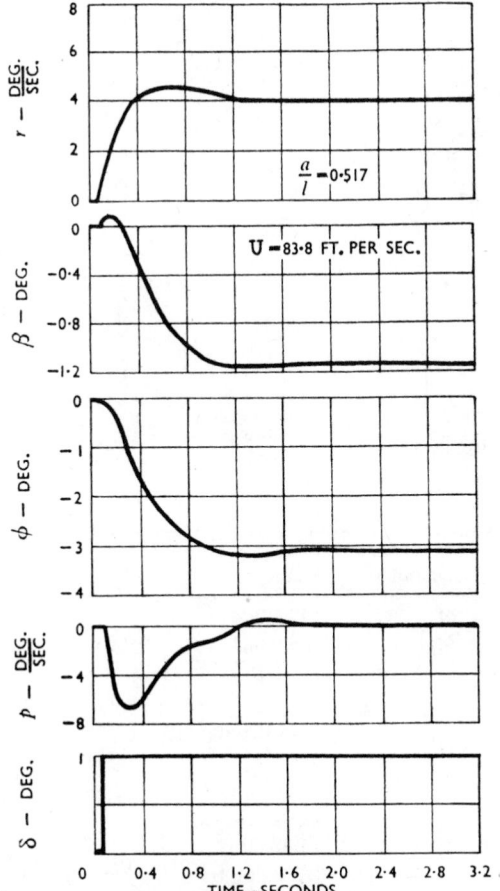

Figure 12.29   Step response of analog-simulated vehicle model.

The system block diagram of Fig. 12.30 shows the vehicle dynamics equations to be identical with those we derived earlier except for three new terms added to account for wind gust input forces and torques. Figure 12.31 shows actual numerical values for the 1968 sedan used in the study while Fig. 12.32 gives some mechanical details. Availability of a valid vehicle dynamics model allowed the use of computer simulation in the preliminary design stages of this study. Thus various concepts could be rapidly tried out at minimum expense, and potential problem areas could be uncovered and resolved before committing resources to a hardware development program. Since the computer model studies showed that a system with significant performance improvements was feasible, hardware development was carried out and the actual system subjected

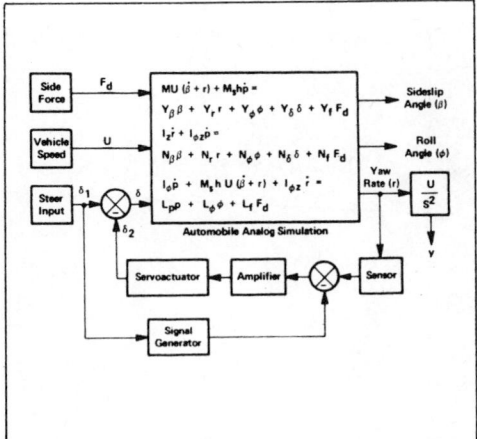

Figure 12.30   Automotive adaptive steering system.

Parameters and stability derivatives	Symbol	Value
Total mass	M	148 slugs
Sprung mass	M_S	128 slugs
Height of sprung-mass center of gravity above roll axis	h	1.183 ft
Side-force stability derivatives	Y_β	$-24{,}020$ lb/rad
	Y_r	0
	Y_ϕ	3497 lb/rad
	Y_δ	12,010 lb/rad
	Y_f	1.0 lb/lb
Moment of inertia about vertical axis	I_z	3674 slug ft^2
Effective product of inertia due to inclined roll axis	$I_{\phi z}$	-58 slug ft^2
Yaw-moment stability derivatives	N_β	5586 lb ft/rad
	N_r	-6170 lb ft sec/rad
	N_ϕ	11,220 lb ft/rad
	N_δ	63,130 lb ft/rad
	N_f	4.0 lb ft/lb
Moment of inertia about roll axis	I_ϕ	446 slug ft^2
Roll-moment stability derivatives	L_p	-2140 lb ft sec/rad
	L_ϕ	$-39{,}010$ ft lb/rad
	L_f	3.0 lb ft/lb
Forward speed of vehicle	U	70 mph

Figure 12.31   Model numerical parameters for 1968 sedan.

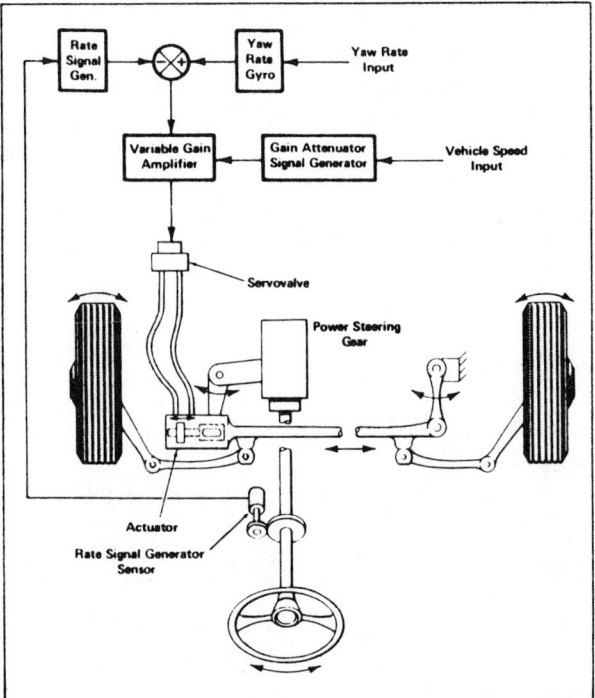

Figure 12.32 Hardware arrangement for adaptive steering.

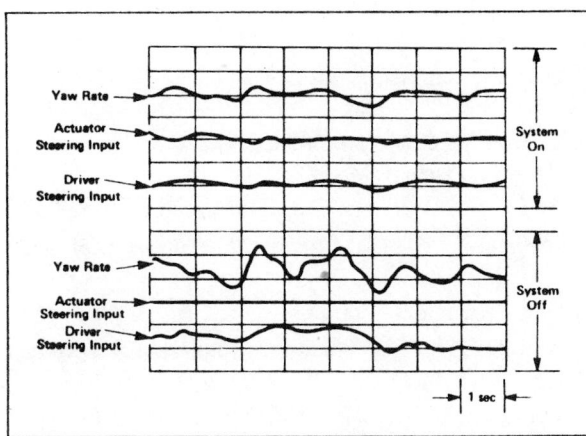

Figure 12.33 Performance improvement due to adaptive steering for rutted-road driving.

to a road-test program. These road tests were also successful; Fig. 12.33 shows the degree of reduced vehicle motion and reduced driver input achieved when driving a heavily rutted road. While the technical feasibility of this proposed new automotive accessory was proven, at the time of this writing it has not appeared on the market, so it appears that economic factors were not propitious for its introduction. The risk involved in funding such research, which may or may not pay off in business profits, is a necessary part of technical progress.

BIBLIOGRAPHY

1   Abkowitz, M. A., *Stability and Motion Control of Ocean Vehicles*, M.I.T. Press, Cambridge, 1969.

2   Basso, G. L., A Methodology for Measurement of Vehicle Parameters Used in Dynamic Studies, National Research Council of Canada Rept. 13497, Ottawa, 1973.

3   Bekker, M. G., *Introduction to Terrain-Vehicle Systems*, Univ. of Mich. Press, Ann Arbor, MI, 1969.

4   Bekker, M. G., *Theory of Land Locomotion*, Univ. of Mich. Press, Ann Arbor, MI, 1956.

5   Bhattacharyya, R., *Dynamics of Marine Vehicles*, Wiley-Interscience, N.Y., 1978.

6   Blader, F. B. and E. F. Kurtz, Dynamic Stability of Cars in Long Freight Trains, ASME Paper 73-WA/RT-2, 1973.

7   Blakelock, J. H., *Automatic Control of Aircraft and Missiles*, John Wiley and Sons, N.Y., 1965.

8   Bradfield, E. N., Experimental Determination of the Moments of Inertia, Product of Inertia, and Inclination of the Principal Axis of Conventional Aircraft by the Spring Oscillation Method, FTC-TIM-71-1001, USAF Flight Test Center, Edwards AFB, Calif., 1971.

9   Bruce, E. L., Regetz, J. D., and R. R. Lovell, SERT II Spacecraft Attitude Response, NASA TM X-2324, 1971.

10   Bundorf, R. T., Directional Control Dynamics of Automobile-Travel Trailer Combinations, Gen'l. Mtrs. Eng. Staff Rept. 3202, 1967.

11   Bundorf, R. T., Pollock, D. E., and M. C. Hardin, Vehicle Handling Response to Aerodynamic Inputs, Gen'l. Mtrs. Res. Lab Rept. GMR-403, 1963.

12   Byrne, R. (ed.), *Railroad Equipment Dynamics*, Am. Soc. of Mech. Eng., N.Y., 1976.

13   Clark, S. K. (ed.), *Mechanics of Pneumatic Tires*, Mono. 122, Nat'l. Bur. of Stds., Wash. D.C., 1971.

14   Chi, C. C., LIM Guidance System Dynamics, Airesearch Corp., Rept. 73-9065, 1973.

15   Coffey, H. T., Colton, J. D., and K. D. Mahrer, Study of a Magnetically Levitated Vehicle, Fed. Rail. Adm. Rept. FRA-RT-73-24, 1973.

16   Coffey, H. T., *et al.* An Evaluation of the Dynamics of a Magnetically Levitated Vehicle, Fed. Rail. Adm. Rept. FRA-ORD and D-74-41, 1974.

17 Cooperider, N. K., Suspension Systems for Tracked Air Cushion Vehicles, General Electric Co. Rept. 69-C-014, Schenectady, 1968.

18 Ellis, J. R., *Vehicle Dynamics*, Business Books, London, 1969.

19 Etkin, B., *Dynamics of Atmospheric Flight*, John Wiley and Sons, N.Y., 1972.

20 Garg, V. K. and K. D. Mels, Lateral Stability of a Six-Axle Locomotive, ASME Paper 75-RT-7, 1975.

21 Goran, M. B. and G. W. Hurlong, Determining Vehicle Inertial Properties for Simulation Studies, Bendix Tech. Jour., Spring (1973).

22 High-Speed Ground Transportation Journal.

23 Journal of Terramechanics.

24 Koch, J. B. C., A Computer Model of Steady-State and Transient Traction Forces and Aligning Moment Developed by Pneumatic Tires, Univ. of Mich. Highway Safety Res. Inst. Rept. UM-HSRI-PF-75-2, 1975.

25 Law, E. H., Hadden, J. A., and N. K. Cooperrider, General Models for Lateral Stability Analyses of Railway Freight Vehicles, U.S. Dept. of Trans. Rept. FRA-OR and D-77-36, 1977.

26 Leatherwood, J. D., Dixon, V. G., and D. G. Stephens, Heave Response of a Plenum Air Cushion Including Passive and Active Control Concepts, NASA TN D-5202, 1969.

27 Lindgren, A. G., Cretella, D. B., and A. F. Bessacini, "Dynamics and Control of Submerged Vehicles," *ISA Trans.*, **6**, No. 4 (1967), pp. 335–346.

28 Lloyd, S. E., Development of a Flat Surface Tire Rolling Resistance Facility, SAE Paper 780635, 1978.

29 McRuer, D., Ashkenas, I., and D. Graham, *Aircraft Dynamics and Auto. Control*, Princeton U. Press, Princeton, N.J., 1973.

30 Meisenholder, S. G. and T. C. Wang, Dynamic Analysis of an Electromagnetic Suspension System for a Suspended Vehicle System, Fed. Rail. Adm. Rept. FRA-RT-73-1, 1972.

31 Moment of Inertia Instruments, Bulletin S8018, Space Electronics, Inc., Meriden, Conn.

32 Nordeen, D. L., Vehicle Directional Equations for an Inclined Roll Axis, Gen'l. Mtrs. Proving Grounds Rept. A-2165, 1969.

33 Nordeen, D. L., Analysis of Tire Lateral Forces and Interpretation of Experimental Tire Data, Gen'l. Mtrs. Res. Lab. Rept. GMR-575, 1966.

34 Nordeen, D. L., Rasmussen, R. E., and J. B. Bidwell, Tire Properties Affecting Vehicle Ride and Handling, Gen'l. Mtrs. Eng. Staff. Rept. 3759, 1968.

35 Nybakken, G. H., Dodge, R. N., and S. K. Clark, A Study of Dynamic Tire Properties over a Range of Tire Constructions, Univ. of Mich. Tire and Susp. Syst. Res. Group, 1972.

36 Orne, D. and T. Schmitz, "Analysis of a Platform for Measuring Moments and Products of Inertia of Large Vehicles," *J. Dyn. Sys., Meas., Control*, **98**, June (1976).

37 Post, T. M. and J. E. Bernard, Response of Vehicles to Pavement Undulations, Univ. of Mich. Highway Safety Res. Inst. Rept. UM-HSRI-76-6, 1976.

38 Proc. First Ship Control Sys. Symp., Nov. 1966, Vol. I, U.S. Navy Marine Eng. Lab., Annapolis, MD.

39 Rasmussen, R. E., Hill, F. W., and P. M. Riede, Typical Vehicle Parameters for Dynamics Studies, Gen'l. Mtrs. Proving Ground Rept. A-2542, 1970.

40 Rasmussen, R. E. and A. D. Cortese, The Effect of Certain Tire-Road Interface Parameters on Force and Moment Performance, Gen'l. Mtrs. Eng. Staff Rept. A-2526, 1969.

41 Richardson, H. H., Captain, K., and W. A. Birch, The Dynamics of Fluid-Suspended Ground Transport Vehicles: A First-Order Analysis, ASME Paper 67-TRAN-58, 1967.

42 Reitz, J. R., *et al.*, Preliminary Design Studies of Magnetic Suspensions for High-Speed Ground Transportation, Fed. Rail. Adm. Rept. FRA-RT-73-27, 1973.

43 Segel, L., On the Lateral Stability and Control of the Automobile as Influenced by the Dynamics of the Steering System, ASME Paper 65-WA/MD-2, 1965.

44 Sewall, J. L., Parrish, R. V., and B. J. Durling, Dynamic Responses of Railroad Car Models to Vertical and Lateral Rail Inputs, NASA TN D-6375, 1971.

45 Siddale, J. N., Dokainish, M. A., and W. Elmaraghy, On the Effect of Track Irregularities on the Dynamic Response of Railway Vehicles, ASME Paper 73-WA/RT-1, 1973.

46 Steeds, W., *Mechanics of Road Vehicles*, Iliffe and Sons, London, 1960.

47 Sweet, L. M., Richardson, H. H., and D. N. Wormley, Linearized Models, Stability Criteria and Experimental Verification for Plenum Air Cushions with Compressor-Duct Interactions, Dept. of Trans. Rept. DOT-FRA-ORD and D-74-40, 1974.

48 Valentine, R. W., Hybrid Computer Simulation of Vehicle Directional Control, SAE Paper 700156, 1970.

49 Whittemore, A. P., A Technique for Measuring "Effective" Road Profiles, SAE Paper 720094, 1972.

50 Wickens, A. H., General Aspects of the Lateral Dynamics of Railway Vehicles, ASME Paper 68-WA/RR-3, 1968.

51 Wolowicz, C. H. and R. B. Yancey, Experimental Determination of Airplane Mass and Inertial Characteristics, NASA TR R-433, 1974.

52 Wong, J. Y., *Theory of Ground Vehicles*, John Wiley and Sons, N.Y., 1978.

53 Zupp, G. A. and H. H. Doiron, A Mathematical Procedure for Predicting the Touchdown Dynamics of a Soft-Landing Vehicle, NASA TN D-7045, 1971.

PROBLEMS

12.1 The rigid body of Fig. 12.1 has its center of mass at the origin of the body-fixed xyz axis set. It is desired to find I_{xz} experimentally by rotating the body at constant speed ω about one of the three axes and simultaneously measuring the appropriate bearing reactions. Explain clearly and completely, using appropriate equations, exactly what such a test involves and give specific formulas and experimental procedures for calculating I_{xz} from measured experimental data. It is necessary to be able to tell whether I_{xz} is positive or negative.

12.2 For the automobile model of Equations (12.46) and (12.47):
(a) Use center-of-pressure concepts from fluid mechanics to include a force input f_w due to wind gusts.
(b) Rewrite the equations using as unknowns ψ and β, rather than ψ and Y_0.
(c) Obtain transfer functions relating ψ, Y_0, and β to δ and f_w.
(d) Derive Eq. (12.48).

12.3 Use center-of-pressure concepts from fluid mechanics to modify the equations of Section 12.2.3 to allow a force input due to wind gusts.

12.4 Manipulate Equations (12.75)–(12.77) to obtain the system characteristic equation and transfer functions relating β, r, and ϕ to δ.

12.5 Modify the model of Section 12.2.3 to include road profile inputs. The road profile is given by specifying the vertical height z_r above a horizontal reference plane as a function of forward displacement x and sidewise displacement $y: z_r = z_r(x,y)$ (see Fig. P12.1). Explain in detail how the equations must be modified to include this effect and actually derive the necessary new terms, giving all laws and assumptions used. Assume wheels never leave contact with the road and neglect tire dynamics.

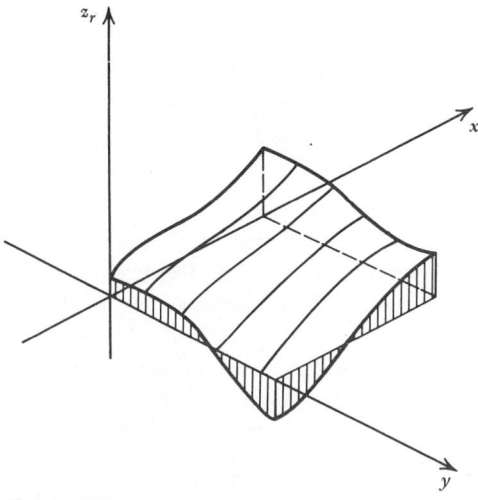

Figure P12.1

12.6 Use the simplified model of Fig. P12.2 to include steering-system dynamics in the equations of Section 12.2.3. Assume the given input will be a steering wheel torque T_s.

   (a) Explain clearly how equations for this new system will be coupled to those derived in the text. What role will front wheel angle δ now play?

   (b) Derive differential equations for the system of Fig. P12.2, explaining all laws and assumptions used. Linearize any nonlinear terms that may appear.

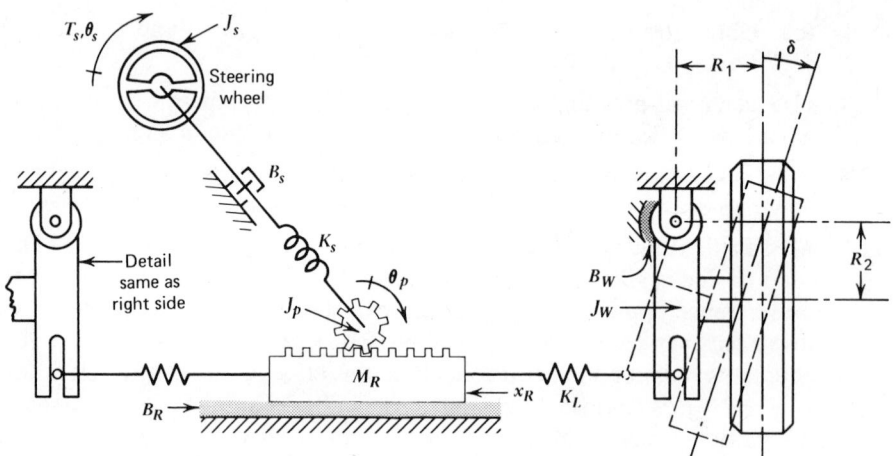

Figure P12.2

12.7 Check the automobile pulse test data of Figs. 12.6 and 12.7 for conformity to good pulse testing standards. Are the time histories (transients) and frequency functions (transforms) consistent with the usual pulse testing criteria? Explain.

12.8 In the model of Section 12.2.3, the *gyroscopic* torque effects of various high-speed rotating objects are treated as negligible. In some proposed flywheel *powered* cars, gyroscopic torque effects of the flywheel may not be negligible. In Fig. P12.3, considering *only* the flywheel and using the rigid-body dynamics equations from the text, discuss the nature and magnitude of the torques exerted on the car due to flywheel gyroscopic action, assuming flywheel spins at constant speed. Take flywheel moments of inertia as $I_{xx} = I_{zz} = I_{yy}/2$. Can these flywheel gyroscopic effects be reduced by mounting flywheel in a different attitude? Explain.

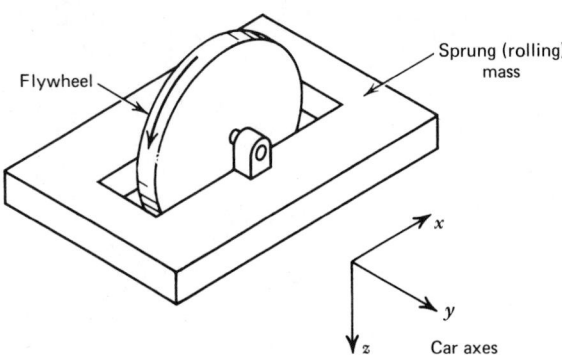

Flywheel

Sprung (rolling) mass

x

y

z Car axes

Figure P12.3

12.9 A car/trailer combination is a rather complex dynamic system even if we restrict our attention to lateral dynamics only. To gain some basic understanding without being overwhelmed by detail, simplified models are generally analyzed first and refinements added later, one by one. The simplest model, Fig. P12.4, which still gives considerable useful and fairly accurate information, treats the car and trailer each as one-wheeled vehicles. Furthermore, the car is assumed to have constant forward speed, no yawing rotation, and only sidewise and forward translation. The trailer has both sidewise translation and yawing rotation, but no rolling. All bodies are considered rigid and the tires produce only a side force proportional to tire slip angle.

$M_c \triangleq$ (car mass)/4, (real car has four wheels)

$M_t \triangleq$ (trailer mass)/2, (real trailer has two wheels)

$J_t \triangleq$ (trailer moment of inertia about COM)/2

$K_{ac} \triangleq$ car tire cornering coefficient, per tire

$K_{at} \triangleq$ trailer tire cornering coefficient, per tire

$U \triangleq$ car/trailer forward speed, assumed constant

$\beta_a \triangleq$ car sideslip angle

$\delta \triangleq$ steering input angle

$V_c \triangleq$ sidewise velocity of car COM

$V_t \triangleq$ sidewise velocity of trailer COM

$\eta \triangleq$ trailer yaw angle

(a) Using small-angle assumptions and space-fixed axes (as in Section 12.2.2), derive two simultaneous linear differential equations describing this system, using V_c and η as the unknowns.

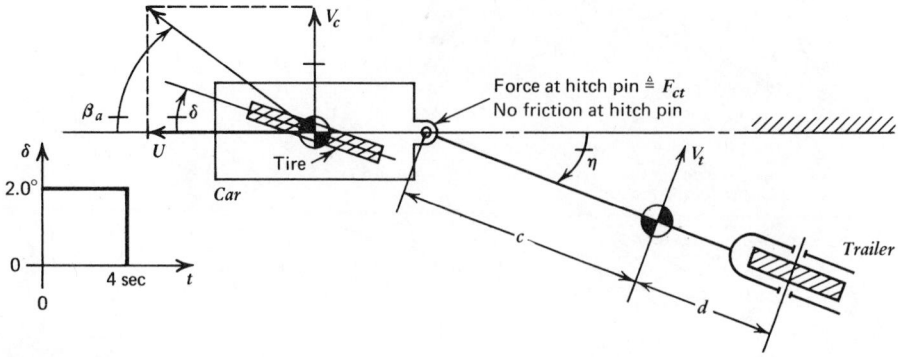

Figure P12.4

(b) Use digital simulation to get numerical results for a lane-changing maneuver in which the driver steers from one lane to another using the δ input of Fig. P12.4. For this maneuver, V_c, sidewise translation $x_c = \int V_c \, dt, \dot\eta$ and η are all initially zero. Get time histories of $\dot V_c, V_c, x_c, \ddot\eta, \dot\eta, \eta$, and also trailer sidewise translation, using numerical values:

$$M_c = 37.5 \text{ slugs}, \quad c = 12.0 \text{ ft}, \quad K_{\alpha c} = 11000 \text{ lb}_f/\text{rad}$$
$$M_t = 62.6 \text{ slugs}, \quad d = 2.0 \text{ ft}, \quad K_{\alpha t} = 14000 \text{ lb}_f/\text{rad}$$
$$J_t = 2250. \text{ slug-ft}^2, \quad U = 88 \text{ ft/sec (60 mph)}$$

(c) Using the equations of part a, obtain the system characteristic equation and investigate stability using Routh criterion.

12.10 We wish to model a car/trailer combination as in Fig. P12.4 but with the following modifications:
(a) Angle η is now *relative* to car.
(b) Car both translates sidewise and yaws.
(c) Car front and rear tires have different cornering coefficients (treat car as four-wheel vehicle).
(d) Trailer has two wheels but these can be considered as one tire with twice the cornering coefficient.
(e) Car and trailer are each subject to wind gust forces.
(f) The hitch pin is still frictionless but now has a torsion spring that exerts a torque proportional to η.
Derive differential equations that describe this system, defining needed symbols and explaining methods as you go. System outputs are to be car sidewise velocity V_c and yaw rate r, and towing angle η. System inputs are car front wheel angle δ, car wind force, and trailer wind force.

CHAPTER 13

Human Factors in Man/Machine Systems

Human beings are involved in the operation of many engineered systems, sometimes in very significant ways. For example, the overall performance of most vehicles depends on a combination of human and machine factors. As engineers strive to place the design of such systems on a more rational basis, it becomes necessary to obtain mathematical models of certain aspects of human response and behavior. While the human mind and body is of course not subject to extensive "redesign," if we understand and can reasonably model the human factors in a man/machine system, hopefully we can design the machine to optimize the overall system behavior within the constraints of what is "humanly" possible.

In this chapter we present two examples of such human factor modeling studies. The first involves the vibration characteristics of the seated human body and is typical of those situations where the thinking and adaptive aspects of human behavior have a minimal effect, and methods that treat the body as an assemblage of inanimate masses, springs, and so on are applicable. In the second example we deal with the behavior of a human pilot or driver responding with steering movements to a visual tracking task. Here the mind is consciously and adaptively involved and experimental test procedures must be designed with this in mind if usable results are to be achieved.

13.1 EXPERIMENTAL MODELING OF HUMAN BODY VIBRATION

In developing mathematical models of human body motion and vibration one can utilize either a strictly experimental approach or else try to apply rational mechanics to a mass/spring/damper representation of the body parts involved. In the latter case, experimentation may still be necessary to obtain numerical parameter values for the "component parts," but once these are given, overall

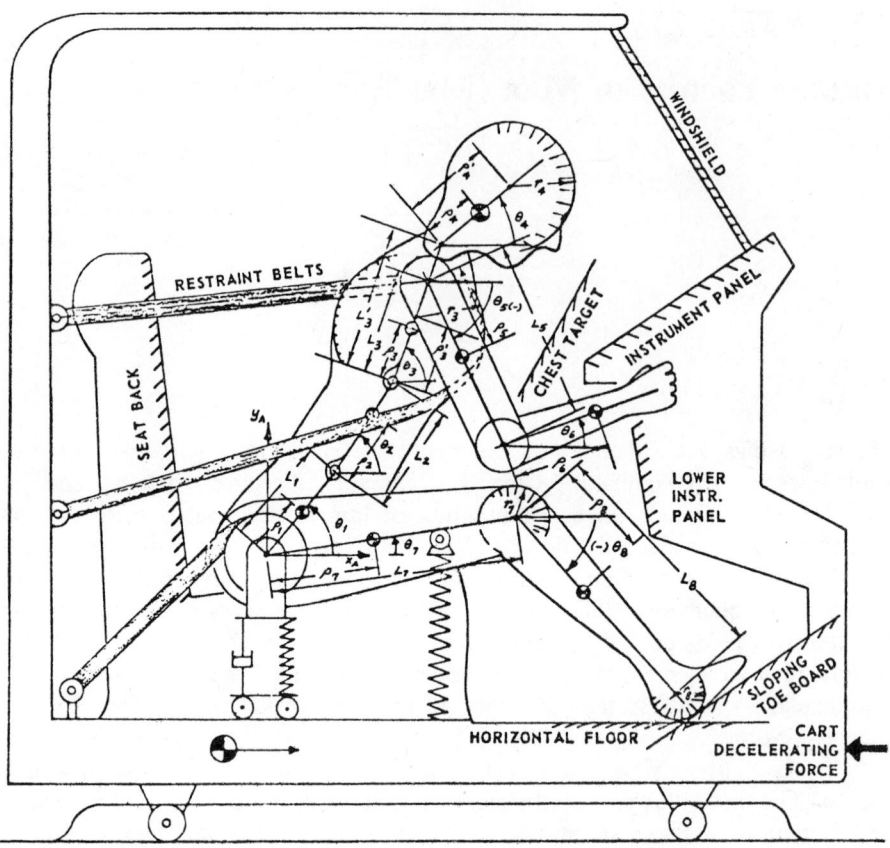

Figure 13.1 Mass/spring/damper model of human body for vehicle crash studies.

model behavior is determined mathematically. Figure 13.1[1] shows a model of this type developed for automobile crash studies.

The example to be discussed in this section utilizes a purely experimental approach to develop a mathematical model for predicting certain aspects of the vibration response of the seated human body. Much of the work on human vibration response has been directed toward the development of safe tolerance levels that can be used by designers of machinery and vehicles to insure that human operators will not suffer excessive fatigue or even permanent physical damage from the vibration environment created by the machine. Figure 13.2,

[1]McHenry, R. R. and K. N. Naab, Computer Simulation of the Automobile Crash Victim in a Frontal Collision—A Validation Study, Cornell Aero. Lab, Rept. YB2126-V, July 1966; Roberts, V. L., Stech, E. L. and C. T. Terry, Review of Mathematical Models Which Describe Human Response to Acceleration, ASME Paper 66-WA/BHF-13, 1966.

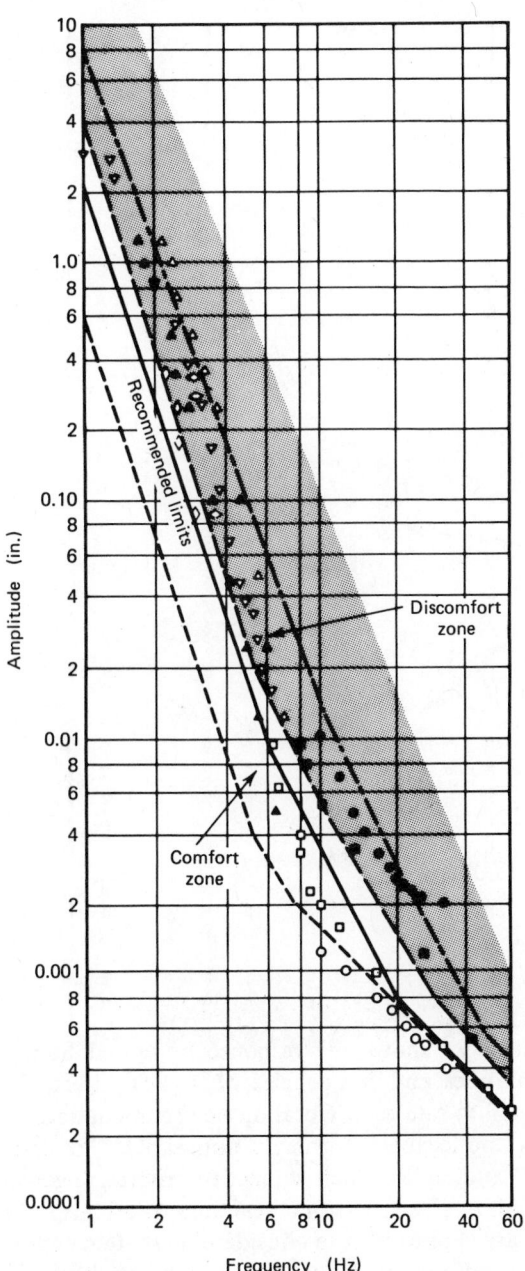

Figure 13.2 Vertical vibration response of seated human subjects.

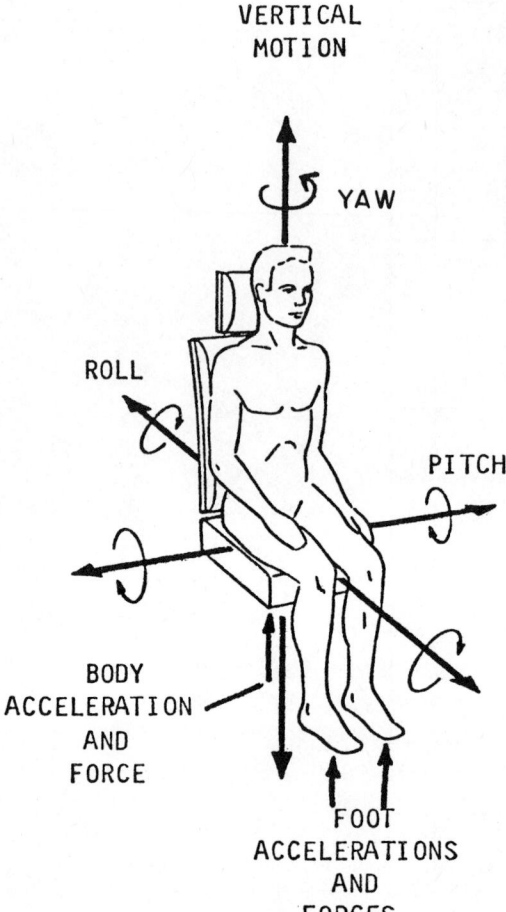

VERTICAL
MOTION

YAW

ROLL

PITCH

BODY
ACCELERATION
AND
FORCE

FOOT
ACCELERATIONS
AND
FORCES

Figure 13.3   Vibration simulator motion axes.

based on data presented by Janeway,[2] shows the response of seated human
subjects exposed to various amplitudes and frequencies of vertical sinusoidal
vibration. The subjects are requested to rate each vibration environment according to qualitative criteria such as "noticeable," "strongly noticeable," "uncomfortable," "very uncomfortable," and so on. Such subjective criteria, together
with the day-to-day variability of a given person's reactions (and also the
variation from person to person), are sure to lead to considerable scatter in data
of this sort; however, useful trends and numerical guidelines can be established.

[2]Janeway, R. N. Human Vibration Tolerance Criteria and Applications to Ride Evaluation, SAE Paper
750166, 1975.

While the single-sine-wave data of Fig. 13.2 are useful, real vibration enviroments always include complex combinations of periodic, transient, and random effects. The *overall* fatiguing or damaging effect of these complex vibration environments is not easily related to the single-sine-wave experimental data. However, various criteria have been proposed and used, the RMS acceleration level being perhaps the most common. That is, it is postulated that when the RMS acceleration of any complex vibration pattern exceeds a certain critical level, then human discomfort will become intolerable, human performance will be degraded below acceptable levels, and/or physical damage will occur. While the RMS acceleration criterion has been of some practical use, many problem areas still exist, leading to a search for some better and more rational criteria. The study we now describe is a part of this search that has lead to some promising results.

The contributions of this investigation [3] fall into two major categories; the development, by carefully controlled experimental testing, of transfer functions for specific input/output points on the seated human body, and the postulation of a new criterion of vibration severity, the *absorbed power*. Both these efforts depend on the availability of a fully instrumented vibration simulator for carrying out the test program. The simulator used is an electrohydraulic servo-system capable of driving a seated subject in four degrees of freedom; vertical translation and pitch, roll and yaw rotation (see Fig. 13.3). Since the simulator accepts electrical commands, convenient electronic signal generators for sinusoidal, transient, or random waveforms may be employed. Also, complete instrumentation providing electrical signals proportional to all force and motion quantities of interest is included to allow easy data processing.

While various output/input transfer functions were studied, we here concentrate on that which relates acceleration at the "seat-of-the-pants" (as input) to head acceleration as output (see Fig. 13.4a). Since use of the transfer function concept requires essential linearity, preliminary tests were run to validate this assumption. The amplitude and frequency ranges over which linearity is required may be deduced from tolerance data such as that of Fig. 13.2. Sinusoidal tests at 3, 5, 20, and 60 Hz for input amplitudes 0 to 0.8g rms showed good linearity (Fig. 13.4b), so complete transfer function testing was initiated using sinusoidal test methods. Using averaged data from groups of ten human subjects, the amplitude ratio curve of Fig. 13.5 was obtained. (Phase angle data are presumably also available, but the reference does not show any.) Statistical analysis of the data used to plot the average curve (see page 533) shows that scatter is not excessive. A final modeling step was to develop an analytical representation of the measured transfer function, using cut-and-try methods on the straight-line asymptotes of the logarithmic frequency-response curves. A

[3]Pradko, F., Lee, R. A., and J. D. Greene, Human Vibration-Response Theory, ASME Paper 65-WA/HUF-19, 1965.

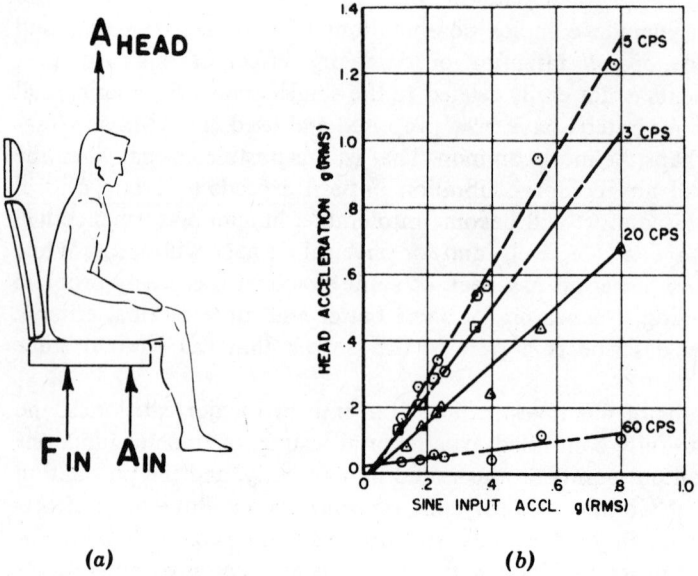

(a)   (b)

Figure 13.4 Experimental linearity check of human vibration response.

good fit was obtained using the following form of function:

$$\frac{A_{\text{head}}}{A_{\text{in}}}(s) = \frac{K(\tau s + 1)\left(\dfrac{s^2}{\omega_{n1}^2} + \dfrac{2\zeta_1 s}{\omega_{n1}} + 1\right)\left(\dfrac{s^2}{\omega_{n2}^2} + \dfrac{2\zeta_2 s}{\omega_{n2}} + 1\right)}{\left(\dfrac{s^2}{\omega_{n3}^2} + \dfrac{2\zeta_3 s}{\omega_{n3}} + 1\right)\left(\dfrac{s^2}{\omega_{n4}^2} + \dfrac{2\zeta_4 s}{\omega_{n4}} + 1\right)\left(\dfrac{s^2}{\omega_{n5}^2} + \dfrac{2\zeta_5 s}{\omega_{n5}} + 1\right)} \quad (13.1)$$

Figure 13.6 graphically compares the measured data with the analytical curve fit.

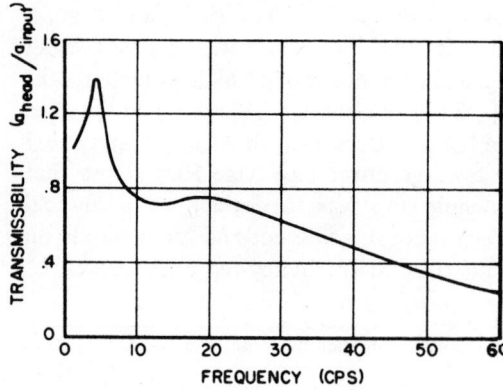

Figure 13.5 Average transfer function based on measurements on ten subjects.

Table For Figure 13.5

FREQUENCY	MEAN g	STANDARD DEVIATION	CONFIDENCE INTERVAL (90%)	
			UPPER BOUND	LOWER BOUND
1	1.011	0.032	1.032	0.989
3	1.182	0.105	1.253	1.111
4	1.389	0.157	1.495	1.282
5	1.298	0.302	1.401	1.195
7	0.901	0.282	1.092	0.710
10	0.76	0.20	0.836	0.684
15	0.74	0.23	0.828	0.652
20	0.76	0.22	0.843	0.677
30	0.63	0.18	0.698	0.562
40	0.49	0.14	0.570	0.410
50	0.35	0.12	0.423	0.277
60	0.25	0.12	0.302	0.198

Another transfer function studied was the input impedance at the seat-of-the-pants, $(f_i/v_i)(s)$. Figure 13.7 shows measured data and the analytical curve fit for this function. A particularly convincing demonstration of the predictive quality of this model is obtained by simulating the analytical function on an analog computer and then "playing" some chosen input simultaneously into both the analog model and the hydraulic seat with a human subject in place. The output "force" of the analog model, and the actual force measured on the human subject, are then displayed side by side on a two-channel recorder, giving, typically, the very good results shown in Fig. 13.8. While it is not our purpose here to explore in detail the practical applications of these human body vibration models, a few words along these lines might be helpful. Figure 13.9 shows in simplified fashion some major aspects of the problems faced by vehicle designers in the area of ride quality. Terrain models derived from studies of actual road (or off-road) profiles define the "path roughness" environment in

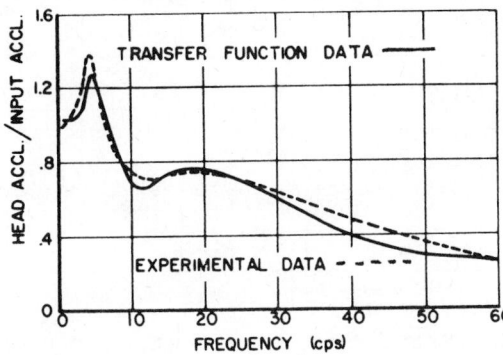

Figure 13.6   Comparison of measured data with curve-fit transfer function.

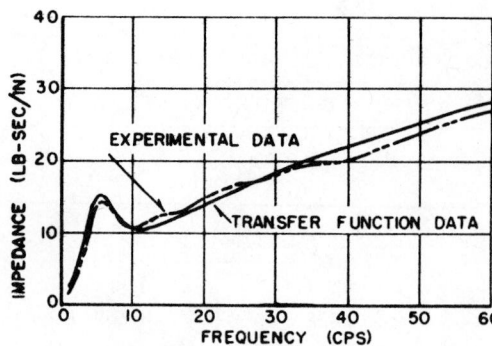

Figure 13.7 Measured input impedance at seat-of-the-pants.

which the vehicle must perform. These profiles serve as inputs to the mass/spring/damper models of the vehicle to produce motion/force outputs at the driver's-seat attachment points. Thinking of the seat as a vibration-isolation device, it may be helpful to consider its dynamic model as separate from the vehicle, perhaps using impedance or four-pole techniques to couple the individual models. The significant contribution of the present study is the capability for now extending rational analysis to the final interface—the seat with the human body—allowing prediction of head motions, vitally important to operator performance because of the critical functions of the brain, eyes, and vestibular sense organs.

We turn briefly now to the concept of absorbed power as a measure of the severity of human body vibration. At the interface between the vehicle seat and the buttocks there is a flow of mechanical power between the machine and the human body. While the human body is a complex mechanical structure, it

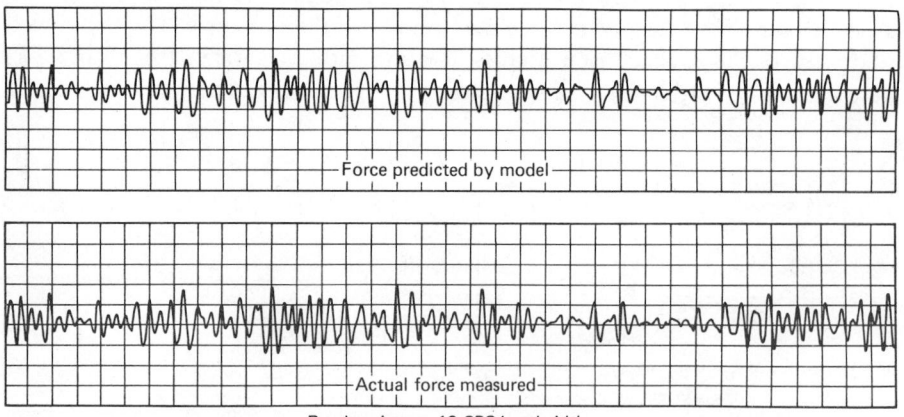

Figure 13.8 Comparison of actual and model-predicted force time histories.

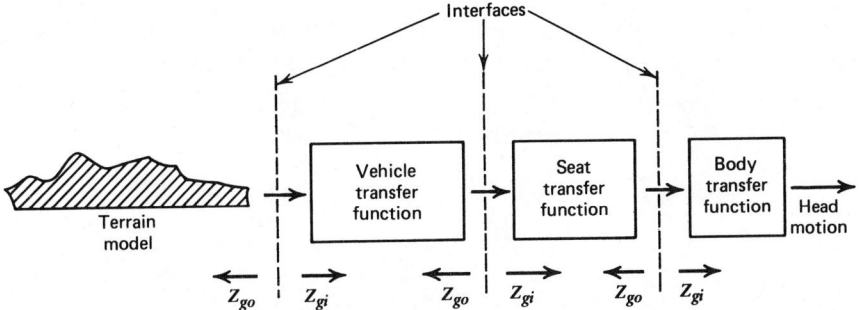

Figure 13.9 Coupling of component transfer functions for overall-system simulation.

certainly contains inertial and elastic energy storing elements and "frictional" dissipative elements. In considering the power flow at the seat/body interface, body inertial and elastic elements may take power from the seat for part of a cycle but will return this power to the seat at a later time. A net power flow—the average power—will however be necessary to supply the body dissipative elements. This net power is called the absorbed power and it can be measured as the time-averaged product of force and velocity at the seat/body interface. The investigators postulated that the absorbed power associated with a certain vibratory motion would be a better measure of the severity of vibration than would the RMS acceleration of that same motion.

To test this hypothesis the following experiment was performed. Twenty-six different "rides" of random waveform with frequency content in the 4–36 Hz range were run with four control subjects to establish the RMS acceleration level of each ride. The rides were then given to each of ten subjects in order of increasing RMS acceleration, and the subjects were asked to rate each succeeding ride as to whether it was more, or less, severe than the previous one. If RMS acceleration were truly a good measure of vibration severity, then none of the rides should be singled out as "misplaced" in the sequence. In fact, the subjects reported nine rides as "out of sequence," showing the unreliability of RMS acceleration. When the absorbed power data was processed for each ride it correlated almost perfectly with the subjective ratings of the human subjects. Figure 13.10 shows some numerical data from this test. The standard deviation columns indicate the variability, respectively, of the measured RMS accelerations and absorbed powers for the ten repeat runs (one for each subject) of each ride. The column headed "No. of Calls" indicates the number of times a ride was judged to be misplaced in the sequence.

While the results of this and other later[4] tests have been quite encouraging to the concept of absorbed power, more experience will be required to thoroughly establish it as the optimum measure of human vibration severity. If this

[4]Lins, W. F. and H. Dugoff, Motion Simulation and its Application to Ride Dynamics Research, SAE Paper 720003, 1972.

RIDE NO.	ACCELERATION RMS g's	STANDARD DEVIATION	NO. OF CALLS	POWER (WATTS)	STANDARD DEVIATION
1	0.430	0.019	0	19.64	2.194
2	0.475	0.015	0	68.64	6.401
3	0.480	0.014	5	17.42	1.665
4	0.512	0.019	0	31.83	3.038
5	0.559	0.032	4	16.35	3.150
6	0.675	0.021	0	209.76	19.766
7	0.672	0.017	6	30.98	2.768
8	0.703	0.028	0	56.49	5.951
9	0.828	0.037	7	26.57	3.240
10	0.850	0.033	1	80.80	7.571
11	0.881	0.024	0	364.16	39.274
12	0.855	0.024	3	228.43	24.615
13	0.924	0.032	5	66.26	5.828
14	0.962	0.032	0	116.32	9.754
15	1.025	0.034	2	66.67	7.358
16	1.121	0.047	5	62.03	6.008
17	1.164	0.034	0	562.28	48.904
18	1.201	0.033	5	167.06	13.748
19	1.205	0.034	0	177.19	13.961
20	1.211	0.046	4	54.15	6.716
21	1.210	0.042	0	92.20	9.990
22	1.503	0.046	0	920.92	84.645
23	1.628	0.050	10	83.24	10.721
24	1.599	0.096	0	172.69	19.024
25	1.854	0.053	0	406.12	46.676
26	1.942	0.073	3	137.36	18.022

Figure 13.10 Experimental data for comparison of acceleration and absorbed power as vibration severity criteria.

can be done in a general way, it could become a very powerful technique since energy is a scalar quantity and thus multi-axial and mixed translational/rotational vibration environments could be simply characterized by addition of the component-mode absorbed powers.

13.2 BASIC ANATOMY AND PHYSIOLOGY OF THE NERVOUS AND MUSCULAR SYSTEMS PERTINENT TO TRACKING AND PILOTING BEHAVIOR

In preparation for our next human factors modeling example, we give a brief survey of some known facts and theories relating to the operation of certain human body systems that are involved when a human being is engaged in tasks such as piloting an aircraft or driving a car. While the engineering approach to the measurement of the desired mathematical models is mainly a "black box" or input/output concept that does not depend heavily on an understanding of internal body functions, an elementary familiarity with these functions is of some help in interpreting the measured results. Recall that *anatomy* deals with the description of body parts while *physiology* explains how they work; we require basic information of both types.

We begin by examining the overall organization of the nervous system. Figure 13.11 shows the system divided into two major sections: the central nervous system and the peripheral nervous system. In a piloting task, for example, a visual input to the eyes (sensors) might show another aircraft approaching dangerously close, calling for an evasive maneuver of some sort. The eye receptors produce an electrical signal that is transmitted along sensory nerve fibers to the brain, where signal processing and decision making occurs. If action is felt to be necessary, the brain sends out through the motor nerves an electrical signal to muscles (effectors) in the hands and/or feet, which actuate the aircraft controls so as to initiate the desired maneuver. Receptors in the hands and feet feed back to the brain signals carrying information related to forces and motions, allowing precise manipulation of the aircraft controls. As the maneuver proceeds, new visual information is integrated with other sensor signals (engine sounds, body tilt and acceleration, etc.) to successfully carry through the desired action.

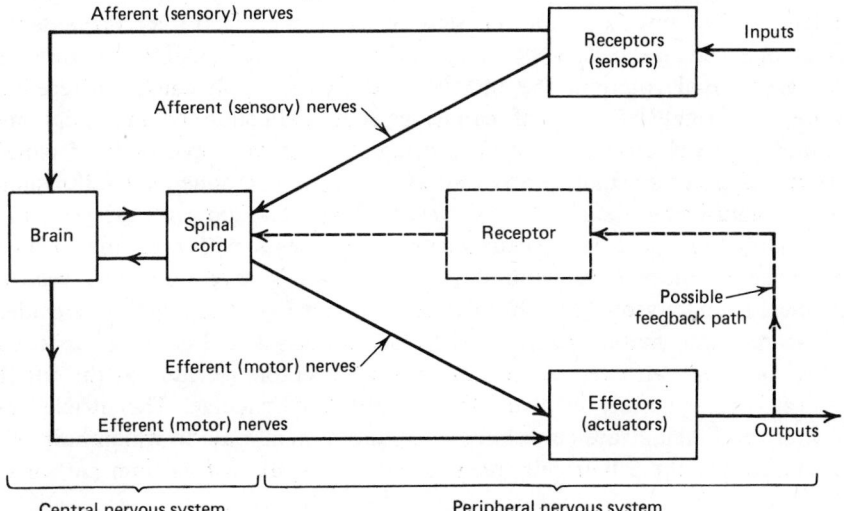

Figure 13.11   Organization of human nervous system.

The peripheral nervous system can be further subdivided into cerebrospinal ("voluntary") and autonomic ("involuntary") sections. The voluntary nervous system consists of the connections between centers in the central nervous system with the body wall by means of cranial and spinal nerve fibers. It includes those parts of the brain concerned with consciousness and mental activity, those parts of the brain, spinal cord, and nerve fibers (both sensory and motor) that control the skeletal muscles, and the end organs (receptors and effectors) of the body wall. The involuntary system consists of the connections between centers in the central nervous system with the viscera by means of the visceral nerve fibers. It controls the "automatic" body functions such as breathing, heartbeat, digestion, elimination, and so on. It is further subdivided into the parasympathetic system (which governs under normal conditions) and the sympathetic system (which takes over under emergency conditions to make the body more "effective").

We next discuss sensors, concentrating on those most important in piloting tasks. The receptors of the *visual sense* are the rods and cones of the retina in the eye; the rods are understood to carry brightness information while the cones discern color and form. A photochemical reaction, involving a purple substance (rhodopsin) that is quickly turned white by light, produces the electrical signal sent to the brain. The visual sense is used both in the perception of spatial relationships and also for instrument reading.

Particularly important in piloting are the *vestibular senses*; those giving information on linear and angular acceleration and body tilt with respect to gravity. The receptors of angular acceleration are the cristae ampularis, located at the semicircular canals of the inner ear. These canals are tubelike structures oriented roughly in three perpendicular planes and partially filled with a liquid (endolymph) that acts as an inertial sensing mass (see Fig. 13.12). An enlargement at one end has cells with hairy endings that are sensitive to pressure. Under accelerated rotation, the inertia of the endolymph causes differential pressure in a right/left pair of canals causing deflection of the hairs and generation of an electrical signal. The orientation of the three pairs of canals allows resolution of a vector rotation into rectangular components, the threshold of detectability being about 1 to 2 deg/sec^2. Those familiar with the construction of multi-axial angular accelerometer instruments will note the striking resemblance of these man-made devices to nature's invention. The relative mechanical simplicity of this vestibular sense organ has encouraged a considerable research[5] into mathematical models for its operation. For linear acceleration and body tilt with respect to gravity, the receptors (located at the utricle and saccule of the inner ear) are the otoliths and maculae. The utricle and saccule are each small sacs containing fluid and a membrane to which hair cells are attached. On these hair cells rests a layer of crystals of calcium carbonate

[5]Meiry, J. L., The Vestibular System and Human Dynamic Space Orientation, NASA CR-628, 1966; Third Symposium on the Role of the Vestibular Organs in Space Exploration, NASA SP-152, 1968.

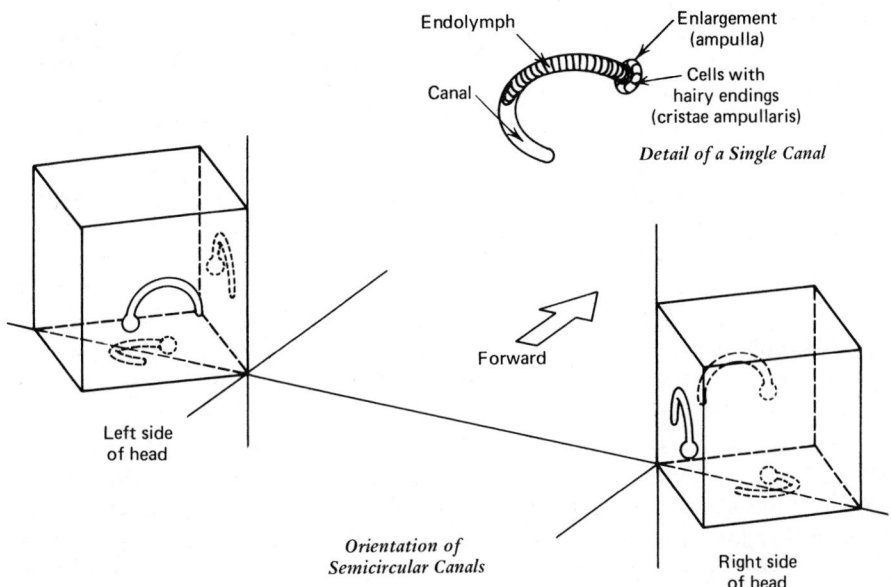

Figure 13.12 Semicircular canals of the inner ear.

called otoliths ("ear stones"). In the utricle the membrane is horizontal; in the saccule, vertical. Tilting or translational acceleration of the head causes gravity or inertial forces on the otoliths and thus a bending of the hair cells, which gives rise to an electrical signal. The threshold of acceleration detection is about 0.002 to 0.02 g's; for head tilting it varies widely, the mean being 2 or 3 degrees.

The *proprioceptive senses* are those inside the body that are activated by movement or force changes of body tissues, including the muscles, while the *tactual (touch) sense* is located in the skin and responds to pressure differences between adjoining skin areas. Proprioceptive senses enable one to control voluntary muscular activities even without the aid of vision. For example, if a pencil is laid on a table and a subject is given a brief look at it, he can reach out and pick it up even with his eyes shut, thus there is apparently some "position feedback" in the body members that indicates their location. Similarly, the forces necessary to lift a dead weight are finely adjusted to the resistance offered, such that the weight is held still and not accelerated away by an unbalanced force. The proprioceptive senses also serve to perceive changes in orientation and equilibrium, either by detecting changes in position of the body members caused by external forces or by detecting those reflex changes in the muscle system that automatically act to maintain posture. Note that the erect human body is "statically unstable," like a broomstick balanced on its end, and would fall over if there were not continuous and automatic corrections applied to keep it vertical. Due to the difficulty of isolating and working with proprio-ceptors, little or no data on thresholds is available. Because of the greater accessibility of the tactual sense, numerical data relative to its behavior are available. Pressure is felt when a differential force deforms the surface of the skin since the difference in pressure between the point of contact and the surrounding skin upsets the neutral equilibrium that existed before the stimulus was applied. If the pressure is *maintained*, equilibrium is restored and the sensation of touch ceases, unless movement occurs. The thresholds for pressure perception vary widely with location in the body; from about 0.02 N/mm^2 at the finger tip to 1.2 N/mm^2 at the thick part of the sole. The time interval for loss of sensation when a steady force is applied varies with the force magnitude, ranging from about 2 sec for a 0.0002 N force to 10 sec for a 0.01 N force for the back of the hand.

Turning now to the nerves, we first consider their structure and intercon-nection. The basic unit of a nerve—the nerve cell—is called a neuron and its function is to receive nerve impulses and convey them to other cells. A nerve in the body is made up of a bundle of neurons, similar to a multiconductor electrical cable. Figure 13.13 shows the main features of a typical neuron and its associated structures. The cell body is the order of 4 to 130 μm in diameter, depending on its location in the body. Radiating out from the cell body are several dendrites and a single axon. The dendrites are the receivers of signals from other neurons whose nerve endings ("end brushes") are in close proximity to the dendrite, the interface between end brush and dendrite being called a synapse.

The nerve fibers (axons, dendrites) are of the order of 2 to 20 μm in diameter in humans. The axon is the transmitter of nerve impulses and may be up to 4 feet in length. While, in a dissected nerve, electrical impulses can be transmitted in either direction, in an operating nervous system transmission is always *from* the end brush of one neuron *to* the dendrite of the other (through the synapse), and thus signals in an axon always go *away* from the cell body. If a nerve is a motor (efferent) nerve, its end brush discharges into either a muscle fiber (as in Fig. 13.13) or into the dendrites of another neuron leading to a muscle. If a nerve is a sensory (afferent) nerve, its end brush discharges into the dendrites of a neuron leading to the central nervous system. Synapses provide the first stage of "sorting" and distribution of impulses. They provide a multi-plicity of possible interconnections between nerves. For example, a single stimulus may be relayed to several different effectors and stimuli from more than one sensor may reach a single effector (see Fig. 13.14). The threshold for signal transmission at a synapse is higher than that in the nerve fiber itself, and this threshold varies from synapse to synapse, thus presenting preferred paths to an impulse.

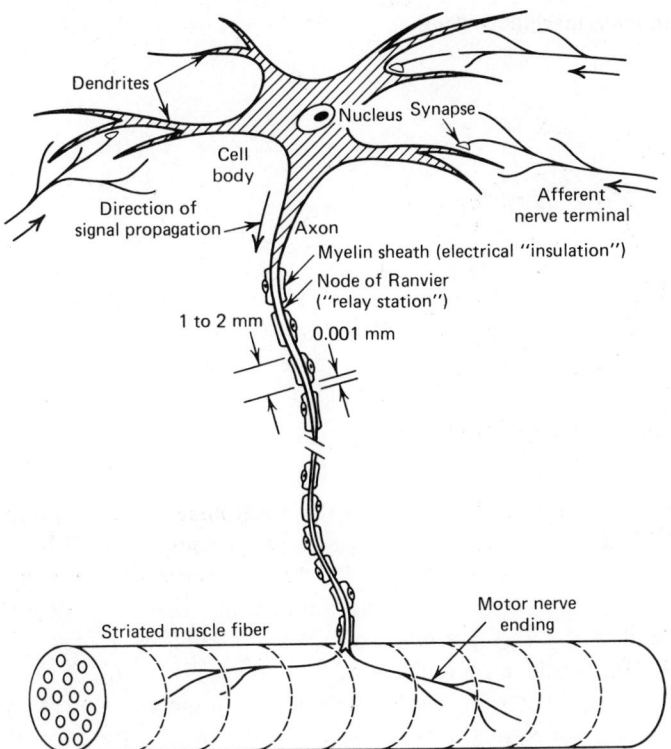

Figure 13.13   Main features of a neuron.

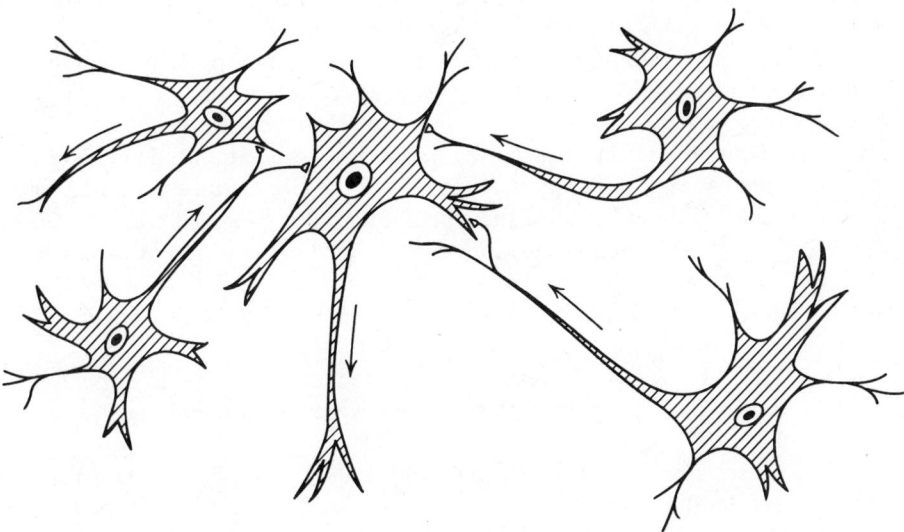

Figure 13.14   Neuron interconnections.

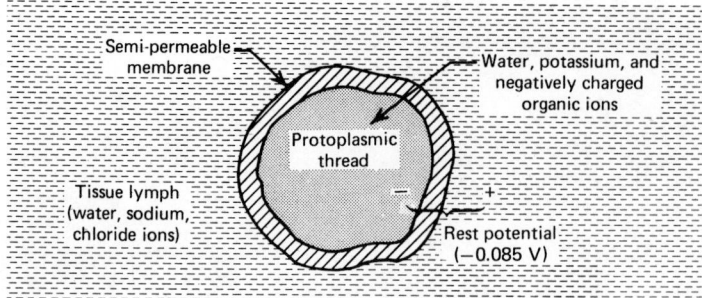

Figure 13.15 Cross section of nerve fiber (axon).

Research has provided considerable detail[6] of the mechanisms of generation and transmission of electrical nerve signals; we here provide only a brief overview. Figure 13.15 shows a cross section of a typical nerve fiber (axon). Experiments show that if the fiber is not being stimulated, there is a steady potential difference, called the *rest potential*, of about −0.085 V across the membrane separating the inside and outside of the axon. This equilibrium condition is supported by a balance between differences in ionic concentration across the membrane and the electrical forces due to the potential difference. If now two tiny electrodes, separated by some distance along the length of the axon, are inserted into the axon, one can feed electrical stimuli into the fiber at one electrode and read out the response of the fiber at the other electrode. The observed behavior is briefly described as follows.

If the applied stimulating voltage is of the same polarity as the rest potential, a voltage pulse of a size roughly proportional to the stimulating pulse is propagated toward the receiving electrode. However the propagated pulse attenuates rapidly with distance, losing about 50% of its amplitude every 1 or 2 mm. This behavior is essentially independent of the size of the stimulating pulse. If we now reverse the polarity of the stimulating pulse so that it tends to reduce the −0.085 V rest potential toward zero, a strikingly different behavior is observed. At first, if the stimulating pulse is small, a behavior similar to that explained above (rapid decay with distance) is observed. However, as the size of the stimulating pulse is increased so that the potential is reduced to about −0.065 V a critical threshold is reached at which an electrochemical reaction occurs and the potential difference across the membrane jumps to about +0.045 V.

This pulse is called the *action potential* and it now propagates down the axon at essentially fixed speed and *without* attenuation. This transmission without attenuation is accomplished by a distributed amplifying mechanism of

[6]Katz, B., *Nerve, Muscle and Synapse*, McGraw-Hill, N.Y., 1966.

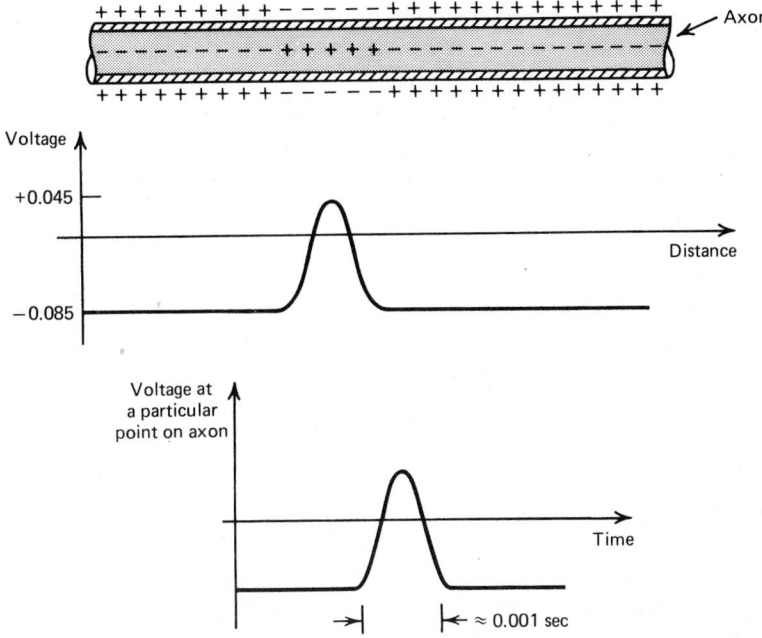

Figure 13.16   Propagation of nerve action potential.

the axon that adds sufficient energy to make up for losses. That is, the axon is not just a passive electrical cable but rather has also the active property of amplification, distributed along its length. Figure 13.16 shows the action potential phenomenon. If one increases the stimulating voltage beyond the critical threshold, the action potential will still be triggered, but its magnitude ($+0.045$ V) will remain the same, regardless of stimulus magnitude. This is called the "all-or-none" law.

Two general types of nerve fibers are found in the body: myelinated and unmyelinated. Figure 13.13 shows a myelinated axon. It has a segmented "insulating" sheath; each segment of the sheath is formed by a cell called a Schwann cell, which wraps itself around the axon in the manner of electrical insulating tape. An unmyelinated fiber is simply a "bare" axon without the myelin sheath. These bare fibers have rather poor electrical signal transmission properties and would have to be of relatively larger diameter than myelinated fibers to carry signals with the same speed and efficiency. (Where the optic nerve leaves the retina, a "blind spot" exists since the space taken up by the nerve is not available for rods and cones. Nature has minimized this blind spot by using the smaller myelinated fibers in this location.) In unmyelinated fibers the amplifying mechanism is distributed more or less uniformly along the length of the fiber. In myelinated fibers the amplifying mechanism is localized at those points, called the nodes of Ranvier, where the bare axon is exposed between the myelin segments. Between the nodes of Ranvier the nerve impulse travels by simple conduction and thus attenuates. However, it does not decay below the firing threshold before it reaches the next node. That is, the distance between nodes, the attenuation rate, and the magnitude of the initial action potential (relative to the threshold level) are such that the impulse can be propagated from one node to the next. The initial action potential is about five times the threshold level, thus a decay of 80% can occur before firing at the next node will be inhibited.

The velocity of propagation of the nerve impulse varies from fiber to fiber according to type and size; larger fibers have higher transmission speeds. For a given fiber the velocity is the same all along it, ranging from 0.3 to 100 m/sec. Also, smaller fibers have somewhat higher triggering thresholds. Nerve fibers exhibit:

1 An absolute refractory phase. This refers to the fact that, just after an action potential has been triggered, if another stimulus (no matter how large) is applied within 1 or 2 milliseconds after the first, no new action potential will be triggered. That is, the fiber is "desensitized" temporarily following an impulse. This behavior limits the possible upper frequency of nerve impulses to the range 500 to 1000 Hz. Actually, observed maximum rates do not exceed about 400 Hz.

2 A relative refractory phase. Following the absolute refractory phase there is a period of a few milliseconds called the relative refractory phase during which the triggering threshold is larger than normal and, if an action potential is triggered, its magnitude is smaller than normal.

3 A hyperexcitable phase. Following the relative refractory phase there is a period of a few milliseconds during which the threshold is reduced, and a smaller-than-normal stimulus will trigger the action potential.

Following the above three phases the fiber recovers to its normal behavior.

Since the size and shape of the action potentials does not vary greatly, the information carried by the electrical nerve signals must be in some kind of time or frequency code. Observation of pulse trains carried by various nerves generally reveals that the frequency increases when the level of the particular stimulus (light, heat, pressure, etc.) increases.

We conclude this section with a brief explanation of the stimulation and response of muscles. The types of effectors found in the body may be classified as:

1 Striated (striped) muscles. These are the skeletal muscles attached to the bony parts by tendons.

2 Smooth muscles. } These automatically regulate continuing body
3 Endocrine glands. } activities such as breathing, heart pumping, and so on.

As far as the body's performance of steering or piloting tasks is concerned, only the striated muscles are of primary interest; thus we concentrate on them.

A striated muscle is stimulated into action by the nerve impulse discharged into it at the synapse that forms the interface between the nerve end-brush and the muscle fiber. The excitation appears to be transferred across this synapse by chemical rather than electrical means. The nerve action potential stimulates the release at the synapse of a chemical called acetylcholine. Muscle fibers react to stimulation in a manner similar to that of nerve fibers. The acetylcholine release results in a reduction of the muscle's rest potential to a value below a trigger threshold. An electrical impulse then propagates along the muscle fiber at a constant speed of a few meters per second, generally slower than in a nerve fiber.

The development of force and motion by a contraction of muscle fibers is brought about by the movement and interaction of two kinds of protein filaments found in the muscle: actin and myosin. Details of the process are not completely understood; however, it is believed to involve the splitting of a chemical (adenosine triphosphate (ATP)) upon contraction and a resynthesis of ATP upon relaxation. The electrical signals transmitted through the muscle fiber initiate and control these reactions governing the conversion of chemical energy into mechanical work and heat. A striated muscle is made up of many muscle fibers (elongated cells) 1 to 40 mm in length and 0.01 to 0.15 mm in diameter. A single fiber, when stimulated, contracts as much as possible or not at all, thus there is an "all-or-none" law, just as in nerve fibers. If a certain threshold stimulus is exceeded, full contraction occurs; smaller stimuli cause no contraction, larger stimuli cause the same contraction. By controlling the number of fibers contracted in the complete muscle, the desired level of force can be obtained. The large number of fibers, together with the averaging effect of connective tissue, give the overall force a smoothly varying nature even though the individual elements are operating in an on-off fashion. Just as in nerve fibers, muscle fibers exhibit absolute and relative refractory phases and hyperexcitability.

Because muscle fibers can only exert tensile (rather than compressive) forces whereas both kinds of forces are needed for operation of the body, muscles are used in pairs called agonists and antagonists. These muscle pairs are connected to bony lever systems in such a way that contraction of one muscle of the pair and relaxation of the other gives a net tension force while the reverse process gives compression. When both members of a given pair are completely contracted the member is rigidly fixed. Movement requires contraction of agonists and relaxation of antagonists. In a slow reciprocating movement both agonists and antagonists are in constant partial tension called a "muscle fixation." As speed increases, periods of tension in opposing muscles overlap less and less until a "resonance" occurs at about 2 Hz where there is no overlap at all. This is called a "ballistic motion." Speed, skill, and accuracy in activities involving motion appear to depend on the replacement of muscle fixation by ballistic movement through training and practice. Increase of frequency above about 2 Hz causes attenuation of motion amplitude until it reaches zero at 8 to 10 Hz for all body members.

Muscular activity may be classified into three types:

1 Reflexes. In a reflex activity the receptors and effectors are connected at the spinal cord; no other part of the central nervous system is involved. There is no conscious control of such activities and there is very small lag in response. Examples are the contraction of the eye's pupil when exposed to bright light and the jerk of the leg when the knee is tapped.

2 Consciously controlled (voluntary) motion. Here the higher levels of the central nervous system are involved and the activity is at all times under conscious control.

3 Learned responses. These are intermediate between reflex and voluntary motions in the sense that once a decision to act has been made, the response is carried out without conscious control. In fact, if conscious control is attempted, performance may be impaired; for example, throwing a ball or tying a necktie.

13.3 EXPERIMENTAL MODELING OF HUMAN TRACKING/PILOTING BEHAVIOR

While a rather detailed physiological understanding (which we only surveyed at an elementary level in Section 13.2) of the underlying body functions is available, an overall model of human behavior in tracking/piloting tasks, suitable for man/machine system design, cannot be deduced from such an approach. Rather, it is necessary to perform input/output experiments on the human operator under conditions that in some way simulate the essential aspects of the real-world situation in which the model will be used. A great deal of research along these lines has been, and continues to be, carried out, and certain useful results are now considered well established. In this section we concentrate on an explanation of the techniques employed in developing these models.

A first necessary step is to describe the general structure of the interactions between man and machine. We use for this a block diagram suggested by McRuer and Krendel.[7] In Fig. 13.17 the inanimate portion of the system is separated into machine dynamics and display dynamics. Using aircraft as an example, machine dynamics refers to the model relating, say, pilot control stick motion to aircraft pitch axis rotation. These dynamics are of course amenable to physical analysis using techniques such as those of Chapter 12. While a small private aircraft might not have sophisticated instrument displays available to the

---

[7]McRuer, D. T. and E. S. Krendel, "The Man-Machine System Concept," *Proc. IRE*, **50**, No. 5, May (1962).

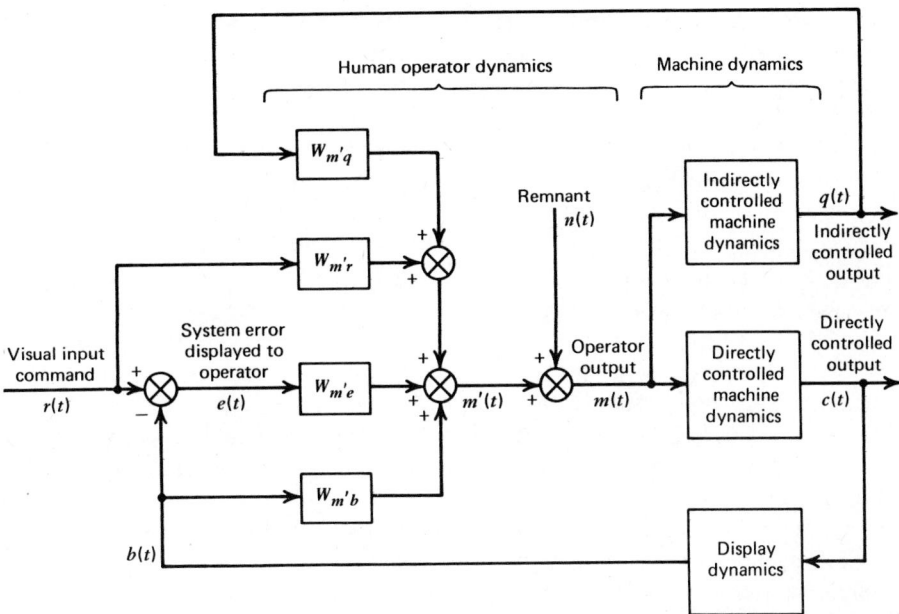

Figure 13.17 Structure of man/machine interactions in steering/piloting tasks.

pilot, advanced military and commercial vehicles make considerable use of such aids. These display systems gather data from various sensors (such as radars, altimeters, radio beacons, etc.), manipulate and integrate the information, and present it visually in numerical and/or graphical form to the pilot, perhaps on a cathode-ray tube screen. Such displays exhibit dynamic characteristics such as unavoidable lags due to information processing delays and also intentional effects such as using derivatives to provide "anticipation."

In representing the human operator for the most general case, a multiplicity of input paths must be provided to account for the various senses (sight, hearing, rotation, etc.) and also for the fact that a single sense (such as sight) can discriminate several different aspects of the same stimulus simultaneously. If our goal is a linear model, we represent these various signal processing paths as separate transfer function blocks whose outputs are additively combined to form the single operator output that is usually a motion (or force) of the hands or feet on the vehicle controls. In Fig. 13.17 a single directly controlled and a single indirectly controlled machine output are shown. However, the diagram is easily modified to provide for multiple control modes, such as a pilot simultaneously controlling aircraft pitch and yaw. Indirectly controlled machine outputs are physical quantities affected by the operator's output but are not the direct objects of control. When these machine variables are sensed by the operator and utilized as components of the control decision, they must be included in the system diagram. For example, when a helicopter pilot is attempting to control the vehicle's forward speed, engine noise may be one of several speed-related sensory inputs taken into account even though the pilot is *not* trying to control engine noise.

In addition to providing for the influence of various indirectly controlled machine outputs (Fig. 13.17 shows one such), the human operator model provides individual signal-processing paths for $r(t)$, $b(t)$, and $e(t)$ since the operator may be able to separately perceive all these aspects of a single visual input. For example, an automobile driver proceeding down a road in good weather can simultaneously perceive the white lines marking the lane ($r(t)$), the car's lateral position ($c(t)$), and the difference between the two, ($e,(t)$). Since any or all of these aspects of the visual signal might be used in making steering decisions, our model should provide somehow for these possibilities. A final feature of the conventional human-operator model, which is not generally found in other models we have considered, is an explicit representation of the difference between the predictions of the linear model and the actual operator output. This difference is given the name *remnant* in Fig. 13.17 and experimental studies generally determine both the "best" linear model (the W transfer functions) and also the remnant $n(t)$, which clearly must be a small quantity if the linear model is a good predictor of actual behavior.

The general structure shown in Fig. 13.17 has to a large extent defied attempts at experimental characterization because of the complexity of the interactions between the various signal-processing paths. Successful models *have* been obtained, however, for *portions* of the overall structure corresponding to situations where certain of the paths are intentionally made inactive by choice of experimental conditions. Fortunately, some of these simplified structures correspond reasonably well with certain real-world application conditions, making the models practically useful. Perhaps the most common of these simplified structures is the so-called *compensatory tracking task*. Here the operator is allowed only to view the "error" signal $e(t)$, no indirectly controlled outputs are sensed, and machine and display dynamics are combined, to give the diagram of Fig. 13.18. Using a control stick with attached potentiometer, an analog computer to simulate W_{cm}, a signal generator to provide $r(t)$, and an oscilloscope

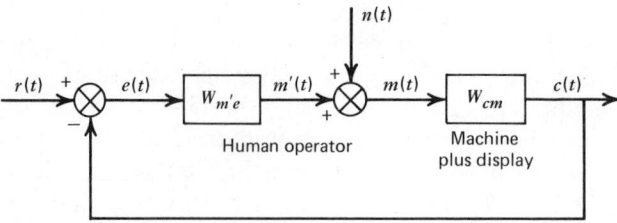

Figure 13.18 Compensatory tracking configuration.

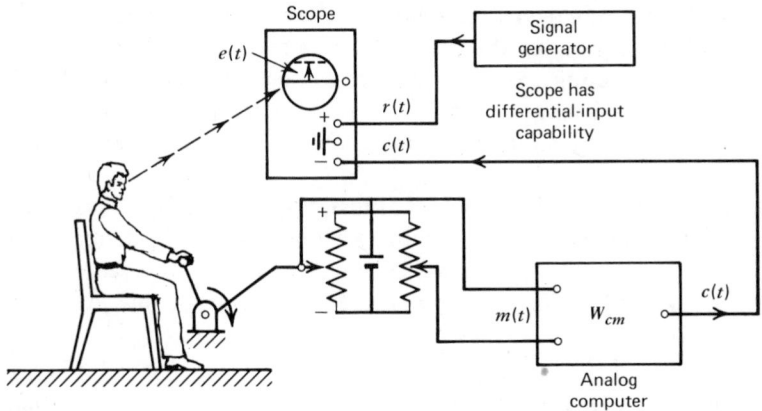

Figure 13.19   Possible apparatus for compensatory tracking.

display, a relatively simple and inexpensive apparatus for such tests can be assembled as in Fig. 13.19. A high sweep speed is used on the scope to display $e(t)$ as a horizontal line that moves above and below a fixed horizontal reference line at $e = 0$.

Whereas inanimate systems (and even human ones such as body vibration, where the mind is not much involved) can be tested, in principle, with any kind of dynamic input with the same results, piloting and tracking models turn out to be quite sensitive to the type of input employed. This is basically due to the fact that the human operator displays *adaptive* characteristics, that is, he changes his performance to suit conditions. This type of behavior is of course nonlinear. However, it is a *desirable* nonlinearity and is in fact one of the main advantages of human operators over machines. This adaptability, however, prevents us from using, say, sinusoidal testing because othe operator quickly recognizes the repeatable nature of a particular sine wave, adapts to it and performs at a higher accuracy level than would be possible with, say, a narrow-band random signal of comparable frequency. Of course, if the real-world application actually *were* sinusoidal, then sinusoidal testing would be appropriate. We thus conclude that testing for piloting/tracking models must be done with test signals similar to those of the actual application; since most real-world situations are quite random, random-signal testing is usually employed.

The adaptability of the human operator extends also to the machine/display dynamics. For the same type of input $r(t)$, different forms and/or numerical values of $W_{m'e}$ will be employed, depending on W_{cm}. Much of the research on human operator models has thus consisted of testing various combinations of $r(t)$ and W_{cm} to discover what forms and numerical values of $W_{m'e}$ are synthesized by the brain to cope with a particular situation. While the number of combinations might at first appear to be boundless, practical constraints on vehicle dynamics plus the physical limitations of the human body itself serve to define operating envelopes that have been thoroughly explored. Careful evaluation of these many results has lead to a relative simple general model, including rules for adjusting its parameters to suit the particular $r(t)$ and W_{cm} of interest.

The task of finding the "best" linear transfer function $W_{m'e}$, based on measurements of $e(t)$ and $m(t)$, is sometimes formulated as an optimum filter problem. We wish to choose $W_{m'e}$ so as to minimize the mean-square error between $m'(t)$, the output of $W_{m'e}$, and $m(t)$, the total operator output. From convolution integral:

$$m'(t) = \int_0^\infty h(\tau)e(t-\tau)\,d\tau \tag{13.2}$$

where $h(\tau)$ is the impulse response of $W_{m'e}$. The error is then

$$E(t) = m(t) - m'(t) = \text{the remnant}$$
$$= m(t) - \int_0^\infty h(\tau)e(t-\tau)\,d\tau \tag{13.3}$$

We desire to select $h(t)$ such that it minimizes $\overline{E^2(t)}$:

$$\overline{E^2(t)} = \overline{[m(t) - m'(t)]^2} \tag{13.4}$$

The $h(t)$ which does this is the solution of the Weiner–Hopf integral equation:

$$R_{em}(\tau) = \int_0^\infty h(u)R_{ee}(\tau-u)\,du \tag{13.5}$$

where

$$R_{em}(\tau) \triangleq \text{cross-correlation function between } e \text{ and } m$$

$$R_{ee}(\tau) \triangleq \text{autocorrelation function of } e$$

While the Wiener–Hopf integral equation may be unfamiliar to the reader, transformation to the frequency domain gives:

$$\phi_{em}(i\omega) = W_{m'e}(i\omega)\phi_{ee}(\omega) \tag{13.6}$$

or

$$W_{m'e}(i\omega) = \frac{\phi_{em}(i\omega)}{\phi_{ee}(\omega)} \tag{13.7}$$

where ϕ_{em} and ϕ_{ee} are, respectively, the cross-spectral density of e and m and the power-spectral density of e. This result is of course identical with our earlier random-signal test methods, which made no mention of remnants or minimization of mean-square errors. We may thus use all our earlier methods and results from random-signal testing here. However, we now also have the additional useful interpretation given by the mean-square error-minimization viewpoint.

While Eq. (13.7) shows $W_{m'e}$ calculated from one cross- and one power-spectral density, examination of the literature shows that generally two cross-spectral densities are used instead. This alternative method is more accurate since it minimizes certain bias errors caused by the presence of remnant effects in the operator input signal $e(t)$. That is, measurements of $W_{m'e}$ *must* be made in the *closed-loop* configuration of Fig. 13.18, since the form of $W_{m'e}$ depends on the nature of the machine dynamics W_{cm}, so a "machine" to be controlled *must* be present. We cannot isolate the portion of the system between e and m for "open-loop" measurements. If we are thus forced to make our measurements in the configuration of Fig. 13.18, it is clear that effects of the remnant signal $n(t)$ will propagate through W_{cm} and appear in $e(t)$. This causes the cross correlation between e and n to be nonzero, which can be shown[8] to cause a bias error in our measurements of $W_{m'e}$ if we use Eq. (13.7). This bias error can be shown[8] to be minimized by using instead the relation

$$W_{m'e} = \phi_{rm}/\phi_{re} \qquad (13.8)$$

While we will not here try to show mathematically why Eq. (13.8) is *preferable* to Eq. (13.7), it is fairly easy to show that Eq. (13.8) can be used to find $W_{m'e}$. Using ordinary block diagram manipulations on Fig. 13.18, it is easy to show that:

$$M = \frac{W_{m'e}}{1 + W_{m'e} W_{cm}} R + \frac{1}{1 + W_{m'e} W_{cm}} N \qquad (13.9)$$

$$E = \frac{1}{1 + W_{m'e} W_{cm}} R - \frac{W_{cm}}{1 + W_{m'e} W_{cm}} N \qquad (13.10)$$

These give

$$\frac{M}{R} = \frac{W_{m'e}}{1 + W_{m'e} W_{cm}} = \frac{\phi_{rm}}{\phi_{rr}} \qquad (13.11)$$

$$\frac{E}{R} = \frac{1}{1 + W_{m'e} W_{cm}} = \frac{\phi_{re}}{\phi_{rr}} \qquad (13.12)$$

[8]McRuer, D. T. and H. R. Jex, "A Review of Quasi-Linear Pilot Models," *IEEE Trans. Human Factors*, Vol. HFE-8, No. 3, Sept. (1967), pp. 231–249.

from which

$$W_{m'e} = \frac{M}{E} = \frac{\phi_{rm}}{\phi_{rr}} \frac{\phi_{rr}}{\phi_{re}} = \frac{\phi_{rm}}{\phi_{re}} \qquad (13.13)$$

Sometimes we wish to find the human operator transfer function, not from controlled laboratory tests, but from data records generated during actual operations. In some such cases the input r is either inaccessible to measurement and/or is very small, the man/machine system being "exercised" by disturbance inputs to the machine. A good example is aircraft piloting when we are trying to maintain a *steady* course $(r(t) \equiv 0)$ in the face of atmospheric turbulence which disturbs the aircraft motion. The man/machine system diagram now appears as in Fig. 13.20 rather than Fig. 13.18. In such a situation Eq. (13.8) cannot be used since r is not available, and we must use Eq. (13.7) even though it is known to have some inaccuracy. Fortunately, when our model is good, the remnant n is small and the previously mentioned bias errors are not great. Also, various data processing modifications based on Eq. (13.7) have been developed[9, 10] improve accuracy.

Returning to the configuration of Fig. 13.18, while the remnant cannot be measured directly, its power-spectral density can be calculated indirectly from measurements on the accessible signals r and m. From Eq. (13.9), using the fact that the power-spectral density of the sum of two signals is the sum of their individual power-spectral densities:

$$\phi_{mm} = \left| \frac{W_{m'e}}{1 + W_{m'e} W_{cm}} \right|^2 \phi_{rr} + \left| \frac{1}{1 + W_{m'e} W_{cm}} \right|^2 \phi_{nn} \qquad (13.14)$$

$$\phi_{nn} = |1 + W_{m'e} W_{cm}|^2 \phi_{mm} - |W_{m'e}|^2 \phi_{rr} \qquad (13.15)$$

We can also calculate for the controlled variable c what portion of it is "caused" by input r and how much is due to remnant n:

$$C = \frac{W_{m'e} W_{cm}}{1 + W_{m'e} W_{cm}} R + \frac{W_{cm}}{1 + W_{m'e} W_{cm}} N \qquad (13.16)$$

$$\phi_{cc} = \left| \frac{W_{m'e} W_{cm}}{1 + W_{m'e} W_{cm}} \right|^2 \phi_{rr} + \left| \frac{W_{cm}}{1 + W_{m'e} W_{cm}} \right|^2 \phi_{nn} \qquad (13.17)$$

The coherence function γ^2 discussed in Sections 6.2 and 6.3 is also useful here

[9]Wingrove, R. C. and F. G. Edwards, A Technique for Identifying Pilot Describing Functions from Routine Flight-Test Records, NASA TND-5127, 1969.
[10]Wingrove, R. C., Comparison of Methods for Identifying Pilot Describing Functions from Closed-Loop Operating Records, NASA TND-6235, 1971.

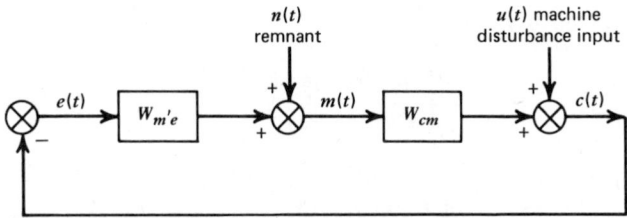

Figure 13.20   Compensatory tracking with disturbance input.

for judging the quality of our measured human operator transfer function $W_{m'e}$. For the signal m we can interpret γ^2 as:

$$\gamma^2 \triangleq \frac{\text{spectral density predicted by } W_{m'e}}{\text{total spectral density}} \tag{13.18}$$

$$\gamma^2 = \frac{\left|\dfrac{W_{m'e}}{1 + W_{m'e}W_{cm}}\right|^2 \phi_{rr}}{\phi_{mm}} = \frac{\left|\dfrac{\phi_{rm}}{\phi_{rr}}\right|^2 \phi_{rr}}{\phi_{mm}} = \frac{|\phi_{rm}|^2}{\phi_{rr}\phi_{mm}} \tag{13.19}$$

If the remnant were zero, then $W_{m'e}$ would predict *all* of the spectral density of m and γ^2 would be 1.0. The closer γ^2, measured from Eq. (13.19), comes to 1.0, the better the model. Experiments show that for random-appearing signals with frequency content not exceeding 0.7 Hz, a linear human operator model explains most of the operator's output and γ^2 is 0.9 or more.[11] When the frequency content extends to about 1.6 Hz, γ^2 drops to about 0.75, and to 0.6 when frequencies go to 2.4 Hz. As the input frequencies get higher, the operator enters a region where a human being is really not capable of performing well. These "unrealistic" demands on the operator appear to lead to confusion and thus to unpredictable nonlinear behavior that causes γ^2 to fall to low values. Fortunately, for those frequency ranges where a human operator is *appropriately* employed, the linear model is quite accurate.

Having discussed in some detail how the measurements are made, we are now ready to present the model that has emerged from the many experiments carried out. Measurements based on Equations (13.7) or (13.8) of course give the usual frequency-response graphs of amplitude ratio and phase angle of $W_{m'e}(i\omega)$ versus frequency. It is then desirable to fit analytical functions to these graphs. While various versions of the model are in use, the most common form that fits

[11]McRuer, D. T. and E. S. Krendel, "The Human Operator as a Servo System Element," *J. Franklin Inst.*, **267**, May and June (1959).

the measured data well is given by:

$$W_{m'e}(i\omega) = \frac{Ke^{-i\omega\tau}(\tau_L i\omega + 1)}{(\tau_I i\omega + 1)(\tau_N i\omega + 1)} \tag{13.20}$$

Of the parameters K, τ, τ_L, τ_I, and τ_N, three (K, τ_L, τ_I) are synthesized by the brain and adjusted to suit the nature of input r and machine dynamics W_{cm}, while two (τ, τ_N) are associated with neuromuscular dynamics and thus essentially fixed, with $0.12 < \tau < 0.18$ sec and $\tau_N \approx 0.1$ sec. For example, if $W_{cm} = 1.0$ (simplest possible machine dynamics), and if $r(t)$ is a random-appearing signal with flat frequency spectrum out to f_{co} (f_{co} not to exceed 0.9 Hz), then experiments show that the adaptive parameters are adjusted as follows:

$$K \approx \frac{2.2}{f_{co}^2}, \qquad \tau_I \approx \frac{K}{9.5}, \qquad \tau_L \approx \tau_N \tag{13.21}$$

Note that setting $\tau_L = \tau_N$ is a "cancellation compensation" similar to what a control system designer might use in an "all-machine" system to offset a lag τ_N.

Experiments using many different combinations of $r(t)$ and W_{cm} have been run and the results compiled.[12] One can thus consult such a catalog and hopefully find a case similar to that needed in a particular practical application. Also, careful study of all the specific cases run has revealed certain rules and trends[13] that can be extrapolated to new situations without the need for new experiments. Studies[14] to relate the measured overall characteristics to physiological details have also been pursued. Figure 13.21 shows a structure suggested by McRuer and Krendel. While most of the experimental effort has been applied to the simpler compensatory tracking task, the so-called *pursuit task*, where r, c and e are all displayed to the operator, is also of practical interest (see Fig. 13.22). The overall test apparatus is identical to that of Fig. 13.19. However, the display now uses a dual-beam oscilloscope. Here a fast sweep speed is again used to make r and c appear as horizontal lines, but a small square wave is now superimposed on $r(t)$ to distinguish it clearly from $c(t)$. While the man/machine model derived from Fig. 13.17 for pursuit tracking is more complex than that for compensatory tracking since now three transfer functions $(W_{m'r}, W_{m'e}, W_{m'b})$ are involved, the human operator performance[15] in the pursuit task is generally better because the display provides more useful information to the operator.

[12]McRuer, D. T. and E. S. Krendel, "The Human Operator as a Servo System Element," *J. Franklin Inst.*, **267**, May and June (1959).
[13]McRuer, D. T., *et al.*, Human Pilot Dynamics in Compensatory Systems, USAF, AFFDL-TR-65-15, 1965.
[14]Magdaleno, R. E. and D. T. McRuer, Experimental Validation and Analytical Elaboration for Models of the Pilot's Neuromuscular Subsystem in Tracking Tasks, NASA CR-1757, 1971.
[15]Reid, L. D., The Measurement of Human Pilot Dynamics in a Pursuit-Plus-Disturbance Task, Univ. of Toronto, UTIAS Rept. 138, 1969.

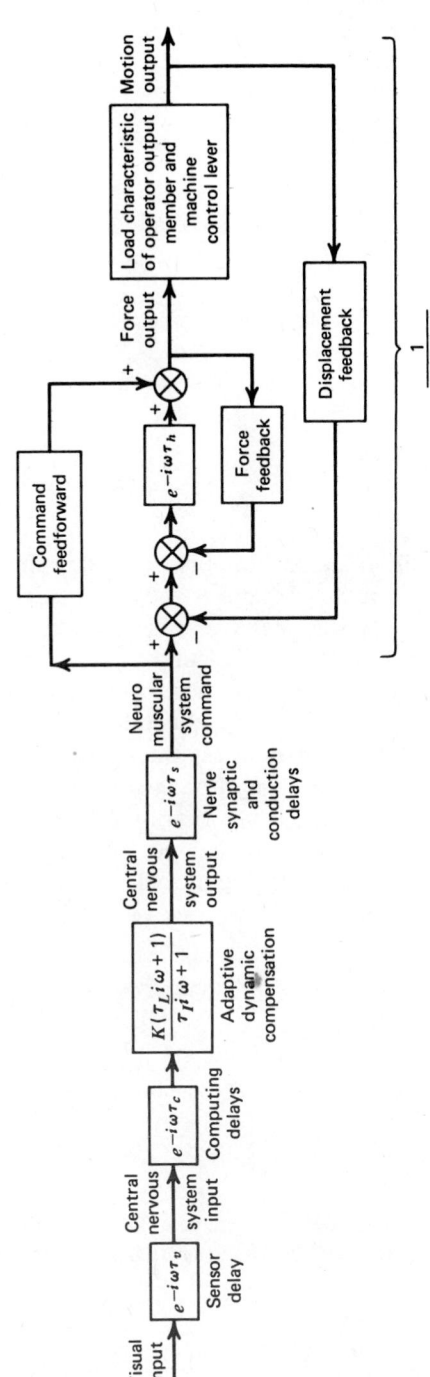

Figure 13.21 Physiological interpretation of human operator dynamics.

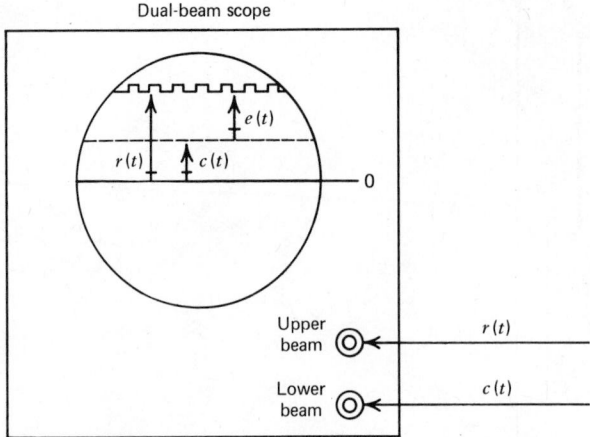

Dual-beam scope

Figure 13.22   Oscilloscope display for pursuit tracking.

Our final reference to the voluminous literature of this field concerns the nature of the human operator's response when the dynamics of the machine undergo gradual or sudden[16] changes. To study such behavior the operator is allowed to control a machine with fixed dynamics for a sufficient length of time to adapt to these conditions. Then, unknown to the operator, the machine dynamics are changed in the desired way to a new set of fixed conditions. We are interested mainly in what strategy the operator uses in his transition from one set of machine dynamics to another and also how long it takes to adapt to the new set of conditions. Practical interest in such situations involves both routine operation's such as changes in vehicle dynamics caused by gradual altitude and speed changes or sudden changes such as staging of rocket vehicles, and also emergency operations induced by failures of portions of the flight control systems.

BIBLIOGRAPHY

1   Allen, R. W. and H. R. Jex, An Experimental Investigation of Compensatory and Pursuit Tracking Displays with Rate and Acceleration Control Dynamics and a Disturbance Input, NASA CR-1082, 1968.

2   Irish, F. J., Potter, R. A., and R. D. McKenzie, Modeling Tools for Design of Air-Cushion Restraint Systems, SAE Paper 710015, 1971.

3   Krause, H. E., Research on a Distributed Parameter Mathematical Model of the Human Body in Dynamic Mechanical Environments, AMRL-TR-70-113, Univ. of Dayton Res. Inst., 1970.

[16]Weir, D. H. and A. V. Phatak, Model of Human Operator Response to Step Transitions in Controlled Element Dynamics, NASA CR-671, 1967.

4 Mahig, J., Effects of Variation of Speed on Vehicle Handling During Wind Gust, ASME Paper 73-ICT-3, 1973.

5 McRuer, D. T. and E. S. Krendel, "The Human Operator as a Servo System Element, *J. Franklin Inst.*, **267**, May and June (1959).

6 McRuer, D. T. and R. Klein, Effects of Automobile Steering Characteristics on Driver/Vehicle System Dynamics in Regulation Tasks, 11th Annual Conf. on Manual Control, NASA Ames Res. Cent., May, 1975.

7 Milsum, J. H., *Biological Control Systems Analysis*, McGraw-Hill, N.Y., 1966.

8 Simmons, G. J., A Lumped Constant Model of CO_2 Exchange in Respiration, Sandia Corp. Rept. SC-R-66-100, Albuquerque, 1966.

9 Stolwijk, J. A. and J. D. Hardy, "Temperature Regulation in Man—A Theoretical Study," *Pflügers Archiv*, **291** (1966), pp. 129–162.

10 Varterasian, J. H. and R. R. Thompson, The Dynamic Characteristics of Automobile Seats with Human Occupants, SAE Paper 770249, 1977.

11 Weis, E. B., Clarke, N. P., and H. E. von Gierke, Mechanical Impedance as a Tool in Biomechanics, ASME Paper 63-WA-280, 1963.

12 11th Annual Conf. on Manual Control, NASA Ames Research Center, 1975.

PROBLEMS

13.1 It has been suggested that pulse testing might be a more efficient method for obtaining the human body transfer functions of Section 13.1. Discuss this proposal and select a pulse shape, duration, and amplitude for A_{in} to be used in a pulse test designed to obtain the transfer function of Eq. (13.1).

13.2 We wish to use the impedance coupling methods of Section 7.4 to calculate seat velocity v_M in Fig. P13.1 from experimentally measured

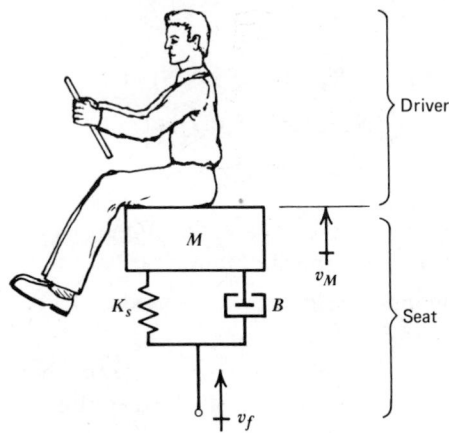

Figure P13.1

human body impedance (Fig. 13.7), a mass/spring/damper model of the driver's seat, and frame velocity v_f, which is assumed known. Derive an expression for $(v_M/v_f)(s)|_{\text{loaded}}$ in terms of $(v_M/v_f)(s)|_{\text{unloaded}}$, the body input impedance of Fig. 13.7, and the output impedance of the driver's seat. Assume that an analytical form for the experimentally measured impedance of Fig. 13.7 is available.

13.3   Equation (13.1) can be put in the form

$$\frac{A_{\text{head}}}{A_{\text{in}}}(s) = \frac{C_1 s^5 + C_2 s^4 + C_3 s^3 + C_4 s^2 + C_5 s + C_6}{s^6 + C_7 s^5 + C_8 s^4 + C_9 s^3 + C_{10} s^2 + C_{11} s + C_{12}}$$

with numerical values

$$C_1 = 11.28 \times 10^1, \qquad C_2 = 10.40 \times 10^4, \qquad C_3 = 38.70 \times 10^6$$

$$C_4 = 16.08 \times 10^8, \qquad C_5 = 44.31 \times 10^{11}, \qquad C_6 = 98.41 \times 10^{11}$$

$$C_7 = 95.34 \times 10^1, \qquad C_8 = 46.20 \times 10^4, \qquad C_9 = 45.12 \times 10^6$$

$$c_{10} = 10.17 \times 10^9, \qquad C_{11} = 64.20 \times 10^{11}, \qquad C_{12} = 98.42 \times 10^{11}$$

Use digital simulation to find the transient response of A_{head} to the pulse input of A_{in} developed in Problem 13.1. Also have the digital simulation program punch cards giving the time histories of A_{in} and A_{head} at proper sampling intervals. Use these cards as input to a pulse-test program that computes $A_{\text{in}}(i\omega)$, $A_{\text{head}}(i\omega)$, and their ratio $(A_{\text{head}}/A_{\text{in}})(i\omega)$. Compare the frequency-response curves thus produced with Fig. 13.6.

13.4   In the literature one may find mechanical models of varying degrees of complexity for representing the vibration of the seated human body. Figure P13.2 shows perhaps the simplest such model. Using the following numerical values:

$$M_1 = 46.2 \text{ kg}, \qquad M_2 = 34.3 \text{ kg}, \qquad K_1 = 1.17 \times 10^5 \text{ N/m}$$

$$K_2 = 0.623 \times 10^5 \text{ N/m}, \qquad B_1 = 1.98 \times 10^3 \text{ N/(m/sec)}$$

$$B_2 = 0.81 \times 10^2 \text{ N/(m/sec)}$$

compute $(A_{\text{head}}/A_{\text{in}})(i\omega)$ and $(f_{\text{in}}/v_{\text{in}})(i\omega)$ and compare with Figures 13.6 and 13.7 with regard to general features, not specific numerical values.

13.5   Using digital simulation, compare the pulse response of the model of Fig. P13.2 with that obtained in Problem 13.3.

13.6   For the system of Fig. P13.2, set up a digital simulation program that would allow one to study the absorbed power and RMS acceleration at the f_{in} location for random inputs of v_{in}.

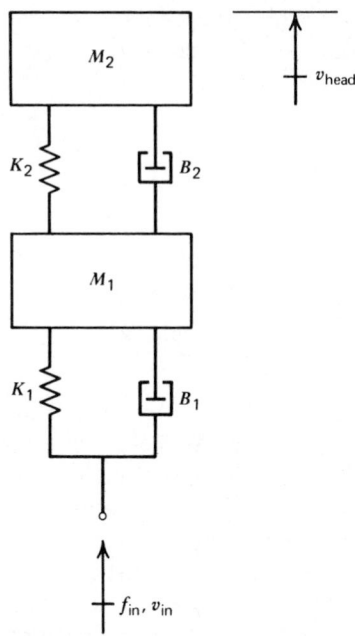

Figure P13.2

13.7 In Section 13.3, in lieu of numerical problems, the author has found that the assignment, for student review, of selected papers from the references or bibliography is useful in developing understanding of the material.

CHAPTER 14
Gas Turbine Dynamics

Combustion engines (spark-ignition and diesel) and gas turbines are important power sources for both stationary and mobile applications. While steady-state performance and economy will always be important considerations in the design and operation of these prime movers, dynamic behavior has become of increasing interest. Many stationary applications (such as electric generator drives), while thought of as nominally constant speed, require governor feedback systems to maintain nearly constant speed in the face of load changes and other disturbances. The design of these and other associated control systems requires knowledge of prime mover dynamics since the control systems are never perfect and small dynamic perturbations occur continuously. Mobile applications such as vehicle powerplants always involve unsteady operation due to the need to maneuver the car, plane, or ship. Transient behavior is thus of interest whether or not feedback control systems are used to improve performance.

In this chapter we introduce some of the basic modeling approaches employed in the gas turbine field. This example is chosen mainly because the area (particularly the aircraft jet engine applications) appears to have been more highly developed and reported in the literature. While jet engine dynamics have been studied since about the 1950s, interest has recently intensified because of developments in vertical take-off craft (including helicopters) where engine dynamics are intimately coupled with flight control and thus become a vital part of overall vehicle design.

Requirements for low pollution and improved fuel economy in automotive reciprocating engines are promoting considerable interest in sophisticated engine control systems that use a microprocessor to integrate data from engine sensors and optimize engine operation under changing conditions. These applications have spurred activity in engine dynamics studies. However, this important area of prime mover dynamics has not yet reached the maturity exhibited by the aircraft gas turbine field and thus a comprehensive open literature is not presently available, preventing "outsiders" from getting a clear and complete view of recent developments.

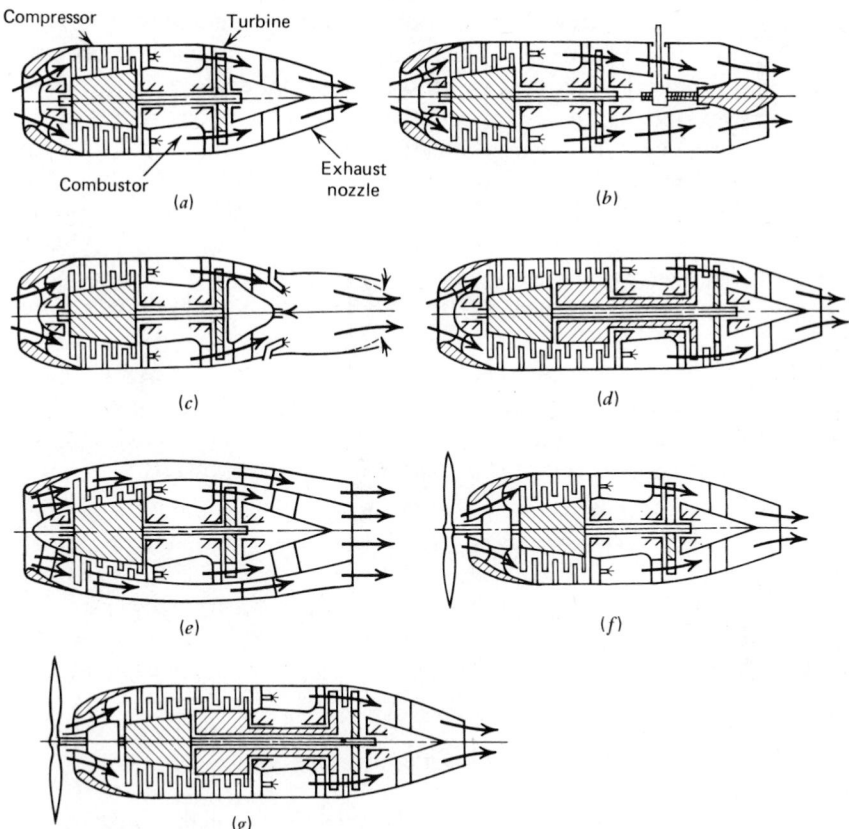

Figure 14.1 Turbine engine types.

14.1 BASIC JET ENGINE TYPES AND OPERATION

Since its original invention, the jet engine has evolved into a variety of configurations suited to particular classes of applications. Figure 14.1, based on an original Russian reference,[1] displays some of the more common versions, with 14.1*a* being the basic single-spool engine in which inlet air is compressed, energy added by fuel combustion, sufficient energy to drive the compressor extracted by a turbine, and the remaining energy converted to propulsive force by expansion in an exhaust nozzle. In 14.1*b* the exhaust nozzle is made of variable area to allow a second control input in addition to the basic control of fuel flow rate. An afterburner is added in 14.1*c* to provide additional control in the form

[1]Zalmanzon, L. A. and B. A. Cherkasov, *Control of Gas-Turbine and Ramjet Engines* (Government Defense Industry Press, Moscow, 1956), NASA TT F-41, 1961.

of short bursts of thrust needed under certain operating conditions. The dual-rotor ("two-spool") type of engine (14.1d) was devised to minimize compressor surge problems present in high-pressure-ratio single-spool engines; its dynamics are more complex since there are two inertias rotating at different speeds. By accelerating air not used for combustion, the ducted-fan engine of 14.1e can obtain additional thrust at relatively low fuel cost. Finally, 14.1f and 14.1g show single- and dual-spool turboprop engines in which the turbine extracts almost all the energy to drive a propellor, with the exhaust jet providing negligible propulsive force. Other variations, not shown in Fig. 14.1, involve compressors and/or turbines whose blades have adjustable angles, giving additional control inputs.

A brief summary[2] of engine steady-state operation will be useful at this point. The basic ideal engine cycle, called the Brayton cycle, is shown in Fig. 14.2. The isentropic compression from state 1 to 3 is accomplished by ram compression in the inlet plus compressor action. For the compressor:

$$\frac{h_3 - h_2}{h_2} = \frac{1}{\eta_c}\left[\left(\frac{P_3}{P_2}\right)^{(k-1)/k} - 1\right] \tag{14.1}$$

where

$$\eta_c \triangleq \text{compressor efficiency}$$

$$k \triangleq \text{specific heat ratio}$$

$$P_2, P_3 \triangleq \text{compressor inlet and outlet pressures}$$

$$h_2, h_3 \triangleq \text{compressor inlet and outlet enthalpies per unit mass}$$

The enthalpy rise $h_3 - h_2$ represents energy added to the fluid by the compressor. For the combustor:

$$\dot{m}_f h_c \eta_b = \dot{m}_a c_{pa}(T_4 - T_3) + \dot{m}_f c_{pf}(T_4 - T_3) \tag{14.2}$$

where

$$\dot{m}_f, \dot{m}_a \triangleq \text{mass flow rates of fuel and air}$$

$$h_c \triangleq \text{fuel enthalpy of combustion}$$

$$\eta_b \triangleq \text{burner efficiency}$$

$$c_{pf}, c_{pa} \triangleq \text{specific heats of fuel and air}$$

$$T_3, T_4 \triangleq \text{burner inlet and outlet temperatures}$$

[2]Sobey, A. J. and A. M. Suggs, *Control of Aircraft and Missile Powerplants*, John Wiley and Sons, N.Y., 1963, pp. 31, 416.

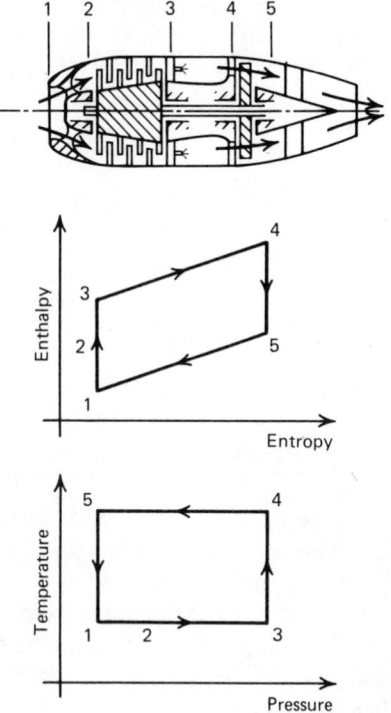

Figure 14.2   Simple single-spool engine and cycle.

The temperature rise $T_4 - T_3$ indicates a further increase in enthalpy across the combustor. Since the turbine must extract sufficient energy from the flow to run the compressor, there is a drop in enthalpy across the turbine:

$$\frac{h_4 - h_5}{h_4} = \eta_T \left[1 - \left(\frac{P_5}{P_4} \right)^{(k-1)/k} \right] \tag{14.3}$$

where

$$\eta_T \triangleq \text{turbine efficiency}$$

$$h_4, h_5 \triangleq \text{turbine inlet and outlet enthalpies per unit mass}$$

$$P_4, P_5 \triangleq \text{turbine inlet and outlet pressures}$$

If the exhaust nozzle is unchoked:

$$\dot{m}_f + \dot{m}_a = CA\sqrt{2\rho(P_5 - P_0)} \tag{14.4}$$

where

$$C \triangleq \text{nozzle discharge coefficient}$$

$$A \triangleq \text{nozzle flow area}$$

$$\rho \triangleq \text{density of exhaust gas}$$

$$P_0 \triangleq \text{ambient pressure}$$

For the more common choked condition:

$$\dot{m}_f + \dot{m}_a = \rho A V_s$$

where

$$V_s \triangleq \text{sonic velocity}$$

For turboprop engines, engine rotary speed is the response variable of major interest since propellor thrust depends directly on it; similarly for turbojet engines since jet thrust depends on gas flow rate, which in turn depends on compressor (engine) speed. For operation at any steady speed, the enthalpy rise across the compressor must just match the turbine enthalpy drop. When turbine enthalpy drop exceeds compressor enthalpy rise, excess shaft power is available to accelerate the engine to a higher speed. Using this energy point of view one can express engine speed changes by the equation

power to accelerate inertia = (turbine power production)

$$\qquad\qquad\qquad\qquad - (\text{compressor power consumption})$$

$$(\text{angular velocity})(\text{net torque}) = (\dot{m}_f + \dot{m}_a)(h_4 - h_5) - \dot{m}_a(h_3 - h_2) \qquad (14.5)$$

$$(\omega)(J\dot{\omega}) = (\dot{m}_f + \dot{m}_a)(h_4 - h_5) - \dot{m}_a(h_3 - h_2) \qquad (14.6)$$

where

$$\omega \triangleq \text{engine angular velocity}$$

$$J \triangleq \text{moment of inertia}$$

Thus a step increase in fuel flow rate results in an increase in turbine power output which accelerates the engine toward a higher equilibrium speed. Then compressor power demand will again match turbine output since the power to drive the compressor increases with compressor speed.

In the above set of equations, only the rotor speed relation is a differential equation, the rest are algebraic and thus assume the variables involved are related on an instantaneous basis. Experience shows[3] that such a viewpoint, which assumes that flow and thermodynamic processes occur much more quickly than rotor speed changes, and thus can be modeled on a quasi-static basis, leads to practically usable engine models for situations where a frequency range up to about 10 Hz is sufficient. When higher frequency ranges (50–100 Hz) are of interest, such as those related to inlet flow and shock-position dynamics, then engine flow, volume-charging, and combustion system dynamics can no longer be neglected. Very comprehensive and complex engine models have been formulated by both engine manufacturers and various government agencies to deal with these more difficult cases. The models developed by engine manufacturers[4] are often proprietary and thus largely inaccessible to a general reader. However, unclassified government studies[5] are available which give a quite detailed explanation of the necessary techniques, including experimental verification of their accuracy.

Development of both the low- and high-frequency analytical models described above is beyond the scope of this short chapter. Rather, in the next sections we explain a semiempirical modeling approach successfully employed to obtain models of existing engines.

14.2 SEMIEMPIRICAL, LOW-FREQUENCY, LINEARIZED, SMALL-SIGNAL MODELS FOR SHAFT-POWER TURBINES

The earliest studies[6] of turbine engine dynamics employed a semiempirical approach still useful in certain applications today. Experimental testing of engine response to step inputs of fuel flow gave the typical response of Fig. 14.3.[7] Such test results, showing obvious single-time-constant trends (especially clear in engine speed) for the response variables, encourage use of the low-frequency model that considers rotor inertia as the major energy storage element. Model development begins by listing all the inputs (main fuel flow, exhaust nozzle area, afterburner fuel flow, compressor blade angle, etc.) and outputs (engine speed, thrust, compressor output pressure, etc.) pertinent to the particular engine. Each output variable is assumed, for steady-state operation, to be some (nonlinear) function of all the input variables. For small changes away from an equilibrium operating point a linearized model is possible using partial

[3]Kerrebrock, J. L., *Aircraft Engines and Gas Turbines*, The MIT Press, Cambridge, 1977, p. 215.

[4]100-PW-100(F100/F401) Digital Dynamic Simulation User's Manual, Rept. PWA FR-3794B, Pratt & Whitney Aircraft, 1974; Durand, H. P., Simulation of Jet Engine Transient Performance, ASME Paper 65-WA/MD-16, 1965.

[5]Seldner, K., Mihaloew, J. R., and R. J. Blaha, Generalized Simulation Technique for Turbojet Engine System Analysis, NASA TN D-6610, 1972.

[6]Boksenbom, A. S. and R. Hood, General Algebraic Method Applied to Control Analysis of Complex Engine Types, NACA Rept. 980, 1950.

[7]Craig, R. T., Vasu, G., and R. D. Schmidt, Dynamic Characteristics of a Single-Spool Turbojet Engine, NACA RM E53C17, 1953.

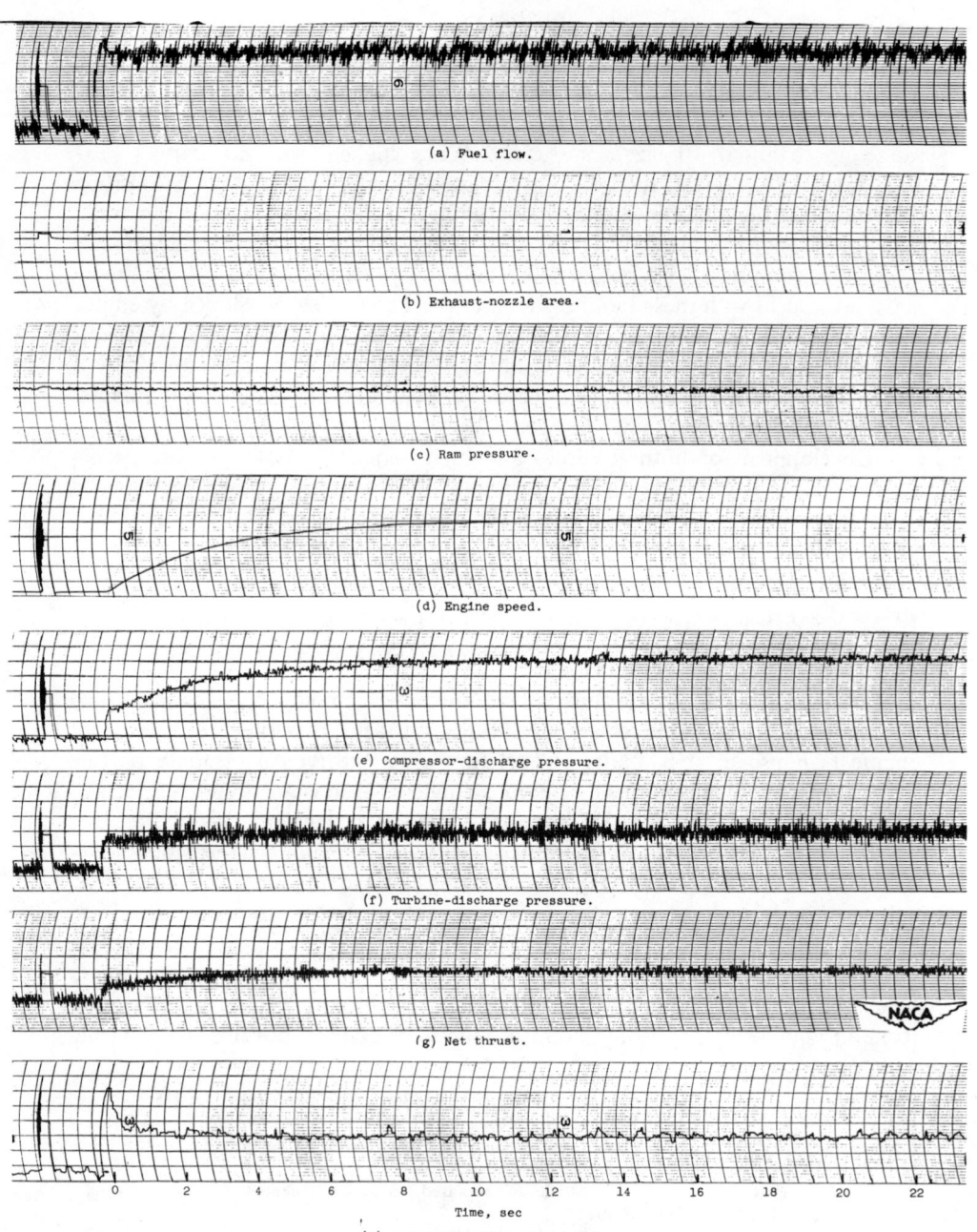

(a) Fuel flow.

(b) Exhaust-nozzle area.

(c) Ram pressure.

(d) Engine speed.

(e) Compressor-discharge pressure.

(f) Turbine-discharge pressure.

(g) Net thrust.

(h) Turbine-discharge temperature.

Time, sec

Figure 14.3 Measured step response of engine to fuel flow input.

derivatives of the chosen output variable with respect to the various inputs. The partial derivatives' numerical values can be obtained from the slopes of experimental graphs relating outputs to inputs for steady-state operation or from physical analysis such as that outlined in Section 14.1.

For engines designed to provide a shaft-power output rather than jet thrust (turboprop, helicopter, marine propulsion drives, etc.), a relatively straight-forward modeling approach[8] is possible. Taking a turboprop engine as example, suppose we are interested in relations between engine speed, turbine-outlet temperature, fuel flow, and propellor blade angle setting. The shaft output torque Q_e produced by the engine is some function of fuel flow $\dot{m}_f$ and engine speed ω:

$$Q_e = Q_e(\dot{m}_f, \omega) \qquad (14.7)$$

The torque Q_p required to drive the propellor depends on blade angle β and propellor speed ω_p,

$$Q_p = Q_p(\beta, \omega_p) \qquad (14.8)$$

where $\omega = N\omega_p$, N being the gear ratio between engine and propellor. When Q_e differs from Q_p/N, an unbalanced torque $Q_e - Q_p/N$ is available to accelerate the total rotary inertia:

$$\Delta\omega = \frac{1}{J_e + J_p/N^2} \int \Delta\left(Q_e - \frac{Q_p}{N}\right) dt \qquad (14.9)$$

where

$$J_e \triangleq \text{engine-shaft inertia}$$

$$J_p \triangleq \text{propellor-shaft inertia}$$

Since torques Q_e and Q_p are each accessible to direct measurement during separate dynamometer testing of the engine and propellor, experimental curves defining the relations given in (14.7) and (14.8) may be obtained for existing hardware.

These experimental curves provide needed numerical values of partial derivatives in the linearized relations:

$$\Delta Q_e \approx \frac{\partial Q_e}{\partial \dot{m}_f}\bigg|_{\omega=\omega_0} \Delta\dot{m}_f + \frac{\partial Q_e}{\partial \omega}\bigg|_{\dot{m}_f = \dot{m}_{f0}} \Delta\omega \qquad (14.10)$$

$$\Delta Q_p \approx \frac{\partial Q_p}{\partial \beta}\bigg|_{\omega_p=\omega_{p0}} \Delta\beta + \frac{\partial Q_p}{\partial \omega_p}\bigg|_{\beta = \beta_0} \Delta\omega_p \qquad (14.11)$$

[8]Ketchum, J. R. and R. T. Craig, Simulation of Linearized Dynamics of Gas-Turbine Engines, NACA TN 2826, 1952.

where ω_0, $\dot{m}_{f0}$, ω_{p0}, and β_0 are the steady-state operating point values about which the small perturbations $\Delta\omega$, $\Delta\dot{m}_f$, and so on take place. We can now express the accelerating torque as:

$$\Delta\left(Q_e - \frac{Q_p}{N}\right) = \frac{\partial Q_e}{\partial \dot{m}_f}\Delta\dot{m}_f - \frac{1}{N}\frac{\partial Q_p}{\partial \beta}\Delta\beta + \left(\frac{\partial Q_e}{\partial\omega} - \frac{1}{N^2}\frac{\partial Q_p}{\partial\omega}\right)\Delta\omega \quad (14.12)$$

Similarly, turbine outlet temperature T_5, which is a function of $\dot{m}_f$ and ω, can be linearized as

$$\Delta T_5 \approx \frac{\partial T_5}{\partial \dot{m}_f}\bigg|_{\omega=\omega_0}\Delta\dot{m}_f + \frac{\partial T_5}{\partial\omega}\bigg|_{\dot{m}_f=\dot{m}f_0}\Delta\omega \quad (14.13)$$

The engine model relating response variables ω and T_5 to inputs $\dot{m}_f$ and β is conventionally represented by the matrix diagram of Fig. 14.4. Here inputs enter vertically, are multiplied by the coefficients in the boxes for the row corresponding to the output of interest, and the resulting terms are summed to produce that output. For example:

$$\left[-\frac{1}{N}\frac{\partial Q_p}{\partial\beta}\Delta\beta + \frac{\partial Q_e}{\partial \dot{m}_f}\Delta\dot{m}_f + \left(\frac{\partial Q_e}{\partial\omega} - \frac{1}{N^2}\frac{\partial Q_p}{\partial\omega}\right)\Delta\omega\right]\frac{1}{J_t s} = \Delta\omega \quad (14.14)$$

which is easily reduced to the form:

$$(\tau s + 1)\Delta\omega = K_{\beta\omega}\Delta\beta + K_{\dot{m}\omega}\Delta\dot{m}_f \quad (14.15)$$

For T_5 we get:

$$\frac{\partial T_5}{\partial \dot{m}_f}\Delta m_f + \frac{\partial T_5}{\partial\omega}\Delta\omega = \Delta T_5$$

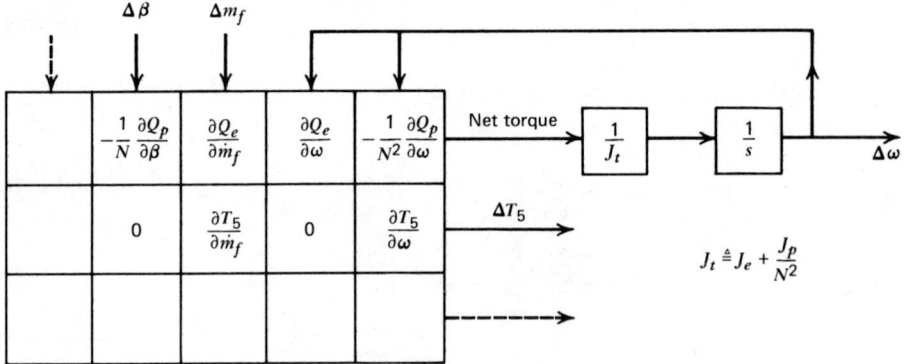

Figure 14.4 Matrix representation of shaft-power engine dynamics.

which, upon substituting for $\Delta\omega$ from (14.15), leads to:

$$(\tau s + 1)\Delta T_5 = K_{\beta T}\Delta\beta + K_{\dot{m}T}(\tau_T s + 1)\Delta\dot{m}_f \qquad (14.16)$$

This scheme is easily extended to include more inputs and/or outputs, as indicated by the dashed lines in Fig. 14.4. Experiments[9] have validated this modeling scheme as a practical tool.

14.3   SEMIEMPIRICAL, LOW-FREQUENCY, LINEARIZED, SMALL-SIGNAL MODELS FOR TURBOJET ENGINES

The modeling technique just explained in Section 14.2 encounters some difficulty when applied to jet engines, where no external power takeoff shaft is accessible for torque measurements; that is, the compressor "internally" absorbs all the turbine torque. While the basic linearization scheme is still usable, a rather roundabout method[10] of implementation is required. Let us take a simple single-spool engine with fuel flow and exhaust nozzle area as the only two control inputs and engine speed and turbine outlet temperature as outputs. The net torque Q available for acceleration still exists and is taken to be some function of fuel flow, speed and exhaust nozzle area A_n:

$$Q = Q(\dot{m}_f, \omega, A_n) \qquad (14.17)$$

However, (14.17) cannot be interpreted, as was (14.7), in terms of a steady-state dynamometer test where Q is measured for various values of $\dot{m}_f$, A_n, and ω since Q is not accessible for measurement and also is zero for any steady-state condition. While Eq. (14.17) is not susceptible to direct experimental evaluation, it is nonetheless a valid relation among the variables and may be used in further manipulations:

$$\Delta Q \approx \left.\frac{\partial Q}{\partial \dot{m}_f}\right|_{\substack{A_n = A_{n0} \\ \omega = \omega_0}} \Delta\dot{m}_f + \left.\frac{\partial Q}{\partial A_n}\right|_{\substack{\dot{m}_f = \dot{m}_{f0} \\ \omega = \omega_0}} \Delta A_n + \left.\frac{\partial Q}{\partial \omega}\right|_{\substack{\dot{m}_f = \dot{m}_{f0} \\ A_n = A_{n0}}} \Delta\omega \qquad (14.18)$$

If we divide this through by the coefficient of $\Delta\omega$, the coefficients of $\Delta\dot{m}_f$ and ΔA_n become:

$$\frac{\partial Q/\partial \dot{m}_f|_{\omega_0}^{A_{n0}}}{\partial Q/\partial\omega|_{A_{n0}}^{\dot{m}_{f0}}} = -\left.\frac{\partial\omega}{\partial\dot{m}_f}\right|_{A_{n0}} \qquad \frac{\partial Q/\partial A_n|_{\omega_0}^{\dot{m}_{f0}}}{\partial Q/\partial\omega|_{A_{n0}}^{\dot{m}_{f0}}} = -\left.\frac{\partial\omega}{\partial A}\right|_{\dot{m}_{f0}} \qquad (14.19)$$

[9]Taylor, B. L. and F. L. Oppenheimer, Investigation of Frequency Response of Engine Speed for a Typical Turbine-Propellor Engine, NACA Rept. 1017, 1951.
[10]Ketchum and Craig, *op. cit.*

The partial derivatives $\partial\omega/\partial\dot{m}_f$ and $\partial\omega/\partial A$ *can* now be numerically evaluated from steady-state experiments.

Equation (14.18) can now be rewritten as

$$-\frac{\Delta Q}{\partial Q/\partial\omega} = \frac{\partial\omega}{\partial\dot{m}_f}\Delta\dot{m}_f + \frac{\partial\omega}{\partial A_n}\Delta A_n - \Delta\omega \qquad (14.20)$$

Newton's law for the rotating inertia J gives:

$$\Delta\omega = \left(\frac{1}{Js}\right)\Delta Q = \left(-\frac{\partial Q/\partial\omega}{J}\right)\frac{1}{s}\left(-\frac{\Delta Q}{\partial Q/\partial\omega}\right) \qquad (14.21)$$

We now define the engine time constant τ by

$$\tau \triangleq -\frac{J}{\partial Q/\partial\omega\big|^{\dot{m}_{f0}}_{A_{n0}}} \qquad (14.22)$$

and since the partial derivative required cannot be directly measured, τ itself must be measured from appropriate *dynamic* testing. Note that for a turboprop engine (Eq. (14.15)) the time constant can be found from *steady-state* measurements only, assuming inertia is known. By combining (14.21) with (14.20), the model for engine speed response is obtained:

$$\Delta\omega = \frac{1}{\tau s}\left(\frac{\partial\omega}{\partial\dot{m}_f}\Delta\dot{m}_f + \frac{\partial\omega}{\partial A_n}\Delta A_n - \Delta\omega\right) \qquad (14.23)$$

$$(\tau s+1)\Delta\omega = K_{\dot{m}_f\omega}\Delta\dot{m}_f + K_{A_n\omega}\Delta A_n \qquad (14.24)$$

For any other desired response variables, such as turbine outlet temperature T_5, the model can be extended by writing

$$T_5 = T_5(\dot{m}_f, A_n, Q) \qquad (14.25)$$

$$\Delta T_5 \approx \frac{\partial T_5}{\partial\dot{m}_f}\bigg|_{\substack{A_n=A_{n0}\\Q=Q_0}}\Delta m_f + \frac{\partial T_5}{\partial A_n}\bigg|_{\substack{\dot{m}_f=\dot{m}_{f0}\\Q=Q_0}}\Delta A_n + \frac{\partial T_5}{\partial Q}\bigg|_{\substack{\dot{m}_f=\dot{m}_{f0}\\A_n=A_{n0}}}\Delta Q \qquad (14.26)$$

In (14.26) the right-most term can be written as

$$\frac{\partial T_5}{\partial Q}\left(-\frac{\partial Q}{\partial\omega}\right)\left(\frac{\Delta Q}{-\partial Q/\partial\omega}\right) = -\frac{\partial T_5}{\partial\omega}\bigg|_{\substack{\dot{m}_f=\dot{m}_{f0}\\A_n=A_{n0}}}\left(\frac{\Delta Q}{-\partial Q/\partial\omega}\right) \qquad (14.27)$$

The term $\partial T_5/\partial\omega$ must be found from dynamic testing, using methods to be explained later. The term $-\Delta Q(\partial Q/\partial\omega)$ is available from Eq. (14.20). In (14.26), $\partial T_5/\partial\dot{m}_f$ and $\partial T_5/\partial A_n$ can be obtained from steady-state tests, where $Q_0=0$.

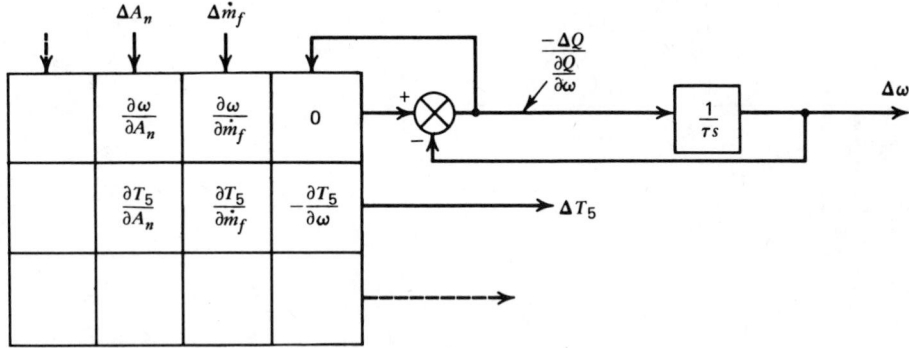

Figure 14.5   Matrix representation of turbojet engine dynamics.

Combining the above equations gives the model for turbine outlet temperature response as

$$(\tau s + 1)\,\Delta T_5 = K_{\dot{m}_f T}(\tau_{Tm} s + 1)\,\Delta \dot{m}_f + K_{A_n T}(\tau_{TA} s + 1)\,\Delta A_n \qquad (14.28)$$

Should the engine model require extension to more inputs and/or outputs, Eq. (14.17) would be augmented with the additional inputs and a new equation of form

$$O = O(\dot{m}_f, A_n, \text{additional inputs}, Q) \qquad (14.29)$$

written for each output. Figure 14.5 shows a conventional matrix representation of such engine models.

To find numerical values for terms in the right-most column of Fig. 14.5, such as $\partial T_5 / \partial \omega$ in the present example, the following method[11] has been found useful. In Eq. (14.28) the time constants τ_{Tm} and τ_{TA} can be expressed in terms of τ by

$$\tau_{Tm} \triangleq a\tau, \qquad \tau_{TA} \triangleq b\tau \qquad (14.30)$$

where the numbers a and b are called *rise ratios*, since for a step input of, say, $\Delta \dot{m}_f$ in (14.28), ΔT_5 would respond as in Fig. 14.6 (see Fig. 14.3 for some actual experimental data). In Eq. (14.28)

$$a = 1 - \frac{\dfrac{\partial \omega}{\partial \dot{m}_f}\dfrac{\partial T_5}{\partial \omega}}{\dfrac{\partial T_5}{\partial \dot{m}_f}} \qquad (14.31)$$

[11]Craig, Vasu, and Schmidt, *op. cit.*

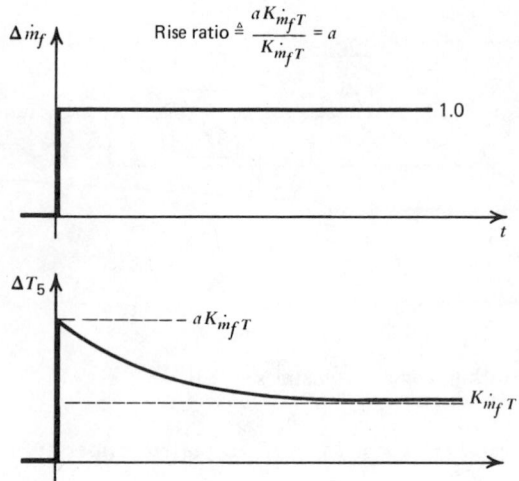

Figure 14.6 Definition of rise ratios.

which can be solved for $\partial T_5/\partial\omega$ as

$$\frac{\partial T_5}{\partial\omega} = \frac{\dfrac{\partial T_5}{\partial\dot{m}_f}(1-a)}{\dfrac{\partial\omega}{\partial\dot{m}_f}} \tag{14.32}$$

and allows calculation of $\partial T_5/\partial\omega$ from measured values of rise ratio "a" since the other two partial derivatives are known.

14.4 CONCLUDING REMARKS

In addition to step-function testing (Fig. 14.3), frequency-response and random-signal methods have been used to study turbine engine dynamics. Figure 14.7[12] shows data for the frequency response of engine speed to exhaust nozzle area. We see that our first-order model predicts amplitude ratio almost perfectly but that phase angle accuracy becomes poor beyond about 2 Hz, requiring addition of a 0.021 sec dead time to match the measured data. Dead times in such engines are attributed to combustion lags, flow transport times, wave propagation effects, and so on. Due to the complexity of combustion phenomena, empirical dynamic models for these aspects of engine operation are usually necessary even in the complex analytical models mentioned in Section 14.1. One

[12]Wenzel, L. M., Hart, C. E., and R. T. Craig, Experimental Comparison of Speed-Fuel-Flow and Speed Area Controls on a Turbojet Engine, NACA TN 3926, 1957.

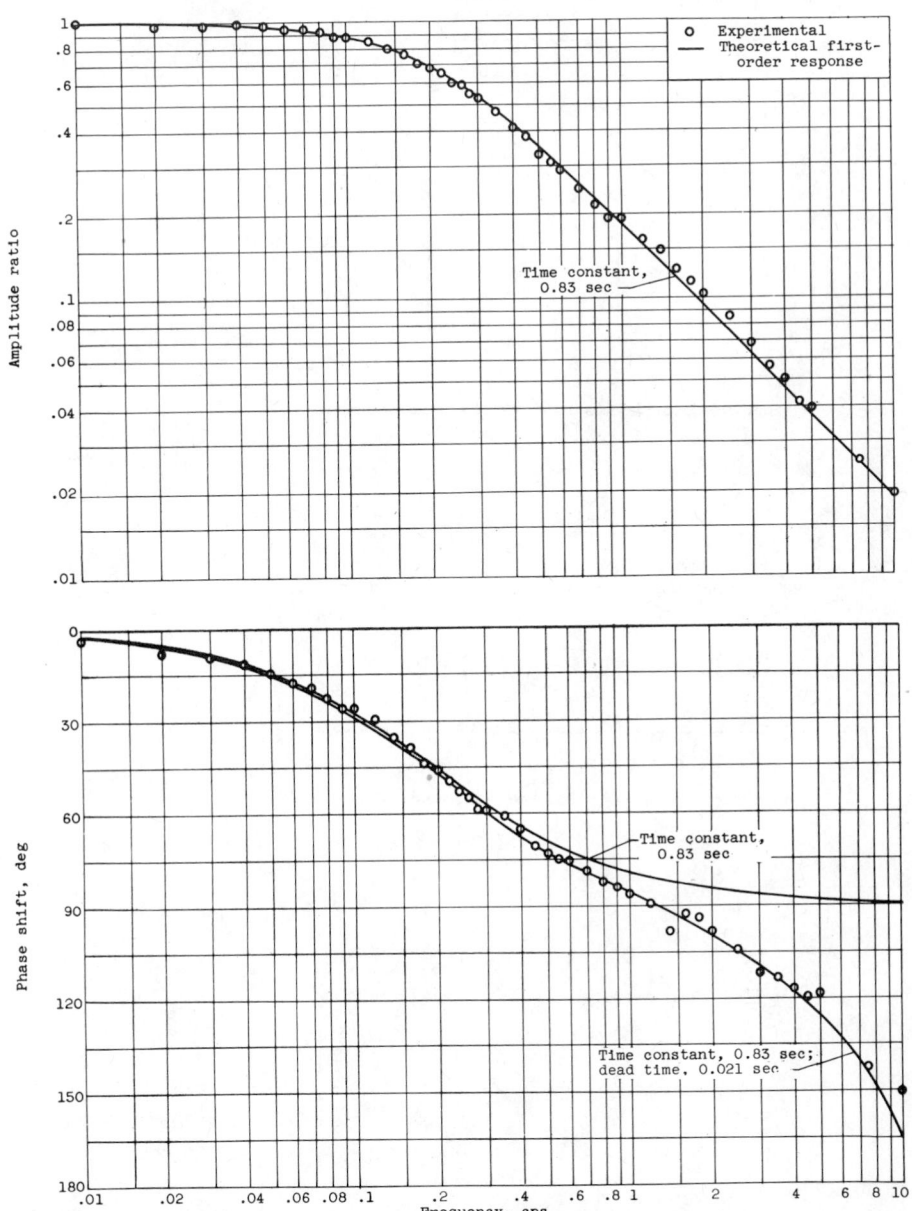

Figure 14.7   Frequency response of jet engine to exhaust nozzle area input.

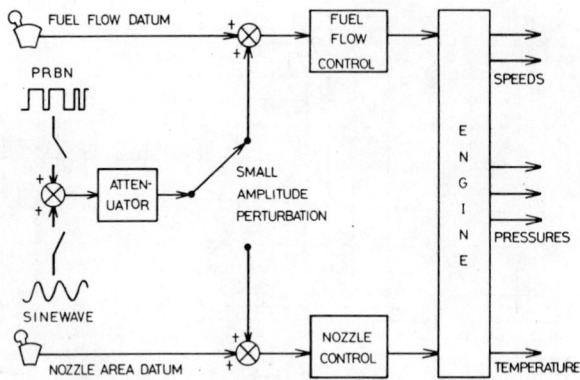

Hardware layout for comparing PRBN with sinewave testing

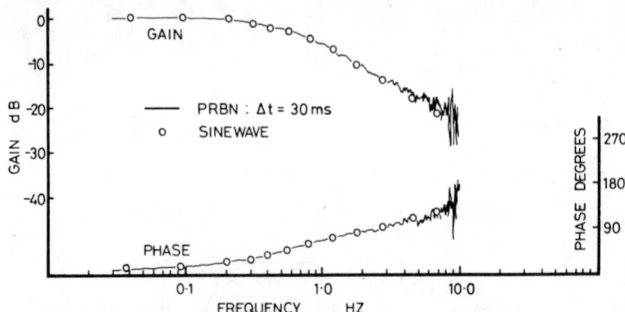

Frequency response of LP rotor speed to actual fuel flow

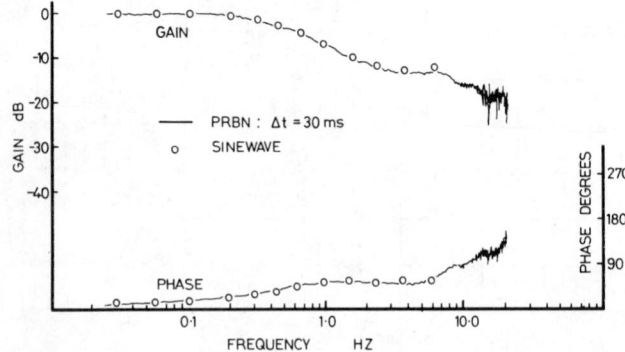

Frequency response of HP compressor delivery pressure to actual fuel flow

Figure 14.8 Comparison of sinusoidal and pseudorandom binary testing of turbine engines.

such study[13] using sweep-frequency test methods (logarithmic sweep rate of one decade per minute, from 0.5 to 130 Hz) found the following empirical relation between combustor pressure P_c and fuel-nozzle pressure P_f:

$$\frac{P_c}{P_f}(s) = \frac{Ke^{-0.0029s}}{(0.0016s + 1)\left(\dfrac{s^2}{(110\pi)^2} + \dfrac{1.6s}{110\pi} + 1\right)} \qquad (14.33)$$

A recent study[14] of marine gas turbine dynamics using pseudo-random binary noise inputs gave results (Fig. 14.8) of accuracy comparable to sinusoidal testing but reduced test time from 30 minutes to 5 minutes. When the detailed nonlinear digital simulation models of an engine are available, they may be used as a starting point for simplified linear models that might be desirable for certain types of engine dynamics studies. Systematic techniques for such linearization and order reduction have been developed.[15]

BIBLIOGRAPHY

1 Arpasi, D. J. and W. M. Bruton, Method of Discrete Modeling and Its Application to Estimation of TF30 Engine Variables, NASA TM X-3443, 1976.

2 Bammert, K. and G. Krey, Dynamic Behavior and Control of Single-Shaft Closed-Cycle Gas Turbines, ASME Paper 71-GT-16, 1971.

3 Cottington, R. V., The Design of a Digital Three-Term Controller as a Turbojet Engine Speed Governor Using Digital Simulation Methods, R. & M. No. 3615, Nat'l. Gas Turbine Establishment, Hampshire, England, 1970.

4 "Gas Turbine Fuel Controls Analysis and Design," *Soc. of Automotive Eng., Prog. in Tech.* Vol. 9, Pergamon Press, N.Y., 1965.

5 Goyal, M. R., Simulation of a Turbocharged Diesel Engine to Predict the Transient Response, ASME Paper 78-DGP-11, 1978.

6 Hart, C. E. and D. J. Arpasi, Frequency Response and Transfer Functions of a Nuclear Rocket Engine System Obtained from Analog Computer Simulation, NASA TN D-3979, 1967.

7 Klarhoefer, C., Optimization of the Idling and Acceleration Characteristics of a Vehicular Gas Turbine by Analog Simulation, ASME Paper 74-GT-103, 1974.

8 Larrowe, V. L., Spencer, M. M., and M. Tribus, "A Dynamic Performance Computer for Gas-Turbine Engines," *ASME Trans.*, Oct. (1957).

[13]Szuch, J. R., Paulovich, F. J., and W. R. Brunton, Study of Turbojet Combustor Dynamics Using Sweep-Frequency Data, NASA TND-6084, 1970.
[14]Cottington, R. V. and C. B. Pease, Dynamic Response Testing of Gas Turbines, ASME Paper 78-GT-31, 1978.
[15]Weinberg, M. S. and G. R. Adams, Low-Order Linearized Models of Turbine Engines, Tech. Rept. ASD-TR-75-24, Aeronautical Sys. Div., Wright-Patterson AFB, Ohio, 1975.

9 Michael, G. J. and F. A. Farrar, Stochastic Regulation of Nonlinear Multivariable Dynamic Systems, UTRC Rept. R76-941620-4, United Tech. Res. Ctr., E. Hartford, Conn., 1976.

10 Sellers, J. and F. Teren, Generalized Dynamic Engine Simulation Techniques for the Digital Computer, NASA TM X-71552, 1974.

11 Scullen, R. S. and R. J. Hames, Computer Simulation of the GM Unit Injector, SAE Paper 780161, 1978.

12 Szuch, J. R., Analysis of Integral Lift-Fan Engine Dynamics, NASA TM X-2691, 1973.

13 Szuch, J. R., Wenzel, L. M., and R. J. Baumbick, Investigation of the Starting Characteristics of the M-1 Rocket Engine Using the Analog Computer, NASA TN D-3136, 1965.

14 Webb, J. A. and M. O. Dustin, Analysis of a Stability Valve System for Extending the Dynamic Range of a Supersonic Inlet, NASA TM X-3219, 1975.

15 Willoh, R., A Mathematical Analysis of Supersonic Inlet Dynamics, NASA TN D-4969, 1968.

16 Zalmanzon, L. A. and B. A. Cherkasov, Control of Gas-Turbine and Ramjet Engines, NASA TT F-41, 1961.

PROBLEMS

14.1 Explain how the techniques of Section 14.2 could be applied to develop a model for a turbine used to drive an electrical generator subject to varying electrical load.

14.2 For Equations (14.15) and (14.16), fill in the steps omitted in their derivation and thus obtain definitions of the various gains and time constants.

14.3 Modify the analysis of Section 14.2 to accommodate an engine with an additional control input in the form of variable-angle compressor vanes and an additional output, compressor discharge pressure.

14.4 Modify the analysis of Section 14.3 to accommodate an engine with an additional control input in the form of variable-angle compressor vanes and an additional output, compressor discharge pressure.

APPENDIX A

LAPLACE TRANSFORM FUNCTION-TRANSFORM PAIRS

	$F(s)$	$f(t)$
1	s	$\delta'(t)$, first derivative of unit impulse
2	1	$\delta(t)$, unit impulse
3	$1/s$	1, unit step, $u(t)$
4	$1/s^2$	t
5	$\dfrac{1}{s^n}$	$\dfrac{1}{(n-1)!}t^{n-1}$
6	$\dfrac{1}{s+a}$	e^{-at}
7	$\dfrac{1}{s(s+a)}$	$\dfrac{1-e^{-at}}{a}$
8	$\dfrac{1}{s^2(s+a)}$	$\dfrac{e^{-at}+at-1}{a^2}$
9	$\dfrac{s+a_0}{s^2(s+a)}$	$\dfrac{a_0-a}{a^2}e^{-at}+\dfrac{a_0}{a}t+\dfrac{a-a_0}{a^2}$
10	$\dfrac{s^2+a_1s+a_0}{s^2(s+a)}$	$\dfrac{a^2-a_1a+a_0}{a^2}e^{-at}+\dfrac{a_0}{a}t$ $+\dfrac{a_1a-a_0}{a^2}$
11	$\dfrac{1}{(s+a)(s+b)}$	$\dfrac{e^{-at}-e^{-bt}}{b-a}$
12	$\dfrac{s+c}{(s+a)(s+b)}$	$\dfrac{(c-a)e^{-at}-(c-b)e^{-bt}}{b-a}$
13	$\dfrac{1}{s(s+a)(s+b)}$	$\dfrac{1}{ab}+\dfrac{be^{-at}-ae^{-bt}}{ab(a-b)}$
14	$\dfrac{s+c}{s(s+a)(s+b)}$	$\dfrac{c}{ab}+\dfrac{c-a}{a(a-b)}e^{-at}$ $+\dfrac{c-b}{b(b-a)}e^{-bt}$
15	$\dfrac{s^2+a_1s+a_0}{s(s+a)(s+b)}$	$\dfrac{a_0}{ab}+\dfrac{a^2-a_1a+a_0}{a(a-b)}e^{-at}$ $-\dfrac{b^2-a_1b+a_0}{b(b-a)}e^{-bt}$

16    $\dfrac{1}{s^2(s+a)(s+b)}$ $\dfrac{t}{ab} - \dfrac{a+b}{(ab)^2} + \dfrac{1}{a-b}\left[\dfrac{e^{-bt}}{b^2} - \dfrac{e^{-at}}{a^2}\right]$

17    $\dfrac{s+a_0}{s^2(s+a)(s+b)}$ $\dfrac{a_0-a}{a^2(b-a)}e^{-at} + \dfrac{a_0-b}{b^2(a-b)}e^{-bt}$

$+ \dfrac{a_0}{ab}t + \dfrac{ab-a_0(a+b)}{(ab)^2}$

18    $\dfrac{s^2+a_1s+a_0}{s^2(s+a)(s+b)}$ $\dfrac{a^2-a_1a+a_0}{a^2(b-a)}e^{-at}$

$+ \dfrac{b^2-a_1b+a_0}{b^2(a-b)}e^{-bt}$

$+ \dfrac{a_0}{ab}t + \dfrac{a_1ab-a_0(a+b)}{(ab)^2}$

19    $\dfrac{1}{(s+a)(s+b)(s+c)}$ $\dfrac{e^{-at}}{(b-a)(c-a)} + \dfrac{e^{-bt}}{(a-b)(c-b)}$

$+ \dfrac{e^{-ct}}{(a-c)(b-c)}$

20    $\dfrac{s+a_0}{(s+a)(s+b)(s+c)}$ $\dfrac{a_0-a}{(b-a)(c-a)}e^{-at}$

$+ \dfrac{a_0-b}{(a-b)(c-b)}e^{-bt}$

$+ \dfrac{a_0-c}{(a-c)(b-c)}e^{-ct}$

21    $\dfrac{s^2+a_1s+a_0}{(s+a)(s+b)(s+c)}$ $\dfrac{a^2-a_1a+a_0}{(b-a)(c-a)}e^{-at}$

$+ \dfrac{b^2-a_1b+a_0}{(a-b)(c-b)}e^{-bt}$

$+ \dfrac{c^2-a_1c+a_0}{(a-c)(b-c)}e^{-ct}$

22    $\dfrac{1}{s^2+a^2}$ $\dfrac{\sin at}{a}$

23    $\dfrac{s}{s^2+a^2}$ $\cos at$

24    $\dfrac{1}{(s+a)^2+b^2}$ $\dfrac{1}{b}e^{-at}\sin bt$

25     $\dfrac{s+a_0}{(s+a)^2+b^2}$ $\dfrac{1}{b}[(a_0-a)^2+b^2]^{1/2}e^{-at}\sin(bt+\phi)$

$$\phi \triangleq \tan^{-1}\dfrac{b}{a_0-a}$$

26     $\dfrac{1}{s[(s+a)^2+b^2]}$ $\dfrac{1}{b_0^2}+\dfrac{1}{bb_0}e^{-at}\sin(bt-\phi)$

$$\phi \triangleq \tan^{-1}\dfrac{b}{-a}, \qquad b_0 \triangleq \sqrt{a^2+b^2}$$

27     $\dfrac{s+a_0}{s[(s+a)^2+b^2]}$ $\dfrac{a_0}{a^2+b^2}+\dfrac{1}{b\sqrt{a^2+b^2}}\times$

$$[(a_0-a)^2+b^2]^{1/2}e^{-at}\sin(bt+\phi)$$

$$\phi \triangleq \tan^{-1}\dfrac{b}{a_0-a}-\tan^{-1}\dfrac{b}{-a}$$

28     $\dfrac{1}{s^2[(s+a)^2+b^2]}$ $\dfrac{1}{a^2+b^2}\left[t-\dfrac{2a}{a^2+b^2}+\dfrac{1}{b}e^{-at}\sin(bt-\phi)\right]$

$$\phi \triangleq 2\tan^{-1}\dfrac{b}{-a}$$

29     $\dfrac{1}{(s+c)[(s+a)^2+b^2]}$ $\dfrac{e^{-ct}}{(c-a)^2+b^2}$

$$+\dfrac{1}{b[(c-a)^2+b^2]^{\frac{1}{2}}}e^{-at}\sin(bt-\phi)$$

$$\phi \triangleq \tan^{-1}\dfrac{b}{c-a}$$

30     $\dfrac{1}{s(s+c)[(s+a)^2+b^2]}$ $\dfrac{1}{c\sqrt{a^2+b^2}}-\dfrac{1}{c[(a-c)^2+b^2]}e^{-ct}$

$$+\dfrac{1}{b\sqrt{a^2+b^2}\,[(c-a)^2+b^2]^{\frac{1}{2}}}e^{-at}\sin(bt-\phi)$$

$$\phi \triangleq \tan^{-1}\dfrac{b}{-a}+\tan^{-1}\dfrac{b}{c-a}$$

APPENDIX B

LAPLACE TRANSFORM THEOREMS

Definition	$\mathcal{L}[f(t)] \triangleq F(s) \triangleq \int_0^\infty f(t)e^{-st}\,dt$
Linearity	$\mathcal{L}[af(t)] = aF(s), \quad a = \text{const}$
	$\mathcal{L}[f_1(t) + f_2(t)] = F_1(s) + F_2(s)$
Derivatives	$\mathcal{L}[f^{(n)}(t)] = s^n F(s) - \sum_{k=1}^{n} f^{(k-1)}(0)s^{(n-k)}$
	$f^{(k)}(t) \triangleq \dfrac{d^k f(t)}{dt^k}, \qquad f^{(0)} \triangleq f(t)$
Integrals	$\mathcal{L}[f^{(-n)}(t)] = \dfrac{F(s)}{s^n} + \sum_{k=1}^{n} \dfrac{f^{(-k)}(0)}{s^{n-k+1}}$
	$f^{(-n)}(t) \triangleq \int \cdots \int f(t)(dt)^n, \qquad f^{(-0)} \triangleq f(t)$
Initial value	$\lim_{t \to 0} f(t) \triangleq f(\epsilon_2) = \lim_{s \to \infty} sF(s) \quad \text{(see Fig. 2.4)}$
Final value	$\lim_{t \to \infty} f(t) = \lim_{s \to 0} sF(s)$
Delayed functions	$\mathcal{L}[f(t-a)u(t-a)] = e^{-as}F(s)$
Periodic functions	$f_1(t) \triangleq \begin{cases} f(t), & 0 \leqslant t \leqslant T \\ 0, & \text{elsewhere} \end{cases}$
($f(t)$ is periodic with period T)	$\mathcal{L}[f(t)] = \dfrac{F_1(s)}{1 - e^{-Ts}}$
Convolution	$\mathcal{L}^{-1}[F_1(s)F_2(s)] = \int_0^t f_1(\tau)f_2(t-\tau)\,d\tau$
Scale change	$\mathcal{L}[f(t/a)] = aF(as)$
Partial derivatives	$\mathcal{L}_y\left(\dfrac{\partial f}{\partial y}\right) = sF(t,s) - f(0,t)$
$f = f(y,t)$	$\mathcal{L}_y\left(\dfrac{\partial^2 f}{\partial y^2}\right) = s^2 F(t,s) - sf(0,t) - \dfrac{\partial f}{\partial y}(0,t)$
	$\mathcal{L}_t\left(\dfrac{\partial f}{\partial y}\right) = \dfrac{\partial}{\partial y} F(y,s)$
	$\mathcal{L}_t\left(\dfrac{\partial^2 f}{\partial y^2}\right) = \dfrac{\partial^2}{\partial y^2} F(y,s)$

INDEX